CONTENTS

FMS in Evolution

FMS – control and communications. The problems and the potential
N. R. Greenwood, General Electric (USA) Automation – Europe... 1

The challenge of quality control in FMS
P. McKeown, Cranfield Unit for Precision Engineering and Cranfield Precision Systems Ltd, UK 9

FMS Beyond Machining

Flexible assembly automation in high volume – a strategic approach
D. C. Fanthorpe, Black & Decker, UK... 15

A flexible manufacturing system (FMS) for PCB production
S. G. Davey, IPL Information Processing Ltd, UK... 29

Integration and control of handling plant in an FMS for surface mounted PCBs
P. D. Parker, GEC Mechanical Handling Ltd, UK... 41

Assembly cell controls using flexible software tools and high level control configuration language
J. Temmes, P. Ruusunen and J. Lempiäinen, Technical Research Centre of Finland (VTT), Finland................ 53

Next generation FMS: Simulation techniques for designing a PCB assembly system
D. A. Cowan, R. H. Davies, T. J. Pontin, The Plessey Company plc, UK.. 63

In-process quality control and corrective feedback in a flexible manufacturing cell
M. Veron, J. Richard and E. Bajic, Centre de Recherche en Automatique de Nancy, CNRS, Université de Nancy, France 75

Systems Experience

The way ahead – the integrated approach
W. H. Horton, Austin Rover Group, UK 85

Advanced integrated manufacturing system (AIMS) for aero engine turbine and compressor discs
M. Butcher, Rolls-Royce Manufacturing Engineering, UK.. 93

Automated manufacture of small compressor blades
D. A. Glew and C. P. R. Hill, Rolls Royce plc, UK... 105

Increase of flexibility and productivity with computer integrated and automated manufacturing
E. Westkämper, Messerschmitt Bölkow Blohm GmbH, West Germany... 121

Mandelli large scale FMS for Volvo car engines
P. Egalini, Mandelli SpA, Italy 127

User project management for Holset FMS
S. Webb, Holset Engineering Co Ltd, UK... 135

Flexible manufacturing cells and systems with computer intelligence
H. Hammer, Werner und Kolb Werkzeugmaschinen GmbH, West Germany 145

CIM – The Integrated Solution

Factors affecting the reliability of real time distributed control systems
P. J. Cornwell, Renishaw Controls Ltd, UK... 159

Economical computer integrated manufacturing
M. J. Henderson, Siemens Ltd, UK... 167

The integration of FMS, Just-in-Time and LAN technology into a common manufacturing system
D. A. Hearn and M. A. Donnellan, Project Managers, Handley Walker Co Ltd, UK.................................. 179

The importance of information technology
M. W. Grant, Istel Automation Ltd, UK.. 193

A new concept in the control of manufacturing systems
P. Anstiss, BAeCAM British Aerospace plc, UK... 201

CIM-OSA, a computer integrated manufacturing based on the open system approach
J. Huysentruyt, Cap Gemini Belgium, Belgium 215

Simulation: Planning Tool for FMS

Simulation for manufacturing systems – a critical review
R. I. Mills, Ingersoll Engineers, UK.. 225

Strategies for material transportation in a computer integrated manufacturing environment
C. L. Moodie, Purdue University and D. BenArieh, AT & T Bell Laboratories, USA.................................... 235

TDL, a task description language for programming automated robotic workcells
A. Adler, Tecnomatix GmbH, West Germany... 247

General theories of flexible integration
J. E. Lenz, CMS Research Inc, USA.. 255

The Technological Base

Tool management system of an FMS of prismatic workpieces
J. Pylkkänen, University of Oulu, Finland 265

Interconnection technology on basis of automatic crane installations
H. J. Roos, Erwin Mehne GmbH & Co, West Germany 275

CIM in sheet metal production
M. Bitz, Trumpf GmbH & Co, West Germany... 285

Present status of sheet metal FMS in Japan
Y. Nakazawa, Mechanical Engineering Laboratory, M. Kiuchi, University of Tokyo, J. Endow, Tokyo Institute of Technology,
M. Shinohara, Tokyo Metropolitan University and S. Matsubara, The Institute of Vocational Training, Japan. 303

Tooling concepts for FMS
P. Tomek, VUOSO, Czechoslovakia... 315

Cells and Systems – Everyman's FMS

Jig boring machines in production with flexible manufacturing systems
Ch. Kesselburg, DIXI SA, Switzerland 327

Flexible manufacturing system for flywheels, bearing housing and gearwheel blanks
M. C. Aldridge, Lister-Petter Ltd, UK 337

Development of a flexible manufacturing system with high power laser
T. Ilar and C. Magnusson, Luleå University of Technology, Sweden 343

FM – the proven approach
F. Popplewell, Deckel Ltd, UK and P. Schmoll, Fr. Deckel AG, West Germany 353

The use of CIM in window manufacture: a case study
T. E. Toth, PA Management Consultants, UK... 375

A modular FMS concept in practice for prismatic parts
R. Tuokko, Valmet Corporation, Finland 385

Decision Making for FMS

Flexible manufacturing systems in Japan – an overview
H. Yamashina, K. Okamura, Kyoto University and K. Matsumoto, Nippon Denso Co Ltd, Japan 405

Justifying FMS to provide a competitive edge
A. A. Hunter, Cummins Engine Co Ltd, UK 417

Flexible manufacturing systems – some fact, some fiction
D. B. Ewaldz, Ingersoll Engineers, USA 427

The integration of FMS within existing factory systems
D. Little, University of Liverpool, UK... 439

Planning for precision in flexible manufacturing systems
R. G. Hannam, University of Manchester Institute of Science and Technology (UMIST), UK......................... 449

Supplementary paper

Flexible media of production automation
W. Apalkow, Instytut Mechaniki Precyzyjnej, Poland 457

Late Papers

European collaboration on FMS – Developments in ESPRIT CIM
P. MacConaill, ESPRIT Directorate, Commission of the European Communities, Brussels, Belgium 471

Management of the Tools Resource in a FMS
L. Borghi, M. Briano and E. Parmeggiani, Comau S.P.A., Modena Division – Italy................................ 477

Understanding the Relationship Between FMS and the Total Computer Integrated Manufacturing Environment
E. A. Herring, Digital Equipment Corporation, USA 491

FMS Integration Projects
M. Dub, Research Institute of Engineering Technology and Economy (VUSTE), Czechoslovakia 499

Robot Cells and System for Small Part Assembly
S. Ljung, ASEA Robotics, AB, Sweden 507

Evaluating FMS by Simulation
Olivier Arnaud and Eric Bloche, Renault Automation, France 513

*Proceedings of the
5th International Conference on*

FLEXIBLE MANUFACTURING SYSTEMS

*3-5 November 1986
Stratford-upon-Avon, UK*

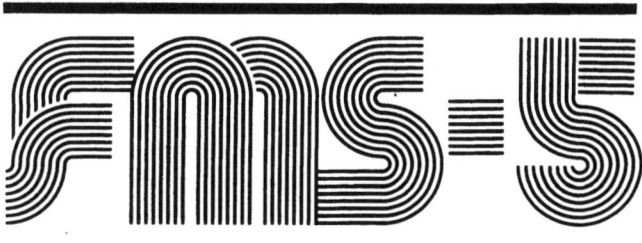

*Edited by
Prof. K. Rathmill*

An international event organised and sponsored by:
IFS (Conferences) Ltd, 35-39 High Street, Kempston, Bedford, UK

Springer-Verlag Berlin Heidelberg GmbH

Proceedings of the
5th International Conference on
FLEXIBLE MANUFACTURING SYSTEMS

The FMS-5 conference is an international event organised and sponsored by:
IFS (Conferences) Ltd, 35-39 High Street, Kempston, Bedford MK42 7BT, UK

CO-SPONSORS

The British Robot Association, UK
The International FMS Club, UK
FMS Magazine, UK
Fraunhofer Institute for Production Automation (IPA), Stuttgart, West Germany
Japan Industrial Robot Association, Japan
The Production Engineering Research Association, UK

FMS-5 PROGRAMME COMMITTEE

Chairman: Professor K. Rathmill, Renishaw Research Ltd
Dr J. Browne, University College, Galway, Ireland
Prof. B. J. Davies, University of Manchester Institute of Science and Technology
K. Gibbs, IFS (Conferences) Ltd
Dr N. R. Greenwood, General Electrical Industrial Automation (Europe)
A. D. Higgs, IFS (Conferences) Ltd
G. M. Hull, Cincinnati Milacron Ltd
A. Kochan, FMS Magazine
C. Lepper, Rolls Royce Ltd
R. I. Mills, Ingersoll Engineers Inc
D. J. Wilcock, TI Machine Tools

British Library Cataloguing in Publication Data

International Conference on Flexible Manufacturing Systems (5th: 1986: Stratford-upon-Avon)
    Proceedings of the 5th International Conference on Flexible Manufacturing Systems, 3-5 November 1986, Stratford-upon-Avon, UK.
    1. Flexible manufacturing systems
    I. Title   II. Rathmill, Keith
    658.5'14   TS155.6

ISBN 978-3-662-37279-1   ISBN 978-3-662-38009-3 (eBook)
DOI 10.1007/978-3-662-38009-3

© Springer-Verlag Berlin Heidelberg 1986
Originally published by Springer-Verlag Berlin Heidelberg New York Tokyo in 1986
Softcover reprint of the hardcover 1st edition 1986

All articles published in this proceedings are protected by copyright which covers the exclusive rights to reproduce and distribute the article (e.g. as offprints) as well as all translation rights. No material published in this proceedings may be reproduced photographically or stored on microfilm, in electronic databases, video disks, etc. without first obtaining written permission from the publisher.
Special regulations for photocopies in the USA: Photocopies may be made for personal or in-house use beyond the limitations stipulated under Section 107 or 108 of US Copyright Law, provided a fee is paid. This fee is US$0.20 per page, or a minimum of US$1.00 if an article contains fewer than five pages. All fees should be paid to the Copyright Clearance Centre, Inc., 21 Congress Street, Salem, MA 01970, USA, stating the ISBN 0-948507-17-9, IFS (Publications) Ltd, the volume, and the first and last page number of each article copied. The copyright owner's consent does not include copying for general distribution, promotion, new works, or resale. In these cases, specific written permission must first be obtained from the publisher.

AUTHOR INDEX

A. Adler 247
   Tecnomatix GmbH, West Germany
M. C. Aldridge 337
   Lister-Petter Ltd, UK
P. Anstiss 201
   BAeCAM British Aerospace plc, UK
W. Apalkow 457
   Instytut Mechaniki Precyzyjnej, Poland
E. Bajic 75
   Centre de Recherche en Automatique de Nancy, CNRS, Université de Nancy, France
D. BenArieh 235
   AT & T Bell Laboratories, USA
M. Bitz 285
   Trumpf GmbH & Co, West Germany
M. Butcher 93
   Rolls Royce Manufacturing Engineering, UK
P. J. Cornwell 159
   Renishaw Controls Ltd, UK
D. A. Cowan 63
   The Plessey Co plc, UK
S. G. Davey 29
   IPL Information Processing Ltd, UK
R. H. Davies 63
   The Plessey Co plc, UK
M. A. Donnellan 179
   Handley Walker Co Ltd, UK
P. Egalini 127
   Mandelli SpA, Italy
J. Endow 303
   Tokyo Institute of Technology, Japan
D. B. Ewaldz 427
   Ingersoll Engineers, USA
D. C. Fanthorpe 15
   Black & Decker, UK
D. A. Glew 105
   Rolls Royce plc, UK
M. W. Grant 193
   Istel Automation Ltd, UK
N. R. Greenwood 1
   General Electric (USA) Automation – Europe
H. Hammer 145
   Werner und Kolb Werkzeugmaschinen GmbH, West Germany
R. G. Hannam 449
   University of Manchester Institute of Science and Technology (UMIST), UK
D. A. Hearn 179
   Handley Walker Co Ltd, UK
M. J. Henderson 167
   Siemens Ltd, UK
C. P. R. Hill 105
   Rolls Royce plc, UK
W. H. Horton 85
   Austin Rover Group, UK
A. A. Hunter 417
   Cummins Engine Co Ltd, UK
J. Huysentruyt 215
   Cap Gemini, Belgium
T. Ilar 343
   Luleå University of Technology, Sweden
Ch. Kesselburg 327
   DIXI SA, Switzerland
M. Kiuchi 303
   University of Tokyo, Japan
J. Lempiäinen 53
   Technical Research Centre of Finland (VTT), Finland
J. E. Lenz 255
   CMS Research Inc, USA
D. Little 439
   University of Liverpool, UK
C. Magnusson 343
   Luleå University of Technology, Sweden

S. Matsubara 303
   The Institute of Vocational Training, Japan
K. Matsumoto 405
   Nippon Denso Co Ltd, Japan
P. McKeown 9
   Cranfield Unit for Precision Engineering and Cranfield Precision Systems Ltd, UK
R. I. Mills 225
   Ingersoll Engineers, UK
C. L. Moodie 235
   Purdue University, USA
Y. Nakazawa 303
   Mechanical Engineering Laboratory, Japan
K. Okamura 405
   Kyoto University, Japan
P. D. Parker 41
   GEC Mechanical Handling Ltd, UK
T. J. Pontin 63
   The Plessey Co plc, UK
F. Popplewell 353
   Deckel Ltd, UK
J. Pylkkänen 265
   University of Oulu, Finland
J. Richard 75
   Centre de Recherche en Automatique de Nancy, France
H. J. Roos 275
   Erwin Mehne GmbH & Co, West Germany
P. Ruusunen 53
   Technical Research Centre of Finland (VTT), Finland
P. Schmoll 353
   Fr. Deckel AG, West Germany
M. Shinohara 303
   Tokyo Metropolitan University, Japan
J. Temmes 53
   Technical Research Centre of Finland (VTT), Finland
P. Tomek 315
   VUOSO, Czechoslovakia
T. E. Toth 375
   PA Management Consultants, UK
R. Tuokko 385
   Valmet Corporation, Finland
M. Veron 75
   Centre de Recherche en Automatique de Nancy, CNRS, Université de Nancy, France
S. Webb 135
   Holset Engineering Co Ltd, UK
E. Westkämper 121
   Messerschmitt Bölkow Blohm GmbH, West Germany
H. Yamashina 405
   Kyoto University, Japan

Late Papers

O. Arnaud 513
   Renault Automation, France
E. Bloche 513
   Renault Automation, France
L. Borghi 477
   Comau S.P.A., Modena Division, Italy
M. Briano 477
   Comau S.P.A., Modena Division, Italy
P. MacConaill 471
   ESPRIT Directorate, Commission of the European Communities, Brussels, Belgium
M. Dub 499
   Research Institute of Engineering Technology and Economy (VUSTE), Czechoslovakia
E. A. Herring 491
   Digital Equipment Corporation, USA
S. Ljung 507
   ASEA Robotics AB, Sweden
E. Parmeggiani 477
   Comau S.P.A., Modena Division, Italy

CORPORATE INDEX

AT & T Bell Laboratories, USA	235
Austin Rover Group, UK	85
BAeCAM British Aerospace plc, UK	201
Black & Decker, UK	15
Cap Gemini, Belgium	215
Centre de Recherche en Automatique de Nancy, CNRS, University of Nancy, France	75
CMS Research Inc, USA	255
Cranfield Unit for Precision Engineering and Cranfield Precision Systems Ltd, UK	9
Cummins Engine Co Ltd, UK	417
Deckel Ltd, UK	353
DIXI SA, Switzerland	327
Erwin Mehne GmbH & Co, West Germany	275
Fr. Deckel AG, West Germany	353
GEC Mechanical Handling Ltd, UK	41
General Electric (USA) Automation – Europe	1
Handley Walker Co Ltd, UK	179
Holset Engineering Co Ltd, UK	135
Ingersoll Engineers, UK	225
Ingersoll Engineers, USA	427
Instytut Mechaniki Precyzyjnej, Poland	457
IPL Information Processing Ltd, UK	29
Istel Automation Ltd, UK	193
Kyoto University, Japan	405
Lister-Petter Ltd, UK	337
Luleå University of Technology, Sweden	343
Mandelli SpA, Italy	127
Mechanical Engineering Laboratory, Japan	303
Messerschmitt Bölkow Blohm GmbH, West Germany	121
Nippon Denso Co Ltd, Japan	405
PA Management Consultants, UK	375
Purdue University, USA	235
Renishaw Controls Ltd, UK	159
Rolls Royce Manufacturing Engineering, UK	93
Rolls Royce plc, UK	105
Siemens Ltd, UK	167
Technical Research Centre of Finland (VTT), Finland	53
Tecnomatix GmbH, West Germany	247
The Institute of Vocational Training, Japan	303
The Plessey Co plc, UK	63
Tokyo Institute of Technology, Japan	303
Tokyo Metropolitan University, Japan	303
Trumpf GmbH & Co, West Germany	285
University of Liverpool, UK	439
University of Manchester Institute of Science and Technology (UMIST), UK	449
University of Oulu, Finland	265
University of Tokyo, Japan	303
Valmet Corporation, Finland	385
VUOSO, Czechoslovakia	315
Werner und Kolb Werkzeugmaschinen GmbH, West Germany	145

Late Papers

ASEA Robotics AB, Sweden	507
Comau S.P.A., Modena Division, Italy	477
Digital Equipment Corporation, USA	491
ESPRIT Directorate, Commission of the European Communities, Brussels, Belgium	471
Renault Automation, France	513
Research Institute of Engineering Technology and Economy (VUSTE), Czechoslovakia	499

FOREWORD

by Professor K. Rathmill
Managing Director of
Renishaw Research Limited

FMS-5 Conference Chairman

FMS-5 represents considerably more than simply the fifth year of an exceptionally successful international conference. These proceedings are, in addition to many other things, a clear indication of the central influence that the concepts and technologies of flexible manufacturing systems philosophy continue to exert in the world's leading industrial societies.

The last two decades have seen fundamental changes in the character of the advanced manufacturing system. Developments in numerous enabling technologies have been complemented by the increasingly sophisticated capabilities of manufacturing engineers to understand, analyse and tackle the challenge of efficient manufacturing system design and operation. Over the last decade, evolving influences in robotics, guided vehicles, sensors, computer control, advanced machine tool design, tooling systems and handling technologies have had a profound impact on thinking in advanced manufacturing system design worldwide. Once revolutionary, and at the time esoteric, concepts implicit in the MOLINS System 24, the Cincinnati Milacron Variable Mission system and the more rigorous work carried out on Group Technology have in recent years received support through the development of a series of progressively maturing and affordable enabling technologies. The earlier exciting, if somewhat daunting, perception of FMS has now given way to a clearer understanding of the widespread potentialities expressed in the title of the session in this Conference which reads "Cells and Systems – **EVERYMAN'S FMS**".

What lessons have we learned from the ever growing applications experience to date? What should industrial managements be doing now to maximise future compatibility with emerging standards in computer integration? How can we best seek to achieve effective quality control in highly automated batch production? Such questions and their responses are at the very heart of this Conference; and they are indeed examples of extremely important issues at the very heart of manufacturing industry itself. Speaking on behalf of the hard working members of the FMS-5 Programme Committee may I express our warm hope and expectation that you will enjoy the Conference proceedings and feel that they have indeed succeeded as an effective channel of thought, experience, information and debate in this vitally important field.

VIII

FMS IN EVOLUTION

FMS – control and communications. The problems and the potential
N. R. Greenwood
General Electric (USA) Automation – Europe

1.0 Introduction:

Over the past few years FMS technology has progressed substantially. Having started out in 1968 with the Molins System 24, FMS is now a natural part of any well founded CIM strategy.

But not only the technology of FMS has been changing, the applicability and overall relevance has also been evolving. Initially FMS was seen as only satisfying the needs of small to mid-size batch manufacturing. With the advances that have been made in for example, computers and computer numerical control systems, the FMS approach can now be applied economically to many different production environments from mass production through to the once-off jobbing shop.

Whilst these general developments have been occurring, the traditional processes used within manufacturing have rapidly advanced to the point where it is now difficult to improve them significantly. So it is the other technologies required for successful implementation of FMS, such as "control" and "communications" which now represent the areas where advancement is likely to hold the key to making FMS easier to implement in the long-term.

2.0 Problems Implementing FMS:

Whilst technological advances have been facilitating the development of successful FMS many of the more strategic problems still remain. Issues such as:

- o How does the FMS fit into the Company's long-term manufacturing and marketing strategy?

- o How should the FMS investment be justified?

- o How should the design and operation of the system be optimized?

- o How should the risks and costs associated with the development of control software be minimized?

- o How should multiple vendor's devices be interfaced into one integrated system?

- o How should components and processes be selected?

The answer to some of these questions lies in technologies such as computer simulation, group technology, broad based long-term planning and common sense. However, the issues of device integration and control software development are particularly interesting since they are areas in which significant progress is likely to be seen during the next couple of years.

All flexible manufacturing systems are, in many ways, alike. They might look significantly different. Factors such as, the manufacturing environment in which the systems are located, the magnitude of the mean time between events, and the complexity of the events all exert a significant influence. However in the most general sense, raw materials go in, and finished parts, hopefully, come out. In fact, the similarities go much further than this, especially in control and communication terms.

All flexible manufacturing systems comprise three major components:

1) A control system, consisting of both computer hardware and software.

2) A communications system. This will consist of some software, a communications medium, and maybe some form of interface hardware at both the computer and the process devices.

3) A number of process devices. These could be machine tools or robots, possibly with a computer numerical control system, or any of a plethosa of other production equipment.

The first two topics are of fundamental importance to all flexible manufacturing systems. Their ability to perform correctly will dictate the success or failure of the whole system. Certainly technically, and possibly commercially as well.

3.0 FMS Control

Typically the computer control system for an FMS accounts for between 15 and 40% of the cost of the total system. In terms of risk, the percentage is probably far higher. Certainly computer hardware has become both less expensive and more powerful, alleviating the situation somewhat, but the control software still remains a major issue. One of the main reasons for this is the current necessity for virtually all such systems to comprise a significant proportion of specially written, application specific software.

The costs and risks associated with the development of this software are immense.

The alternatives to writing application oriented software are essentially threefold:

1) Write software in reusable modules

2) Develop an appropriate software environment which facilitates the generation of application specific systems

3) Develop generic software systems capable of controlling a wide variety of flexible manufacturing systems

Writing software in reusable modules sounds simple enough, but in practice it is rather more difficult. One of the main problems is having both the time and money to sit down and design all the interfaces that might be needed between all the software modules. However, for flexible manufacturing systems which are very similar the approach does have merit. Certainly amongst machine tool manufacturers who have supplied FMS largely comprising their own equipment, a sufficient degree of commonality does exist, making this approach particularly attractive.

Alternatively, development of an appropriate software environment to facilitate generation of FMS control systems represents an awesome task. It requires the reduction of all FMS control requirements into a number of well defined high level functions. These functions may then be programmed and combined in a certain way to produce the required application specific control system. This approach is not unlike that which might be used to create a complex simulation model from a specialized simulation language.

There are some substantial disadvantages to this technique. The cost of creating such a control system development environment is extremely high. Also, the level of skill needed to produce the control systems and the costs/risks still associated with debugging, all combine to make the concept less than attractive to all but the wealthiest of customers.

The third alternative, namely development a generic package capable of controlling a wide variety of flexible manufacturing systems has not really been practical until recently. What was needed to make this approach a viable possibility was a thorough understanding of the problems associated with controlling flexible manufacturing systems in widely differing production environments, and the resources and computing power to design, and implement such a generic control system in a manner in which it eventually could be marketed economically.

This approach is without doubt gaining favour, particularly in the U.S. where some products already exist. However, as yet such systems do appear to have somewhat limited capabilities. This situation will no doubt change in the not too distant future.

4.0 FMS Communications

Until recently, communications systems within FMS were probably the least talked about aspect of the system. Over the past few years this situation has changed radically, almost entirely due to the development of the Manufacturing Automation Protocol (MAP).

Like FMS control systems, communication systems within FMS are remarkably similar, regardless of the actual FMS in which they are installed.

For example the main purpose of all FMS communications systems is to fulfill the following three requirements:

1) To enable files to be sent and received from the devices within the FMS (part, programs, tool offset tables, robot parameter files, PLC recipes, etc.)

2) To carry control instructions to the devices (cycle start, enable optional stops, etc.)

3) To carry status information from the devices (device busy, device idle, alarm, etc.)

Admittedly the individual bias towards any of the three areas might vary significantly according to the nature of the production environment in which the FMS is located. For example, if small components are being produced, part programs are likely to be short. If aircraft wings are being machined, the part programs are likely to be lengthy. Interestingly, the level of status change activity within two such systems could well be the same. In the first instance because parts are being finished quickly, and in the second, because comprehensive quality control data has to be sent frequently.

Where one of the main problems has arisen with FMS communications is with regards the interfacing of many manufacturers device control equipment into one integrated system. It is hoped that the MAP efforts wil eventually lead to the elimination of this problem.

MAP is based on the full OSI (Open Systems Interconnection) seven layer model. This is an established model which identifies generic communication requirements and provides a uniform nomenclature, thus facilitating system comparison.

5.0 The Potential

With the advent of both generic control systems and MAP, the future for the on-going development of FMS is significantly better than it was a couple of years ago. With MAP standardizing communications, and generic controllers reducing the cost and risks associated with developing FMS control software, it will become easier to implement FMS economically. Advances such as PC simulation packages (such as Modelmaster) will continue to enhance the design process, ensuring a higher level of confidence in a system's eventual operation.

It is likely that the trend towards cellular manufacturing will continue. This is likely to help accelerate the rate of take-up of the MAP subnet system. This could well result in the full MAP system being used mainly by the larger manufacturers who might still feel it justified to have say 100 machine tools all controlled from a central MRP/DNC host computer.

To conclude, the control and communications problems associated with the design and implementation of FMS are now being addressed. Reinforcing the opinion that the flexible manufacturing systems of to-day admirably demonstrate the basis of the technology which will be used to manufacture the majority of the products of tomorrow.

14 July 1986   Dr. N.R. Greenwood

The Seven Layers of the OSI Model are as follows:

Layer 7 - APPLICATION

    Provides all services directly understandable by application programs.

Layer 6 - PRESENTATION

    Transforms data to and from agreed standardized formats.

Layer 5 - SESSION

    Synchronizes and manages data.

Layer 4 - TRANSPORT

    Provides reliable, transparent data transfer from end node to end node.

Layer 3 - NETWORK

    Performs message routing for data transfers between non-adjacent nodes.

Layer 2 - DATA LINK

    Detects errors in messages moved between adjacent nodes.

Layer 1 - PHYSICAL

    Encodes and physically transfers messages between adjacent nodes.

Unfortunately MAP's noble aims have resulted in what, for the present at least, is quite a complicated and expensive communications system.

In an effort to reduce this adverse impact, alternative approaches to the full MAP being developed, frequently called MAP subnet. These are likely to be less generic, but both cheaper to implement and of higher performance.

The challenge of quality control in FMS
P. McKeown
Cranfield Unit for Precision Engineering and Cranfield Precision Systems Ltd, UK

Quality improvement of products and services - and of the whole company performance is frequently the strategy which can produce the earliest and maximum return on investment. Results obtained by some leading UK manufacturing companies from Quality Improvement Programmes are outlined. Within QIP strategies, implementing systems for control of quality in automated manufacture is discussed, including the full development of 'deterministic metrology' philosophies and techniques.

1.0 STRATEGY FOR SUCCESS IN ADVANCED MANUFACTURING

Any engineering company aiming to improve its competitiveness and profitability in national and international markets can make a three-pronged attack on this critically important challenge. They can

1.1 Improve PRODUCTIVITY

(Production at higher rate and at reduced overall unit cost)

1.2 Implement planned, successful INNOVATION programmes for the products and services it offers

(A carefully planned programme of product and service innovation managed against the life cycles of its existing products. Note that the opposite of innovation is *obsolescence* and *stagnation*).

1.3 Improve QUALITY of those products and services.

This glimpse of the obvious is nothing less than the formula for increased wealth creation, for halting economic decline and achieving economic growth. All three targets are to be addressed in some detail in this conference, but I suggest that the one which offers the greatest potential for bottom line improvement of company performance in the short to medium term is IMPROVED QUALITY.

Quality may be defined as providing *Fitness for purpose with value for money*. Perhaps a better definition of product or service quality is "fully conforming to the agreed requirements of the customer"; he is concerned with:

- the specification ... his expectation
- conformance to specification ... what he receives
- reliability ... how well it performs
- price ... how much he pays during the lifetime use of the product or service provided
- delivery ... how long he waits

By dedicating themselves to achieving improved quality, top managements must persuade their entire workforces to accept that "we either deliver the goods defect-free and competitively or the customer will find someone who can," and to further accept that "only a quality business can deliver a quality product". *This is the underlying philosophy of TOTAL BUSINESS QUALITY (TBQ)*.

2.0 QUALITY IMPROVEMENT PROGRAMMES

In the last three years, the author has served on the adjudicating panel for the British Quality Awards. This has been a stimulating and educational experience. Each year companies from a wide range of manufacturing and service activities are invited to submit detailed cases. Each is required to demonstrate methods used and benefits achieved over the previous three years in implementing a Quality Improvement Programme. All the winners have been able to demonstrate outstanding achievements and benefits:

- dramatic reduction of defects in their products at all stages of production and at final customer site;
- improvement in overall product liability;
- thus, dramatic decreases in the costs of quality (Fig. 3);
- increases in customer satisfaction;
- increases in overall market share

Figures 4, 5, 6 and 7 are typical of the improved operating efficiencies achieved by three of the award winning companies by implementing successful Quality Improvement Programmes involving all in their organisations. Dramatic improvements in overall efficiency have been achieved in relatively short times. There is great scope for many more companies in this and other industrialised nations of the world to practise the simple, obvious, business logic of TBQ.

Quality Improvement Procedure is described in References 1, 2, and 3. Two of the winning companies have overtaken their Japanese sister companies in Quality achievements. Quality is thus not purely dependent on culture; it is a matter of personal commitment of the top management and the total organisation, leading on to good team performance through effective leadership.

3.0 QUALITY ASSURANCE AND QUALITY CONTROL

3.1 Quality Assurance

The first essential step in achieving an improved and acceptable quality performance in the company is to commit everyone in the organisation to a detailed company operating policy embodied in a comprehensive Quality Manual. BS.5750 Quality Systems provides a detailed guide not only to writing the Quality Manual, but also on how to set up "all those systems by which the company can ensure the continued satisfaction of its customers at a profit".

3.2 Quality Control

Quality control is an important sub-set of overall quality assurance. *Control means measuring and applying corrective action.* Quality cannot be "inspected in"; it can only be designed and built into the product. *Prevention* of defects of any type is the essential feature of modern quality control. The sequence of actions to control quality can be likened to a closed-loop, error-feedback servo-system. *Maximum efficiency of the control of quality can be achieved when the speed of response of this servo is faster than the rate of change of those processes/parameters which cause errors.* Thus, the more closely the control loop is applied to the point of manufacture the greater the efficiency.

We are now seeing in modern manufacturing more effort, time and money being devoted to the automatic control of quality than ever before. Increasingly, quality control is being planned into manufacturing systems at the time of designing those systems. Production line balance can only be ensured by matching the measurement and error feedback control processes to the manufacturing throughput cycle time. This accounts for today's rapidly increasing development and application of:

- immediate post-process (measurement and) control
- in-process measurement and control

and
- geometric adaptive control

In-line measurement and control of quality must be part of the production planning function. When immediate post-process control or in-process control can be used, line balancing is more easily achieved.

3.3 Deterministic Performance of Machines

It is essential that manufacturing engineers thoroughly understand that machines under automatic control are totally deterministic in performance. Central to the control of quality in automated manufacture is the principle of determinism. This states that all machines operate in a deterministic manner. Loxham (Ref. 6) expressed the concept as "an automatic machine may be classified as operating perfectly. It may not be doing what is required and if this is so, it is because it has not been suitably arranged". In other words, all errors of such a machine are systematic right down to a level set by the uncertainty principle.

Portas of CUPE puts it "random results are the result of random procedures". The application of statistics to the performance of machines has been tersely referred to as "a cloak of respectability for bad metrology". Bryan (Ref. 7) uses the term "deterministic thinking" as the correct philosphy in the design and performance analysis of precision machines and goes on to say "the probabilistic approach to a problem is only a tool to allow us to deal with variables that are too numerous or expensive to sort out properly by common sense and good metrology ... it may lead to an implicit assumption about the necessity of ignorance".

3.4 Developments in Dimensional Measurement and Quality Control

3.4.1 *conventional inspection*, - that is sorting good from bad after manufacture (now a totally outmoded and inefficient philosophy and practice.

3.4.2 *operator quality control* - a technique to make the operator responsible for the quality he produces by provision of measurement/gauging equipment with the aim of determining trends and thus preventing defects from occurring.

3.4.3 *accuracy capability studies on production machinery* - an implicit acceptance, if not full recognition of deterministic performance of production machines leading to selection of "accuracy capable" machines and probable use of *statistical sampling techniques* (SQC)

3.4.4 *post process measurement, analysis, and feedback control* of the production process frequently based on strictly sequential measurement of parts (from which patterns of performance can more readily be recognised)

3.4.5 *in-process measurement and control* in which the speed of response of the closed-loop error-feedback servo-system is at its highest (but used almost exclusively in cylindrical grinding)

3.4.6 *adaptive control (including geometric-adaptive)* which can compensate for relatively rapid trends/parameter changes such as grinding wheel wear etc..

3.4.7 *deterministic metrology* - in which full advantage is taken of the deterministic nature of production machines and in which all of the manufacturing sub-systems are optimised to maintain deterministic performance within acceptable quality levels. Here, *system processes* are monitored, e.g. temperature, pressure, flow, force, vibration, acoustic "fingerprinting", these sensors being fast and non-intrusive. Here new techniques such as 3D error compensation by CNC and expert systems will come into use, leading to fully adaptive control. This is a new philosophy of system control in which part measurement is replaced by process measurement.

The presentation will illustrate some current practice and future techniques in the development of deterministic metrology applied to future automated manufacturing systems. Mention will also be made of recent developments in very high precision manufacturing machinery and control systems illustrating how some deterministic metrology techniques are now being applied to individual machine tools today to achieve microtechnology and nanotechnology accuracies.

REFERENCES

(1) McKeown, P.A., "Implementing Quality Improvement Programmes". Annals of CIRP, Vol. 34/2/1985

(2) Ward, J.M., "A Total Quality Improvement Programme", p 1-12, IFS Publications Proceedings of Conference on Automatic Inspection and Product Control. 1985

(3) Huckett, J.D., "Implementing a Winning Quality Improvement Programme", Cranfield Prestige Lecture, April 1985

(4) McKeown, P.A., and Perkins, D.R., "Adaptive Control Grinding of High Precision Cam Profiles", p.253, IFS Publications, Proceedings on Automatic Inspection and Product Control, 1985

(5) McKeown, P.A., "High Precision Manufacturing and the British Economy", James Clayton Lecture, Institute of Mechanical Engineers, Proceedings 1986, Vol. 200 No. 76.

(6) Loxham, J., "The Commercial Value of Investigations into Repeatability", CUPE, Cranfield Institute of Technology, 1975. (unpublished)

(7) Bryan, J.B., "The Power of Deterministic Thinking in Machine Tool Accuracy", 1st International Machine Tool Engineers Conference, Tokyo, Nov. 1984, JMBTA pp 3-17.

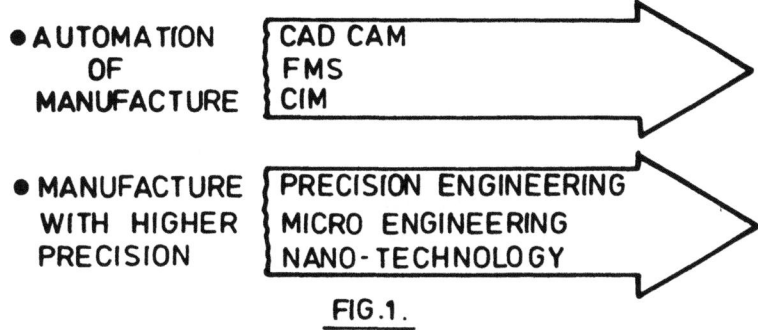

THE MAIN THRUSTS IN MANUFACTURING ENGINEERING

- AUTOMATION OF MANUFACTURE: CAD CAM, FMS, CIM
- MANUFACTURE WITH HIGHER PRECISION: PRECISION ENGINEERING, MICRO ENGINEERING, NANO-TECHNOLOGY

FIG.1.

QIP philosophy - commitment persuasion and training

- 'Quality is free' - it pays for itself through
    - reduced operating costs
    - increased sales through satisfied customers

- 'Zero defects' must be achieved
    - Acceptable Quality Levels (AQL's) are no longer good enough
    - defect free products are possible

- 'Prevention is better than inspection'
    - quality must be 'built in', not 'inspected in'
    - operation and process variables can be eliminated

- 'Everyone is a link in the quality chain'
    - total employee involvement is essential
    - teamwork is the basis of success

- 'Only a quality business can deliver a quality product'
    - a total quality approach to the whole business is the best strategy
    - satisfying the next 'customer' down-the-line throughout the business is a powerful personal motivation to all staff.

Fig. 2

THE COSTS OF QUALITY

1. PREVENTION - the cost of stopping failures from occurring
2. APPRAISAL - measuring whether parts or products conform
3. INTERNAL FAILURE - cost of scrap, rework, etc.
4. EXTERNAL FAILURE - cost of warranty claims and rework after delivery of products
5. EXCEEDING REQUIREMENTS - cost of work which is unnecessary
6. LOST OPPORTUNITIES - lost business due to failure to meet customer requirements

Fig. 3

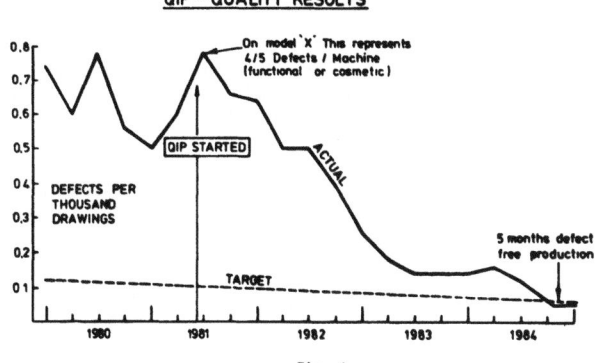

Fig. 4

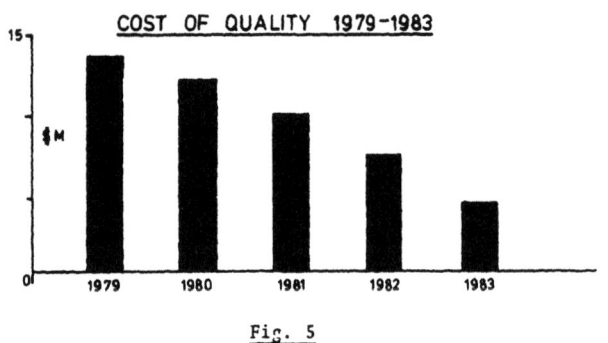

Fig. 5

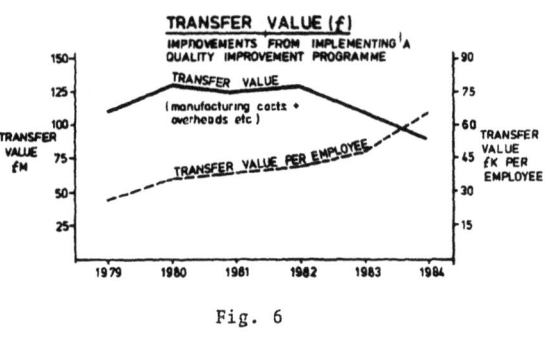

Fig. 6

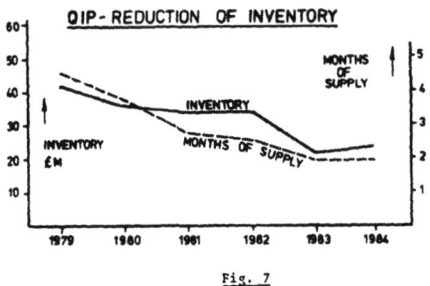

Fig. 7

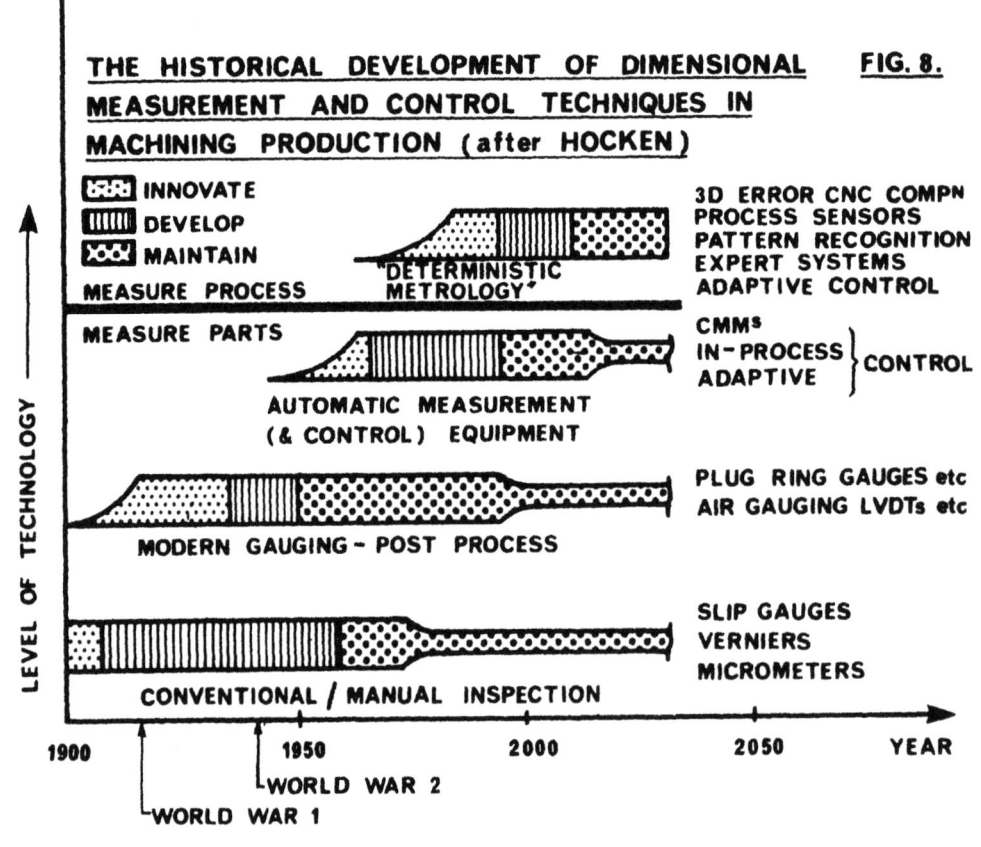

FMS BEYOND MACHINING

Flexible assembly automation in high volume – a strategic approach
D. C. Fanthorpe
Black & Decker, UK

This paper describes a Flexible Assembly System for high volume consumer power tools. It examines the strategic process by which the system was developed including market issues and design for automation and identifies pitfalls for the unwary. The system is described before concluding that high volume flexible assembly automation is possible and brings advantages of cost, quality and additional range to the market place. The underlying theme, however, is that it is only possible to succeed in FAS if the automation is put into the considered context of the whole Manufacturing Business.

FLEXIBLE ASSEMBLY AUTOMATION IN HIGH VOLUME - A STRATEGIC APPROACH

"THERE ARE THREE WAYS TO THROW AWAY YOUR MONEY, ON WOMEN, ON GAMBLING OR ON COMPUTERS" - PRESIDENT POMPIDOU OF FRANCE.

THIS PAPER WILL DEAL WITH THE APPLICATION OF FLEXIBLE ASSEMBLY SYSTEMS WITHIN A TOTAL MANUFACTURING STRATEGY. I WOULD BE ADAMANT THAT TO CONSIDER THE APPLICATION OF FAS, OR INDEED ANY SIGNIFICANT AUTOMATION WITHOUT FIRST CONSIDERING ITS PLACE WITHIN A **DETAILED** MANUFACTURING STRATEGY WOULD BE A DERELICTION OF DUTY ON THE PART OF SENIOR MANUFACTURING MANAGERS AND ENGINEERS, LEADING TO INEVITABLE FAILURES AND DISILLUSIONMENT. I WOULD SINGLE OUT SENIOR MANUFACTURING MANAGERS AND ENGINEERS FOR THAT HEAVY RESPONSIBILITY, BECAUSE IN MANY CASES BOARDS OF DIRECTORS DO NOT HAVE THE SPECIALIST SKILLS AND EXPERIENCE TO DEMAND A JUSTIFICATION IN THE CONTEXT AND PERSPECTIVE OF THE WHOLE BUSINESS, WORSE, THEY WILL PROBABLY HAVE BEEN MATERIALLY MISLED BY MEDIA HYPE ON AUTOMATION GENERALLY, AND BY THE IMMEDIATE INVENTION OF AN ACRONYM TITLE TO EVERY ISOLATED PIECE OF DEVELOPMENT IN MANUFACTURING TECHNOLOGY. THESE BURGEONING ACRONYM TITLES WOULD SUGGEST TO THE UNWARY DIRECTOR THAT HERE IS A STAND ALONE MANUFACTURING CONCEPT, THAT IN ISOLATION WILL TRANSFORM HIS MANUFACTURING PERFORMANCE AND PULL HIS PRODUCTION FACILITIES, KICKING AND SCREAMING, INTO THE 20TH CENTURY. RARELY, IF EVER, WILL THAT BE THE CASE UNLESS IT IS A CONSIDERED APPLICATION OF THE TECHNOLOGY WITHIN THE CONTEXT OF A WHOLE MANUFACTURING AND BUSINESS STRATEGY, WHICH IN ALL LIKELIHOOD WILL CHANGE EVERY FACET OF THE MANUFACTURING BUSINESS.

SO LET ME DESCRIBE SOME OF THE FLEXIBLE ASSEMBLY SYSTEMS, INSTALLED OVER THE PAST FIVE YEARS AT THE BLACK & DECKER U.K. MANUFACTURING PLANT IN SPENNYMOOR, COUNTY DURHAM, AND JUSTIFY MY ORIGINAL PROPOSITION BY ILLUSTRATING THE STRATEGIC CONSIDERATIONS WHICH LED TO THEIR CHOICE AND SUBSEQUENT SUCCESS. FIRST, THE STRATEGIC CONSIDERATIONS.

A MANUFACTURING STRATEGY TO TAKE US INTO THE 1990'S WAS COMMISSIONED IN 1980. THE STRATEGY WAS DEVELOPED OVER 9 MONTHS OF INTENSE STUDY, DEBATE, AND PROJECTIONS OF MANUFACTURING TECHNOLOGY AND MARKET TRENDS. AT THE END OF THE DEBATE, THERE WAS A PLAN WHICH WAS COMPREHENSIVE, COVERED EVERY FACET OF THE BUSINESS, AND WITH COMMITTED CONCENSUS, WOULD REVOLUTIONISE THE MANUFACTURING BUSINESS, AND EFFECT EVERY SINGLE EMPLOYEE TO A VERY SIGNIFICANT DEGREE. WITHIN THAT STRATEGY WERE FLEXIBLE ASSEMBLY SYSTEMS AND THEIR IMPACT ON THE MARKET, MARKET RESPONSE TIMES, COST, QUALITY, NEW PRODUCT DEVELOPMENT, AND PEOPLE.

THE PROFILE OF THE OPERATION IN 1980 CAN BE EASILY RELATED. THERE WAS A VERY TRADITIONAL MACHINE SHOP MAKING METAL COMPONENTS FOR POWER TOOLS - SHAFTS, SPINDLES, GEARS, FOR EXAMPLE. THE SHOP WAS ARRANGED IN DISCREET WORK CENTRES, AT ONE TIME 20 IN ALL. THERE WAS A MOTOR WINDING SHOP, ALREADY AUTOMATED AND LINKED, MAKING AND TESTING MOTOR ARMATURES AND, SEPARATELY, MAKING MOTOR FIELDS. THERE WAS THEN A VAST, HIGHLY LABOUR INTENSIVE, LOW CAPITAL ASSEMBLY SHOP WITH GOOD, BUT BASIC, JIGS AND FIXTURES AND A VERY LOW CAPITAL BASE. SO THE MAIN THRUST OF AUTOMATION AND CAPITAL INVESTMENT HAD BEEN MOTOR WINDING, WITH THE BEST OF TRADITIONAL MACHINE SHOP TECHNOLOGY A SECOND PRIORITY AND ASSEMBLY VERY MUCH AS YOU MAY HAVE OBSERVED 25 YEARS AGO. THE ASSEMBLY TECHNOLOGY, OR LACK OF IT, WAS NOT AN ACCIDENT OR AS A RESULT OF NEGLECT, BUT WAS LABOUR INTENSIVE AND NON CAPITAL INTENSIVE FOR TWO VITAL REASONS:-

A. TO GIVE MAXIMUM FLEXIBILITY AND SPEED OF RESPONSE IN A HIGHLY VOLATILE CONSUMER MARKET.

B. BECAUSE UP TO 1980, FLEXIBLE ASSEMBLY AUTOMATION, IF IT EXISTED AT ALL WAS NOT FINANCIALLY VIABLE NOR, IF TRUTH BE TOLD, TECHNICALLY VIABLE FOR HIGH VOLUME PRODUCTION.

THE FACTORY WAS MAKING A WIDE RANGE OF CONSUMER POWER TOOLS BASED ON MOTORS MADE FROM ONE LAMINATION SIZE BUT IN MANY DIFFERENT CONFIGURATIONS. THE MARKETING INPUT TO THE STRATEGIC STUDY INDICATED:-

A. INCREASING COMPETITION FROM GERMANY AND JAPAN.

B. AN INEXORABLE **INCREASE** IN PRODUCT RANGE TO PREVENT NICHES FOR COMPETITION, AND AS A RESULT OF AN AGGRESSIVE NEW PRODUCT DEVELOPMENT PROGRAMME.

C. COST WOULD BE CRITICAL IN MAINTAINING MARGINS TO FINANCE NEW PRODUCT DEVELOPMENT AND LAUNCHES.

D. MANUFACTURING RESPONSE TIMES TO MARKET DEMAND WOULD REQUIRE TO BE SIGNIFICANTLY REDUCED AS THE RETAIL OUTLETS, OUR CUSTOMERS, GOT MORE AND MORE SOPHISTICATED IN STOCK MANAGEMENT, AND AMBITIOUS IN TERMS OF J.I.T. SUPPLY.

E. CUSTOMER SERVICE AND QUALITY WOULD BE MAJOR FACTORS IN GROWTH AND MEETING COMPETITION.

THE STRATEGY RECOGNISED FROM THIS INPUT THAT IN COST, THERE WERE MAJOR OPPORTUNITIES IN FURTHER AUTOMATION OF MOTOR WINDING BUT PRINCIPALLY IN THE AUTOMATION OF ASSEMBLY AND THE INTEGRATION OF MACHINE SHOP PROCESSES INTO MOTOR WINDING AND ASSEMBLY. **BUT** WE HAD TO PROJECT FAR FASTER MARKET RESPONSE TIMES, MORE DEMAND VOLATILITY, A RAPIDLY EXPANDING PRODUCT RANGE, IN SHORT FOR **MORE** MANUFACTURING FLEXIBILITY, NOT LESS. THE SUPERFICIAL ANSWER AND THE ONE MOST FASHIONABLY TURNED TO WOULD SAY PROGRAMMABLE ROBOTIC ASSEMBLY. HOWEVER, EVEN A SUPERFICIAL EXAMINATION OF ROBOTICS AS THE ROUTE WOULD DEMONSTRATE IT AS A DEFINITE, NON VIABLE ROUTE, BECAUSE:-

A. WHATEVER THE PUBLICISTS SAY, THE CURRENT ROBOT TECHNOLOGY WOULD NOT SUPPORT SMALL TOLERANCE, HIGH VOLUME ASSEMBLY OF SMALL COMPONENTS INCLUDING WIRING.

B. THIS TYPE OF F.A.A. WOULD HAVE TO BE REPEATED UNIQUELY OVER MANY DIFFERENT PRODUCT GROUPS RANGING ACROSS, SAY STRINGTRIMMERS TO DRILLS, TO HEDGECLIPPERS, IN THEIR FRACTIONED VOLUMES.

C. THE QUESTIONABLE FINANCIAL AND TECHNICAL VIABILITY WOULD MORE THAN DISCOUNT ANY LABOUR SAVING.

NO, THAT ROUTE WOULD BE STRICTLY FOR THE ENJOYMENT AND SATISFACTION OF TECHNOLOGISTS, AND WOULD ACHIEVE NOTHING BUT A NET ONCOST ARISING FROM AN INCREASED OVERHEAD BURDEN FOR THE BUSINESS.

SO HOW ARE THE CONFLICTS BETWEEN THE VIABILITY OF FLEXIBLE ASSEMBLY AUTOMATION, THE NEED FOR INCREASED MARKET FLEXIBILITY AND NET PRODUCT COST REDUCTION RESOLVED. THE ANSWER LIES IN DELVING DEEP WITHIN THE WHOLE OF THE MANUFACTURING STRATEGY.

MARKET FLEXIBILITY - THE MANUFACTURING CONSEQUENCES

HOW FLEXIBLE IS FLEXIBLE? - AN EXAMPLE

THREE TYPICAL, VOLUME CONSUMER PRODUCTS IN THE BLACK & DECKER RANGE IN 1980 WERE DRILLS, JIGSAW AND STRINGTRIMMER. (SEE FIGS. 1, 2 & 3). FROM A MARKET PERCEPTION THREE TOTALLY DIFFERENT PRODUCTS IN FUNCTION SHAPE AND SIZE. HOWEVER, LETS LOOK AT THE THREE PRODUCTS FROM A MANUFACTURING VIEWPOINT, WITHOUT COMPROMISING THE CUSTOMERS PERCEPTION OR REQUIREMENTS.

ONE HIGH VOLUME WINDING FACILITY WILL PROVISION ALL THREE PRODUCTS AT THEIR ANNUAL SALES LEVEL. HOWEVER, THAT WOULD DEMAND A HIGH DEGREE OF FLEXIBILITY FROM THE AUTOMATED WINDING LINES. BUT HOW MUCH? A FUNDAMENTAL REVIEW OF POWER REQUIREMENTS SHOWS THAT ALL THREE PRODUCTS COULD BE POWERED BY THE SAME FRAME SIZE MOTOR, WITHOUT COMPROMISING PERFORMANCE IN ANY WAY. THE MOTOR WINDING LINES WOULD THEN REQUIRE THE FLEXIBILITY TO PRODUCE MOTORS WITH DIFFERENT WIND PATTERNS, NUMBERS OF TURNS AND WIRE SIZE. THAT TECHNOLOGY EXISTS AT MINIMUM PREMIUM OVER MORE STANDARD WINDING TECHNOLOGY, SO BY FIRST EXAMINING THE NEED FOR PROLIFERATION IN FRAME SIZES AND KEY RELATIVE DIMENSIONS (E.G. BEARING CENTRE DISTANCES, RELATIVE TO LAMINATION STACK), THEN RATIONALISING, THE FLEXIBILITY REQUIREMENT IS CONSTRAINED TO WITHIN VIABLE, CURRENTLY AVAILABLE TECHNOLOGY. IN FACT THE WINDING LINES ARE FULLY INTEGRATED, COMPUTER CONTROLLED, AND ENCOMPASS SUB ASSEMBLY OF SHAFT AND SECONDARY INSULATION, FINISH SHAFT GRINDING, AUTOMATIC GAUGING AND RESET FEEDBACK, INSULATION FLASH TESTING, SELECTION AND PRESS FITTING OF LAMINATIONS TO SHAFT, PRESS FITTING COMMUTATOR, SLOT INSULATION, WINDING (INCLUDING COMMUTATOR CONNECTION), SLOT WEDGING OVER THE WINDING, WELDING COMMUTATOR CONNECTION, TRICKLE IMPREGNATING, FINISH TURNING THE COMMUTATOR, PRESS FITTING THE FAN, DYNAMICALLY BALANCING IN TWO PLANES, FUNCTIONAL AND SAFETY TESTING.

ALL OF THESE OPERATIONS ARE FULLY LINKED AND FLOWING AT 600 PIECES PER HOUR, WITH ON BOARD COMPUTER CONTROL AND FAULT DIAGNOSIS, RESETTABLE FOR WIND PATTERN AND NUMBER OF TURNS AT THE PRESS OF A BUTTON, AND DATA COLLECTION OF LINE PERFORMANCE IN TERMS OF DOWNTIME, THROUGHPUT, QUALITY AND FINISHED PRODUCT INTEGRITY. SO HOW FLEXIBLE IS FLEXIBLE IN SUB ASSEMBLING, WINDING AND TESTING A MOTOR ARMATURE? NOT VERY, BUT WHO NEEDS IT. THREE FUNDAMENTALLY DIFFERENT PRODUCTS WITH THREE FUNDAMENTALLY DIFFERENT MOTORS START TO BE PERCEIVED IN MANUFACTURING TERMS AS MORE SIMILAR THAN DIFFERENT AFTER STRATEGIC RATIONALISATION OF FRAME SIZE AND STANDARDISATION OF KEY RELATIVE DIMENSIONS. SUDDENLY, THE FLEXIBILITY REQUIREMENT OF THE MOTOR MANUFACTURING PROCESS IS TRANSFORMED INTO IMMEDIATELY AVAILABLE, VIABLE HIGH VOLUME TECHNOLOGY WITH NO COMPROMISE IN PRODUCT PERFORMANCE AND NO IMPOSED CONSTRAINTS ON THE MARKET. IN FACT THE AUTOMATION OF MOTOR MANUFACTURING BRINGS A NEW LEVEL OF INTEGRITY TO THE MARKET IN CONSISTENCY AND 100% TESTING OF THE POWER SOURCE, AND INCREASING VALUE FOR MONEY. LET ME NOW MOVE ON TO THE ASSEMBLY OF FINISHED PRODUCT, WITH A SIMILAR STRATEGIC APPROACH.

AUTOMATIC ASSEMBLY OF FINISHED PRODUCT - A STRATEGIC APPROACH TO DESIGN FOR AUTOMATION

WHEN IS A DRILL NOT A DRILL?

I HAVE DESCRIBED THE STRATEGIC APPROACH AND THE CONSEQUENTIAL TECHNOLOGY FOR ARRIVING AT THE COMPONENTRY FOR THE HEART OF ANY POWER TOOL, THE MOTOR. BUT THE KEY TASK REMAINS TO ASSEMBLE THE ARMATURE, FIELD AND TRANSMISSION INTO A RECOGNISABLE FINISHED POWER TOOL. THE STRATEGIC APPROACH SAYS FIRST LOOK TO THE DESIGN FOR AUTOMATION. THE STARTING POINT IN ESTABLISHING THE MANUFACTURING PLAN FOR AUTOMATION WAS THE EXISTING DESIGN FOR MANUAL ASSEMBLY AND BY WAY OF EXAMPLE I WILL EXAMINE THE DRILL IN SOME DETAIL. (SEE FIG. 4). THE ORIGINAL DESIGN IS REPRESENTED HERE. LET ME CRITIQUE IT. FOUR MAJOR PLASTIC PARTS, TWO

OF WHICH MUST BE MOULDED IN HIGH VOLUME TO DEMANDING TOLERANCE LEVELSBECAUSE THEY CONTAIN MOULDED TO SIZE LOCATIONS FOR BEARINGS, THE MOTOR FIELD AND MOTOR BRUSH GEAR. VERY COMPLEX MOULDINGS WHICH WOULD BE CHALLENGING IN HIGH VOLUME WITHOUT THE ADDITIONAL CONSTRAINTS OF TIGHT TOLERANCES, UPON WHICH THE INTEGRITY OF THE FINISHED PRODUCT DEPENDS. FROM AN ASSEMBLY AUTOMATION POINT OF VIEW AN ABSOLUTE NIGHTMARE - AXIAL ASSEMBLY OF SUB COMPONENTS E.G. THE ARMATURE & FIELD INTO THE MOULDING, RADIAL ASSEMBLY OF SUB COMPONENTS, E.G. BRUSHES INTO MOULDED BRUSH LOCATIONS, FLEXIBLE MOTOR WIRING IN VIRTUALLY ALL PLANES, ORIENTATION OF 4 SUB ASSEMBLIES, AND THEIR ASSEMBLY, SCREWS IN TWO PLANES AND FROM OPPOSING DIRECTIONS ON ONE AXIS. ONLY THE UNWARY DIRECTOR, OR AN ENGINEER IN LOVE WITH ROBOTIC TECHNOLOGY FOR ITS OWN SAKE WOULD CONTEMPLATE AUTOMATIC ASSEMBLY OF THIS DESIGN.

ANALYSIS OF TOTAL ASSEMBLY TIME SHOWS:

TRANSMISSION SUB ASSEMBLY	20
MOTOR SUB ASSEMBLY, INCLUDING MOTOR WIRING	20
SWITCH WIRING	15
RUN IN AND TEST	20
PACKAGING	25
	100%

A MAJOR PART OF ASSEMBLY TIME IS THEREFORE MOTOR SUB ASSEMBLY AND MOTOR WIRING DUPLICATED ACROSS ALL OF THE PRODUCT ASSEMBLY LINES, FOR EXAMPLE THE JIGSAW AND THE STRINGTRIMMER. AUTOMATING THIS PART OF THE FINISHED PRODUCT ASSEMBLY WOULD THEREFORE HAVE TO BE DUPLICATED ON EACH INDIVIDUAL PRODUCT ASSEMBLY LINE, WHATEVER ITS VOLUME UNLESS A STRATEGIC DESIGN DECISION COULD BE TAKEN TO DIFFERENTIATE THE MAJOR MOTOR BEARING, BRUSH AND WIRING LOCATIONS FROM THE MAIN PRODUCT HOUSING. LEAPING STRAIGHT TO THE CONCLUSION OF THAT STRATEGY DEBATE, THE DESIGN FOR AUTOMATION WAS A REVOLUTIONARY CONCEPT IN VOLUME POWER TOOLS, THE SELF CONTAINED, FINISHED, RUN AND TESTED MOTOR MODULE AND MODULARISED TRANSMISSION ENCLOSED BY NON STRUCTURAL CLAMSHELL MOULDINGS. (SEE FIG. 5).

LET ME NOW CRITIQUE THIS DESIGN. THE MOTOR IS ASSEMBLED FROM RATIONALISED AND STANDARDISED ARMATURES AND FIELDS IN VERY HIGH VOLUME, BECAUSE THE FINISHED MOTORS NOW SUPPLY THE COMBINED VOLUMES OF THREE PRODUCT RANGES. THE ASSEMBLY OF THE RUNNING MOTOR IS RELATIVELY EASY TO AUTOMATE, AND VERY VIABLE BECAUSE THE CAPITAL INVESTMENT IS SUPPORTED BY A VOLUME IN THE ORDER OF 2 MILLION PER YEAR. THE INTEGRITY OF THE MOTOR IS ENHANCED BY A WHOLE ORDER, BECAUSE IT IS ASSEMBLED WITH MACHINE PRECISION, AND CAN, FOR THE FIRST TIME, BE 100% RUN AND TESTED AS A MOTOR RATHER THAN AS A FINISHED PRODUCT WHEN ALL VALUE HAS BEEN ADDED. THE FLEXIBILITY REQUIREMENT OF DIFFERENT WIRING REQUIREMENTS FOR DIFFERENT WORLD MARKETS CAN BE EASILY ACCOMMODATED AND PROGRAMMABLE WITHIN THE AUTOMATION. A MAJOR PART OF ANY FINISHED PRODUCT ASSEMBLY TIME IS REMOVED FROM INDIVIDUAL PRODUCT LINES AND CONSOLIDATED INTO ONE PIECE OF AUTOMATION, REMOVING THE DUPLICATED CAPITAL COST OF MOTOR ASSEMBLY FROM EACH INDIVIDUAL PRODUCT LINE. THE FINISHED MODULARISED MOTOR IS ALSO ROBUST, AND EASILY MANIPULABLE IN SUBSEQUENT AUTOMATED ASSEMBLY. COMBINING THE MODULARISED MOTOR WITH VARIOUS MODULARISED TRANSMISSIONS ALSO GIVES A RANGE OF PRODUCTS. FOR EXAMPLE, THOUGH IT IS BARELY RECOGNISABLE TO THE CONSUMERS EYE, A MOTOR MODULE AND A MODULARISED DRILL GEARBOX COMBINED, IS FUNCTIONALLY A DRILL, AND A MOTOR MODULE AND A MODULARISED JIGSAW TRANSMISSION IS FUNCTIONALLY A JIGSAW. (SEE FIG. 6). THEY THEN ONLY REQUIRE TO BE PLACED IN AN AESTHETIC, ERGONOMIC SKIN, CABLED, RUN IN AND TESTED TO BECOME FINISHED PRODUCT OF EXCELLENCE IN THE CUSTOMERS PERCEPTION.

TO CONCLUDE THE CRITIQUE OF THIS STRATEGIC DESIGN FOR AUTOMATION, THE MAJOR CLAMSHELL MOULDINGS ARE NO LONGER CRITICAL IN MOULDED DIMENSIONAL TOLERANCES, THERE ARE ONLY TWO, AND THERE IS SUBSTANTIAL MATERIAL COST SAVING IN CHANGING FROM GLASS FILLED NYLON IN THE OLD DESIGN, REQUIRED FOR LOAD BEARING AND HEAT RESISTANCE, TO MUCH CHEAPER GRADES OF PLASTIC ASSOCIATED WITH A ROLE ONLY IN AESTHETICS AND ERGONOMICS. FURTHER, REDESIGN OF A PRODUCT RANGE FOR AESTHETICS, STYLE OR ERGONOMICS DOES NOT INVOLVE THE DESIGNER IN ANY OTHER CONSIDERATION THAN THE PROJECTED MARKET FASHION SINCE ALL MAJOR FUNCTIONAL LOCATIONS ARE CONTAINED WITHIN THE MOTOR AND TRANSMISSION MODULE. ENGINEERING RESOURCE, AND REPLACEMENT PRODUCT LEADTIMES ARE THEREFORE MUCH REDUCED, VITAL IN AN INCREASINGLY 'FASHION' CONSCIOUS AND COMPETITIVE MARKET. VARIATIONS IN MARKET PREFERENCES FOR STYLE CAN SIMILARLY BE EASILY ACCOMMODATED. FOR EXAMPLE, THE CONTINENTAL EUROPEAN PREFERENCE IN DRILLS TENDS TO BE THE PISTOL GRIP CONFIGURATION, WHEREAS THE U.K. PREFERENCE TENDS TO BE THE MID HANDLE CONFIGURATION. (SEE FIG. 7). THE MANUFACTURING AUTOMATION HARDLY CARES, SINCE IT ONLY HAS TO COPE WITH VARIATIONS IN THE EXTERNAL AESTHETIC SKIN, ALL THE REMAINDER OF THE INTERNALS BEING COMMON, AND ASSEMBLED ON THE SAME MACHINE. (SEE FIG. 8). THIS DESIGN APPROACH, THEREFORE, BRINGS SIGNIFICANT ADDITIONAL FLEXIBILITY TO THE MARKET PLACE WHERE PREVIOUSLY IT DID NOT EXIST.

WHEN IS A DRILL NOT A DRILL? WELL, FROM A DESIGNERS POINT OF VIEW, OR THE AUTOMATIC ASSEMBLY EQUIPMENTS POINT OF VIEW WHEN ITS A JIGSAW ....
    OR A STRIMMER ....
    OR ANY OTHER HAND HELD POWER TOOL.

    (SEE FIG. 9)

FLEXIBLE ASSEMBLY AUTOMATION IN HIGH VOLUME - THE PRACTICAL REALITY

THIS STRATEGIC APPROACH TO AUTOMATION, STARTING WITH THE CUSTOMER, THE MARKET AND WORKING ALL THE WAY BACK THROUGH THE MANUFACTURING PROCESS AND THE DESIGN, ISOLATES THE TRUE REQUIREMENT FOR FLEXIBLE ASSEMBLY, MINIMISING WASTEFUL AND UNNECESSARY SOPHISTICATION, AND THROUGH THE RELATIVE SIMPLICITY OF THE SOLUTION, OFFERS A BETTER GUARANTEE OF SUCCESS, WITH OPTIMUM CAPITAL INVESTMENT AND PRODUCT COST. WHAT OF THE REALITY, HOWEVER, THE APPLICATION OF THIS MANUFACTURING STRATEGY. IN THE TIME ALLOCATED I WILL DESCRIBE THE INTEGRATED MANUFACTURING PROCESS FOR THE GROUP OF PRODUCTS USED FOR EXAMPLE ABOVE, WITH THE COMMON DENOMINATOR OF MOTOR FRAME SIZE, THOUGH THE SAME PRINCIPLE HAS BEEN APPLIED ACROSS THE WHOLE PRODUCT RANGE MANUFACTURED AT BLACK & DECKER, SPENNYMOOR.

(REFER FIG. 10).

THE ARMATURE AND FIELD LINES, PREVIOUSLY DESCRIBED ARE SOLIDLY LINKED TO THE MOTOR MODULE ASSEMBLY MACHINE.

THE MOTOR MODULE ASSEMBLY MACHINE RECEIVES TESTED ARMATURES AND FIELDS VIA CONVEYORS AND ALL OTHER COMPONENTRY - END MOULDINGS, BEARINGS, HEAT SINKS, WASHERS, BRUSHES, BRUSH BOX PRESSINGS, ETC., FROM A VARIETY OF BOWL FEEDERS OR BANDOLIERS, ALL WITH A MINIMUM OF 8 HRS STORAGE CAPACITY. AN IN LINE POWERED AND FREE CONVEYOR CARRIES THE ACCUMULATING SUB ASSEMBLY THROUGH THE MACHINE ON ASSEMBLY PALLETS, WHICH ARE RETURNED TO THE START POSITION ON A RETURN CONVEYOR WHICH ALSO INCORPORATES A REPAIR STATION IN THE LATEST GENERATION. THE MACHINE INCORPORATES SELECTIVE ASSEMBLY FOR ARMATURE END FLOAT WITHIN BEARING CENTRES, AND CULMINATES BY APPLYING A RANGE OF ELECTRICAL SAFETY TESTS AT HIGH VOLTAGE AND A FUNCTIONAL RUN TEST TO MEASURE NO LOAD CURRENT AND WHOLE UNIT INTEGRITY. ON BOARD COMPUTERS MONITOR ALL OPERATIONS AND

ON FAILURE OF ANY ONE OPERATION, WILL PASS THE PALLET THROUGH THE REMAINING OPERATIONS WITHOUT FURTHER ATTEMPTED WORK. ON THE LATEST GENERATION OF THIS MACHINE, SUCH A RECORDED FAILURE, FOR EXAMPLE, A FAILURE TO LOAD A SCREW, WILL RESULT IN THAT PALLET BEING RETURNED COMPLETE TO THE BEGINNING OF THE PROCESS, FOR A SECOND ATTEMPT, AFTER WHICH, IF STILL A FAILURE, IT IS PARKED AT THE REPAIR STATION FOR MANUAL INTERVENTION.

THE LATEST GENERATION ALSO HAS THE PROGRAMMABLE CAPABILITY TO FIX A VARIETY OF SUPPRESSION COMPONENTS TO SUITE A WHOLE RANGE OF WORLD TEST BOARD REQUIREMENTS, AND TO CHOOSE FROM A VARIETY OF MOTOR/TRANSMISSION INTERFACES.

THE ON BOARD COMPUTER MONITORS PERFORMANCE OF THE LINE ON VIRTUALLY ANY HISTORIC BASIS, AND PROVIDES ANY OPERATING ANALYSIS BY STATION OR COMPONENT, AND A FULL QUALITY ASSURANCE ANALYSIS.

FOR THE PRESENT, THE RUN, TESTED MOTOR PACKS ARE THEN TRANSFERRED MANUALLY BETWEEN THE MOTOR PACK MACHINE AND THE VARIOUS PRODUCT ASSEMBLY LINES. THIS MISSING LINK AWAITS THE IN HOUSE DEVELOPMENT OF A COMPREHENSIVE MANUFACTURING PLANNING AND CONTROL SYSTEM BEFORE WE WOULD WISH TO CONTEMPLATE THE INSTALLATION OF AN AUTOMATIC, FLEXIBLE LINKAGE SYSTEM, PROBABLY A.G.V.'S.

FOR TIME CONSTRAINTS, I WILL FOLLOW THE ASSEMBLY PROCESS FOR THE DRILL ONLY. IN THIS CASE THE MOTOR PACK IS PRESENTED TO THE DRILL TRANSMISSION ASSEMBLY MACHINE WHICH IS A ROBOTIC ASSEMBLY CELL DESCRIBED IN EXCELLENT DETAIL BY JACK HOLLINGUM IN VOLUME 6 OF ASSEMBLY AUTOMATION MAGAZINE, FEBRUARY, 1986. THIS CELL ASSEMBLES THE MODULAR DRILL TRANSMISSION, WITH FINAL ASSEMBLY OF THE TRANSMISSION TO THE MOTOR PACK. THE WHOLE IS THEN RE-TESTED IN THE RUNNING MODE, GUARANTEEING ACCUMULATING INTEGRITY OF THE EVENTUAL FINISHED UNIT. FROM HERE THE PROGRESSIVE SUB ASSEMBLY ENTERS THE FINAL ASSEMBLY LINE. THE EXTERNAL CLAMSHELLS ARE MANUALLY PLACED IN PALLETS, AND THE MOTOR/TRANSMISSION PLACED MANUALLY INTO ONE HALF OF THE CLAMSHELLS. SWITCH AND CABLE PLACEMENT ARE MADE MANUALLY AND THE TOP CLAMSHELL COVER PLACED OVER THE WHOLE. THE PALLET THEN ENTERS THE AUTOMATIC IN LINE SYSTEM FOR SCREW SELECTION, PLACEMENT AND TORQUE DOWN, FOLLOWED IN AUTOMATIC SEQUENCE WITH LIVE RUN-IN AND A COMPREHENSIVE RANGE OF ELECTRICAL TESTS. THE CHUCK IS FITTED AUTOMATICALLY AND THE DRILL IS THEN READY FOR PACKAGING. CARTONS ARE AUTOMATICALLY ERECTED, THE DRILL PLACED TO THE CARTON MANUALLY, THEN THE CARTON IS CLOSED. BAR CODE AND LABEL APPLICATION THEN FOLLOWS, BEFORE CARTONS ARE COLLATED INTO GROUPS FOR AUTO SHRINKWRAPPING. SHRINKWRAPPED CARTONS ARE THEN TAKEN FROM THE END OF THE PACKAGING LINE BY A ROBOT PALLETISER WHICH FILLS AN AUTO FED PALLET, BEFORE CALLING UP AN A.G.V. FOR DELIVERY OF THE PALLET, VIA AUTO PALLET STRETCHWRAP TO THE LOADING DOCK OR STORAGE LAYDOWN AREA WHICHEVER IS PROGRAMMED.

SO THE WHOLE OF THE DRILL RANGE MANUFACTURING PROCESS IS AUTOMATED AND LINKED IN SIGNFICIANT ISLANDS OF AUTOMATION, WITH THE EXCEPTION OF TWO MANUAL OPERATIONS, THE COLLATION OF PARTS TO THE AUTO ASSEMBLY LINE, AND PLACEMENT OF THE FINISHED, TESTED DRILL TO THE CARTON. THE COLLATION OF PARTS TO THE AUTO ASSEMBLY CARRIERS IS LARGELY AUTOMATABLE WITHIN AVAILABLE TECHNOLOGY, EXCEPT FOR SWITCH AND CABLE WIRING. A HARD WIRING DESIGN OPTION FOR SWITCH TO MOTOR IS FEASABLE BUT THERE WOULD STILL REMAIN THE PROBLEM OF CABLE CONNECTION. THIS IS SURMOUNTABLE BY EXTENDING THE HARD WIRING SOLUTION ALL THE WAY FROM THE MOTOR, THROUGH THE SWITCH TO A CABLE PLUG-IN CONNECTION ON THE DRILL HANDLE. HOWEVER, THIS IS NOT ACCEPTABLE IN VARIOUS TEST BOARD MARKETS AND CURRENTLY WOULD INCREASE UNIT COST. THE CURRENT MANUAL METHOD OF CABLE AND SWITCH WIRING REMAINS AS THE ONLY VIABLE METHOD AVAILABLE.

PLACEMENT OF DRILL TO ERECTED CARTON IS ALSO TECHNICALLY FEASIBLE, BUT NOT FINANCIALLY VIABLE. MANUAL INTERVENTION AT THIS POINT ALSO OFFERS THE OPPORTUNITY FOR FINAL AESTHETIC INSPECTION BEFORE THE DRILL IS APPRAISED BY THE CRUEL AND DISCERNING EYE OF OUR INDIVIDUAL CUSTOMERS.

EPILOGUE

FLEXIBLE ASSEMBLY OF HIGH VOLUME CONSUMER PRODUCT IS TECHNICALLY AND FINANCIALLY FEASIBLE, AND CAN EVEN ENHANCE YOUR MARKET OFFERING IN TERMS OF COST, QUALITY AND RANGE. HOWEVER, THE VITAL COMPONENT FOR SUCCESS IS THE STRATEGIC THINKING WHICH PRECEDES THE EXECUTION, STARTING WITH THE CUSTOMER AND HIS NEEDS, NEVER COMPROMISING THE MARKET FOR THE SAKE OF MANUFACTURING OBJECTIVES AND WORKING THROUGH DESIGN FOR AUTOMATION FOR OPTIMUM CAPITAL AND PRODUCT COST. OUR EXPERIENCE WOULD INDICATE THAT THE HARDWARE TO FOLLOW THROUGH YOUR STRATEGY IS READILY AVAILABLE FROM A DEDICATED AND PROFESSIONAL CAPITAL EQUIPMENT INDUSTRY, WHICH IS JUST WAITING TO TRANSFORM GOOD STRATEGIC THINKING INTO HARDWARE REALITY ON AN INTEGRATED PARTNERSHIP BASIS, RATHER THAN THE TRADITIONAL ADVERSORIAL BASIS. I CANNOT SPEAK TOO HIGHLY OF THE CONTRIBUTION MADE BY OUR CAPITAL EQUIPMENT SUPPLIERS TO THE OVERALL INITIATIVE AND ITS SUCCESS. AFTER THAT, ALL THAT IS NEEDED IS THE FINANCE, CONSTANT DIRECTION, REVIEW AND REFINEMENT OF THE PLAN, DEDICATED ENGINEERS, AND ABOVE ALL, THE COURAGE AND FAITH TO ENTER THE SECOND GREAT INDUSTRIAL REVOLUTION. PRESIDENT POMPIDOU WAS ONLY TWO THIRDS RIGHT.

FIG. 1 - A BLACK & DECKER DRILL CIRCA 1980 - PRE AUTOMATION

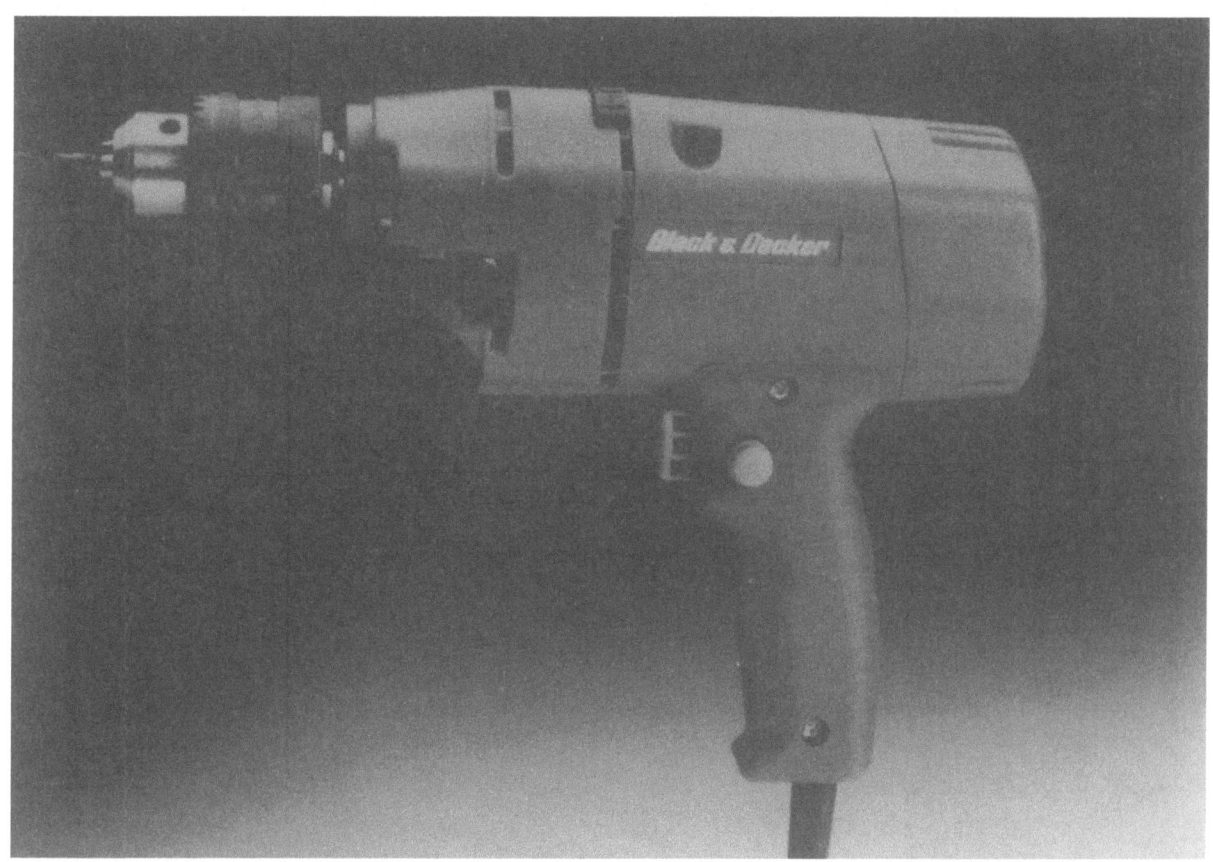

FIG. 2 - A BLACK & DECKER JIGSAW CIRCA 1980 - PRE AUTOMATION

FIG. 3 - A BLACK & DECKER STRIMMER CIRCA 1980 - PRE AUTOMATION

FIG. 4 - DRILL DESIGNED FOR MANUAL ASSEMBLY

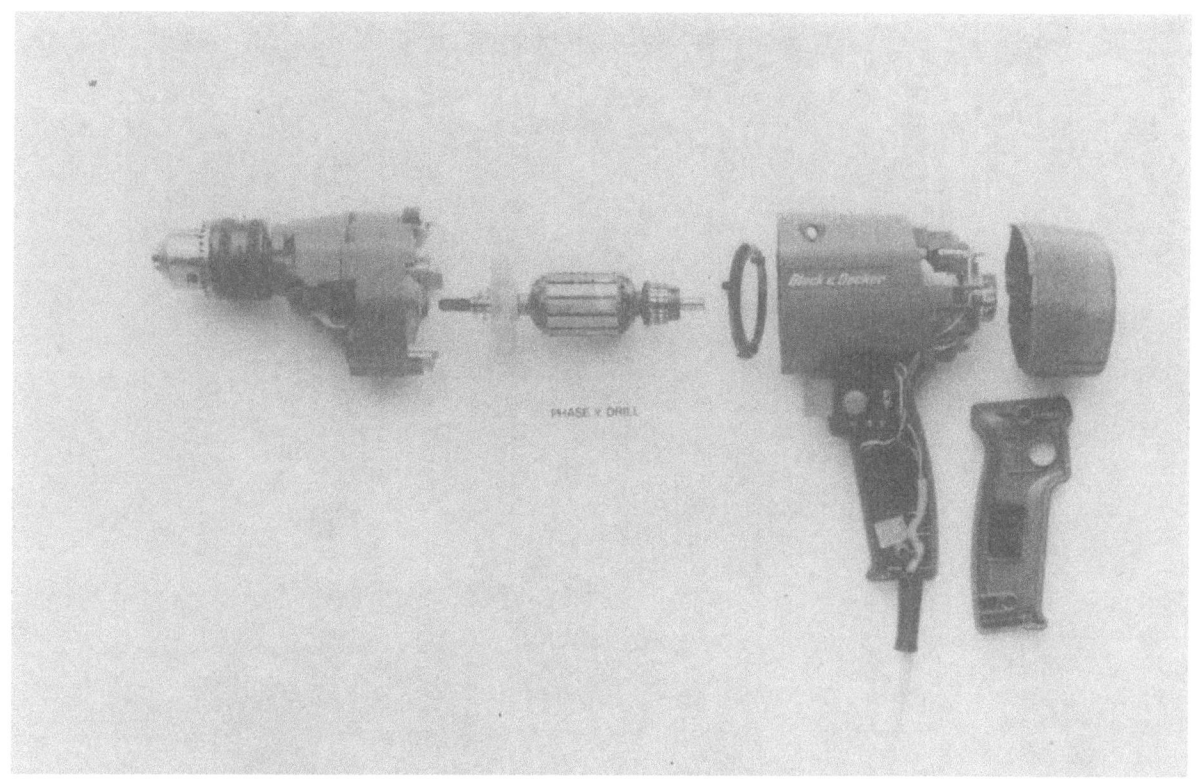

FIG. 5 - DRILL DESIGNED FOR AUTOMATIC ASSEMBLY

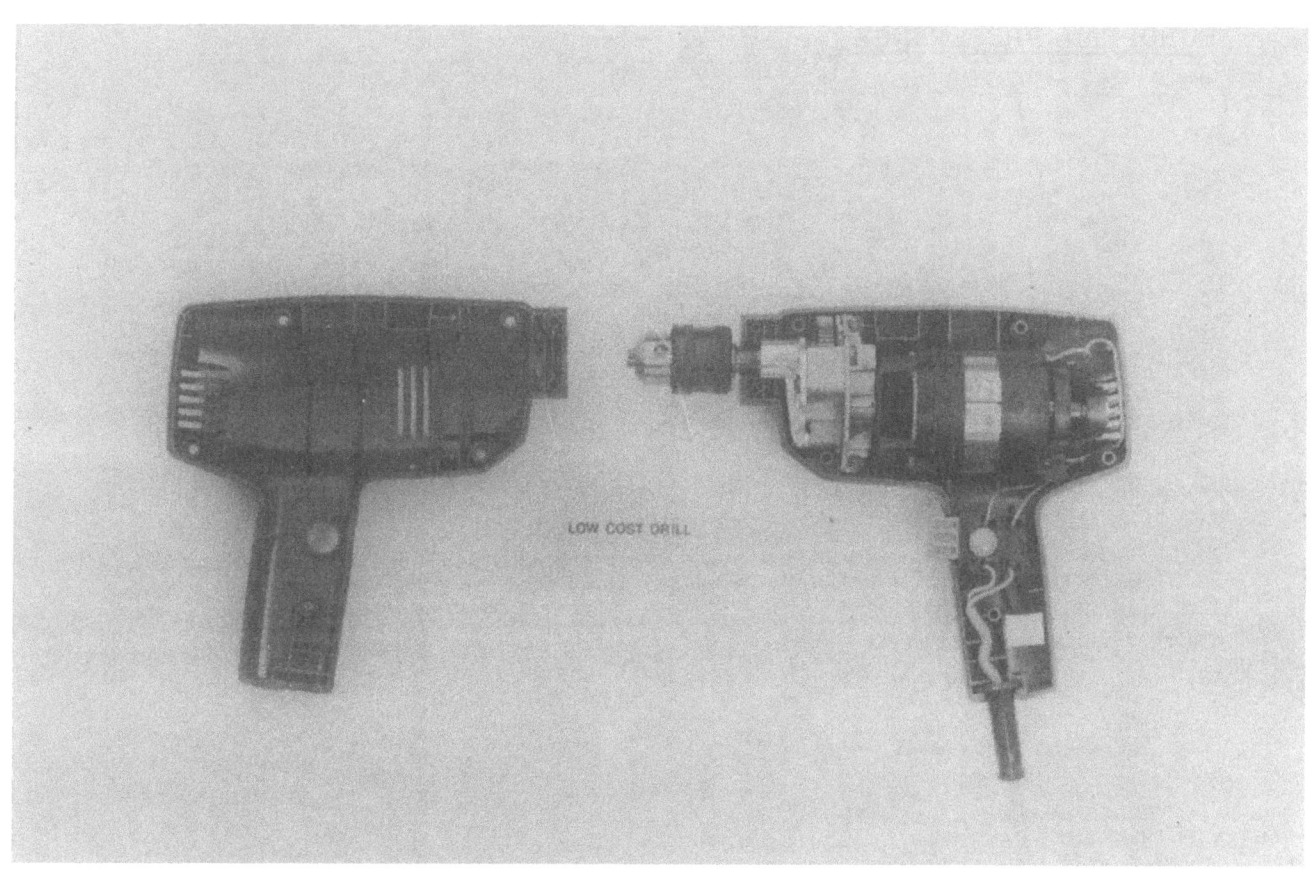

FIG. 6 - JIGSAW DESIGNED FOR AUTOMATIC ASSEMBLY - UNDRESSED

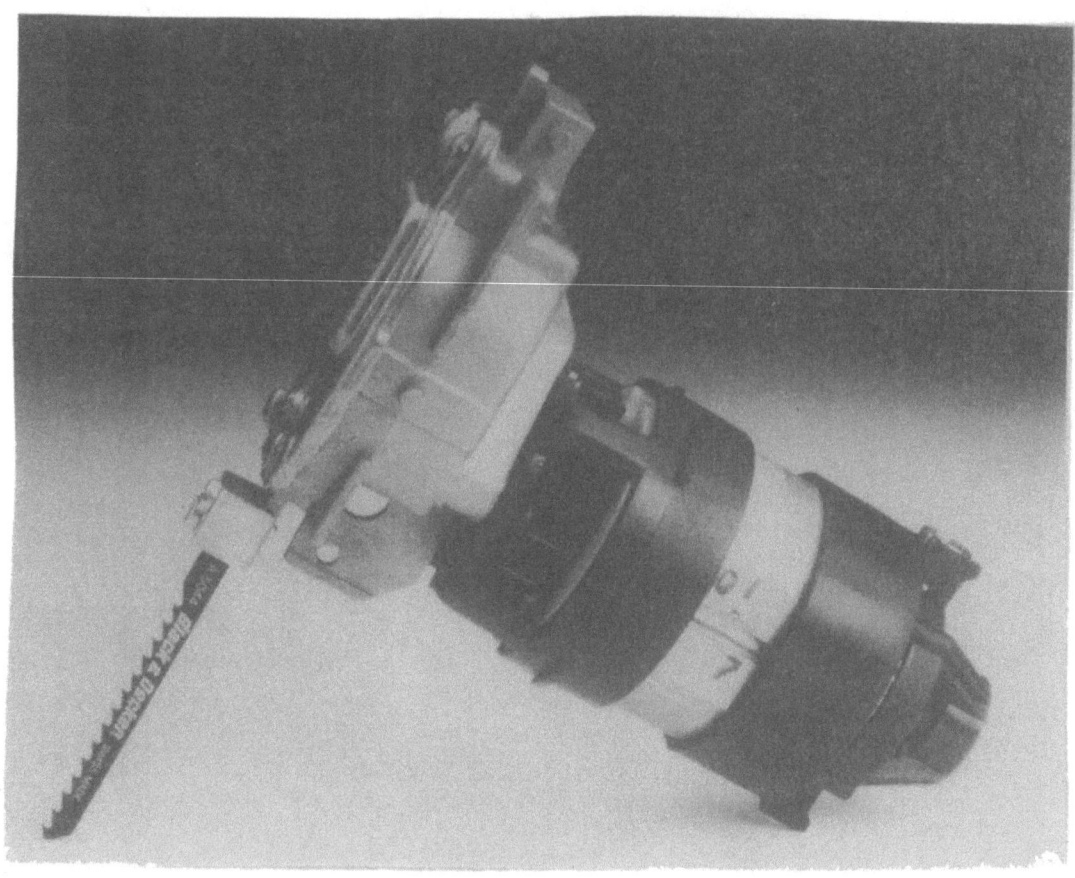

FIG. 7 - 330W RANGE DRILLS - U.K. PREFERENCE AND CONTINENTAL EUROPEAN PREFERENCE

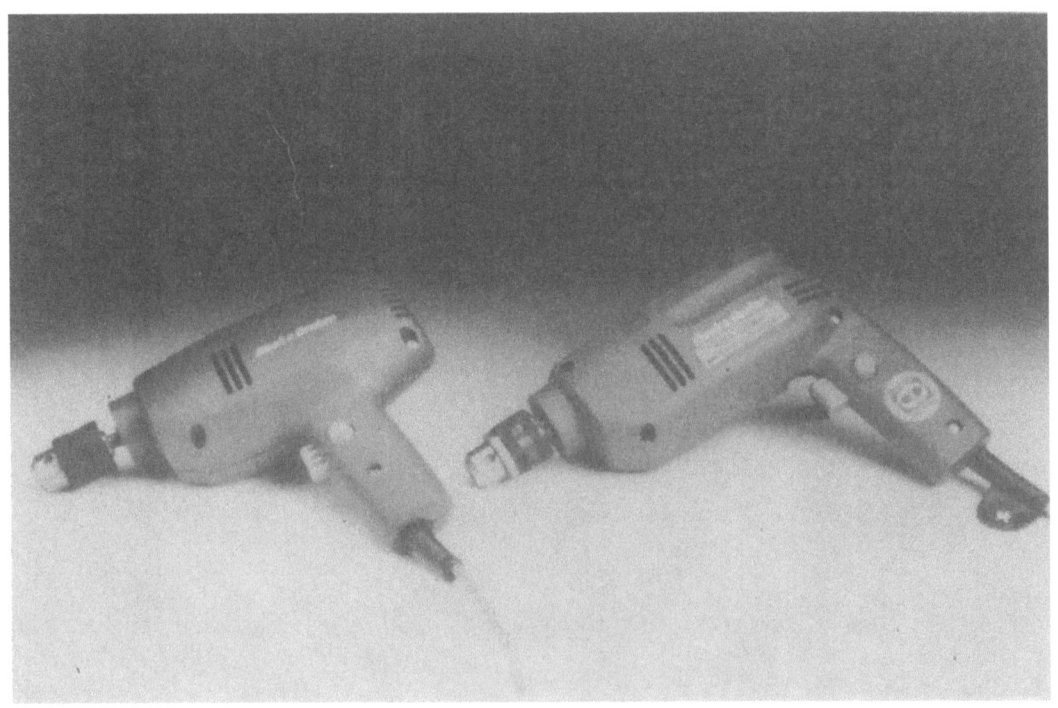

FIG. 8 - SAME DRILL INTERNALS - ONE DRESSED FOR THE U.K. MARKET - ONE DRESSED FOR THE EUROPEAN MARKET

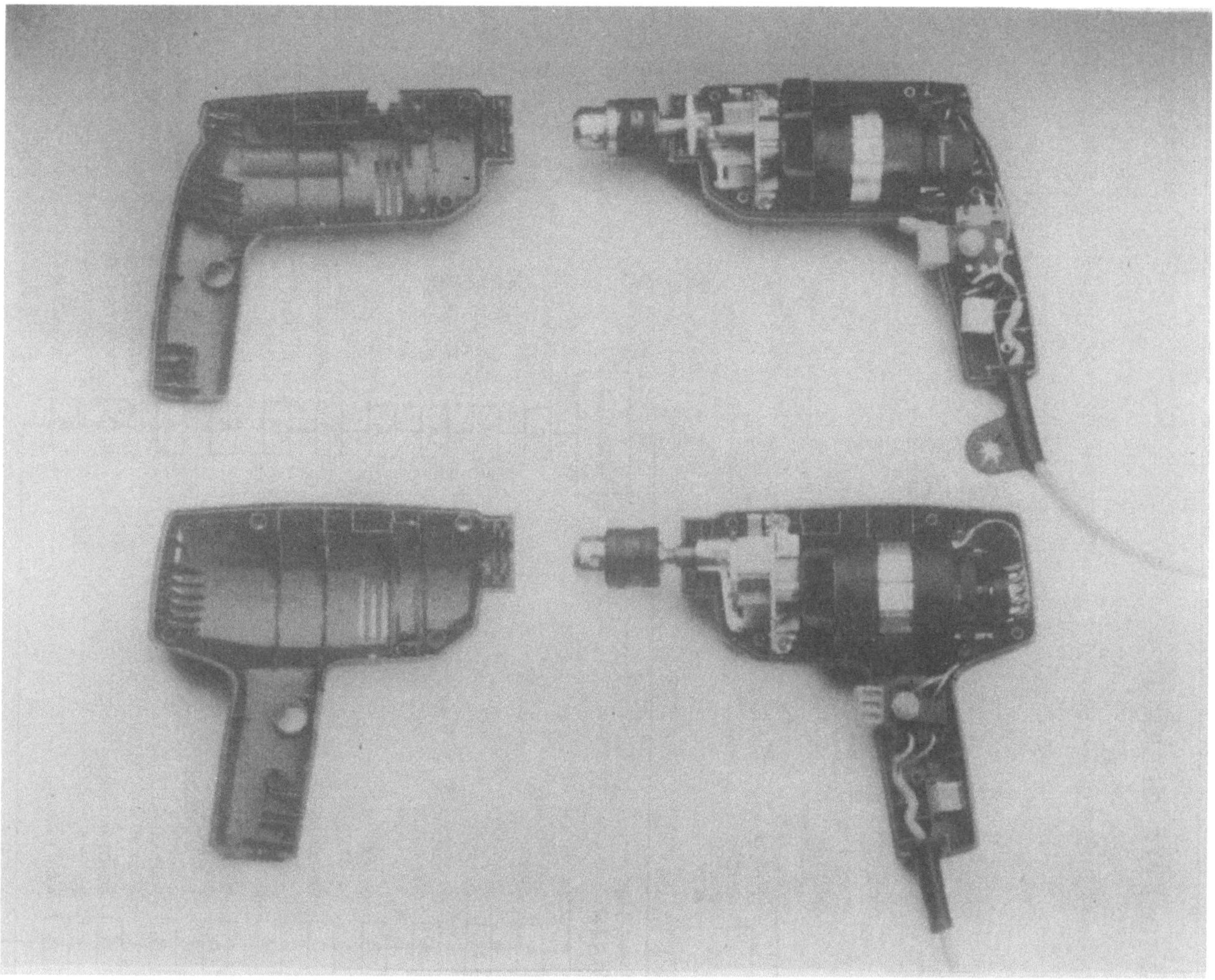

FIG. 9 - AUTOMATIC ASSEMBLIES FOR A DRILL, A JIGSAW AND A STRIMMER

FIG. 10 - SCHEMATIC FOR THE AUTOMATIC MANUFACTURE AND ASSEMBLY OF A RANGE OF HIGH VOLUME POWER TOOLS

KEY: M - MANUAL OPERATION   A - AUTOMATIC OPERATION

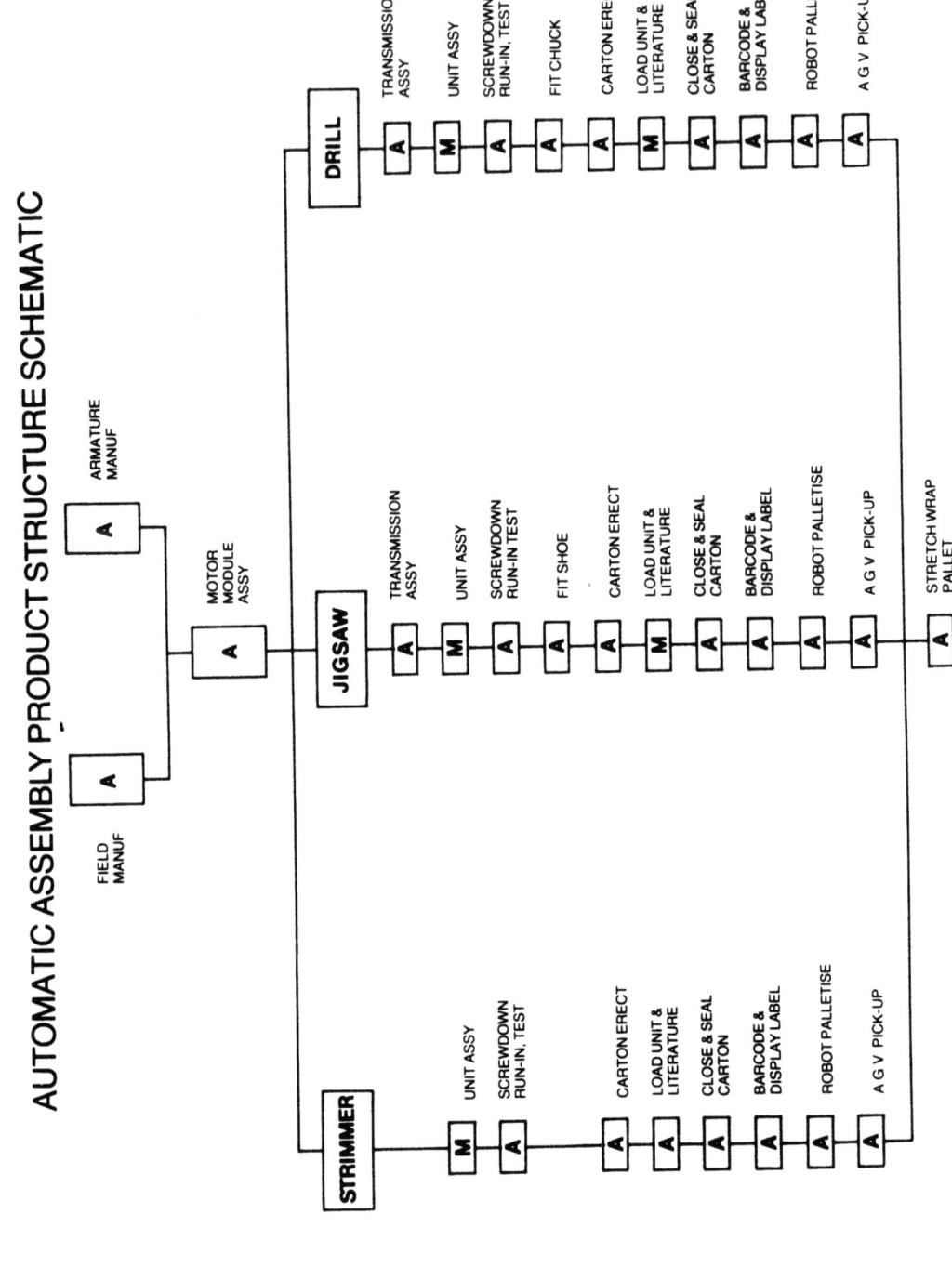

A flexible manufacturing system (FMS) for PCB production

S. G. Davey

IPL Information Processing Ltd, UK

ABSTRACT

Surface mounted printed circuit board (PCB) assembly equipment is expensive and therefore for it to be cost effective it must be properly utilised. If batch sizes are small, the large set-up times of such equipment may not be compatible with this requirement. The design and high level control of a simple flexible assembly system for surface mounted PCB manufacture, which overcomes this apparent conflict, is presented. This is achieved by equipment designed for fast mechanical reconfiguration and reprogramming via a host mini-computer system. This FMS exhibits full vertical integration with management and electronics design computer systems.

1. INTRODUCTION

Surface mounted printed circuit boards (PCBs) offer a considerable space saving over conventionally assembled boards. This reduced board area is particularly important in the confined environments of weapons systems. British Aerospace Naval Weapons Division has a requirement, therefore, for the production of small batches (< 50 units), of surface mounted PCBs.

Microprocessor based 'intelligent' equipment is becoming increasingly popular as a cost effective method of assembling PCBs quickly and accurately. However, such equipment is expensive and may not be fully utilised if product batches are small because of the potentially long set up times. This is the conflict that faced British Aerospace when they decided to set up a new facility for small batch manufacture of customised printed circuit boards utilising the latest generation of surface mounted devices. To overcome this problem the British Aerospace FMS was designed with the following principal features :-

(a) A central processor (host computer) capable of offline data preparation; fast downloading of configuration instructions to production line equipment; and line control, monitoring, and fault correction.

(b) Complete vertical integration with the site mainframe computer and electronics computer aided design (CAD) system for forward load control, and part-data acquisition.

(c) A local to line stores and inventory control system with facilities for offline job kitting.

(d) Fast mechanical reconfiguration of line machines.

(e) An efficient mechanical handling system for materials transport.

(f) A cellular line structure with facilities for independent operation of individual cells; in addition to normal integrated operation.

(g) A typical batch size of fifteen units and a maximum throughput of four hundred boards per operating day.

2. LINE DESIGN AND OPERATION

As surface mounted PCB production requires an unvarying series of processing steps a simple serial line design was adopted. However, to accommodate re-work and to add further flexibility, the line is divided into three processing cells, in addition to a stores cell. The function of the processing cells is as follows:

Cell One: Substrate Pre-clean, Print, and Assembly.

Cell Two: Assembly Bake, Solder, and Deflux.

Cell Three: Assembly Test.

Figure 1 shows a three dimensional representation of the finished line, and figure 2 an equipment layout plan. Figure 3 shows line control and materials flow.

2.1 Stores Cell

The stores cell provides the following facilities:

(a) Semi-Automatic kitting of jobs by an operator under control of the supervisory host computer.

(b) Job kitting feasibility studies.

(c) Stock forward planning and inventory control.

(d) Part extraction and re-booking facilities.

(e) A bar coded stock labelling system.

The stores hardware consists of four carousel storage units which operate under host computer control. The stores operator is provided with an industrial visual display unit (VDU), keyboard, and bar code reader.

Parts arriving are booked into the stores system via a comprehensive man-machine interface using the bar code reader and VDU. All parts have printed bar codes. Part quantities and locations are held in a database on the host computer system. Jobs may only be kitted when the database shows that there are sufficient parts available in the stores for manufacture of the complete job. During kitting, parts are extracted by the stores operator under the supervision of the host computer. Bar coding facilitates correct part identification and substrate traceability.

Each substrate, which is uniquely identified by serial number, is fixed in a tooling plate as it is extracted from the stores for kitting. Tooling plates are batched into cassettes containing up to seven plates. A given job may therefore consist of one or more cassettes.

Extracted components are kitted onto pallets for use by the pick and place machine in cell one. The positioning of parts on the pallets is monitored by the host computer for later compilation of the pick and place program.

2.2 Cell One: Assembly

The following processes are carried out in cell one:

(a) Substrate pre-clean (de-grease) in an ultrasonic boiling solvent cleaning system.

(b) Screen printing of the substrates with solder cream.

(c) Assembly of the PCBs by placement of components onto the substrates.

When a job enters the cell the line operator is instructed by the host computer to configure the pick and place machine with the correct pallet set (containing components), and load the cassettes (containing substrates) on the line input conveyor. A unique print screen for the part to be manufactured is loaded into the screen printer as part of the configuration process. The pick and place machine and screen printer programs are automatically downloaded by the host computer as part of the configuration sequence. Cell set-up is therefore a simple task and can be achieved in minutes rather than hours.

When configuration is complete the host computer allows the cell to start processing the job. This proceeds without host intervention unless a processing error occurs.

Cassettes are picked up from the input conveyor by an overhead transport unit which transports the batch of substrates through the three-tank pre-clean process. On completion of cleaning, the tooling plates are individually extracted from the cassettes by a robot and then placed in the solder screen printer. The screen printer prints solder cream onto the contact pads for the surface mounted devices on the substrate. The robot then moves the plate to the pick and place machine. This unit picks up to one hundred and twenty different component types from feeders mounted on the machine pallets and places them onto the substrate in an optimised program sequence. Assembled boards are ejected at the other side of the machine where they are removed by a second robot and re-batched into a waiting cassette. When the cassette is complete it may be moved into cell two, if the cell is free.

2.3 Cell Two - Solder

The following processes are carried out in cell two:

(a) Cassette pre-heat and solder cream cure.

(b) Vapour phase soldering.

(c) Cassette Cooling.

(d) Three stage boiling solvent deflux.

(e) Drying.

This cell is controlled by a microprocessor based single board computer which supervises the cell materials transport system and controls the operation of the processing stages. The host computer downloads processing parameters (eg. processing times and temperatures) as each cassette enters the cell. Up to four cassettes, each with a different processing requirement, may be processed simultaneously. As cassettes move through the cell, monitored parameters such as temperatures are generated and uploaded to the host computer for archive. Thus it is possible to extract the actual processing details of a particular substrate should the need arise. Transport through the cell is effected by two intelligent overhead transportation units which have the capability to pick up cassettes, transport them and leave them in any of the processing stages.

Cassettes entering the cell are picked up by one of the overhead transport units and placed in the first of two pre-heat ovens. When the cassette has reached a pre-set temperature and a specified 'soak' period has been observed, it is moved to the second pre-heat oven where a similar sequence of events takes place. Cassettes are then transported to the vapour phase reflow solder tank where condensation of an inert fluid, at a temperature of 215 degrees C, causes rapid heating and thus soldering in an airless environment. On removal from this stage, after a precise pre-programmed dwell period, the hot cassette is transported to the cooler which allows controlled solidification of the soldered joints in a cool air stream. The final stages of the process are a three stage solvent defluxing system, followed by hot air drying. After drying the cassettes are placed on an outstation for transfer into cell three.

2.4 Cell Three - Test

This cell contains the automatic test equipment, (ATE). Substrate tooling plates are individually removed, by a robot, from the cassette and placed in the automatic test equipment. Testing takes place via a 'bed of nails' test jig which makes contact with the test pads on the circuit board. On completion of testing substrates are returned to the cassette from whence they were taken.

The host computer downloads test programs to the ATE as part of the cell configuration sequence. Each type of substrate requires a unique test jig, which is mounted by the operator on orders from the host computer. The test jigs are barcoded to prevent incorrect jig usage. Test results are uploaded to the host computer for archive and interpretation by the line monitoring software.

3. LINE CONTROL AND INTEGRATION

The high level control of the FMS line is carried out entirely by the host mini-computer system. The production cycle, from job kitting to manufactured assembly testing, is coordinated with complete operator visibility by this machine. The host system has an extensive man-machine interface which requires only two operators and a production controller to support the production process. Each of these individuals is provided with a computer terminal, (see figure 3).

The host computer is also a principal node in the vertical integration of the FMS with the site mainframe computer and CAD systems. Therefore it is possible, for instance, for board designs to be processed giving raw assembly data which is downloaded to the host for compilation into placement programs, (see figure 4).

3.1 Production Control

A forecast jobs list of orders for the FMS is regularly downloaded from the site mainframe computer, (see figure 5). This list is sorted and displayed to the production controller in "due-date" order, to emphasise job priorities. Before a particular job may be manufactured it is necessary for the required manufacturing data and tooling to be available. When these conditions are satisfied, the production controller may select the job for manufacture. This causes a request to be sent to the mainframe computer for the download of production data; consisting of a kitting list, placement program and test program for the job. The job then enters the jobs awaiting manufacture (JAM) queue.

Jobs in the JAM queue are displayed to the operator and dispatched to the line in "due-date" order, provided there are sufficient components in the stores to complete manufacture. However, the production controller has facilities to override this order if necessary.

Once in manufacture, jobs are monitored by the host computer giving the production controller complete visibility of job status at any time. Stock usage data, process-monitored data and test results generated by the manufacturing cycle are uploaded to the site mainframe for part ordering and archive respectively.

3.2 Manufacturing Line Coordination

Cell configuration and error management is coordinated with the line operator via a VDU, keyboard and bar code reader. Owing to the stand alone capability of the line equipment, once configured, manufacture takes place with little interaction from the host computer until the job is completed by the cell, or an error occurs. (See section 2.)

Downloaded assembly data is compiled into optimised placement programs for the pick and place machine. These part programs are maintained on the host computer for possible future use.

Processing parameters uploaded from the cell two equipment and test results uploaded from cell three are archived on the host computer. This data may be examined by the production controller to aid process development and study product reject rates. The host software also carries out line error analysis, which may be used by the production controller for process optimisation and MTBF studies on the line equipment.

4. CONCLUSIONS

The British Aerospace Surface Mount FMS was installed at Filton during the first six months of 1985, and became fully operational in November 1985 after commissioning and integration. The 'bespoke' software system provided for the host computer system has operated reliably since installation.

5. ACKNOWLEDGEMENT

The author would like to thank the staff of British Aerospace PLC for assistance in the preparation of this paper, and for permission to publish the facts presented.

6. DISCLAIMER

The opinions expressed here are those of the author and not necessarily those of British Areospace PLC.

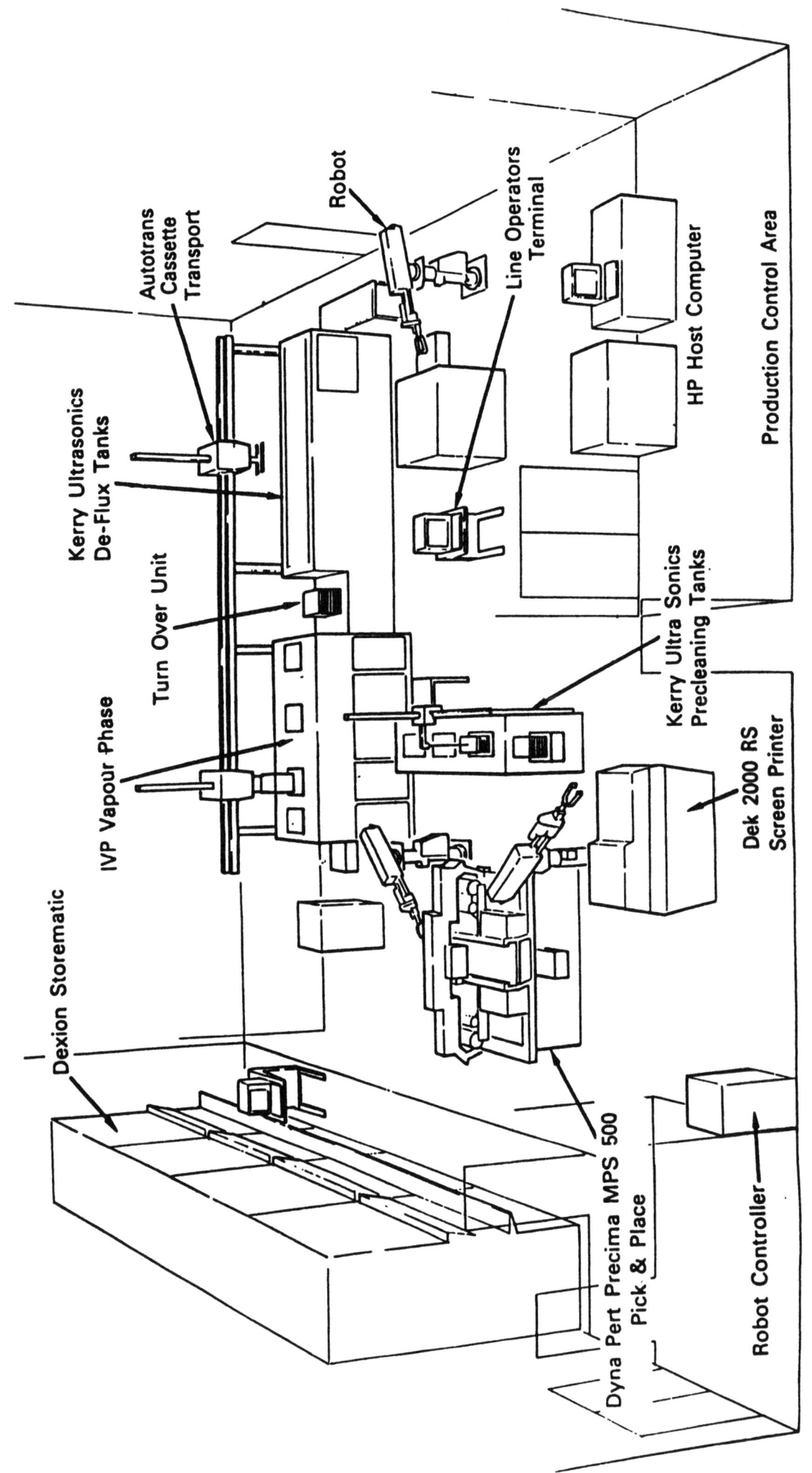

Figure 1 - British Aerospace PCB Flexible Assembly System

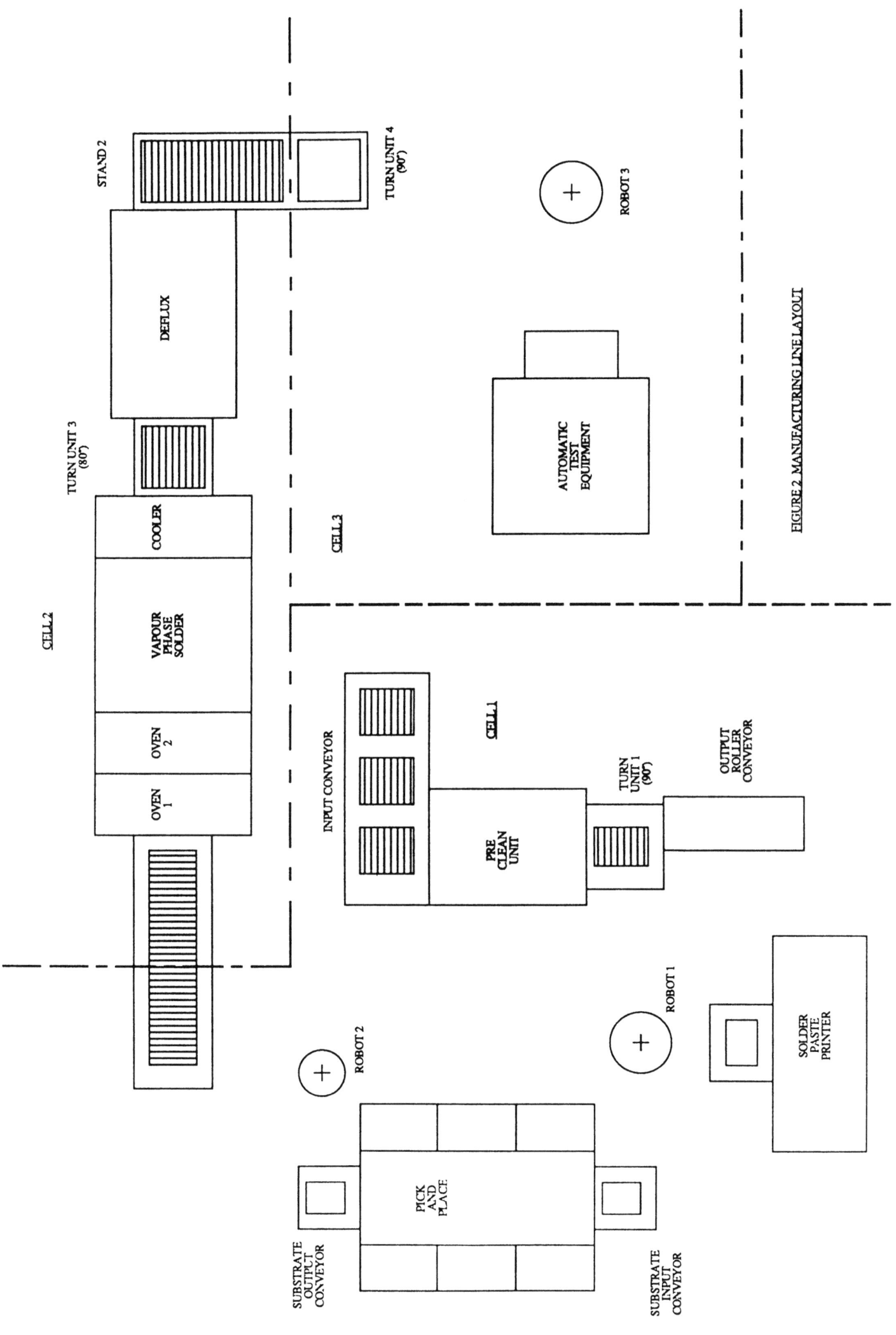

FIGURE 2 MANUFACTURING LINE LAYOUT

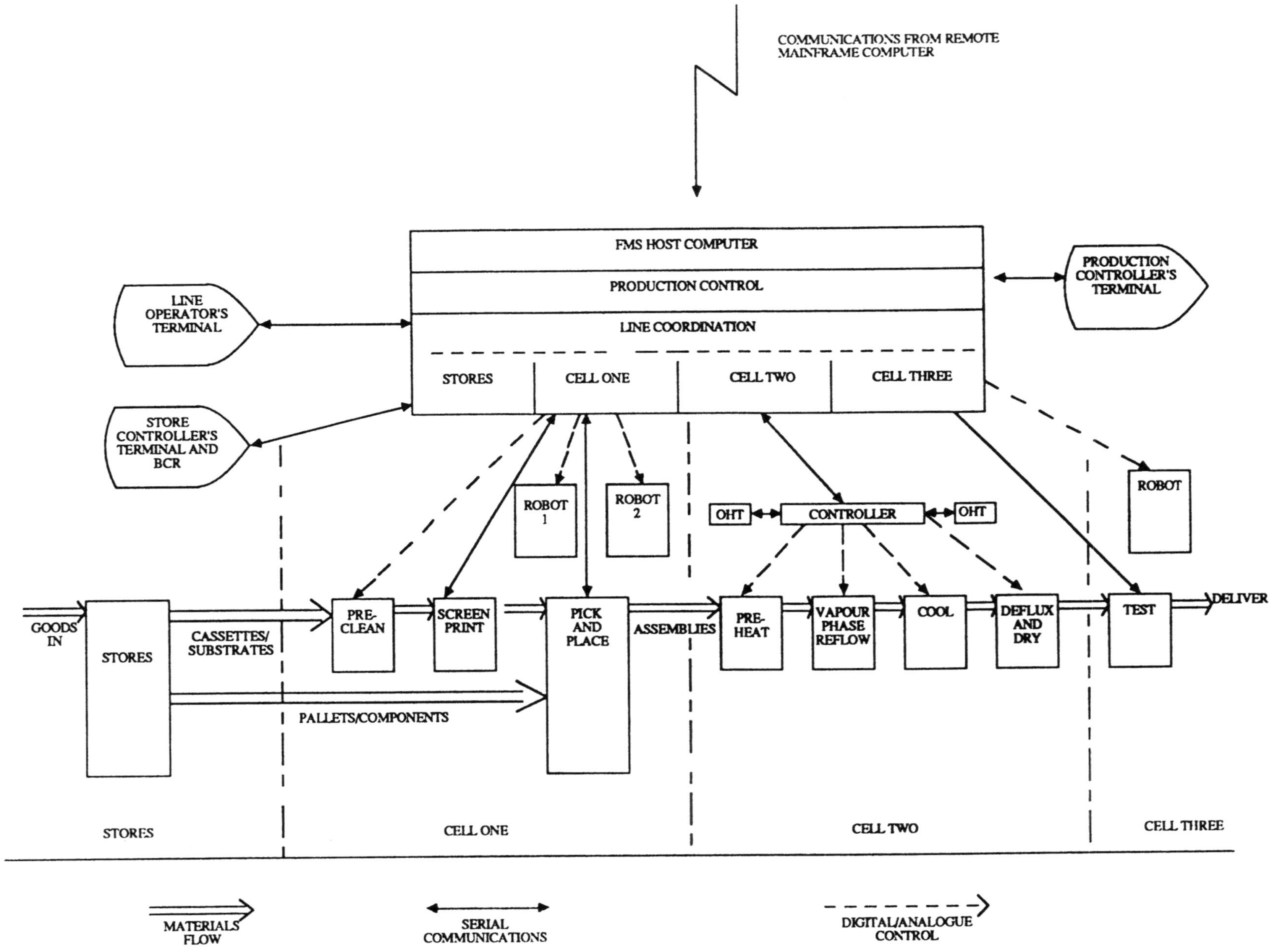

FIGURE 3 MATERIALS FLOW, CONTROL AND COMMUNICATIONS

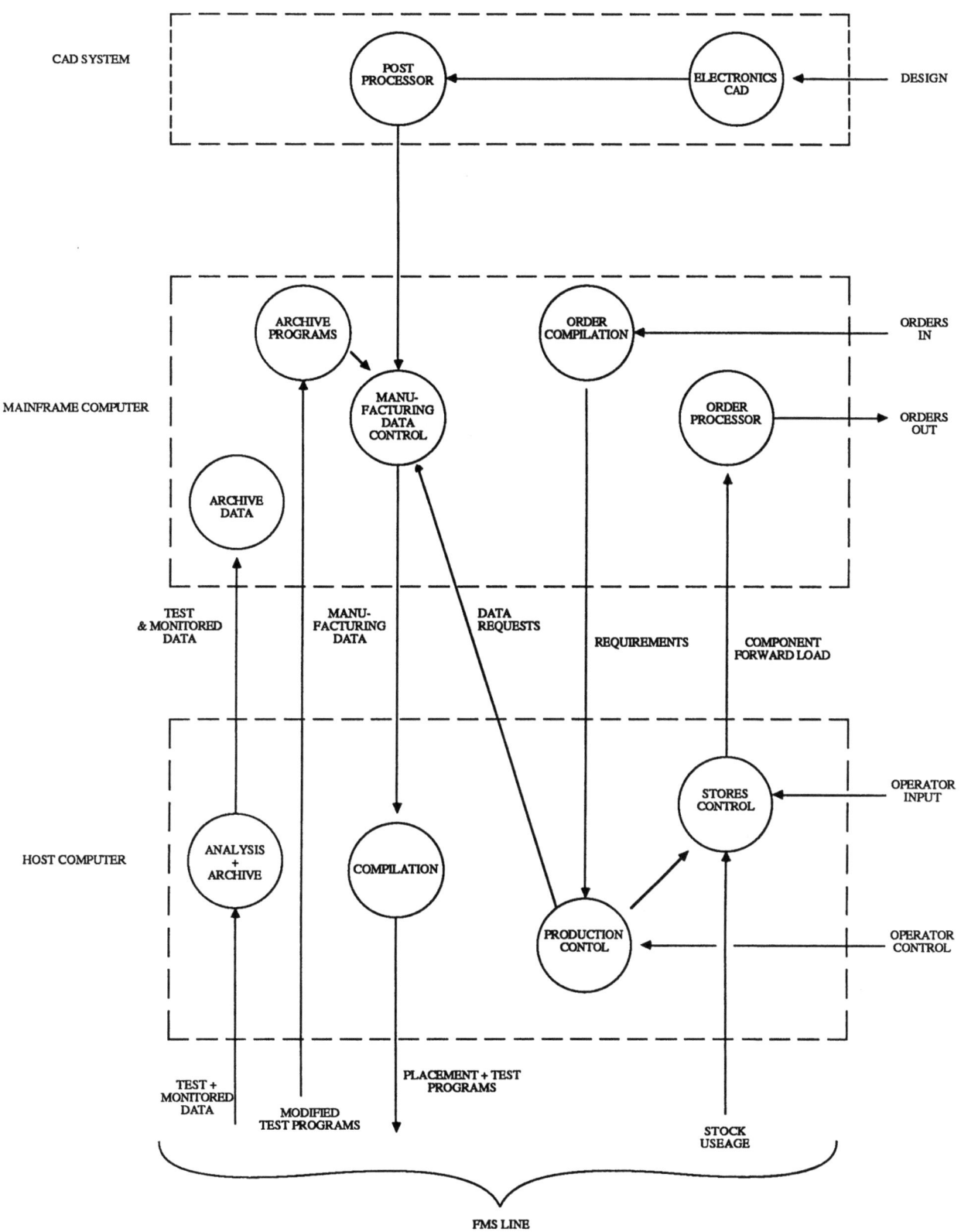

FIGURE 4 : LINE VERTICAL INTERATION

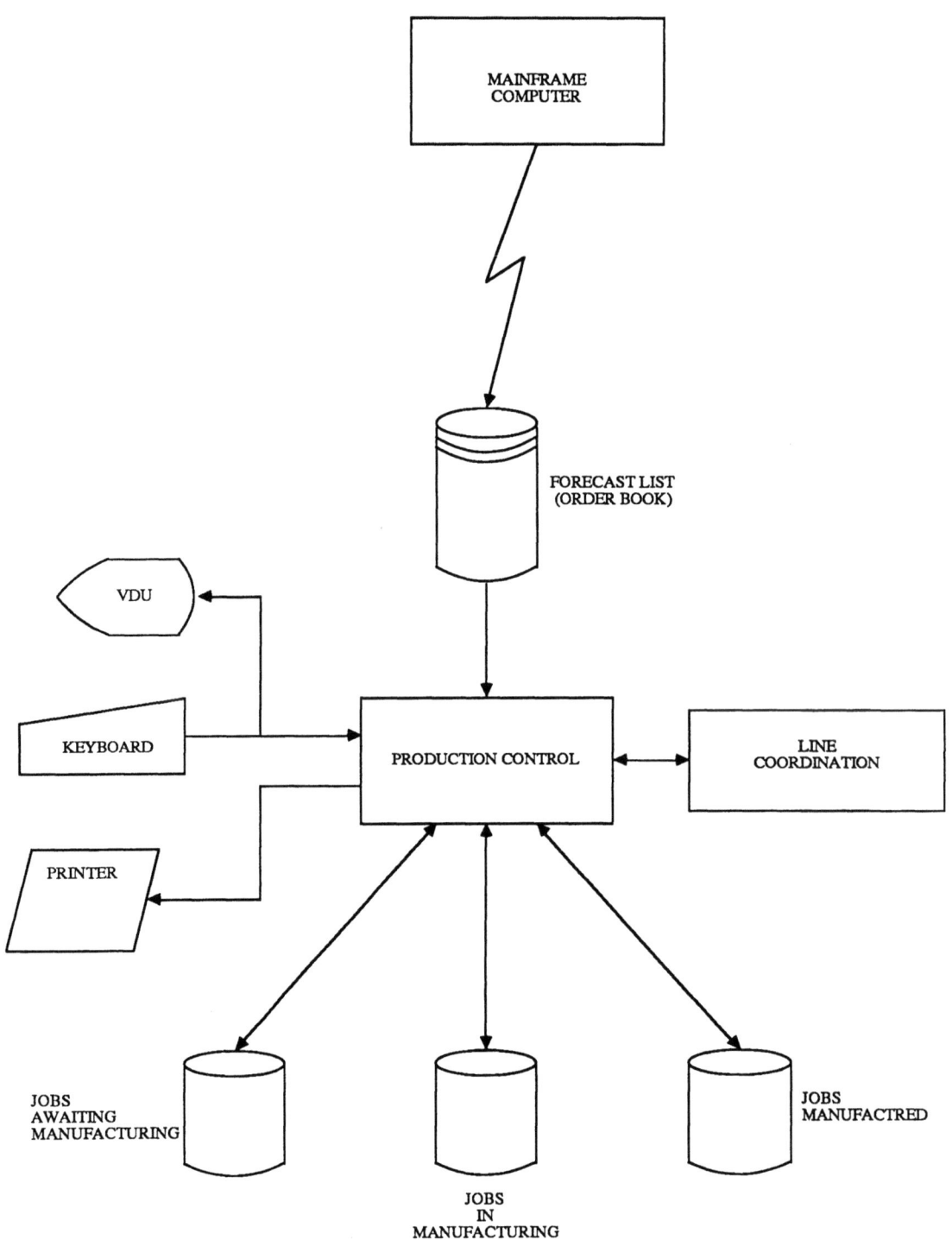

FIGURE 5 PRODUCTION CONTROL

Integration and control of handling plant in an FMS for surface mounted PCBs
P. D. Parker
GEC Mechanical Handling Ltd, UK

This paper gives an overall view of the system integration requirements and the solutions adopted for control of the various machines used in the production of Surface Mounted PCBs. Production machines with different interfacing requirements, both mechanical and electrical, were integrated into a working FMS using robotics in combination with purpose designed plant. The paper shows how the control systems for the robots were also used for machine interfacing. The control network between the machines and the robot systems using both digital and data links to a mini computer is illustrated. The contrast between safety and operating requirements is discussed.

SURFACE MOUNTED PCBs

Most electrically operated items in use today employ printed circuit technology and semiconductors, microchips etc., somewhere within their control systems. During the last few years the individual transistor has largely been replaced by microcircuits in the familiar Dual in Line packages. These range in size from six pin to 40 pin but the ever increasing complexity of Very Large Scale Integrated circuits has led to the need for larger carriers to connect the elements. The physical limitations of the Dual in Line approach were themselves becoming a limiting factor on the performance of the circuit design which could be achieved. Thus a new breed of carrier for VSLI was needed which enabled the connections to be made much closer together and closer to the actual semiconductors within the microcircuit chips. This new breed of microcircuit is known by the general designation Surface Mount as distinct from Dual in Line.

To go with the new circuits there was a need for the existing Dual in Line packages to be made available in the new format and most of the familiar standard TTL and CMOS types are now available in Surface Mount form. Along with the microcircuit development has come the requirement for other components such as resistors and capacitors to be made in Surface Mounting form and a large variety of all the circuit elements which go to make up a working system are now available.

This development has gone hand in hand with the development of multilayer circuit boards. Printed Circuit Boards for Dual in Line packages are in the main double sided but the new Surface Mounting chips are often used on boards of up to twelve layers. The packing density can be further increased by placing the Surface Mounted chips on both outer surfaces of the multilayer board. The increase in packing density is of the order of ten times that which can be achieved with conventional Dual in Line packages (Fig.1). Moreover circuits can now be developed which were not possible before the introduction of Surface Mount technology.

The use of Surface Mounting components has in turn led to the need for automatic assembly machinery. Automatic machines for the placement and insertion of Dual in Line packages have existed for some time; also specialist machines to insert other components such as resistors and capacitors. In the main these have been separate items of plant because the shape of these components was different. However with Surface Mount components the need for separate plant items for assembling the components to a circuit board disappears. A number of manufacturers can now offer Pick and Place machines for Surface Mount components which can handle all the components used on many circuit cards. The placement of components by hand on such circuit cards is not a feasible economic concept.

Surface Mounting has also changed the methods used for soldering the components to the circuit card. A number of different systems are in use of which the Vapour Phase system was chosen for the plant described in this paper.

DEVELOPMENT HISTORY

When our customer first asked for assistance with this project, the specific task was to link mechanically five proprietary machines and liaise with the appointed host computer software consultants on communication requirements.

Space and cost constraints led to the appraisal of many plant layouts, and an early decision was made to use robotic transfer for three main reasons:-

(a) A cost appraisal showed that it would cost no more than designing fixed transfer equipment.

(b) It offered flexibility of plant layout without the need to consider the special design of transfer equipment for each change of plant configuration.

(c) The input/output system of the robot's programmable controller could be used to control ancillary equipment, and in addition, communicate both with proprietary machines and with the host computer. Thus additional programmable logic controllers were not required. The robots to perform the transfer requirements were built to design specifications, prepared by the author's company, which covered also the gripper arrangements and the proprietary programmable controllers used to control the robot.

The various proprietary items of PCB Surface Mount manufacturing and assembly plant were chosen by our customer as the best solution for the processes to be performed. The choice made was not constrained by any interfacing or handling requirement. The decision to transfer using robotics required additional special safety features to be incorporated in the design of the plant and its control system.

Mechanical linking naturally required the integration of electrical control and power. Perhaps in a few years' time when MAP or similar specifications are established, the electrical linking and interfacing of various pieces of equipment will be a simple matter, but as will be seen from Appendix A, the integration of control became complex.

Inevitably, when the decision was taken to provide an automatic handling system to link the various machines into a Flexible Manufacturing System, there were a large number of apparently incompatible controls and logic, all of which needed to be integrated into a workable system. Because the various processes for manufacture brought together machines from different suppliers it was not surprising to find that each supplier had different interfacing requirements, both mechanical and electrical.

Machines were found to operate with internal and interfacing controls at different voltages ranging from 5V TTL DC levels through 24V AC and DC to 110V AC. The power requirements for individual items of plant varied from a few watts to about 100kW and careful planning was required to ensure that the switching of large loads did not cause interference to other machines, the controlling computer, or microprocessors.

During the development phase it became clear that the power supply available within the building would not be of sufficient capacity or of low enough impedance to ensure satisfactory operation. The opportunity was therefore taken to plan a new electrical distribution system and to have a supply installed with spare capacity for future likely extensions.

A plan was drawn up of the control requirements for each machine and jointly agreed between each machine supplier, our customer and the host computer software consultants. Table 1 lists the principal suppliers of plant. It can be seen that the task of getting together an agreed scheme was not inconsiderable.

Each machine, except the Automatic Testing Equipment (ATE), included as part of its design a system for handling the PCBs. However the pick-up and put-down positions varied in height and spatial orientation. In addition, some machines (the Solder Pasting Printer, the Component Placing machine and the ATE Table) handled individual PCBs, whereas the Cleaning, Defluxing and Vapour Phase Soldering plants were designed to deal with a cassette containing many PCBs simultaneously. The ATE system was a test system only and required both "bed of nails" jigs to connect to the PCBs and a handling system to place and remove the PCBs into and from the "bed of nails".

DESCRIPTION OF PLANT

The FMS described in this paper was installed to provide a production facility for Surface Mounted circuit cards. The cards themselves are manufactured elsewhere in the building and are brought to the plant to be fitted with the components. The whole plant is contained in a room approximately 10m x 16m. The room is fully air conditioned and humidity controlled. "Clean" room conditions are observed. The cards follow a prescribed route through the various items of plant but depending on whether the card is to have components on one or both sides, this route is variable. Cards are manufactured in batches but the batch size can be from one to many hundreds. A typical sequence for a single sided Surface Mounted multilayer circuit card would be as follows:-

The PCBs are batched in special carriers in handling cassettes.
Cassettes containing PCBs are put through a cleaning process.
Individual PCBs are taken, one at a time, from the cassette and solder pasted.
All the Surface Mounting components are placed upon the PCBs.
Prior to batch soldering, the PCBs are replaced in a cassette.
The PCBs in cassettes are put through the soldering process.
The cassettes containing the PCBs are de-fluxed and then cooled.
The cassettes are placed into a Turning Unit.
PCBs are taken, one at a time, tested and replaced in the cassette.

For most of the above processes, individual proprietary machines were used.

The final plant layout (Fig.3) required the specific design of twelve items of equipment to meet the process and machine transfer requirements.

One of the key items was the cassette to hold the PCB carrier plates. The PCBs are transported from process to process in carrier plates, all of which are of fixed size. Many sizes of PCB are able to be accommodated in these plates but the use of a fixed size of carrier enabled the design of the system to be simplified. The cassette to hold these carriers had to be able to withstand the temperatures of the soldering plant and be able to pass through the defluxing and pre-cleaning plants. The cassettes were manufactured from stainless steel, laser profiled to minimise distortion. The accuracy of the cassettes was important because the robots were required to be able to withdraw and replace individual PCB carriers in the cassettes at various stages in the production sequence. A further feature of these cassettes was that they had to be capable of holding the PCB carriers vertically for cleaning, at an angle of 5 degrees to horizontal for soldering, and horizontal for robot unload/load purposes.

Three turning units for the cassettes were designed, two of which turn through 90 degrees and one which turns through 85 degrees. These turning units present the PCB carriers horizontally to the robot gripper jaws. The cassettes are transported through the pre-clean, soldering and deflux plants from overhead conveyors supplied with and built as part of the plant items. Fig.2 shows a cassette in one of the turning units.

Three robots were designed to withdraw the carrier plates from the cassettes and place these onto the entry conveyors or tables of the solder pasting machine, the component pick and place machine or the automatic testing equipment (Figs.4,5). Two of the robots are used to replace the PCB carriers back into the cassettes. All the robots were designed to use common parts as far as possible, the gripper design being identical on each. The robots have a servo driven vertical axis and use air cylinders for the other movements. This is because the reach, rotational and gripper movements can operate between fixed stops where incremental control is not required. Because of the clean conditions of the room, all the exhaust air is piped away. One robot has an additional rotational axis and moves to two 90 degree positions either side of a central position. Both the reach and gripper actions are in two stages, each separately controlled, the former being telescopic. The gripper can hold the carrier plate loosely or tightly and this enables gentle handling of the plates when putting them down onto a horizontal surface. The robot accuracy and repeatability was designed to be better than 0.5mm spatially. All the movements of the robots are equipped with feedback monitoring and each movement has been allocated a maximum time. Should a movement not be completed within the allotted time then the controller will send a signal to the mini computer to indicate an error.

The robot controllers are a proprietary design and are basically servo axis controllers but having an extensive facility for the connection of additional inputs and outputs. These facilities enabled the system to be put together without further Programmable Logic Controllers.

The automatic testing equipment had to be provided with a connection system and means of getting the PCBs into and out of this system. This was achieved by the design of a purpose built loading table. The ATE load table consists of a platform which is raised and lowered by a screw jack to precise heights. The robot controller for the number three cell controls this table in conjunction with the associated robot and turn unit. A dialogue takes place between the ATE system and the robot controller to ensure that proper contact in the test jig is made. If the ATE system does not receive "good" signals then the robot is instructed to withdraw the PCB on its carrier and replace it for another try. A comparison is made before the table is raised to ensure that the actual carrier containing the PCB to be tested is one from the batch that the test jig will accept and that the ATE system expects. Because, from time to time, the test system can be used manually for different purposes, it is essential to ensure that no damage can occur to a PCB or the test jig by attempting to test an incorrect type of PCB. Both the PCBs produced by this system and the test jigs are expensive items. The checking is done by an automatic purpose built code reader attached to the loading table, which sends data to the ATE where the comparison is made. A bar code reader is used manually each time a test jig is placed on the system and in this way the mini computer can ensure that the program sent to the ATE system is the correct one for the particular PCB to be tested. This checking dialogue takes place even when the ATE table is being used in a manual mode for out of sequence testing.

There are of course a number of other features in this plant to prevent and report errors. Most of the detection of PCB carriers throughout the plant is done by proximity detectors or by photocell units. Most of these are interfaced into the system via the robot controllers.

SYSTEM INTEGRATION

The plant is configured as 3 "operating cells" (Fig.3). The purpose designed cassettes are manually loaded with up to 7 PCB carriers and placed on the Cassette Input Conveyor in Cell 1. Up to three such cassettes may be loaded and are automatically positioned sequentially to be picked up by the overhead conveyor system of the Precleaning plant. The PCBs are carried through the cleaning plant in a vertical position. The plant itself consists of boiling solvent tanks and a cooling tank. At the end of the cleaning process the cassette is placed into the 90° turn unit which allows the robot in Cell 1 to extract the PCB carriers horizontally, one at a time, and place them on the table of the Solder Pasting Printer. The Solder Pasting Printer is a proprietary silk screen printer adapted to print the PCBs with a solder cream. The PCBs are taken from the robot gripper jaws into the machine, printed and then returned to the table. Following the printing process, the same robot picks up the PCB carrier and places it onto the entry conveyor to the Component Placing machine.

The component pick and placing machine forms the heart of the FMS physical operations. This machine is programmed from the mini computer for the particular design of PCB to be handled. The machine has a number of component magazines containing the microchips, resistors, capacitors to be used. Two "robotic" grippers place the components onto the PCB. The machine has x, y, z and ∅ actions which allow any component to be placed in any desired position. It also incorporates a tooling bank so that components requiring different pick up methods can be handled. The chosen machine is capable of placing components at a rate of approx. 3000 per hour. Because of the programmable facility this machine can handle PCBs of different design in any sequence, so long as its component storage cassettes are suitably equipped.

The Component Placing machine unloads the PCB, now with its electronic components held in position by the sticky solder cream and still sitting in the carrier plate, onto an exit conveyor where Robot No.2 in Operating Cell 1 collects the carrier, placing it into an empty cassette. This cassette is then transferred from Stand 1A to Stand 1B to be ready for collection by another overhead conveyor and put through the various tanks and ovens forming the Vapour Phase Soldering line.

When passing through the Vapour Phase Soldering plant the PCB carriers are held at 5 degrees to horizontal to enable the fluids used to access the PCBs without any air pockets forming. This angle also facilitates the draining off of the fluids when the cassette is withdrawn from the tanks, thus minimising the waste of expensive fluids. A turning unit in the middle of this line changes the cassette attitude such that the PCBs are again vertical when passing through the Defluxing plant.

At the end of the Vapour Phase Soldering and Defluxing line forming operating Cell No.2 there is a further Stand onto which the cassettes of PCB carriers are placed.

Operating cell No.3 contains a further Turning Unit to place the PCB carriers horizontal, and a third robot which extracts the carriers one at a time to place them into the purpose designed Automatic Test Equipment Load Table.

This ATE Load Table which is driven by the controller for Robot No.3 lifts the PCB carrier into contact with the Bed of Nails jig and allows the ATE system to perform tests on the PCB. After testing, the robot collects the PCB in its carrier from the ATE table and returns the PCB carrier to the same slot in the cassette before removing the next one for testing. When all the PCBs have been tested the Turning Unit returns to its original position and a "job finished" signal is given to the mini computer.

Not shown in Fig.3 for clarity are an automated stores area, which has its own VDU, keyboard and line printer, and the control office which houses the mini computer and further VDUs, etc.

DATA SIGNALLING

Fig.6 shows the interfacing for the whole FMS plant.

A number of the items forming the FMS communicate with the mini computer over serial data links. In order to reduce the length and number of such links at the mini computer, Multiplexers are provided into each of which up to eight serial data links can be connected. These data links are used for sending configuring programs for control of the Component Placing machine, the Automatic Testing equipment, the Vapour Phase Soldering plant and the Solder Pasting Printer.

Data links are also used for receiving information from the Bar Code Readers. A separate data link is used in addition to send and receive overall management information between the customer's mainframe computer and the mini computer.

The Multiplexers are placed amongst the plant such that the RS232 signals can be within the maximum permitted lengths of cable for such circuits. Within a group of RS232 type circuits on a single multiplexer it was found to be necessary to have circuits operating at different Baud rates. This was because of the use of proprietary designs of machines where the selection of Baud rates was not compatible.

The automatic storage system was only provided with 20mA current loop signalling and thus a converter was used to translate these signals to RS232 type signals. The stores cell comprises in addition a local operator's VDU and printer together with a Bar Code Reader. The stores cell comprises the only operational items of plant where the actual commands for operation are given by the mini computer over a data link. In all other cases the instructions from the mini computer are used to set up or program the plant item which then carries out its function by using the handshaking routines on the interface logic to its neighbours. The mini computer is therefore only concerned with such as batch quantity, progress and machine errors. It does not actively control the individual plant items. It can however, issue "stop" or "hold" instructions at certain positions.

LOGIC SIGNALS

Fig.6 also shows the interfacing logic pathways for the whole FMS plant. Inside each safety cell there is a microprocessor controller. This unit is primarily a positional controller for the vertical axis of the robot. However because these controllers had considerable spare interfacing capacity and advantage was taken of this to control all the other functions within each cell. These controllers use optically isolated inputs capable of operating from 5V DC TTL levels up to 28V dc. The outputs are in the form of normally open relay contacts which are rated at up to 5A at 110V ac.

The controller itself is powered directly from the mains supply to the cell via the main isolator for the cell but remains active when the robot and plant motive power is isolated. The control program software is retained in battery backed RAM but a facility is included to have this software operating from an EPROM. Apart from some of the hardwired safety logic all the logic signals to and from the other items of plant are handled by the I/O field of these controllers. The controllers also have I/O connections directly to the mini computer and can therefore give and receive predetermined messages to or from the mini computer.

Appendix A contains a selection, copied from the maintenance manual for this installation, of the function of some of the logic circuits shown in Fig.6. It will be seen that a large range of system voltages are involved and that care had to be taken to decide which end of a logic circuit should be "passive" or "active". In general inputs to an item are "active" and outputs from an item are "passive", but this logic could not be followed in all cases due to the nature of the interface on some plant items.

Appendix A and Fig.6 show as a "logic" signal some 415V power supplies where these operate a motor without any other circuitry being involved. Strictly these are of course not logic signals but unless they were shown as such the diagram would not have shown that any electrical connections were made to some of the linking conveyors.

ELECTRICAL POWER

The plant layout was constrained by the nature of the building and changes were required to the ceilings to accommodate the overhead transports of the Cleaning plant and the Vapour Phase plant. These modifications affected in turn the already installed lighting and air conditioning systems.

As already noted, the power requirements exceeded the spare capacity (approximately 100 Amps per phase) of the available supply in the room. For fire safety reasons it was necessary to operate all the plant in the room from a single source feeder. Thus the provision of a new electrical feed had to be engineered, and this had to come, for capacity reasons, from a different high voltage substation. The new feeder was to be installed without losing production in the adjacent area in which existing PCB manual assembly was taking place. Temporary arrangements were therefore made to feed the new local distribution switchgear from the existing distribution switchgear in the room until the new main supply was available. This enabled the new plant to be installed but not fully commissioned. In due course this temporary feed cable was used in the opposite direction to power the original distribution system from the new distribution cubicle.

The new plant required overall approximately 250 Amps per phase and 100 Amps per phase was deemed sufficient for the original manual plant, including all the lighting and air conditioning. In addition it was known that a further building extension was to be erected alongside the PCB plant (this is now in course of erection) and provision was therefore made to run a 400A per phase supply, which allowing for some diversity of loading would be large enough for all the loads foreseeable.

Partly because of the long run of the feeder cable and partly to provide a lower than normal source impedance, an oversize cable was selected. With the Vapour Phase Soldering plant requiring at full power some 115 Amps per phase, care was required to avoid the sudden switching of too large a load either on or off, which might have caused transient interference with microprocessor systems connected to the common feeder. Oversize cabling was used to reduce the size of the transient voltage fluctuations but in addition the design of the switching of the loads within this plant was arranged to be progressive under software control. In this way the largest step of load was limited to approximately 6kW, spread across the three phase supply.

All the plant forming this FMS is fed from individual overload circuit breakers incorporating residual earth leakage current protection and situated in a common supply distribution cubicle. Some of the individual items such as Bar Code Readers, VDU terminals, the mini computer, etc., were of such low power consumption that the provision of individual circuits was not justified and these items were powered from ring main circuits but using non standard 13A outlets. This prevented these items from being affected by other incidental loads such as floor polishers, vacuum cleaners, etc. Such services were catered for by further ring main circuits using conventional 13A outlets.

SAFETY

This description of the safety aspects of this plant is a brief outline only of the typical problems and the solutions adopted. The safety aspects are complex and a full description would require a separate paper to be given.

The decision to supply robotics for the transfer of PCBs between the processes meant that safety cells were necessary. However these safety cells overlapped with the overall concept of operating cells for control purposes. Fig.6 shows the control concept for signalling (both logic and data) and also shows the operating and safety cell boundaries. For each safety cell there is only one electrical feed and all plant forming the safety cell is controlled via a single "on load" isolating switch. Emergency Stop pushbuttons operating directly in the power control circuitry serve to stop all electrically driven moving parts instantly and also releases the compressed air from any air cylinder. However the microprocessor controllers in each safety cell remain operational when an Emergency Stop is pushed. This requirement arises because it is necessary to allow manual (semi-automatic) control of machines for setting up purposes and for experimental production. Also, if an Emergency Stop button is pushed by accident the loss of time to reset the computer system and pick up the sequence would otherwise be considerable. The retention of all the microprocessor memory enables a resumption of work in such circumstances without a total resetting operation.

The manual operating requirement meant that the Solder Paste Printer and the ATE system had to be capable of being operated with the robot parked safely and held inoperative and with the safety gates open allowing access. In these conditions each machine had to be interlocked such that it was safe for an operator to use. This requirement contrasted with the need for the machines to work in an automatic mode with their individual guards open when the robot was in use.

All of these features were catered for by purpose designed circuitry using hardwired techniques and not relying on software or electronic circuits. The safety cages are provided with a system of access gates which are controlled by a locking system which does not require any electrical connections to be made to the gates or the associated locks.

The whole method of safe operation relies on a system of keys and key interlocked switches built into the robot controller cubicles. These interlocked switches ensure that motive power ceases and that the robots are parked and mechanically locked before a key is available to unlock a safety gate. Manual operation of a machine inside a safety cell is achieved by removing a further key previously trapped in the gate and using this inside the safety cell to restore motive power to the item to be operated. The robots however, cannot be moved under such conditions and remain mechanically locked and electrically isolated.

The Stores cell has no special additional safety requirements, as each storage machine is equipped with its own safety devices.

CONCLUSION

The foregoing description gives an outline of the way in which the logic and control system has been developed. Design of the system commenced in the Autumn of 1984 for installation in the Summer of 1985 and has been in operation from September 1985. The design is such that a number of changes to the operating sequences are possible by changes made to the local software of the robot controllers. Inevitably some of the original ideas on how the system should operate have had to be changed due to the operational experience. However, only one change has been made to the logic circuitry and this was to ensure that the Solder Pasting Printer had to be started by pushbutton after manual intervention for replenishing the solder cream instead of automatically from the robot controller.

The author would like to express his thanks for the help given by the staff and management of British Aerospace and for permission to publish this information, including the photographs.

TABLE 1

Plant suppliers:

Cassettes for PCB Carriers	-	GEC Mechanical Handling Ltd.
PCB Board Pre-cleaning	-	Kerry Ultrasonics Ltd.
Cassette Turning Units	-	GEC Mechanical Handling Ltd.
Solder paste printing	-	DEK Printing Machines Ltd.
Conveyors (powered & roller)	-	GEC Mechanical Handling Ltd.
Component Placing	-	Dynapert-Precima Ltd.
Vapour Phase Soldering	-	I.V. Products Ltd.
Defluxing Plant	-	Kerry Ultrasonics Ltd.
Automatic Testing Equipment	-	Marconi Instruments Ltd.
Bed of Nails jigs	-	Everett Charles Ltd.
ATE Load Table	-	GEC Mechanical Handling Ltd.
Automatic storage equipment	-	Dexion Ltd.
Mini computer and VDUs	-	Hewlett Packard UK Ltd.
Bar Code Readers	-	Hewlett Packard UK Ltd.
Robots and controllers	-	GEC Mechanical Handling Ltd.

APPENDIX A: SELECTION OF LOGIC AND DATA SIGNALLING CIRCUITS WITH REFERENCE TO FIG.6

I/O SIGNALLING (excluding RS232/423 and 20mA current loop signals)

NAME OF SIGNAL	POWER SOURCE	VOLTAGE	EFFECT
Cassette present at lift position.	Junction Box	12VDC	Informs Host Computer that a cassette is available for processing.
Roller conveyor empty	Cell 1 Console	24VDC	Informs Cell Controller to allow processing of cassette and issue Finished Cassette signal to pre-clean.
Hooks clear (Input conveyor)	Cassette Input Conveyor	110VAC	Normally closed contact in pre-clean allows operation of input conveyor. Contact is opened to inhibit conveyor.
Finished Cassette (Pre-clean)	Pre-clean	15VDC	Signal from Controller output to allow overhead conveyor to remove empty cassette.
Not clear to start (Pre-clean)	Pre-clean	15VDC	Normally closed contact in Host Computer is opened to halt Pre-clean at start position of sequence, over input conveyor.
Emergency Stop (Cell 1)	Cell 1 Console Silk Screen Printer	110VAC 12VDC	Double pole break latching pushbutton operates to stop Cell 1 motive power and stops Silk Screen Printer independently.
Go I/P 1	Silk Screen Printer	5VTTL	Signal from Controller to printer to initiate operation.
Cover safe (O/P 3)	Cell 1 Console	24VDC	Signal given by open collector 7406 output from silk screen printer but powered from 24VDC line in Cell 1 Console with return taken to earth in printer unit.
Run Conveyor (Substrate Input Conveyor)	Cell 1 Console	415VAC	Power feed to 3 phase motor to drive conveyor. [There are no limit switches or other circuitry on this conveyor]
Board Sent [(Pick & Place to Cell 1)] Send Board [(Pick & Place to Cell 1)]	Pick & Place	24VDC	[This is a 2 pair (4 wire) circuit used to handshake] [between the Pick & Place and Cell 1 Controller.] [Both rising and falling edges of the DC levels are] [employed giving a total of 4 signalling conditions.]
Got Substrate (Cell 1)	Junction Box	12VDC	Informs Host computer that Robot 1 is processing a substrate. Signal is on from picking up until collected by Pick & Place.
Not Clear to Start (Cell 1)	Junction Box	12VDC	When asserted by Host computer, holds Robot at start of cycle.

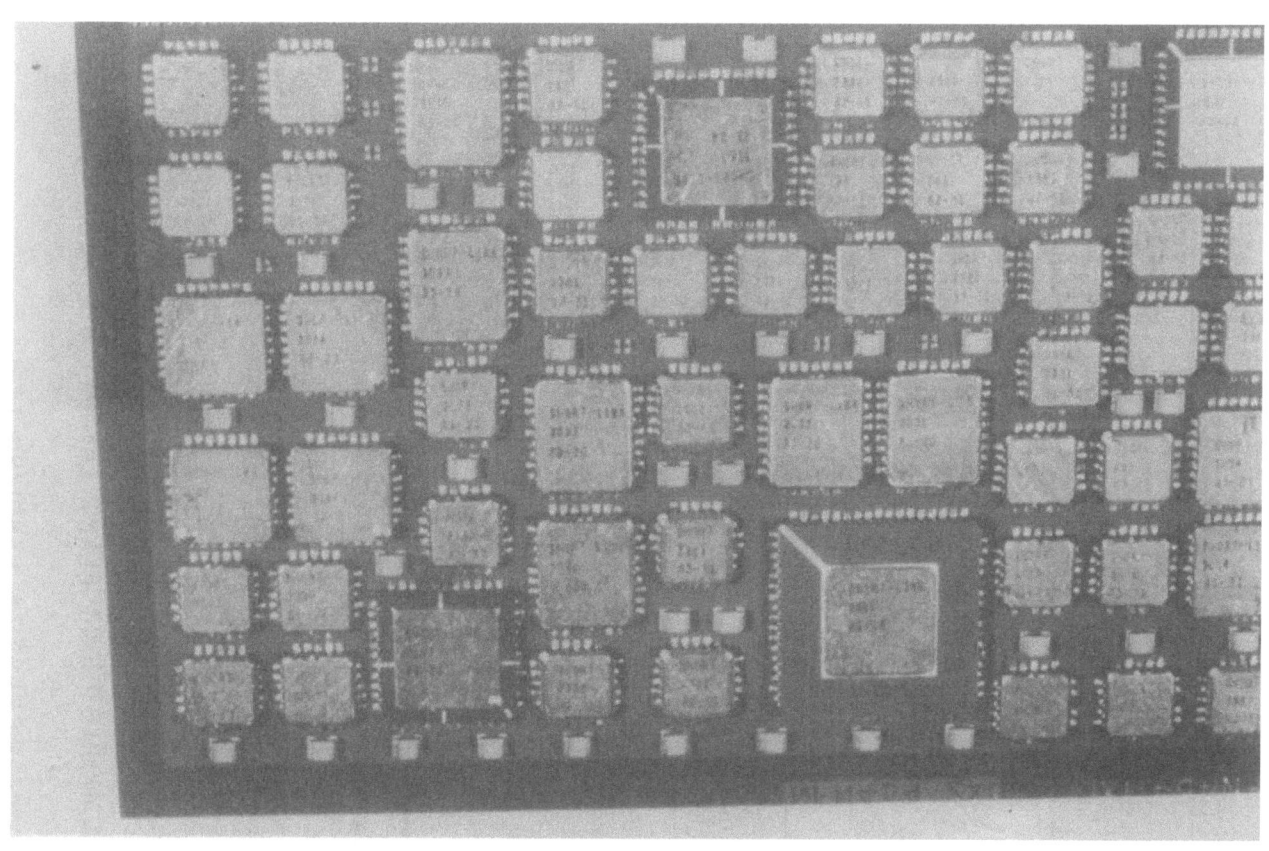

Fig.1 Part of a Surface Mounted PCB

Fig.2 Turning Unit with Cassette

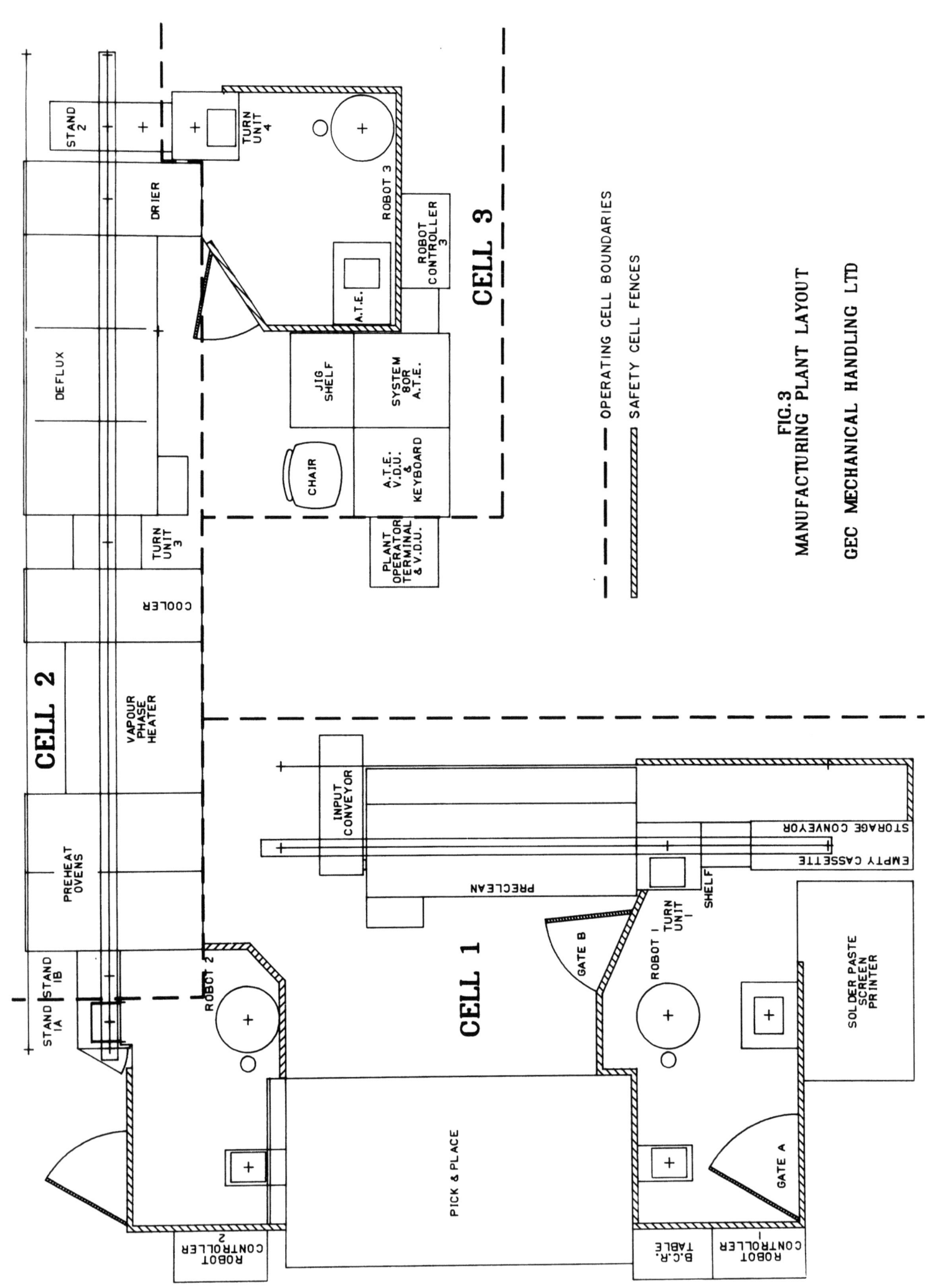

FIG. 3
MANUFACTURING PLANT LAYOUT
GEC MECHANICAL HANDLING LTD

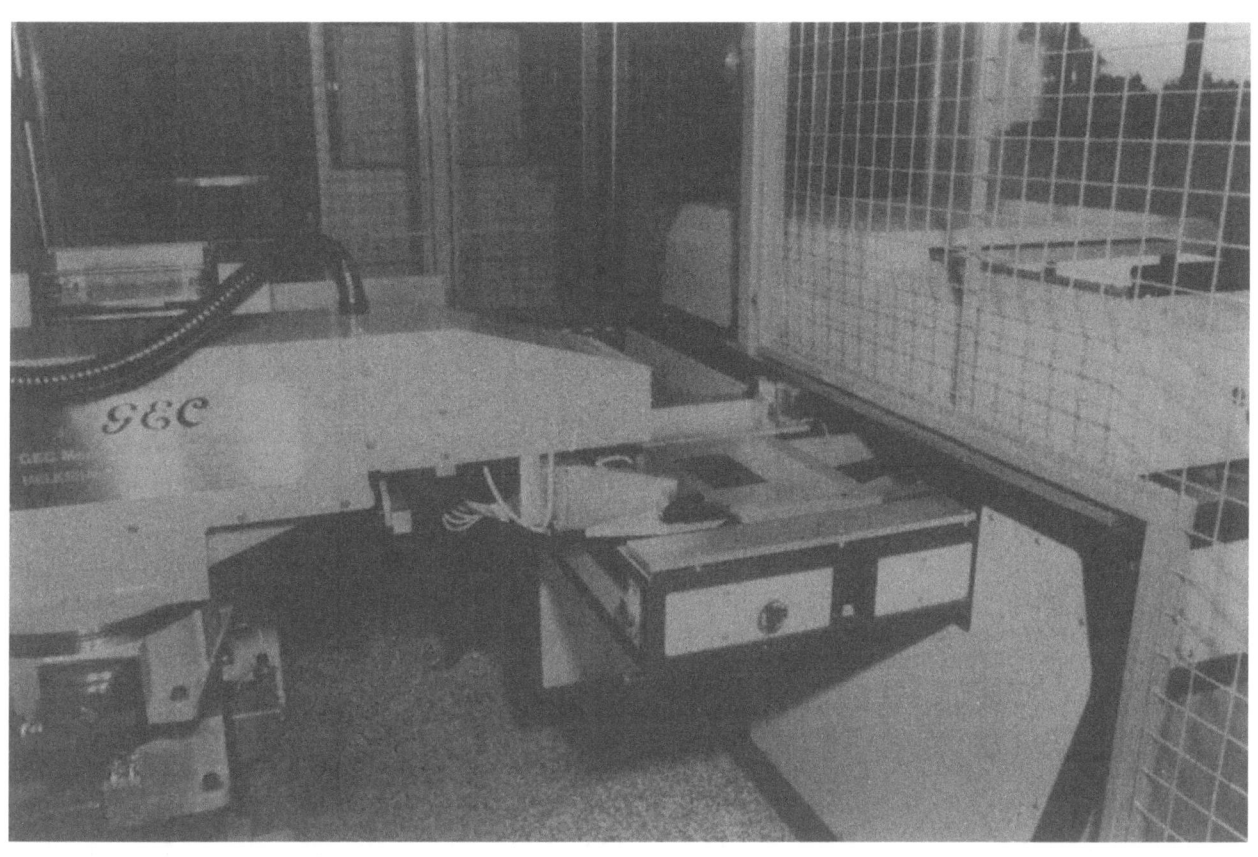

Fig.4   Robot placing PCB into Solder Pasting Printer

Fig.5   Robot withdrawing PCB from ATE Loading Table

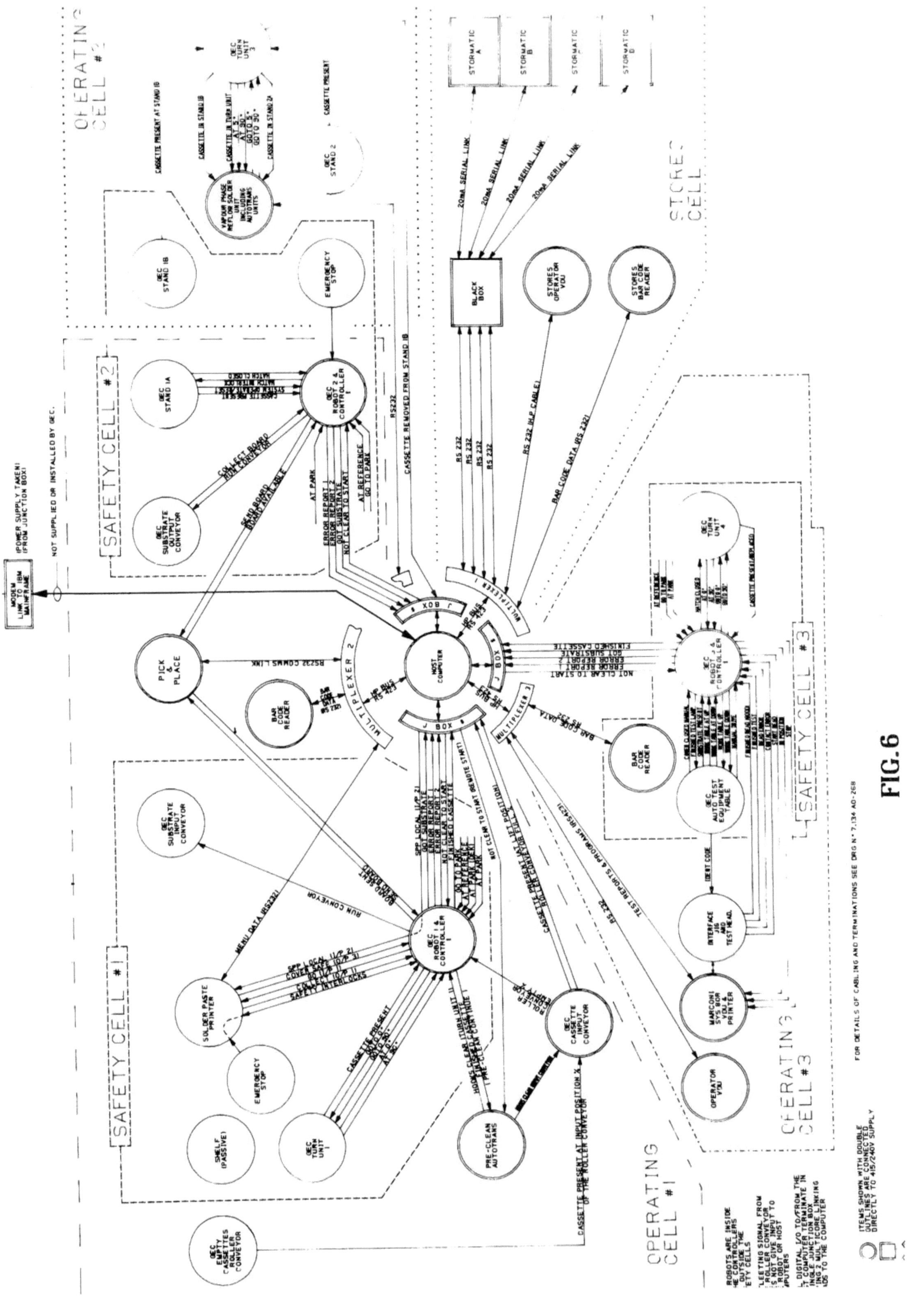
OPERATING CELL #1
OPERATING CELL #2
OPERATING CELL #3
SAFETY CELL #1
SAFETY CELL #2
SAFETY CELL #3
STORES CELL
HOST COMPUTER
J BOX
MULTIPLEXER 1
MULTIPLEXER 2
MULTIPLEXER 3
MODEM LINK TO IBM MAINFRAME
(POWER SUPPLY TAKEN FROM JUNCTION BOX)
NOT SUPPLIED OR INSTALLED BY GEC.
PICK & PLACE
RS232 COMMS LINK
BAR CODE READER
GEC ROBOT 1 & CONTROLLER 1
GEC ROBOT 2 & CONTROLLER 1
GEC SUBSTRATE INPUT CONVEYOR
GEC SUBSTRATE OUTPUT CONVEYOR
SOLDER PASTE PRINTER
EMERGENCY STOP
SHELF (PASSIVE)
GEC TURN UNIT 1
PRE-CLEAN AUTOTRANS
GEC CASSETTE INPUT CONVEYOR
GEC EMPTY CASSETTES ROLLER CONVEYOR
GEC STAND 1A
GEC STAND 1B
GEC STAND 2
GEC TURN UNIT 3
VAPOUR PHASE REFLOW SOLDER UNIT INCLUDING AUTOTRANS UNITS
BLACK BOX
STORES OPERATOR VDU
STORES BAR CODE READER
STORMATIC A
STORMATIC B
STORMATIC C
STORMATIC D
20mA SERIAL LINK
RS 232
RS 232 (H.P CABLE)
BAR CODE DATA (RS 232)
GEC ROBOT 3 & CONTROLLER 1
GEC TURN UNIT 4
GEC AUTO TEST EQUIPMENT TABLE
INTERFACE JIG AND TEST HEAD
MARCONI SYS 80R VDU & PRINTER
OPERATOR VDU
FINISHED CASSETTE
GOT SUBSTRATE
ERROR REPORT 1
ERROR REPORT 2
NOT CLEAR TO START
AT PARK
AT REFERENCE
GO TO PARK
CASSETTE REMOVED FROM STAND 1B
COLLECT BOARD
RUN CONVEYOR
SEND BOARD
BOARD AVAILABLE
MENU DATA (RS232)
CASSETTE PRESENT AT INPUT POSITION X OF THE ROLLER CONVEYOR
TEST REPORTS & PROGRAMS (RS423)
IDENT CODE
FOR DETAILS OF CABLING AND TERMINATIONS SEE DRG N° 7,134 A0-268
ITEMS SHOWN WITH DOUBLE OUTLINES ARE CONNECTED DIRECTLY TO 415/240V SUPPLY

FIG.6

Assembly cell controls using flexible software tools and high level control configuration language

J. Temmes, P. Ruusunen and J. Lempiäinen
Technical Research Centre of Finland (VTT), Finland

ABSTRACT

In the Technical Research Centre of Finland a flexible, automatic assembly cell has been built for research and development of automatic assembly systems and for feasibility studies of industrial assemblies. The main components of the cell are two industrial robots with dedicated microprocessor controllers, a conveyor system and a hydraulic press with multi-purpose microprosessor controllers and a vertical carousel storage with Programmable Logic Controller (PLC). In addition the cell includes auxiliary equipment like automatic tool changers, feeders and screwdrivers which are mainly controlled by binary signals. The cell controller is a common 80186-based controller system for automation with some software extensions to cover also manufacturing cell controls. The control software in the cell controller consists of a real time database with software blocks to handle controls, machine communications, operator communications and data-management.

In order to manage many small production tests with different assemblies and equipment with minimal effort the cell control software must be parameter based so that new configurations can be defined without changes in the software. Control definitions i.e. control sequences, error handling, operator displays and syntax for machine communications are given by configuration tools using high level configuration language. Communications between the cell computer and different machines are carried out using serial links and line-by-line character based communications (terminal type communications). By setting parameters and describing communication strings in a description file it is possible to cover most of the machines having some kind of terminal connection for the operator. With the terminal type communications it is easy to realize the Virtual Terminal Protocol i.e. to connect directly the operator's terminal to the machines. The VTP and the terminal type communications together with the description files give us an opportunity to build up hierarchical control systems with almost unlimited number of control levels.

1 INTRODUCTION

In the Technical Research Centre of Finland (VTT) a flexible, automatic assembly cell has been built during the years 1985 - 1986. The main purposes of the cell are development of new equipment for flexible assembly systems and feasibility studies of industrial assemblies.

The research work connected to the cell is a part of a large research programme which was contributed by the Technology Development Centre of Finland (TEKES). The research programme in the field of flexible manufacturing consists of several pilot FM-systems and assembly cells. Technically the pilot systems are designed to cover the main manufacturing functions in a plant, although they are located in different cities of Finland. Volume of the four year programme is 3-4 million $, including about 50 man-years and remarkable industrial research volumes. The assembly cell described here is aimed for final assembly. The construction work of the cell has been realized in the Metals Laboratory and the Electrical Engineering Laboratory of VTT.

Due to the small amount of large-scale manufacturing industry in Finland there are only a very limited number of fixed assembly automation systems in use. Now as programmable automation technique is making automation flexible also to small batch manufacturing, industry is eager to seek new solutions to assembling problems.

2 EQUIPMENT OF THE SYSTEM

Figures 1 and 2 are overviews of the cell.

2.1 Conveyor And Assembly Stations

The components of the cell are placed around a modular conveyor forming a rectangular loop with five assembly stations (manuf. Bosch). Assemblies are built on pallets which are transported on belts from one station to another. The size of a pallet is 240 x 400 mm.

The conveyor system is controlled by a MC6801 based microcomputer system Mechatronic (manuf. Arlacon). The controlling algorithms are programmed in VTT using McBasic which is a dialect of Basic for machine control purposes. The microcomputer is controlling the conveyor with 32 binary I/O-signals. The conveyor controller is connected to the cell controller by using the terminal connection of the microcomputer. Through the terminal connection it is possible either to do all normal Basic programming operations or to give parameters to a running Basic program. The conveyor control program is fixed and controlled by high-level commands given by the cell controller. Communication speed is 2400 baud.

2.2 Workpiece Handling

Component and workpiece handling is done mainly by a PUMA 760 robot (manuf. Unimation) which is situated in the middle of the conveyor loop. The robot has a large working space and is able to carry loads up to 10 kg. The robot is equipped with a commercial tool changer (manuf. Nokia) to make it possible to handle various kind of workpieces. The gripper and the exchanger limits the weight of the workpiece to app. 6 kg depending on the type of the gripper.

Robot programming is done with the earlier version of VAL, VAL-I, using programming tools in the robot controller. All programs are some basic operations in the sense of assembly tasks and they are activated when needed by the cell controller. The robot is connected to the cell controller by using the terminal connection because there is no special computer connection available. Through the terminal connection it is possible both to do all normal VAL programming operations and to give parameters to a running VAL program. Communication speed is 2400 baud.

2.3 Assembly

Assembly tasks are done by an assembly robot Seiko RT-2000. The high accuracy of the robot makes it suitable for tasks requiring precise operation. In one of the assembly cases the robot has been equipped for screwing. The screwing heads can be changed automatically. Several types of screws can be used in one assembly. Screws like other volume parts are fed by vibratory bowl feeders. The programming language of the Seiko robot is DARL, a Basic like robot programming language. The robot is programmed like the Puma robot to execute basic assembly operations that are activated by the cell controller. The robot is connected to the cell controller by additional serial port making it possible to use also customized command strings instead of normal operator terminal commands. Communication speed is 2400 baud.

2.4 Storing And Feeding

Storing and feeding of large and heavy parts is realized by a vertical carousel storage. The storage is a common commercial storage (manuf. Electrolux) with some improvements for the use with Puma robot. The storage is two-sided: one side for manual loading and unloading and the other for robot loading and unloading. To gain the precision needed for robot picking of workpieces there were some modifications to be made. We are using extra limiters on the sides of the shelves, low speed positioning and a level sensor in the robot gripper. With these arrangements sufficient positioning precision for Puma was gained. The storage includes 8 shelves, 100 kg capacity and 2000 x 320 mm dimensions each. The storage is controlled by a Programmable Logic Controller Autolog 36 (manuf. FF-Elektroniikka). The storage is controlled by binary signals and equipped with a small manual control panel. The connection to the cell controller is made using a serial port. The storage control program is fixed and controlled by upper level commands coded with single small ASCII characters. Communication speed is 300 baud.

2.5 Fittings

In metal shops there is a need for very tight fittings eq. when assembling bearings on shafts. For tight fittings the cell is equipped with a hydraulic press capable of 50 kN axial force. The press has changeable upper and lower tools. The press is served by the Puma robot. Reliability of operation is achieved using several sensors and a microcomputer controller. Sensors include optical encoder, pressure measurement and several on/off sensors. The controller is the same type as in the conveyor system with the same characteristics.

3 REQUIREMENTS FOR THE CELL CONTROLS

One of the most important features that was considered during the design of the cell was flexibility. To achieve system flexibility there must be flexibity both in the mechanics and control. The control system should be able to totally utilize the flexibility offered by the mechanical construction.

In our case the cell controller must be able to handle automatic set-ups (like robot or machine program downloading) in the run-time and allow us to achieve even batch size one. Also the controller must be able to handle the cell level manufacturing information like material flows, assembly status for products, production statistics and manufacturing request status. Because of the cell will be used for tests of different products and equipment we had some special requirements concerning the cell configuration:

- possibility to connect a great variety of manufacturing machines like robots, CNC-machines, microcomputer controllers and PL-controllers to the cell controller without large hardware or software work,
- efficient software tools for configuring the cell controls for new products without deep understanding of computer programming,
- generality of control software to cover a great variety of assembly systems.

4 CELL CONTROL HARDWARE

The cell controller is a common 80186-based multiprocessor controller system Nokia 1000 (manuf. Nokia Electronics). It is a family of computer equipment meant to be used in industrial control tasks and includes a large variety of I/O and control capabilities. The controller hardware includes 512 kB memory, 10 serial (RS-232C or current loop) ports, 16/16 digital I/O, floppy disk, fixed disk, terminal and printer. Digital I/O is used only for some operators pushbuttons and some common sensors. The operating system in the cell controller is Intel RMX. Together with the RMX there is a general software package for automation control made by the manufacturer of the cell controller. In order to cover all control needs that are common in small manufacturing systems some extensions have been programmed to the package in VTT. The extensions handle communications to the machines and information management tasks.

Figure 3 is a schematic description of the cell controls hardware architecture. As described earlier every machine has a processor-based controller. Most of the controllers are different: two robot contollers, one PL-controller and two microprocessor controllers. Facing the fact that the controllers are differing from each other and have no standardized upper level computer connection alredy available, the connections were realized using serial terminal connections. The cell controller handles all control functions and equipment concerning the total operation of the cell. All task specific equipment and operations are handled in the lower level controllers.

5 CELL CONTROL SOFTWARE

The software architecture of the cell controller is presented in the figure 4. The software is based on a common real time database located in the main memory. In the database there is system status information, control definitions, type definitions, display definitions, report definitions, I/O configurations and parameters for operations. There are also definitions for communications. The software that uses the database can be divided into five major parts: configuration tools to make changes to the database definitions, operator interface to form a hierarchical display system for the operator, software blocks for machine communications, control blocks for sequencing and supervising the operations and software blocks for production information management.

During production the operator is using hierarchical displays for supervising and controlling. Should a fail occur it is also possible to give commands directly to each lower level controller trying to recover the failed machine.

Cell programming for new products and equipment is handled by using configuration tools in the cell controller and using programming tools in the lower level machine controllers, in our case in robot controllers. Figure 5 describes the user interfaces in different phases of operation.

5.1 Configuring The Controls

Configurations are made using a special line-by-line configuration language. By using the configuration language it is possible to change definitions in the database and also to read and set values in the database. All references to database objects are presented with symbolic names. In the figures 6 and 7 there are examples of configuring a new operators display to the system and configuring a new assembly task to the system.

5.2 Operators Displays

After defining the operators displays to the database the operators interface software forms the given displays, updates them according to the values in the database and interprets the operators inputs for run-time manual control of the cell. There are capabilities also for graphics and colors in the displays. Operators input tools include commanding by cursor positioning, by several soft-keys and by direct fixed function keys. Figure 8 gives an example of an operators display.

5.3 Control Blocks

The real time controls of the system (mainly sequencing of operations) are defined using control primitives called blocks.

Typical control blocks are:

- sequence step,
- chain of sequences,
- group of chains,
- robot controller blocks,
- storage controller block,
- program up- or downloading,
- etc.

The control tasks are created by using the basic control blocks and the control chains by grouping the tasks. The control tasks are connected to each other by directing some block outputs as inputs for another block in another task. Task chains are then executed in a cycle or by request.

5.4 Communications

The electrical connection is RS-232C or current loop type and data coding is the most common TTY-protocol. Message exchange is line-by-line ASCII-character based communication. It is possible to use echoed characters and response strings to quarantee the validity of transferred commands.

For each machine there is a fixed number of automatic operations available. All character strings for these operations are described in communications description lists in the database. In addition to the group of automatic operations to the machines there is a possibility to direct connection from the operators terminal to the machines. In our case the direct connection (Virtual Terminal Protocol as it is called in MAP definition) is very straightforward to realize because of the terminal type protocols we are using. The group of automatic operations and the direct connection makes it possible to build a multilevel system of such controllers by using line-by-line configuration language to control lower level controller. Of course there will be limits to the number of control levels because of the communication speed, but anyhow the concept is open for such ideas.

5.5 Upper Level Connection

As the cell is included in the national research programme in the field of manufacturing automation with several FM-systems around the country, there will also be a central production control system for all the FM-systems and cells. The cell controller will thus have a connection to a central production control computer.

6 CONCLUSIONS

First, when building a small FM-system or a cell with requirements for automatic set-ups and production management, connections between the machines and the controller becomes often a major problem. There are two types of solutions available. Either you have to buy new system from one manufacturer or you have to do a lot of additional work to make the communications work. This is the case specially with older machines. New standardization efforts, especially MAP, are trying to ease the problem. MAP is still far from the factory floor and will probably not lower the connection costs inside small systems. In our case with the terminal type protocols and the communication strings in the database, we can cover many different machines and microprosessor controllers without changing the software. In this case it is also easy to realize direct connection from the operator's terminal to the machines.

Secondly, by building the controls using a real time database and general control primitives to execute the controls according to parameters in the database, it is possible to make a control system that is easy to adapt to new products and different production systems.

7 REFERENCES

Ä1Ä Lempiäinen J., Temmes J., "Flexible assembly cell realization for feasibility studies of industrial assemblies." Proc. of the 7th Int. Conference on Assembly Automation, Zurich 4 - 6 Feb. 1986, pp. 227 - 232. IFS (Publications) Ltd, 1986.

Figure 1. A photograph of the cell.

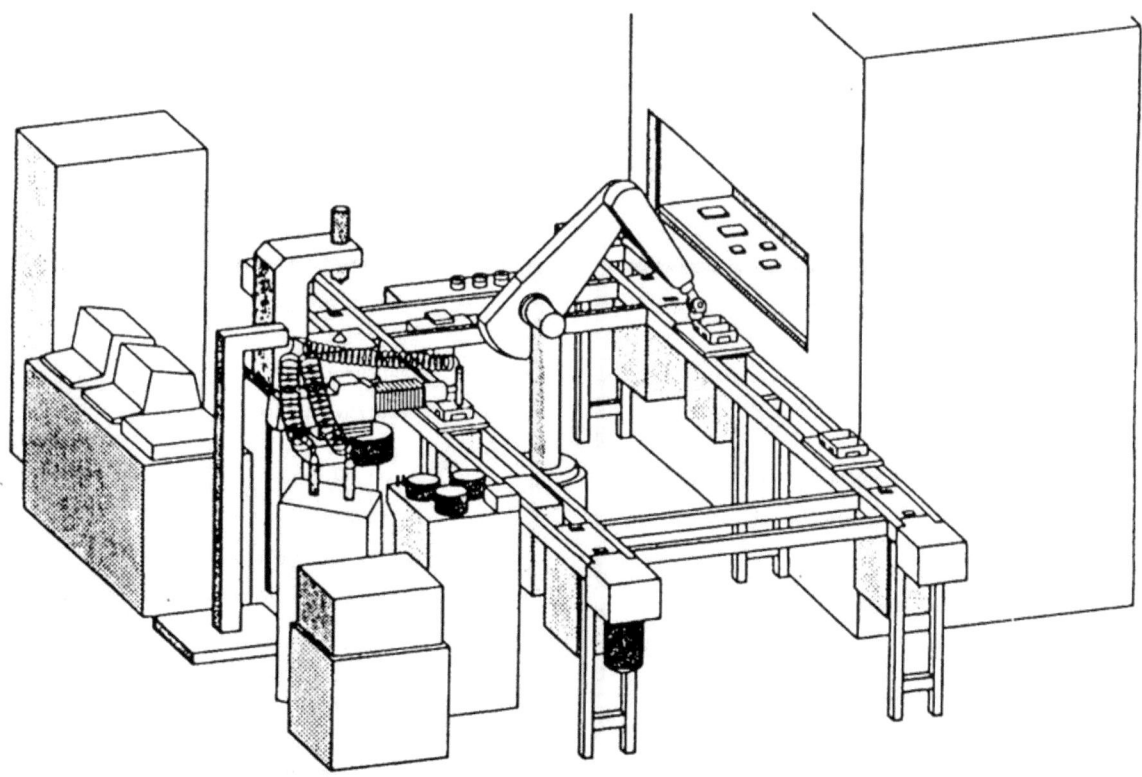

Figure 2. A schematic picture of the cell.

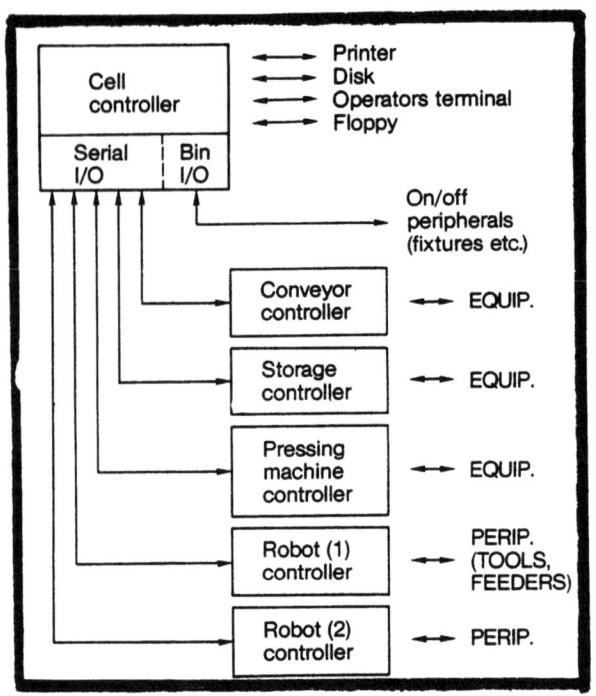

Figure 3. Hardware architecture of the cell control system.

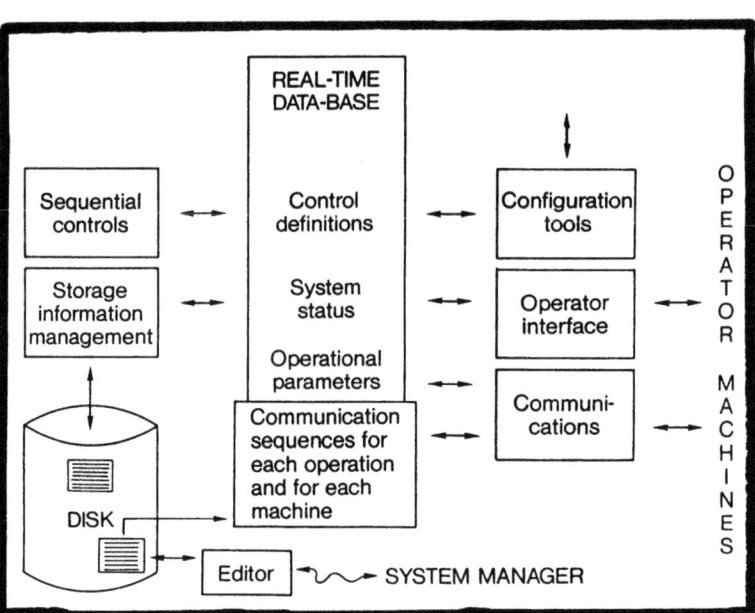

Figure 4. Software architecture of the cell controller.

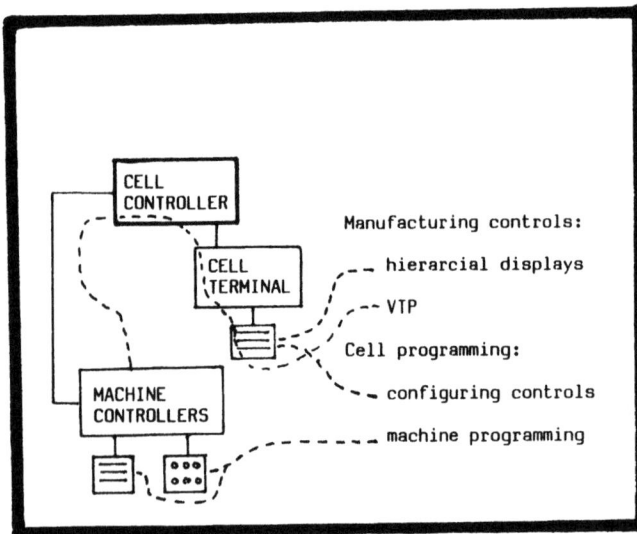

Figure 5. User interfaces of the cell controller.

```
*CREATE-DISPLAY
 Display:           CELL_OVERVIEW
 Type (M):
 Window (GEN):
 Text:              OVERVIEW OF THE CELL
 User/System (U):

- CLEAR
- XY: 40,4
- LITERAL: 'SYSTEM STATUS',13,LEFT
- IF: SYSTEM_RDY = 1
- THEN:
- LITERAL: 'READY',5,RIGTH
- ELSE:
- LITERAL: 'NOT READY',9,RIGTH
- ENDIF:
- XY: 40,10
- STATE: PUMA_CONT.RDY,8,3,UPDATE
- XY: 40,12
- STR: PUMA_CONT.PROG_NAME,20,3
- E
```

Figure 6. Example of configurating an operator's display.

```
    *CREATE-TASK
     Name:          TM_PHASE_1
     Type:          ROBOCON
     Unit Code:
     Window (GEN):
     Text:          GET 1:ST PART FOR ASSEMBLY
     User/System (U):

    LOGIC:
    - START = STEP2.STR
    - PROGNAME = PUMA(1)
    - OPERATION_TYPE = TM_OP_TYPE
    -

    Command: E

    --- Task for TM_PHASE_1 created ---
```

Figure 7. Example of configuring a control task.

```
--------------------------------------------------------------------------
VTT ASSEMBLY                    CELL DISPLAY                      05.10.1986
CELL                                                   PAGE 1/1    12:05:24
--------------------------------------------------------------------------


    CELL STATUS:  alarm

        STATISTICS: cumulative time: 120:23:12
                    cumulative run time: 52:12:22
                    no of stops: 23


        SCREWING STATUS: ready

        PRESSING STATUS: running

        PICKING STATUS: alarm


        PRODUCT: SHAFT_GB_1
            STATUS: waiting
            PHASE: SECOND_BEARING_FETCH


--------------------------------------------------------------------------
** PUMA NOT READY **
--------------------------------------------------------------------------
SCREWING   PRESSING   PICKING   PRODUCTS      STATIONS
--------------------------------------------------------------------------
```

Figure 8. Example of a operator's display

Next generation FMS: Simulation techniques for designing a PCB assembly system

D. A. Cowan, R. H. Davies, T. J. Pontin
The Plessey Company plc, UK

ABSTRACT

FMS applied to electronic assembly offers an opportunity for a significant advance in the state of the art, particularly if the cell can concurrently assemble mixed format PCBs. The methodology to design, and use a robotic PCB assembly simulation model for a total assembly system approach, including the product profile quantified in terms of mix, volume and phasing reflected by the production master schedule is described.

INTRODUCTION

Productivity problems assembling small batches of complex PC boards with inserted components are further compounded by new design trends. To get more functions per unit area at lower cost, electronic circuit designs are increasingly using SMT (Surface Mount Technology) and VLSI (Very Large Scale Integration). The resultant mixed format PCBs presently require discrete assembly operations for the different types of conventional, odd-form and surface mounted components.

This paper describes a system to assemble mixed componentry cost effectively. It can be integrated into a CIM strategy and the enabling technologies of CAD, CAE, CAM, CAT, CAPP etc.

The total system has been simulated and the process tested using entity cycle diagrams.

SYSTEM OBJECTIVES

First, the system should place components correctly, failure rates being measured in a few parts per million. Next it should place components cost effectively at a high overall rate and system up-time, avoiding the problems of setup and changeover of components and tooling associated with automatic component insertion equipment and dedicated robots.

The system should be software driven so that it can be integrated to obtain the benefits from CIM technologies.

FINAL SYSTEM DESIGN

At Plessey we have designed a universal robotic assembly system which can cope with batch quantities down to one and can select from over 2,000 components. With an average component placement cycle time of less than three seconds, it is competitive up to large batch quantities with conventional automatic equipment. [1]

Because the system interfaces over the broad spectrum from CAD through production, sales, material control and purchasing it is important all functional relationships and data paths are clearly identified and optimised. We achieved this using IDEF methodology.

The cell is designed as a modular system, and to achieve a high system up-time, place components while the rest of the system is running. For short batches a large number of different components may be required over a few hours, so the cell is designed to give the robot immediate access to over 2,000 different components. This is achieved by computer scheduling and on-line simulation to obtain optimum loading of components to the cell.

A high system up-time with small batch quantities requires minimising robot programming time which we achieve by incorporating automatic generation of component loading programs in the computer aided design of the printed circuit boards. The system is data-driven and software-controlled.

Components need feeding in many ways and the system takes account of tape, tray, tube and bulk delivery. A vision system is necessary to get good placements with minimum failure.

The assembly system in effect starts afresh with each board delivered to it, calling up the program and component data according to a bar code identification. The batch quantity is therefore irrelevant to the functioning of the system.

The mechanical layout for a single robot system is shown in Figs. 1 and 2. Boards are loaded to platters which can be stored in a stacker system before entering the assembly cell. Boards are identified to the cell either by a bar code on the board or a bar-coded label on the platter.

Alignment of boards to the robot system is achieved by mounting all boards on primary and secondary datum location pins on the transport pallets. When the pallet is docked in the work area, two locating plungers enter the same location pins in the pallet to remove board positional tolerance errors.

To obtain access to 2,000 different components, the component delivery systems are mounted on up to seven carousel conveyors arranged at 30 degree intervals around the robot, each with a fixed dispensing point within reach of the robot gripper. The carousels are substantial, each being 15ft long, with a 3hp motor drive and between 200 and 300 positions on to which magazines, tapes or other modular dispensing systems are attached.

Different types of robot grippers may be obtained either by gripper changing or an indexing head on the robot arm. Operations such as preforming of axial leads and combing of dual in-line component legs are performed before insertion. With a two-robot system these operations, with initial pick-up, are performed by the first robot. The second robot aligns and places the components.

As the board passes the reader, the code initiates the data required for that board, indicating which components go where. All information is stored, and as each board passes the bar code reader it downloads the data for that board only. Changeover time between entirely different boards is usually less than five seconds, and the overall system time is less than three seconds per component on boards containing over 100 components on average. The premium to pay for complete flexibility is therefore very small.

All the boards are handled within the period of the carousel loading. This is done to a master schedule, so the board/component combination can be treated as a repetitive flow operation sequence, and just-in-time advantages can be achieved in inventory, work in progress and cycle time.

For components requiring insertion, the variables are; component presentation, lead forming, lead straightening including compensation for lead material and cross section, lead cutting, insertion and clinching.

These tasks are best addressed by two robots performing concurrently and employing vision systems with insertion and clinching performed by the second robot. The vision system, with a rapid integrated feed back for robot guidance, has the same language and is used for location of components, component leads and board artwork. Other vision systems inspect the components, boards and final assembly.

Components may be mounted to both sides of the PCB. This is achieved with a board turnover unit attached to the transport platter. Surface mounted devices requiring retention on the board before soldering have an adhesive applied at the pin correction station before placement on the board. Conventional axial leaded components are preformed to the required pitch and shape at one of the intermediate stations.

The grippers and ancillary equipment are of minimum number and maximum flexibility, with the ability to compensate for body variation and also variations between body and leads which can be quite substantial in odd form components. Feedback, both positive and negative, to determine good and bad conditions is required.

Because the whole system is data driven and software controlled it is possible to have data collection and quality analysis performed automatically via the relevant feedback loops. If a part is not formed correctly or misplaced or a robot is malfunctioning the statistical analysis identifies the cause.

METHODOLOGY

There are five principle steps in the methodology: Business Objectives, Problem Definition, PCB Profile Analysis, Generic Simulation Model, Detailed Solution Model.

BUSINESS OBJECTIVES

The Business must respond more flexibly to market demands, reducing costs and offering shorter lead times whilst maintaining high quality.

DEFINITION OF THE PROBLEM

Many vendors supplying military and other specialised markets have a PCB profile containing the following six features:

Features	Major Influence
Low Volume	Difficult financial justification
High Mix	Demands high flexibility
Low Commonality	Large setting element between batches
Small Batch Sizes	Increased number of settings
High Number of Engineering Changes	High change and subsequent rework content
Stringent Quality Requirements	High inspection content and configuration control

Emphasis is focused on the complexity of controlling both material and information flows.

Technology Options

There are four alternative approaches to assembly:

Manual methods although offering flexibility and capability to absorb many inefficiencies which cannot be tolerated by automated systems, are error prone and as a result demand a large inspection element.

Semi-automatic approaches, which involve a manual operation to effect each placement, require a relatively low capital investment. Generally these systems are inflexible and yield poor utilisation.

Dedicated plant, automatic inserters and onserters, have a proven track record but in addition to suffering from inflexibility, generally prove financially inviable for low volumes.

Robotic technology provides the desired flexibility for both inserted and surface mounted components cost effectively, particularly when "intangible" benefits are taken into account.

Because of the number and complex interrelationship between parameters, a method of modelling the problem was demanded such that each option could be evaluated and compared both technically and financially.

PCB PROFILE ANALYSIS

The first step defined a generic representation of a workstation, for all assembly technologies, to identify the system variables, eg the number and mix of component feeds, grippers etc. This is shown using IDEF methodology in Figure 3.

Quantifying these variables required extracting relevant PCB profile information. We performed a series of purpose written routines on a database generated using files captured from the MRP system reformatted to permit efficient sorting. Essential component shape details, such as the pitch and number of leads, were added.

Output from this analysis were a number of distributions covering relevant profile features, e.g.

System Variable	Profile Feature	Sort Routine
Number and mix of component feeds	Part number variety of each component shape	Distribution of part number varieties for each component shape contained on each board

GENERIC SIMULATION MODEL

We simulated a generic workstation, using FORRSIGHT, with system requirements quantified by the profile analysis. The model was run for an equivalent one year workload for alternative technologies and specific equipment types using:

o   Equipment Technical Specifications

o   Stated, Measured or Estimated Operation Times

o   Actual Batch Size Distributions

o   PCB Profile Database

o   Range of Market Forecasts

Performance data was input into a financial model to produce a DCF analysis for each iteration.

Key Conclusions

The results highlighted three major areas of inefficiency:

o   Kitting   - Component selection, preform and marshalling

o   Setting   - Loading and setting component feeds and gripper tooling

o   Inspection   - Verification of correct component geography, polarity and physical placement

The combined effect of these, together with the problem of controlling the variety of batches in the system, introduces a massive queuing element.

Rework normally generated by manufacturing and/or component defects, engineering design changes and shortages, greatly increase the problem because assemblies repeat the queue-activity cycle.

Proprietary Equipment

Automation of PCB assembly for a high mix, low volume profile potentially reduces inspection but component marshalling and workstation setting become more acute. For a given volume these problems are inversely proportional to batch size and commonality between assemblies.

Proprietary automated equipment, generally designed for higher volumes and a limited mix, do not adequately tackle these problems, particularly component marshalling.

DETAILED SOLUTION MODEL REQUIREMENTS

Locate all components required over a given time period in the feeds reducing kitting and setting operations to maintaining levels in each cassette.

All component shapes, including "odd form" and surface mounted devices, be accommodated by grippers mounted on an indexable head to reduce tool setting.

The system should process a batch size of one without cost penalty.

Prototype assemblies to be processed rapidly without sacrificing production output.

The generation of programs directly from CAD, and issue changes through software driven procedures, to ensure that the degree of data integrity essential for successful operation is maintained.

To minimise the risks a detailed simulation of the solution was written, using OPTIK, designed to meet two major requirements:

o   As a system design and performance analysis tool. Other businesses can prove a particular configuration against their own PCB profile, production rules and market forecasts.

o   As a scheduling tool after implementation.

Why Simulation?

Simulation is a "what if" technique which does not return an optimum mathematical solution like some Operational Research methods, notably linear programming. It possesses three principal strengths:

o   Virtually unlimited complexity can be modelled accurately, an attribute which is particularly valuable when evaluating FMS.

o   System parameters and input data can be changed interactively. Many iterations may be tried to find a satisfactory solution.

o   Graphical output enables the analyst to validate a model with greater confidence, and provides a means of proving solutions to senior management or a customer.

Entity Cycle Diagrams

The fundamental building blocks of a simulation model are entity cycle diagrams. These represent the problem by expressing system entities, eg carousels, robot, live assemblies, in terms of their events and states. Events are either conditional or bound. the former require at least one condition to be true, whilst bound events are scheduled to occur at a given time. States are treated as attributes or each entity, e.g. carousel waiting for refill, operator fixing feed failure. A simple entity cycle diagram is shown in Figure 4.

Carousel Sequencing

The technique is illustrated by focusing on sequencing the carousels. Consider some key issues:

o   Carousel Specification

How many carousels?
What is the size of each carousel?
What packaging forms must be facilitated?

o   Stock Level Policy

For what time period are the carousels loaded?
How and when are the carousels loaded?
How are components distributed in the carousels?

o   Cell Control

What index speeds are required to balance the cell?
Are one or two robots required?
How many operators are required?

o   Sensitivity - What are the effects of:

Operation failures?
Component failures?
Component shortages?
Engineering changes?

Master Assembly Schedule

A decision process is required which evaluates cell status after every component pick and determines the optimum action for each carousel. A Master Assembly Schedule (MAS) was developed for this purpose.

The MAS lists all components, in placement order, required for assembly within a given time window. As a component is picked it is deleted from the MAS. Up to six thousand components, or approximately forty assemblies, have been shown as possible.

When a component is selected a conditional event, "Carousel Actions", is called. This subroutine first determines the cell status, eg availability of operators, then interrogates the MAS to find the next components to be picked or refilled on each carousel. A decision that optimises the process is actioned - not always the most obvious course.

Output

Performance data is presented by tables and histograms which can be scrutinised any time during the simulation. Because system parameters are changed interactively, change impact can be studied immediately.

Conclusions

This model provides the means to generate and prove design specifications for the cell elements. In addition, the system control is proved and software requirements clearly defined. By linking the output to a financial model, cost, contribution and profit implications for the Business can be evaluated, Fig. 5.

Modelling Strategy

IDEF 0 is used to model the functions within a system. Material and information flows are represented statically. IDEF 1 models the logical data structure. This is also a static representation. Simulation provides a dynamic analysis of both information and materials. Results are input to a financial model which performs capital appraisal, profit and contribution analysis.

REFERENCE

[1] Cowan, D.A. "A totally flexible system for PC board assembly". Proceedings of the 9th British Robot Association Conference, Stratford-upon-Avon, UK, pp 71-80 (May 86)

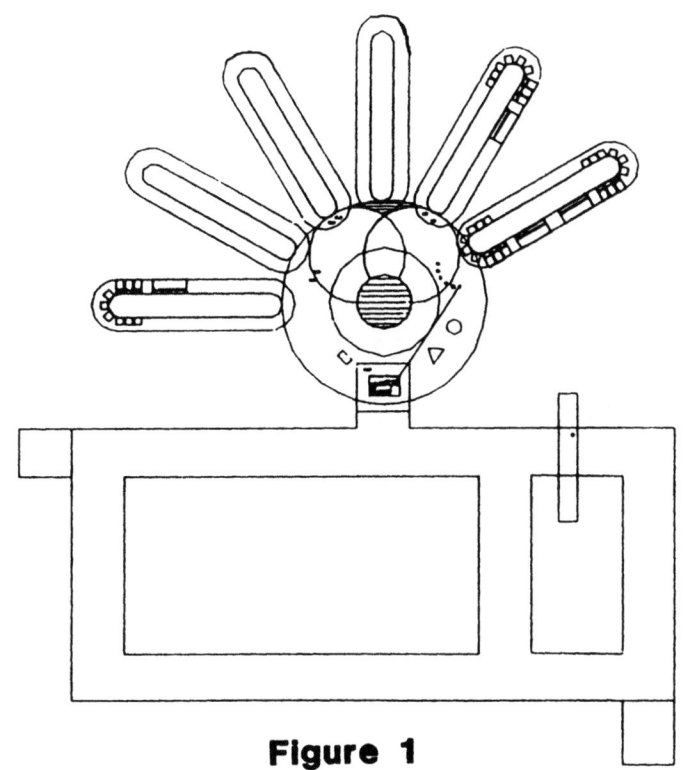

Figure 1

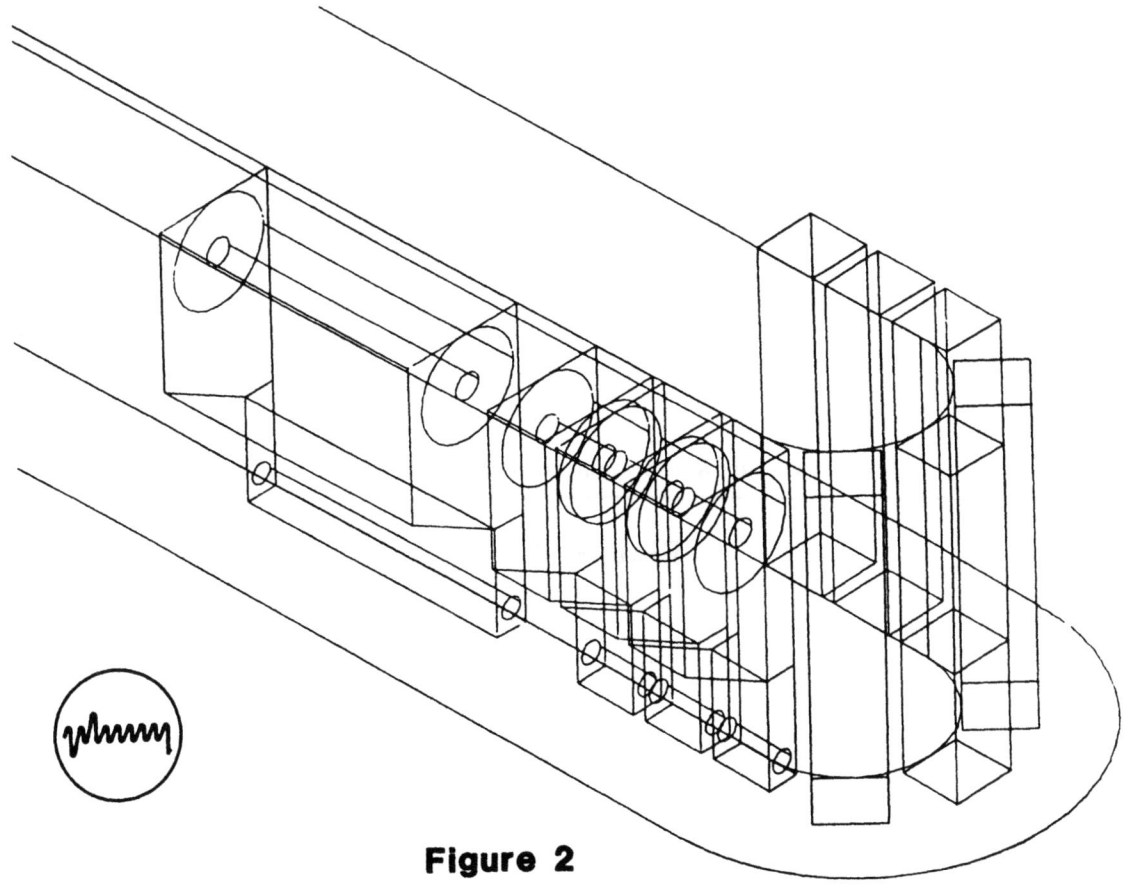

Figure 2

EXECUTE WORKSTATION OPERATIONS

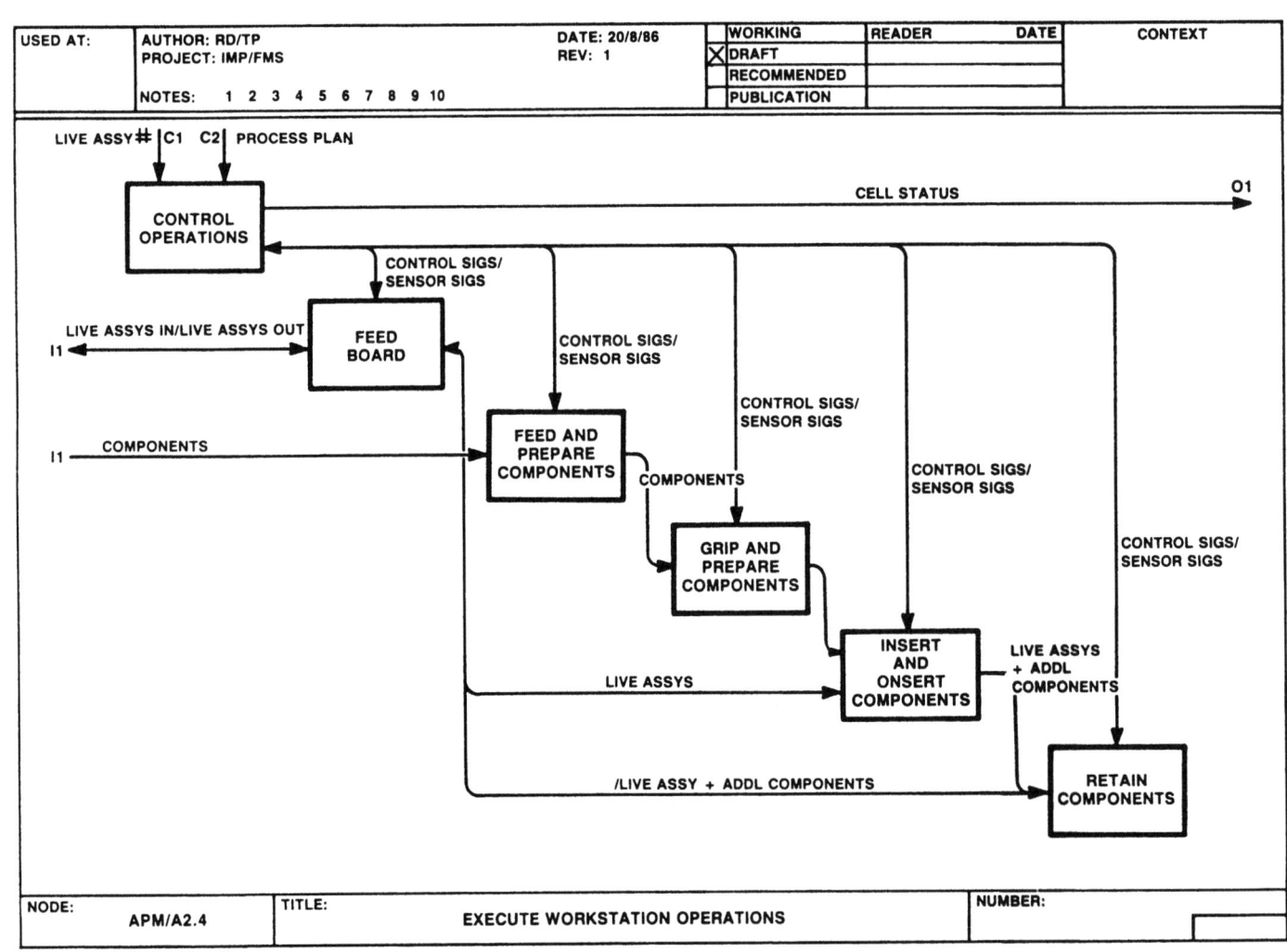

Figure 3

MANUAL ASSEMBLY OPERATOR

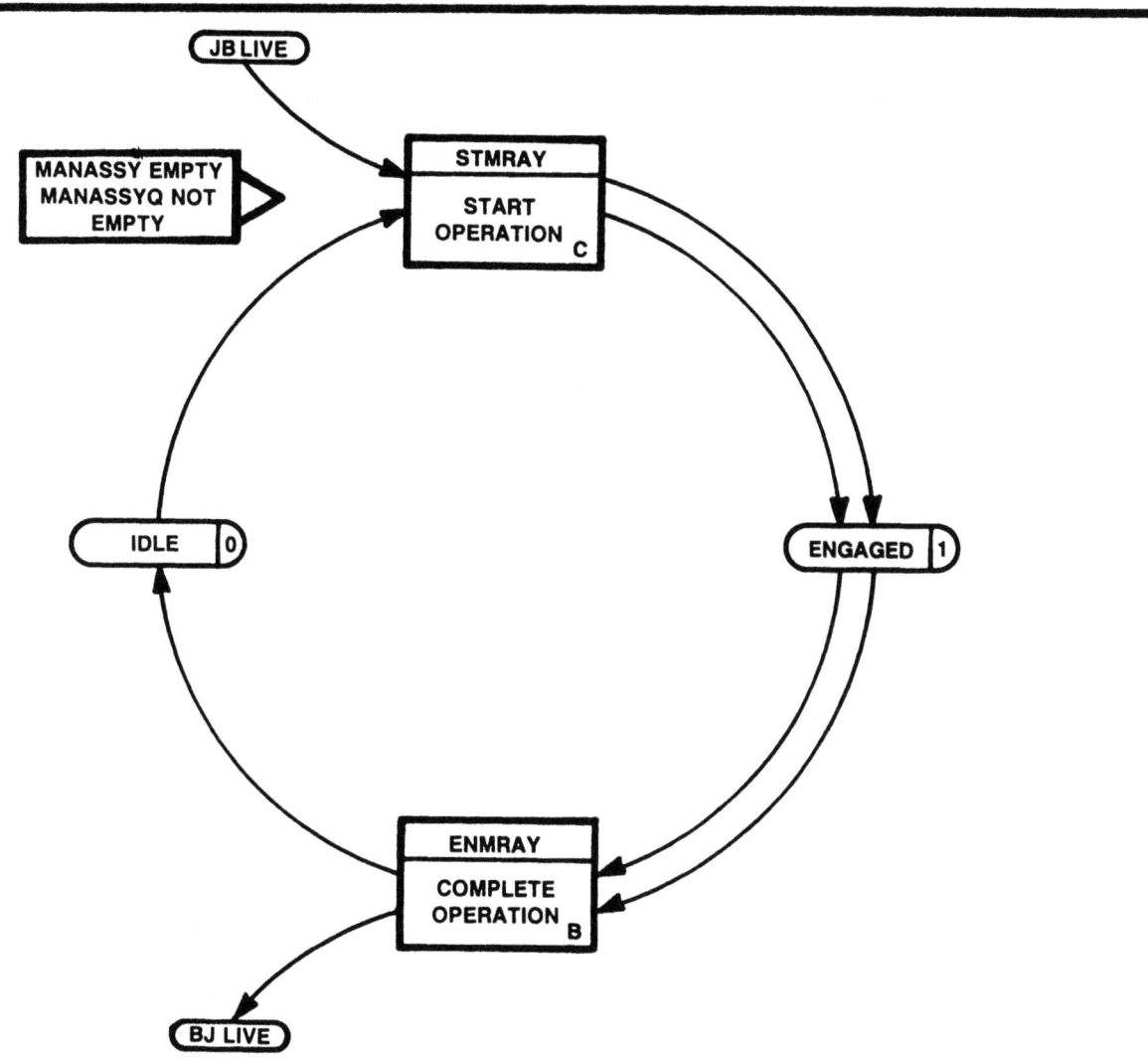

Figure 4

MODELLING STRATEGY

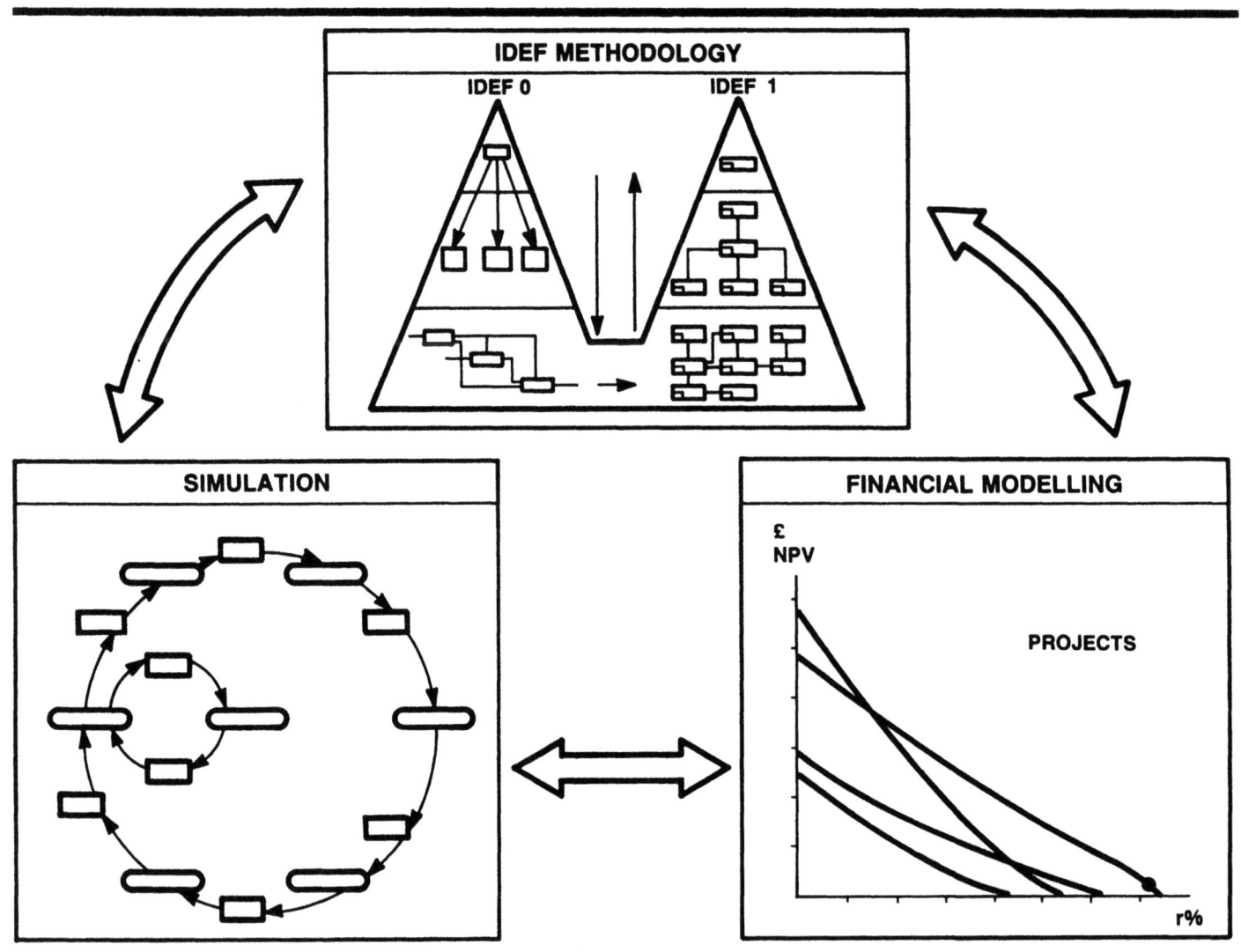

Figure 5

In-process quality control and corrective feedback in a flexible manufacturing cell

M. Veron, J. Richard and E. Bajic
Centre de Recherche en Automatique de Nancy, CNRS, Université de Nancy, France

ABSTRACT

A flexible manufacturing cell with integrated quality control is being developed by the LACN. Dimensional control is realized on a coordinate measuring machine. Integrated inspection allows an appropriate feedback on the manufacturing process with the following benefits :

- measurement analysis ensures on-line corrections of N.C. machine tool programs. Thermal drift and tool wear are predicted during processing. For this purpose an auto-regressive model identifies dimensional fluctuations and then corrections values are estimated through a Kalman filter;
- dimensional results allow storage by the cell robot according to quality classification.

Quality determination for part scheduling and automatic correction planning of N.C. programs are defined in the cell program.

INTRODUCTION

As a priority of flexible automation, high productivity is inextricably bound to the concept of quality [1]. In a flexible manufacturing system quality control must be considered not only as an inspection function but also as a prevention function, because the best way to decrease the cost of quality is to decrease the number of rejected parts. This concept of prevention means that quality control should be used as a feedback control on the manufacturing process. In practise, this requires the use of :
- in-process integrated inspection ;
- measurement analysis and statistical process control to prevent failures.

To maintain a high level of flexibility, "integrated in-process quality control" must be entirely programmable. Ideally, this should be programmed at the same time as the production planning.

The LACN is currently engaged in research on an automated manufacturing cell which integrates quality control and automatic N.C. program correction. This cell includes classical N.C. machine tools, a computerized measuring machine and a gantry robot for handling parts.

Applications are envisaged in small industries as a further step in automation with a minimum of capital investment. To reach this goal our research project includes the following fields :
- development of specific interfaces to ensure easy connection between N.C. machines and the cell computer through a reliable local area network;
- integration and control of gantry robot for parts handling;
- integration of a numerical controlled coordinate measuring machine (CMM) in the automated manufacturing cell;
- statistical process control and corrective feedback to maintain optimal production quality;
- integration of real-time quality control within cell programming.

The first two items above have been described in [2] and [4]. This paper is primarily concerned with the last three items.

INTEGRATION OF THE COMPUTERIZED MEASURING MACHINE

The first step in achieving our goal of an automated flexible manufacturing cell was to integrate the N.C. machines to create a level II DNC system managed by a central computer. A local area network (LAN) allows information transfer for controlling the component machines. The LAN was designed with system controllability as a major objective [3]. Given a relatively small number of stations (less than 20) and the capabilities of a typical industrial process control computer (such as the SOLAR 16/65), a star topology with distributed access was chosen for its low cost [4]. Because existing N.C. machine controllers are not capable of all the functions necessary for integration within the system (keyboard simulation, auxilliary device control, communication protocol, etc...), it was necessary to develop a modular programmable interface (MPI).

The coordinate measuring machine is tired to the system through such an interface (Fig.1). This link plays an important role in the manufacturing process by allowing dimensional information to be fed back to the central computer for on-line analysis and correction of N.C. programs. The CMM is controlled locally by an HP-9825-A running under the "Promesur" measurement software.

This conversational software is "closed" to users and allows only teaching mode programming.

To maintain accuracy the CMM must be kept in an air-conditionned and dust-free room.

The connection of the CMM to the LAN, realized through a MPI, achieves the following functions :
- the usual network interface unit functions ;
- new functions such as simulation of keyboard related to automatic operation needed to interface to the measuring machine software;
- programmable controler functions such control of parts clamps, measuring machine room lock, or other devices.

The measuring machine has two operating modes :
- DNC level I ; all functions of the machine are controlled by an operator who has conversationnal access to the cell computer. This mode is usefull to produce or to adjust measuring programs in teaching mode.
  The measuring machine may be also used as a tool setting system where tool data are obtained and stored in files on the cell computer.
- DNC level II or automatic mode ; the CMM, entirely controlled by the cell computer, receives orders and transmits measuring results without operator intervention.

Finally, when the machine is disconnected from the LAN, operation in local mode is possible.

STASTISTICAL PROCESS CONTROL

Determination of N.C. program correction by geometric results requires sophisticated statistical analysis. Knowledge of only the last measure is insufficient to determine trends in dimension fluctuation. However this trend can become very important over several parts and can exceed the tolerances.

Thermal drift and tool wear are the main causes of dimension fluctuations.

Tool wear can be estimated by calculation, or measured during machining by means of specific sensors, however it is very difficult to identify thermal deformations of the machine, the tool or the part by mean of temperature measurement.

If thermal drift and tool wear are approximatively constant during machining time for one part, N.C. programs can be corrected by means of measurement of the preceeding parts [5]. Our application deals with this case.

Experimentally, a dimensional value of a part series is a stochastic process i.e., the superposition of a random dispersion and a fluctuation. These two processes may be separated by a moving average method, but this method cannot be used to predict corrections. By hypothesis we consider that the random component serie is generated by a Gaussian process. The maximum entropy method identifies this stochastic process by means of an auto-regressive model excited by white noise :

$$v_n = y_n + b_1 \cdot y_{n-1} + \ldots + b_k \cdot y_{n-k}$$

where :

v_n is the Gaussian white noise ;

$y_n, \ldots y_{n-k}$ is the sequence of the $k+1$ last measurements ;

b_i are the coefficients of the auto-regressive model.

The b_i coefficients are estimated by the Burg-Levinson algorithm [6] with the $k+1$ auto-correlations of y :

$$M_{i,n} = E\{y_n \cdot y_{n-i}\} \quad i = 0, \ldots, k$$

The predicted value of y computed at time n for time n+j : y_{n+j}, can be deduced from the b_i coefficients as :

$$\check{y}_{n+1} = b_1 \cdot y_n + b_2 \cdot y_{n-1} + \ldots + b_k \cdot y_{n-k+1}$$
$$\check{y}_{n+j} = b_1 \cdot y_{n+j-1} + \ldots + b_k \cdot y_{n-k+j}$$

To test this method we machined a batch of 1000 parts on a N.C. lathe with rotary sensors on the screw. The machining duration was 7 minutes and the inspected dimension was an exterior cylinder tooling with a finishing cut of 0.2 mm. In these conditions tool wear is on the same order but slightly less than thermal drift.

The best results were obtained with an one-step prediction computed by a third order A.R model. The auto-correlations were estimated using the 7 last measurements. In these conditions error variance is maintained at $(5\mu)^2$, this variance should be $(7\mu)^2$ with a sample correction (i.e., 0 order correction).

However, two hypothesis must be satisfied :
- the number of measurement must be larger than the order M of the model;
- the stochastic process must be stationary.

These two conditions are only satisfied in first approximation. The recursive Kalman algorithm increases the prediction accuracy. The following equation estimates the state vector from the sum of last predicted vector and the product of prediction error and filter gain :

$$\hat{Y}_n = \check{Y}_n + K_n \cdot (y_n - \check{y}_n)$$

with :

$$\hat{Y}_n = [\hat{y}_n, \hat{y}_{n-1}, \ldots \hat{y}_{n-k}]^T$$
$$\check{Y}_n = [\check{y}_n, \check{y}_{n-1}, \ldots \check{y}_{n-k}]^T$$

The prediction value is obtained at time n by :

$$\check{y}_{n+1} = B^T_k \cdot \hat{Y}_n$$

where :

$$B_k = [b_1 \ldots b_k]^T$$

The filter gain Kn is given by the following equation :

$$K_n = [E_n][E_n + W_n + M_{0,n} \cdot A_n]^{-1}$$

Where :

$$E_n = E\{(y_n - \check{y}_n)^2\}$$
$$W_n = E\{w_n^2\} : \text{variance of measure noise.}$$
$$A_n = E\{\Delta b_i^2\} : \text{variance of } b_i \text{ coefficients.}$$

The A_n value is typical of machining process stability.

With the Kalman recursive algorithm, error variance on our experiment is maintained at $(5\mu)^2$ for a three steps forward prediction. This method essentially increases the stability of the predictor. The advantage of this stastistical method can be very important if we want to exploit the machine close to the limits of accuracy. For instance the rejected parts number decreases from 150 with a sample correction to 50 with our algorithm for 1000 parts of our experiment if the tolerance on the geometric dimension is $\pm 10\mu$.

THE CELL PROGRAMMING PROCEDURE

The production of part batches is completely programmable from an independant processor which gathers descriptive manufacturing informations introduced by an operator. This processor is connected to the supervisor computer through the local area network.
The quality control programming is an important characteristic of the cell programming procedure. The operator has to determine :
- the quality classification of workpieces for corresponding storage by the cell robot;
- the planning of machining programs corrections by mean of dimensional measurement feedback analysis.

A graphical and interactive process planning software for a Macintosh micro-computer assumes the cell programming in three steps :
- Description in a "Grafcet" [7] formalism of the tooling sequences and control operations on the machines for one part;
- Simulation of the cell driving software for machining one or several simultaneous part batches;
- Determination of qualities for scheduling parts and determination of the planning for automatic correction of N.C. programs.

The three steps of the cell program functionning are described in three steps called pages.

1) First page : the phases chaining description
Organizing the cell production of part batches requires to know the machining jobs to be done on a part and their chaining for manufacturing one part. This chaining forms the specific description series for a given type of part.

With the help of a rolling menu, the operator defines the chronological stages succession which is automatically represented on a screen in a "Grafcet" formalism, in which a step is a job and the transitions correspond to global movements of the robot between two N.C. machines (Fig.2 a).

A completely defined series begins with an "unmanufactured parts feeding" stage and it ends with one or several "achieved parts storage" stages. So a description series is a graphical comprehensive representation explaining the successive operations to be done on a part (Fig. 2 b).

In case of multiple storages, the operator will define removing conditions for each storage stage according to different qualities of manufactured parts. This will be done in the third page.
During the definition of the description series, the possible stage menu is automatically managed to avoid impossible configurations, as several "feeding" stages in the same series or several "storage" stages without defined a "control" stage before.

1-1) **Stages definitions and representations.** Each stage is defined by the reference of the N.C. program that determines the work to be done on the station and by the physical situation of the affected station.

These N.C. programs that can be learned or programmed beforehand, are stored in the supervisor

computer data bank and are composed of two parts :
a) The ready-to-run N.C. program (programmed sequences, measuring programs, ...);
b) A comment area reserved for :
- some further informations on the used resources tools and part clamping devices references, comments for the operator, execution duration time ...;
- references of supplied local robot moves programs, for loading and unloading the station and their execution duration times.
- adresses in N.C. programs of the parameters to be automaticaly corrected. Such a parameter may be a tool dimension measurement (M1) or any other numerical value (M2) :

$$M1 = (T1, R)$$
$$M2 = (N2, X)$$

Two kinds of stations on which stages are realized may be distinguished :
- active station : for instance, a N.C. machine-tool on which an automatic processing is implemented;
- passive station : for instance, a storage buffer, its main characteristic is its physical position in the cell.

Also two sorts of stages are required :
- stages in which processes are realized by an N.C. program on the active stations (Fig 3. a);
- stages in which processes are local moves of the robot, for instance "feeding" and "storage" on passive stations (Fig 3. b).

When the operator selects from the menu a stage to put in a description series, he has to determine the N.C. program reference, the estimated or real execution durations of this program and of the associated robot programs.

A symbolic reference name is assigned to the description series, this name is common to all the needed files for the parts batch production. The operator can define one or several description series, one for each forecasted kind of parts, that will be simulated one by one or in association in the second page.

All types of N.C. machines-tool or programmable machines (washing machine, etc...) can easily be integrated in the cell structure because of the presented description method.
The stage modelization method allows the operator to link different description series by identifying the "storage" and "feeding" stages of them and also to program the recovery of scrapped parts identified thgrough the quality control.

Let us notice that the "robot special" stage in the rolling menu is used for special purpose as a reversal part stage between two works on a machine-tool and is described by the type of stage corresponding to a passive station.

2) <u>Second page : simulation</u>
The simulator running on the cell programming processor is based on a discrete events and temporal simulation method which summarizes the cell driving software organization, representing thus a model of the production process.

The cell driving software realizes :
- the flow control of manufactured parts by managing an identification card for each part coming in the cell;
- the production optimization regarding to robot movements and to engaged stations rates. The optimization strategy is the one selected by the operator on the programming processor.

- the manufacturing ordering. The realization of a stage is split into elementary actions to be sequentially executed by the machines of the cell :
    + global movement to station I;
    + clamping of part K on post A of station I with gripper P;
    + global movement to station J;
    + placing part k on post J of station J;
    + request for operating the stage on station J.

After defining technological conditions as physical situation of the stations, reference and type of the gripper, pallet settings, ..., the operator can choose between two simulation modes :

1/ Simulation with constraints analysis. After consulting the data bank, execution duration times are set for each stage according to the practicable program reference affected on the station. The technological compatibilities about clamping materials, pallet settings, robot gripper, tools, ..., are cross-checked;

2/ Simulation without constraints analysis. The simulator will operate on the duration times determined in the first page . This is useful to estimate a faced parts manufacturing.

The simulator is composed of :
- a module generating parts progress in cell. It recovers the prematured states of work-done on stations, and then creates the next stage request, according to the affected description series. The stage requests are stored in a waiting queue.
- an optimisation module based on the scheduling strategy choosen by the operator. The stage to be processed is selected between all of those contained in the waiting queue.
- a synchronization module managing the robot tasks when a phase is selected.

Simulation as evaluation means, informs the operator about :
- occupation ratio of the robot,
- duration of production cycles,
- duration of total production,
- on-station parts in progress ratio,
- ...

Once the description series has been simulated and validated both by the operator and the processor, an object file defining the software and handware cell configuration is generated to be sent to the supervisor computer for initialization of the cell driving software.

3) <u>Third page : parts quality control and feedback on the machining process</u>
The operator assignes a dimensional quality value to every "storage" stage inclued in the choosen description series. A quality value is defined with one or several measure results and their tolerances. Every measure result is adressed by an operation number in the measurement listing. For instance in the following example two qualities are defined :

$$Q1 : 12.078 =< \quad OP1 \quad < 13.251$$
$$Q2 : 50.025 =< OP2 + OP3 < 52.745$$

To do that, the operator uses, on the screen, two windows : a window showing the result measures of a reference measurement listing, obtained in teaching mode, and a window for input of qualities values.
Others quality parameters may be created by logical or arithmetical operations on preceeding qualities.

At this step, the manufacturing series is entirely defined and it is possible to process parts with automatic N.C. programs corrections. The feedback control is programmed by the operator who establishes relations between N.C. programs parameters and measurement results. Then the object file generated in the second page is complete and ready to be sent to the cell driving software on operator order.

The feedback software is resident in the supervisor computer and called by the cell driving software. During measurement analysis information may be sent to a manufacturing observation console in case of faults (i.e. tool break, ...) or correction overflows. So the operator has the responsability and the choice to rectify the fault (i.e. tool replacement, ...) and restart, or to cancel the analysis and continue.

CONCLUSION

The cell programming software allows the user to consider the cell as a super production unit, entirely programmable from an unique station. This software ensures a maximum of flexibility by its graphical and interactive conception.

The in-process quality assurance is used as a manufacturing process sensor, and the local network as a closed loop. This feedback process, programmed by the operator, optimizes the quality of the production by correcting N.C. programs.

This research study has been supported by A.D.I. (Agence De l' Informatique) of french ministry of industry.

REFERENCES

[1] J. HATVANY, M.E. MERCHANT, K. RATHMILL, H. YOSHIKAWA
"World survey of C.A.M." C.A.D. special publication, Butterworths, 1983.

[2] R. VOGRIG, P. BARACOS, P. LHOSTE, G. MOREL, B. SALZEMANN
"Flexible manufacturing shop operation."" Presented at 18th C.I.R.P.M.F.S. - S., June 1986. Stuttgart, F. R. of GERMANY.

[3] B. SALZEMANN, E. BAJIC, F. CORBIER, F. MUNERATO
"Ilot automatisé de production. Mise en oeuvre d'une modélisation de contrôle-commande et rétro-action par contrôle-qualité." In Proc. IMACS-IFAC Symposium pp 463-466, June 1986. Villeneuve d' Ascq FRANCE.

[4] F. LEPAGE
"Proposition d'un réseau local industriel hétérogène. Application à un îlot automatisé de production." Thèse de doctorat d'état, Université de Nancy I, 10 Mars 1986.

[5] J. RICHARD
"Contrôle dimensionnel et suivi de production dans un îlot automatisé de fabrication de pièces mécaniques. Analyse des mesures et prédiction des corrections."
Thèse de doctorat d'état, Université de Nancy I, 7 Février 1985.

[6] J.P. BURG
"Maximum entropy spectral analysis." In Proc. 37th Meet. Soc. Exploration Geophysicists, 1967; Stanford Thesis, 1975.

[7] "GRAphe de Commande Etape Transition" - NF C03-190 - ADEPA AFCET

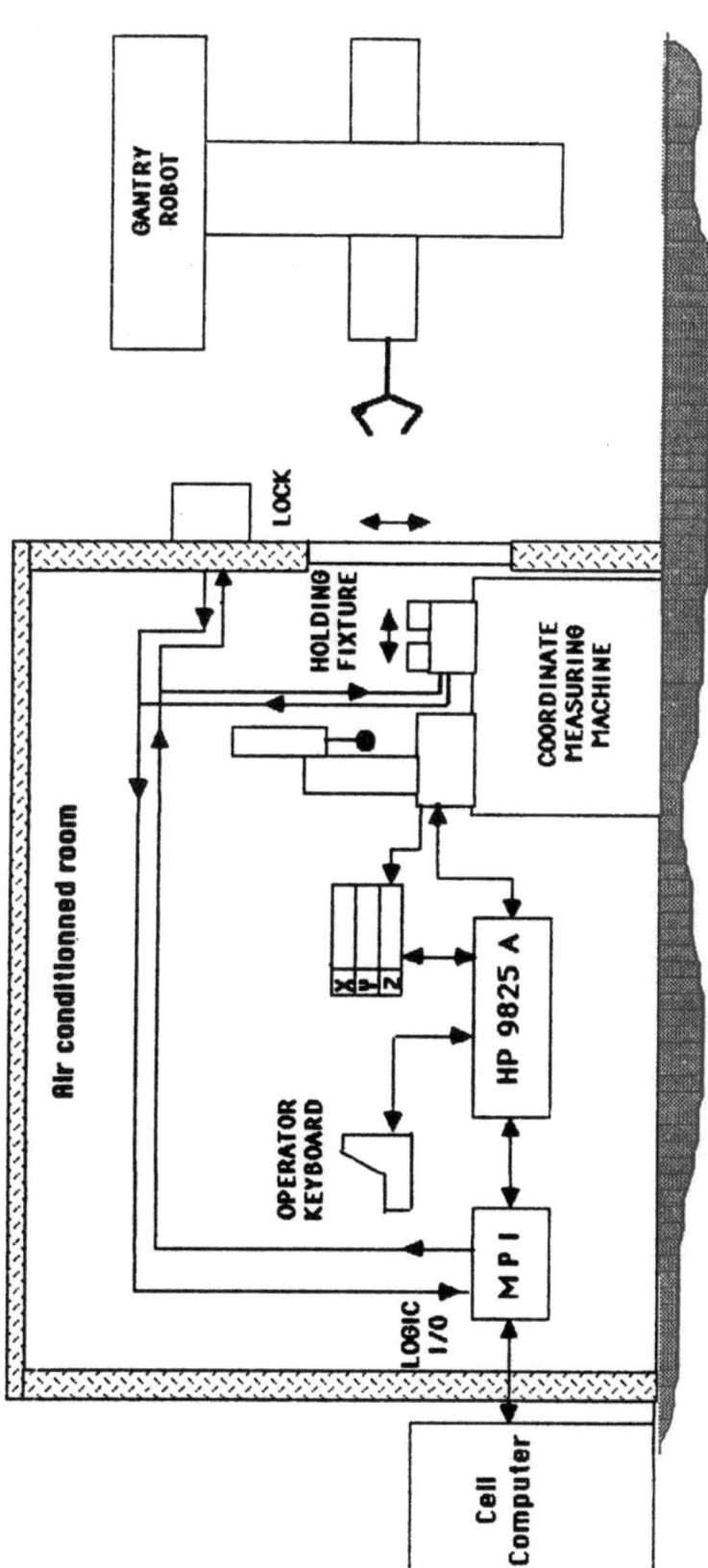

Figure 1.

INTEGRATION OF THE MEASURING MACHINE
IN THE AUTOMATED MANUFACTURING CELL

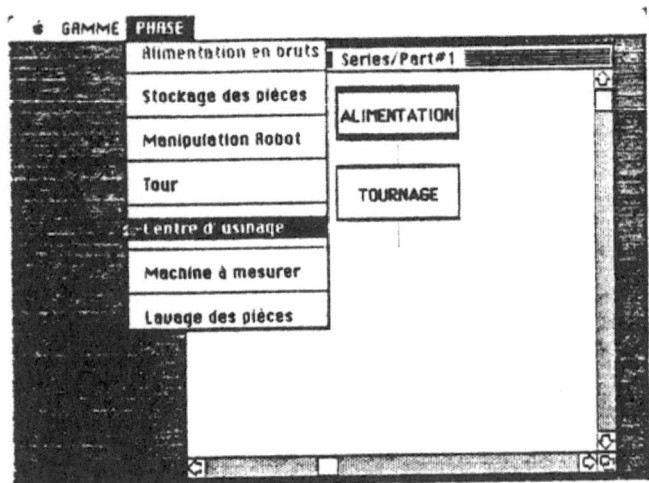

Figure 2a. Determination of the job succession to be executed for to produce a part.

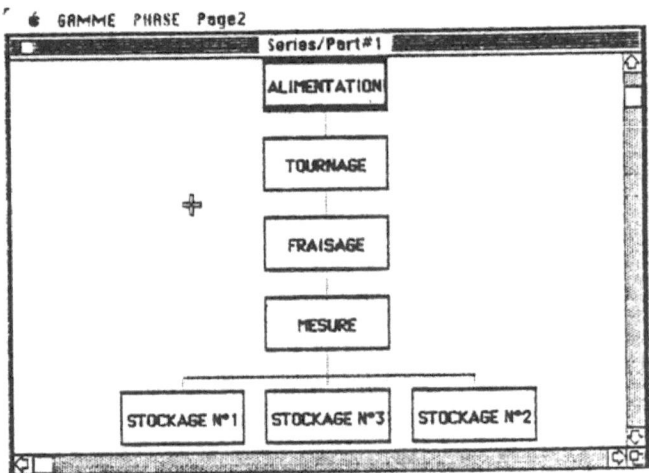

Figure 2b. A complete serie resumes the successive operations.

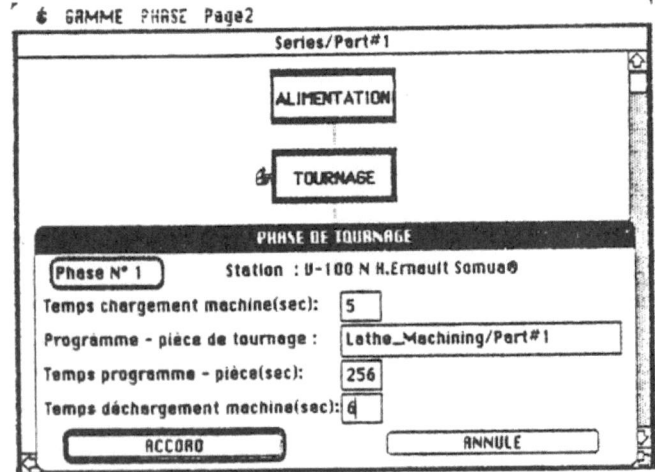

Figure 3a. Definition of an active station stage.

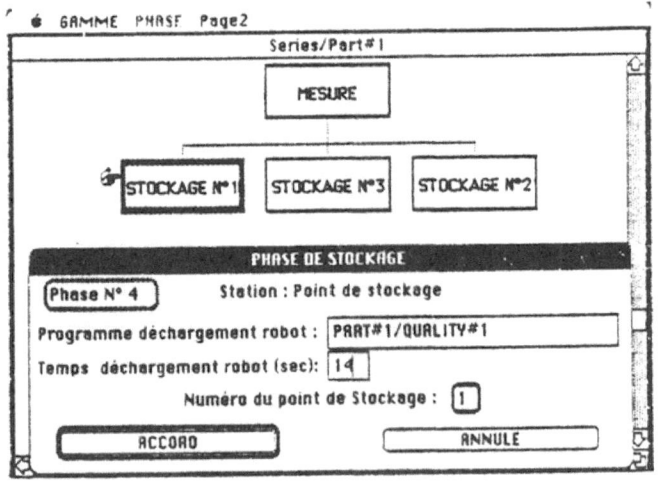

Figure 3b. Definition of a passive station stage.

SYSTEMS EXPERIENCE

The way ahead – the integrated approach
W. H. Horton
Austin Rover Group, UK

Austin Rover regards the Flexible Manufacturing as the total system that covers the Planning, Design, Manufacture and delivery of its products to the customer.

The paper will cover the design and implementation of this Strategy with practical examples of the logic and benefits.

THIS INTERNATIONAL CONFERENCE WILL PROVIDE THE MEANS FOR DISCUSSION ON THE CURRENT STATE OF THE ART OF FLEXIBLE MANUFACTURING SYSTEMS. THERE ARE BOUND TO BE MANY VIEWS OF WHAT IS REPRESENTED BY FLEXIBLE MANUFACTURING SYSTEMS.

THE AUSTIN ROVER VIEW IS THAT IT IS THE TOTAL SYSTEM BY WHICH WE PLAN, DESIGN, MANUFACTURE AND DELIVER OUR PRODUCTS TO THE CUSTOMER.

IN SUCH A SYSTEM, DATA ESTABLISHED AT THE DESIGN STAGE HAS TO MAINTAIN ITS INTEGRITY THROUGHOUT THE TOTAL PROCESS AND HAS TO BE THE DRIVER FOR THE PROCESS. THE KEY TO THE SUCCESS OF THE SYSTEM IS THEREFORE A FIRM INTENT TO REDUCE MANUAL MANIPULATION OF THAT DATA IN PRODUCING THE PRODUCT. THE COMPUTER AIDED DESIGN SYSTEM, THE DATABASE AND THE MANUFACTURING FACILITIES ARE ALL PART OF THE FLEXIBLE MANUFACTURING SYSTEM.

WITHIN THE TIME ALLOTTED ME, LET ME INDICATE TO YOU SOME OF THE DETAIL OF HOW THIS ALL WORKS.

FIRSTLY, THERE IS THE POWERFUL SINGLE SOURCE COMPUTER AIDED DESIGN SYSTEM WITH 330 WORK STATIONS COVERING ALL FUNCTIONS OF THE COMPANY, AND SECURELY NETWORKED BETWEEN ALL OF OUR SITES.

RESPONDING TO THE CONCEPTUAL PLAN FOR A VEHICLE, THE SYSTEM GIVES US THE ABILITY TO OFFER UP A 3 DIMENSIONAL MODEL OF THE COMPONENTS LISTED AS ELEMENTS OF THE PROPOSAL.

THERE IS FURTHER CONCEPTUAL MODELLING IN RESPECT OF THE INTERIOR OF THE VEHICLE, SUFFICIENTLY FIRM TO PROVIDE A BASIS FOR THE INITIAL DESIGN OF THE STRUCTURE OF THE BODY.

AS PART OF THAT DESIGN PROCESS, IT IS NOW POSSIBLE TO ACCURATELY LOCATE THE DRIVER'S EYE, AND FROM THERE WE ARE ABLE TO ACCURATELY CHECK A HOST OF FEATURES, INCLUDING VISABILITY ON INSTRUMENTS AND REARWARD VISION IN BOTH INTERIOR AND EXTERIOR MIRRORS.

SIMILARLY WE CAN LOOK THROUGH THE WINDSCREEN AND ENSURE THAT WE MEET LEGISLATION IN RESPECT OF THE SWEEP AREA OF THE SCREEN AND THAT WE HAVE ACCURATE VISIBILITY AROUND THE WINDSCREEN PILLARS IN ALL SIZES OF DRIVER, UNDER ALL CONDITIONS.

WHEN THESE AND MANY MORE OFFERINGS ARE COMPLETE BODY ENGINEERING CAN THEN BEGIN.

BUT BEFORE THE DESIGN IS APPROVED IT IS USUAL TO PRODUCE A FULL SCALE CLAY MODEL REPRESENTATIVE IN EVERY RESPECT OF WHAT IS PROPOSED. THIS IS A SLOW AND LABORIOUS OPERATION AND WHEN COMPLETED AND APPROVED IT HAS TO BE METICULOUSLY DIGESTED FOR ENGINEERING AND MANUFACTURING WORK TO TAKE PLACE.

THE COMPUTER AIDED DESIGN SYSTEM IS NOW BEGINNING TO OFFER US AN ALTERNATIVE.

FIRSTLY, WE ARE LEARNING TO DESIGN ON THE SCREEN IN 3D. THE OFFERED TECHNOLOGY IS SO FAR ADVANCED AS TO ENABLE US TO ADD COLOUR, SHADING, LIGHT, REFLECTION AND REFRACTION.

THE MODEL CAN BE REPRESENTED TO FULLY SATISFY THE STYLIST, TO THE POINT WHERE HE HAS CONFIDENCE OF ACCEPTABILITY. THE FINAL STAMP OF APPROVAL WILL HOWEVER ONLY COME FROM A 3 DIMENSIONAL FULL SIZE MODEL.

WITH ALL THE DATA ENCAPSULATED IN THE COMPUTER AND WITH THE SURFACES COMPUTER SMOOTHED, WE NOW HAVE A MEANS OF MACHINING THE MODEL AS OPPOSED TO CONSTRUCTING IT IN CLAY.

THROUGH RELATIVELY SIMPLE SOFTWARE THE ACCUMULATED INFORMATION IS NOW BEING FED DIRECTLY INTO A 5 AXIS MODEL MAKING MACHINE TO PRODUCE A FULLY REPRESENTATIVE 3 DIMENSIONAL MODEL THAT CAN BE ASSESSED AND MODIFIED UNTIL APPROVAL IS COMPLETE.

THE DEVELOPMENT OF THIS PARTICULAR PIECE OF TECHNOLOGY IS BEING VIGOROUSLY PURSUED AND TODATE IS GIVING VERY ENCOURAGING RESULTS.

SUBSEQUENT TO THIS APPROVAL THE INTERIOR AND EXTERIOR SURFACES CAN BE RELEASED TO THE REST OF ENGINEERING, THROUGH THE INTEGRATED COMMON DATABASES, FOR FURTHER DETAILED DESIGN WORK TO TAKE PLACE. BECAUSE THE SMOOTHED SURFACES ARE ALL DEFINITIVELY DESCRIBED, EACH AREA CAN BEGIN ITS WORK SIMULTANEOUSLY, AND BECAUSE EVERY PIECE OF DATA COMES FROM A CENTRAL DATABASE EACH OF THE AREAS WILL BE AWARE OF THE PROGRESS MADE, ONE TO ANOTHER.

BY WORKING THIS WAY WE MAKE TWO IMPORTANT GAINS.

FIRSTLY, IN RESPECT OF QUALITY, ALL THE ENGINEERS ARE WORKING FROM A SINGLE CONSISTENTLY UPDATED MASTER MODEL AND THERE IS THEREFORE NO OPPORTUNITY FOR ANYONE TO WORK AGAINST OBSOLETE DATA. SECONDLY WE ACHIEVE A GAIN IN LEAD TIME.

SIMULTANEOUS WITH ALL THIS THE PANELS DESCRIBED AND MACHINED ON THE STYLING MODEL IN RESPECT OF THE EXTERNAL SURFACES CAN BE TRANSMITTED DIRECTLY TO THE TOOL DESIGN AREAS SO THAT WORK CAN BEGIN IN PARALLEL WITH THE OTHER DEPARTMENTS.

SINCE THE PANELS HAVE BEEN SO WELL DEFINED, THE TOOL ENGINEER SIMPLY ADDS THE RUN OUT SURFACES REQUIRED TO GRIP THE METAL DURING PRESSING, AND THEN THE SUPPORTING STRUCTURE TO PRODUCE THE MALE DIE FORM.

THIS PICTURE OF THE COMPUTER MODEL IS TAKEN DIRECTLY FROM THE SCREEN. THE REST OF THE PRESS TOOL, SUCH AS THE SLIDES AND RELEASE MECHANISMS ARE DESIGNED AS STANDARD FITMENTS AROUND THE DIE SET.

ONCE THE DIES ARE DESIGNED THE SAME SOFTWARE USED FOR MACHINING THE FULL SCALE MODEL CAN BE FED INTO 5 AXIS DIE SINKERS TO MACHINE THE DIE.

EARLY DEVELOPMENT WORK, USING THIS PROCESS HAS PRODUCED PANELS FOR OUR NEW ROVER 800 VEHICLE TO SUCH A POINT, THAT APART FROM AUTOMATIC POLISHING, AGAIN UNDER THE CONTROL OF THE COMPUTER, WE HAVE AN "UNTOUCHED BY HAND" TECHNOLOGY AND WE FAITHFULLY PRODUCE THE SURFACES ORIGINALLY STYLED BY THE STYLIST. HIS WORK BECOMES THE INSTRUCTION TO THE MACHINE TOOL.

AGAIN FLOWING FROM VERY CLOSE ACCURACY TO WHICH THE TOOLS ARE PRODUCED THERE IS TOTAL ELIMINATION OF THE TRADITIONAL TRY OUT TIME WHILE MALE AND FEMALE TOOLS HAVE TO BE BEDDED IN TO OVERCOME TOLERANCES IN THE MACHINING PROCESS.

BY THE SPRING OF 1987 THIS TECHNOLOGY WILL BE FULLY COMMISSIONED AND INSTALLED IN OUR TOOL ROOMS AND WE CONFIDENTLY FORECAST A VERY SIGNIFICANT SAVING, BOTH IN TIME AND MONEY IN THE PRODUCTION OF TOOLS. WE WILL ALSO ACHIEVE A LEVEL OF QUALITY PREVIOUSLY UNPRECEDENTED.

WHAT IS EQUALLY IMPORTANT HOWEVER IS THAT THE PANEL PRODUCED IS IDENTICAL TO WHAT HAS BEEN DESIGNED. THE DATA IS THE SAME!

ALTHOUGH UP TO THIS POINT I HAVE TALKED ENTIRELY ABOUT BODY DESIGN AND DEVELOPMENT AND ILLUSTRATED THE ADVANTAGES OF COMPUTER INTEGRATED ENGINEERING THE SAME LEVEL OF EXCELLENCE EXISTS THROUGHOUT, AND I WOULD LIKE TO GIVE YOU ONE FURTHER EXAMPLE.

THE FULL MODEL OF THE ENGINE IS PRODUCED BY POWER TRAIN ENGINEERING AND THIS SLIDE SHOWS HOW THE DATA FOR THE CYLINDER HEAD OF THE NEW M16 ENGINE FOR THE ROVER 800 MODEL WAS TRANSMITTED FROM PRODUCT ENGINEERING TO MANUFACTURING.

UNDER THE SKIN OF THIS MODEL THERE IS ALL THE ENGINEERING DATA, FULLY DIMENSIONED IN THE COMPUTER AND MORE IMPORTANTLY ALL THE WORKING SURFACES SUCH AS INLET AND EXHAUST PORTS AND COMBUSTION CHAMBER ARE FULLY AND PRECISELY DEFINED, AS THESE ARE CRITICAL FOR THE NEW LEAN BURN TECHNOLOGY BEING ADOPTED TO REDUCE EXHAUST EMISSIONS.

WE CAN NOW EXTRACT ONE VERY COMPLICATED FEATURE OF THAT DESIGN, WHICH IS A WIRE FRAME DIAGRAM OF THE COMBUSTION CHAMBER AND THE PORTING, WHICH ARE A VERY CRITICAL PART OF THE DESIGN OF THE CYLINDER HEAD.

THE INFORMATION THAT WE HAVE ON THIS SLIDE CONTAINS EVERY SINGLE PIECE OF DETAIL THAT WE WISH TO KNOW ABOUT THIS PART OF THE PROPOSAL. IT WILL HAVE BEEN SIMULATED TO THE POINT WHERE THE ENGINE DESIGNER CAN PREDICT WITH GREAT ACCURACY THE PERFORMANCE THAT HE WILL GET FROM AN ENGINE BUILT IN THIS WAY. HE IS WORKING TO OVERCOME SUCH PROBLEMS AS LEAN BURN TECHNOLOGY CAN BRING, AND HIS FINAL CONFIGURATION IS VITALLY IMPORTANT!

THE MANUFACTURING ENGINEER HAS NO ALTERNATIVE BUT TO COMPLY WITH WHAT IS ASKED AND HE WILL SEEK MEANS WHICH BEST GUARANTEE THE REPRODUCTION OF THIS HIGH QUALITY.

PREVIOUSLY THE BEST THAT WE WOULD HAVE GOT WOULD HAVE BEEN A FULL SIZE LAYOUT ON PAPER, WITH MAYBE 6 OR 7 SECTIONS TAKEN AT VARIOUS POSITIONS IT WOULD HAVE BEEN LEFT TO THE PATTERN MAKER TO BLEND THE SECTIONS INTO THE FINAL SHAPE. UNDER SUCH CONDITIONS THE ENGINE DESIGNER COULD HAVE LITTLE CONFIDENCE OF HIS FINAL PART OR OF THE RESULTING PERFORMANCE OF HIS ENGINE.

FROM THAT BUSY BUT INFORMATIVE DIAGRAM THE ENGINEER CAN NOW OBTAIN FROM THE SYSTEM A SOLID MODEL DIAGRAM WHICH GIVES HIM A 3 DIMENSIONAL PICTURE OF THE SHAPE OF THIS PART OF THE TOOLING.

HE NOW NEEDS TO FAITHFULLY REFLECT THE INFORMATION THAT IS GIVEN IN THE FORM OF TOOLING TO MANUFACTURE PARTS, A PROCESS TO MANUFACTURE THE PART FROM THAT TOOLING AND A PRODUCTION SYSTEM THAT WILL PRODUCE THE PART TO THE ACCURACY REQUIRED ON A CONTINUOUS BASIS.

A SIMPLE BUT WELL DEVELOPED FLEXIBLE MANUFACTURING SYSTEM CAN MACHINE THE TOOLING EXACTLY AS REQUIRED WITH THE CREATION OF THE NECESSARY SOFTWARE TO TRANSLATE THAT GRAPHICAL INFORMATION INTO MACHINING INSTRUCTIONS. SUCH SOFTWARE NOW EXISTS AND GUARANTEES ABSOLUTELY THAT THE TOLERANCES WILL BE MET, AND THAT THE FULL REQUIREMENTS OF THE DESIGNER IN RESPECT OF TOTAL CONFIGURATION IS ACHIEVED.

THIS IS A SLIDE OF OUR HIGH TECH CENTRE WHICH IS NOW IN PLACE IN OUR LONGBRIDGE PLANT AND CAPABLE OF MANUFACTURING TOOLING AND PROTOTYPE PARTS ON THE BASIS DESCRIBED.

THERE IS NO MANUAL INTERVENTION AND THEREFORE NO CORRUPTION. THE DESIGNERS INPUT TO THE DATA BASE BECOME THE INSTRUCTIONS TO THE MACHINE TOOL, UNDER THE TOTAL CONTROL OF THE COMPUTER.

DEVELOPMENT SO FAR HAS BEEN HIGHLY REWARDING AND AUTOMATED GUIDED VEHICLES ARE NOW BEING ADDED TO THE SYSTEM.

IT WILL THEN BE CAPABLE OF RUNNING ON A CONTINUOUS BASIS 7 DAYS PER WEEK, WITH MUCH OF THAT TIME BEING UNMANNED.

THE ADMINISTRATION OF THE DEPARTMENT IS IN THE HANDS OF A SMALL NUMBER OF HIGHLY TRAINED MULTI-FUNCTIONAL ENGINEERS.

IN ONE VIEW, THIS SETS THE PATTERN OF TOOL ROOMS FOR THE FUTURE.

THE MANUFACTURING ENGINEER CAN THEN TAKE THE TOTAL CYLINDER HEAD CASTING AND SIMULATE IT THROUGH A NUMBER OF MACHINING OPERATIONS. THIS PICTURE TAKEN FROM THE SCREEN DEPICTS SUCH A PROCESS AND AT THIS STAGE REMEMBER NO CYLINDER HEAD HAS YET BEEN MADE. WE CAN EASILY DETERMINE THE NUMBER OF MACHINING OPERATIONS NECESSARY, THE TYPE OF TOOLING REQUIRED, AND THE TIME THAT IT TAKES US TO PERFORM ALL THE TASKS.

THE PROCESS OF MACHINING AND ASSEMBLING THAT CYLINDER HEAD WILL OF NECESSITY REQUIRE THE USE OF ROBOTS. WE CAN AUTOMATICALLY CHOOSE THE RIGHT ROBOT FOR THE JOB FROM THE LIBRARY OF INFORMATION THAT THE DATABASE HOLDS.

WE CAN MANIPULATE THAT ROBOT THROUGH THE WHOLE OF ITS PROCESS AND PROGRAMME IT TO PERFORM THE OPERATIONS REQUIRED.

BY THE TIME THAT THE ROBOT ARRIVES ON SITE IT WILL HAVE BEEN FULLY PROGRAMMED READY FOR USE AND THUS REPLACE A SYSTEM OF TIME ABSORBING MANUAL PROGRAMMING.

WE CAN THEN TAKE ALL THESE PROCESSES, OF WHICH I HAVE JUST TALKED, AND WE CAN COMBINE THEM INTO A SIMULATION TO COVER THE TOTAL INSTALLATION THAT WE ARE INTERESTED IN.

THIS SLIDE IS AN INSTANT SHOT OF THAT DYNAMIC SIMULATION IN RESPECT OF A FLEXIBLE MANUFACTURING SYSTEM. IT IS A MODEL OF EQUIPMENT BELIEVED TO BE NECESSARY, BUT IN A DYNAMIC OPERATIONAL MODE. AT ANY POINT IN TIME WE CAN SEE THE MACHINES COLOURED BLUE, WHICH AT THAT MOMENT ARE INOPERATIVE. WE CAN SEE THE QUEUE OF WORK AGAINST THOSE MACHINES THAT ARE OPERATIVE, WE CAN FOLLOW THE ACTIVITY OF AUTOMATIC GUIDED VEHICLES AND WE CAN MEASURE THE EFFECTIVENESS OF ROBOTS.

WE CAN THEREFORE, OPTIMISE THE INSTALLATION IN EVERY RESPECT INCLUDING THE COST.

WE CAN THEN TAKE THE OPTIMISED SIMULATION AND LAY OUT IN A 3 DIMENSIONAL MODE THOSE MACHINE TOOLS THAT ARE CALLED FOR. WE CAN PLOT WITH GREAT ACCURACY THE SPACE REQUIRED BETWEEN MACHINES, BECAUSE AS PART OF THE SIMULATION WE HAVE DETERMINED HOW MUCH "WORK IN PROGRESS" WE NEED. WE CAN PLAN THE ROUTES FOR AUTOMATIC GUIDED VEHICLES AND WE CAN SAFELY POSITION ROBOTS WITHOUT FEAR OF A CLASH, OR RESTRICTION WHEN THE PLANT STARTS TO OPERATE.

SUCH A FLEXIBLE MANUFACTURING SYSTEM PLANNED BY THESE MEANS IS NOW IN PLACE AT OUR LONGBRIDGE FACTORY. IT IS NOW BEING COMMISSIONED FOR THE MANUFACTURE OF A NUMBER OF DIFFERENT CYLINDER HEADS. IT IS PROGRAMMED

AND SCHEDULED ENTIRELY BY MEANS OF THE C.A.D. SYSTEM.

IT IS CAPABLE OF RUNNING ON A 7 DAY, 3 SHIFT ROUND THE CLOCK OPERATION. SOME OF THE TIME IT WILL BE UNMANNED.

THE ROBOTIC ASSEMBLY OF THE HEAD IS PART OF THE F.M.S.

NOW ALL THESE SYSTEMS REQUIRE A CASTING OF VERY HIGH QUALITY. THEY REQUIRE THAT LEVEL OF QUALITY TO BE CONSISTENT AND TOTALLY PREDICTABLE. WE HAVE TO CAST MUCH OF THE DETAIL TO SIZE IN ORDER TO MINIMISE CAPITAL.

THIS IS PART OF THE TRADITIONAL MEANS OF CASTING, THE MELTING AND POURING OPERATION. IT HAS NO PLACE IN MODERN TECHNOLOGY, BECAUSE IT FAILS TO MEET IN A CONSISTENT FASHION ANY OF THOSE PRE-SET REQUIREMENTS OF CONSISTENT HIGH QUALITY. THE ENVIRONMENT IS BAD WHICH IN TURN CONTAMINATES THE MELT.

WHAT WE HAVE BEEN DRIVEN TO DO IS TO DEVELOP A NEW COMPUTER BASED PROCESS THAT WILL GUARANTEE THOSE REQUIREMENTS, AND HERE YOU SEE THE IMPROVEMENT IN JUST ONE SECTION OF THE PROCESS, THE MELT.

IT IS NOW TOTALLY ENCLOSED AND AGAIN COMPUTER CONTROLLED. THEN LOOK AT THE IMPROVEMENTS TO THE ENVIRONMENT.

THE AUTOMATED FOUNDRY IS NOW BEING INSTALLED IN OUR LONGBRIDGE PLANT AND IS NOW ON-STREAM FOR THE CASTING OF ALUMINIUM CYLINDER HEADS FOR ROVER 800.

HERE IS AN IMPRESSION OF THE TOTAL INSTALLATION.

THE INSTALLATION AGAIN WILL BE CAPABLE OF CONTINUOUS RUNNING 7 DAYS PER WEEK, 24 HOURS PER DAY WITH MOST OF THE PROCESS RUNNING ON AN AUTOMATIC BASIS.

IN THE ASSEMBLY OF THE CAR WE ARE TAKING FULL ADVANTAGE OF AUTOMATION TO COVER THOSE AREAS OF OPERATION WHERE CONSISTENT QUALITY IS DIFFICULT TO ACHIEVE USING TOTALLY MANUAL OPERATIONS.

THE TOP VIEW OF THIS SLIDE SHOWS THE CURRENT SYSTEM WITH A MANUALLY OPERATED PACED CONVEYOR.

FROM THAT SYSTEM WE HAVE BEEN ABLE TO DEVELOP A NUMBER OF ON-LINE OR OFF-LINE AUTOMATED TOTALLY FLEXIBLE CELLS, EACH OF WHICH WILL CONTAIN ROBOTS OR VISION GUIDED SYSTEMS IN A FASHION THAT WILL GUARANTEE CONSISTENCY IN QUALITY, FOR SO LONG AS THE TRACK OPERATES.

WE HAVE A MEANS OF ADDING FURTHER CELLS AS NEW TECHNOLOGY IS MASTERED, SO MORE AND MORE THE PROCESS OF ASSEMBLING THE CAR BECOMES COMPUTER CONTROLLED.

IN SUCH A CELL GLAZING THE VEHICLE AUTOMATICALLY IS ALREADY IN PLACE.

WHEN WE INSTALLED METRO A FEW YEARS AGO, THE MOST ADVANCED MEASURING SYSTEM AVAILABLE TO US WAS THIS. IT IS A RELATIVELY LONG PROCESS ENSURING ACCURACY, BUT WITH THE LIMITATION OF ONLY MEASURING 3 OR 4 BODIES PER SHIFT.

WITH THE APPEARANCE OF LASER BASED VISION SYSTEMS WE NOW HAVE A FAST AND ACCURATE MEANS OF DOING THAT SAME CHECKING WITHIN THE CYCLE TIME OF THE TRACK, AND THIS INSTALLATION CURRENTLY NOW COMMISSIONED FOR

ROVER 800 GIVES US THE ABILITY OF AN ABSOLUTE READ OUT OF ALL CRITICAL DIMENSIONS ON EVERY SINGLE BODY THAT IS BUILT. QUALITY THEREFORE IS THAT MUCH MORE CERTAIN.

THE FULLY AUTOMATED BODY BUILD SYSTEM SUCH AS IS INSTALLED AT THE LONGBRIDGE METRO PLANT IS A DEDICATED LINE. IT WAS THE BEST LEVEL OF TECHNOLOGY THAT WAS AVAILABLE TO US WHEN THE LINE WAS INSTALLED. ALTHOUGH IT GUARANTEES SUPERB QUALITY IT IS INFLEXIBLE.

WHAT IS NOW AVAILABLE AS AN ALTERNATIVE IS A SERIES OF AUTOMATIC GUIDED VEHICLES COMPUTER PROGRAMMED TO TAKE THE BODY THROUGH THE SERIES OF AUTOMATED ROBOTIC OPERATIONS GUARANTEEING THE SAME LEVEL OF QUALITY BUT INFINITELY MORE FLEXIBLE.

WE THUS STYLE OUR STRATEGY AS THE "ENGINE OF CHANGE", BECAUSE IT FORCES US IN A VERY DISCIPLINED WAY TO COME UP WITH PROCESSES WHICH MATCH THE MUCH IMPROVED EXCELLENCE THAT EMINATES FROM THE DESIGN AND PRODUCT ENGINEERING OPERATION.

THE SYSTEM PERMITS MANY FUNCTIONS, TO WORK IN PARALLEL WITH ONE ANOTHER INSTEAD OF ONE WAITING FOR THE OTHER ONE TO FINISH ITS TASK.

IT GUARANTEES A UNIFORM MEASURE OF EXCELLENCE AND IT PRODUCES A TOTAL INTEGRATION OF THE WHOLE. BECAUSE THERE IS ONLY ONE SERIES OF INFORMATION AND THAT CONTROLS THE TOLERANCES TO WHICH WE HAVE TO WORK.

BUT A.R.G. USES A TREMENDOUS NUMBER OF SUPPLIERS AND THEY FORM A VERY IMPORTANT PART OF OUR STRATEGY. BY MEANS OF SEMINARS WE ARE DEMONSTRATING OUR STRATEGY.

WE ARE INDICATING THE POSSIBLE TRENDS DEVELOPING NOW FOR THE TRANSFER OF DATA AND THEN THE ADVANTAGES TO ALL OF US IF THEY WERE CAPABLE OF USING THAT DATA TO PARTNER US IN CONCEPTUAL DESIGN AND DEVELOPMENT WORK. WITH SMALL COMPANIES, WE HAVE DEVELOPED PROGRAMMES TO REDUCE CONSIDERABLY THE LEVEL OF SOPHISTICATION OF EQUIPMENT REQUIREMENTS.

AT ALL LEVELS WE HAVE UNDERTAKEN TO SHARE EXPERIENCES TO AVOID CUMBERSOME AND EXPENSIVE DUPLICATION.

THE LEVEL OF ACCEPTANCE OF SUCH PRINCIPLES HAS BEEN ENCOURAGING AND A NUMBER OF SUCH PROGRAMMES ARE NOW BEGINNING TO TAKE SHAPE.

THE NEW TECHNOLOGY REPRESENTS A TREMENDOUS CHALLENGE IN THE MANAGEMENT OF OUR COMPANY. FORTUNATELY AND OF INFINITE BENEFIT TO THE STRATEGY THERE IS A TOTAL COMMITMENT THAT WE HAVE FROM OUR BOARD, WITHOUT IT I DOUBT IF ANYTHING WOULD BE POSSIBLE.

THE STRATEGY ITSELF CAN ONLY SUCCEED IF THERE IS A TOTAL PARTNERSHIP BETWEEN ITS EMBODIED TECHNOLOGY AND THE PEOPLE WHO HAVE TO OPERATE IT.

THE SUCCESS OF THAT PARTNERSHIP REQUIRES MAJOR CHANGES IN EDUCATION AND TRAINING THROUGHOUT OUR RANKS.

THE SOPHISTICATION OF THE PROGRAMME DEMANDS THE PLANNING OF VERY ADVANCED EDUCATIONAL PROGRAMMES. WE ARE ACHIEVING THESE THROUGH OUR CLOSE COLLABORATION WITH THE UNIVERSITY OF WARWICK, A DECLARED CENTRE OF EXCELLENCE FOR ADVANCED MANUFACTURING TECHNOLOGY.

JUST A SAMPLE OF THE INITIATIVES THAT WE HAVE TAKEN RANGES FROM SENIOR MANAGEMENT TECHNOLOGY TRAINING ON THE CAMPUS IN A RESIDENTIAL PROGRAMME THAT COVERS ALL FUNCTIONS.

OUR INTEGRATED GRADUATE DEVELOPMENT SCHEMES AND TEACHING COMPANY PROGRAMMES ARE THE MEANS OF PROVIDING NEW BLOOD, HIGHLY QUALIFIED YOUNG PEOPLE TO OUR EMPLOYMENT.

TO SUPPLEMENT THIS WE HAVE NOW TAKEN THE INITIATIVE TO SPONSOR AN ADVANCED TECHNOLOGY CENTRE ON THE CAMPUS AS A MEANS OF WORKING CLOSELY WITH A SELECTED GROUP OF ACADEMICS, OUR OWN ENGINEERS AND WHERE NECESSARY STAFF FROM OUR SUPPLIERS IN THE EVOLUTION AND DEVELOPMENT OF NEW TECHNOLOGY FOR USE IN OUR PRODUCT AND IN THE PROCESS OF MANUFACTURING IT.

THE CENTRE IS DUE FOR OCCUPATION IN THE SUMMER OF THIS YEAR.

WE REGARD THESE ACADEMIC LINKS AS A VITAL CONSTITUENT TO RELATING NEW TECHNOLOGY TO OUR BUSINESS AND EQUALLY IMPORTANT IN RELATING OUR PEOPLE TO THE NEW VITAL TECHNOLOGY.

    SO SUMMING UP THEREFORE, WHAT HAS THE STRATEGY DONE FOR US.

    FIRSTLY, IT GIVES US CONSISTENT HIGH QUALITY COMBINED WITH RELIABILITY.

    THEN WE OPTIMISE DESIGN OF BOTH PRODUCT AND PROCESS.

    WE ACHIEVE SHORTER LEAD TIMES.

    SO WE ARE MORE EFFICIENT.

    ULTIMATELY ALL OF THIS REDUCES COST.

WHAT I HAVE TALKED ABOUT IS THE FLEXIBLE MANUFACTURING SYSTEM. IF YOU ASK ME IF IT HAS BEEN EASY I WOULD HAVE TO SAY NO. BUT PROVIDING THERE IS SUCCESS, THEN THERE IS ENTHUSIASM AND VIGOUR TO OVERCOME THE PROBLEMS.

AS YOU PROGRESS YOU REALISE THAT THERE IS REALLY NO ALTERNATIVE!

-ooOoo-

Advanced integrated manufacturing system (AIMS) for aero engine turbine and compressor discs

M. Butcher
Rolls-Royce Manufacturing Engineering, UK

ABSTRACT

AIMS is a flexible manufacturing system for the production of aero engine disc components. These are the high value, rotating components which require structural integrity together with a high degree of precision during the manufacturing process.

The prime objective for the project was to cut operating costs and lead times and to reduce the value of work-in-progress to a level which would enable cost savings in inventory to finance the project capital cost.

The paper details concept strategy, examines equipment requirements, implementation aspects and computer hardware/software support.

INTRODUCTION

Within the aero engine industry, the task of Manufacturing Engineering is to produce products of high value and low volume whilst retaining the ability to be responsive to design changes. Low volume manufacture dictates conventional manufacturing processes with low levels of automation. This in turn demands labour intensive operations with batch sizes large enough to justify traditionally accepted high set times. Large batches create queues between processes which stretch lead times and cause large volumes of high value work-in-progress (WIP). The result is capital tied up in large inventory carrying costs.

This paper describes a flexible manufacturing system for Turbine and Compressor discs, recently installed at Derby, England, which was designed to reduce lead time and WIP and to provide reductions in inventory from which the new system could be self-financed. The project is called A.I.M.S - The Advanced Integrated Manufacturing System.

THE COMPONENTS

The parts selected for the new system of manufacture were the high-value major rotating disc components which hold the turbine and compressor blades. These are particularly suitable targets for cost reduction, because their high material costs and long manufacturing lead times had become, over many years, endemic.

(See figure 1)

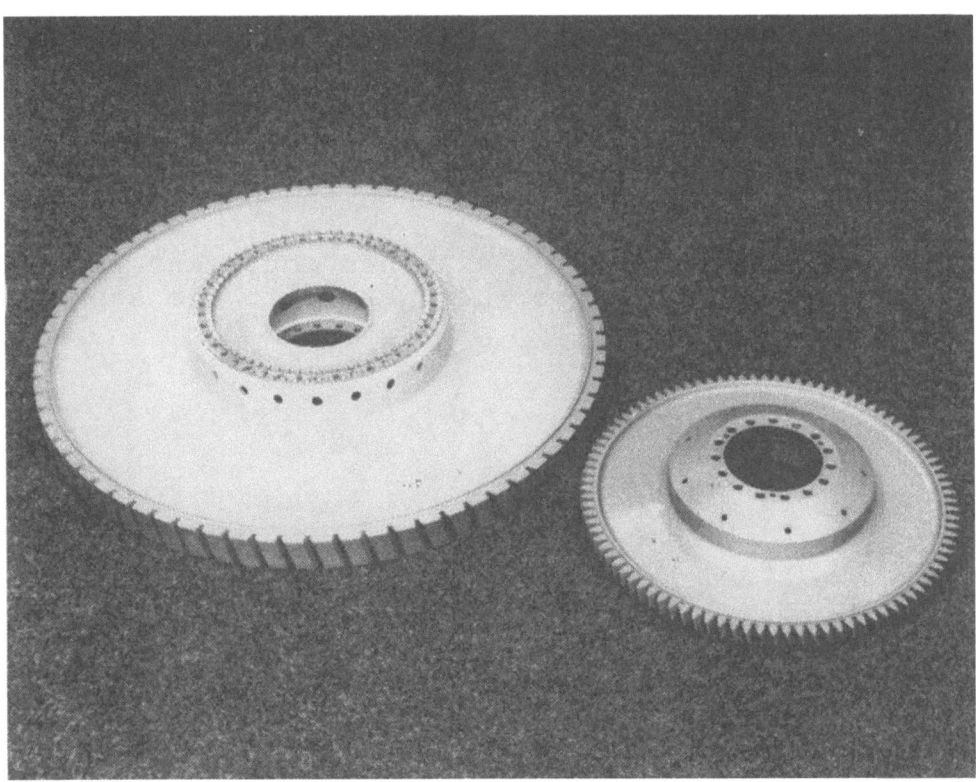

FIG 1 RB211 and Adour turbine discs.

Aero engine discs are highly stressed components which operate under severe conditions, demanding the use of exotic materials such as Titanium, Waspaloy and corrosion resistant steels. The cost of these materials comprise 75% of the value of the finished part, which effects the commitment of the largest proportion of component inventory costs at the very start of the manufacturing cycle. There are 70 different discs and disc assemblies in the large civil engine product family, ranging in value from £2,000 for the smallest disc to £40,000 for the large Electron-Beam welded assemblies.

(See Figure 2)

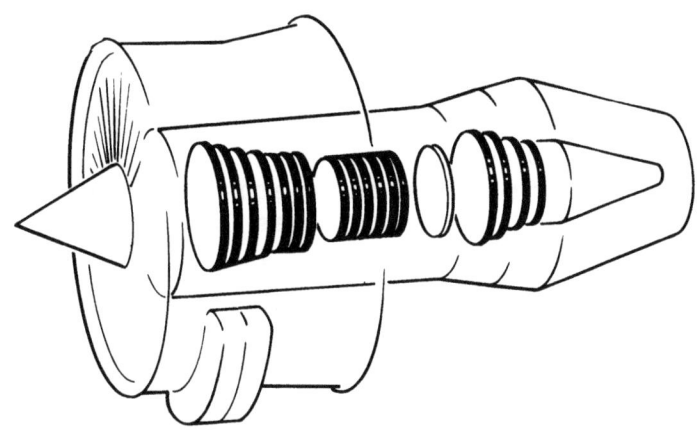

FIG 2   RB211 Major rotating modules.

OBJECTIVES

The objectives set for AIMS were to cut work-in-progress by two thirds, compress production lead times from 26 weeks to six and increase manpower productivity by over 40%. These objectives were to be achieved by deploying 27 machine and process cells in six elderly factory shops covering a total floor area of 9,300 square metres. Two computer systems would instruct and control the movement of parts between the cells and an automatic racking bay. Components and fixtures would be transported by eight Automatically Guided Vehicles (AGV's).

Prior to launch of the final phase of the project in June 1984, a number of 'ground rules' set by senior management had to be incorporated in the project plan. These were as follows:

a) Advanced Manufacturing Technology (AMT) should be applied to satisfy company strategic manufacturing objectives.

b) Existing buildings, plant and equipment should be used to accommodate the new technology.

c) Machine movements, structural changes and equipment installation should take place within the existing working environment without disruption to output. This required total commitment and full co-operation from the workforce.

d) The complex software packages which make up the Central Control System (CCS) should be developed in house.

MANUFACTURING STRATEGY

The prime goal, as defined by the Company Director of Manufacturing Engineering, is become more responsive to our customers requirements. This implies improved quality standards and reduced manufacturing lead times while providing a stable yet competitive and profitable manufacturing base from which to operate in the future.

Senior management boards were created in the mid 1980's to determine, at an early stage in the product design, where and how the components should be made. A procedure was also established which enables Manufacturing to formally accept design schemes and detailed drawings, thereby ensuring a match of Process Capability with the Design Intent.

The essence of the Manufacturing Strategy was to define those parts which should be made domestically at Rolls-Royce and then to apply the appropriate economic levels of technology to manufacture these strategic parts as quickly as possible. Not all of the strategic parts require the same degree of advanced manufacturing technology. In some instances the technology has been developed in-house; in others, development has been carried out in conjunction with specialist tool suppliers; and in some cases 'off the shelf' technology has been utilised.

A series of guidelines was laid down to identify the criteria to be satisfied before a new manufacturing concept was committed and these guidelines were used as the basis for the following ten point AIMS strategy.

- Parts should be organised into family groups.
- Parts should be planned for a common manufacturing method with a standardised sequence of operations.
- Tools should be rationalised and standardised.
- Set-up times should be reduced to provide a 'batch of one' capability.
- Work holders should be standardised.
- In-cycle inspection should be provided.
- All manufacturing processes should be integrated into the workflow and handling system.
- Non-machining operations should be examined with a view to applying AMT wherever viable.
- Workflow and handling should be automated.
- Quality should be improved by matching engineering specification to process capability.

THE ADVANCED INTEGRATED MANUFACTURING SYSTEM (AIMS)

Early in 1981, Rolls-Royce, together with a team of consultants, examined several techniques for disc manufacture available at that time and the outcome of this study became the project plan now called AIMS. The review included a new machining development at Derby which utilised a number of advanced technology turning centres, installed and in operation since 1979. These machines provided the means to reduce the total number of operations required to complete the discs, which in itself was an important step in the FMS development.

In retrospect, the adoption of this new machining technique enables us to clearly define five discrete stages in the evolution of the project.

a) Method Development (1978-1980)
b) Purchased Material - condition of supply (1981)
c) Refining the method (1981-onwards)
d) Integration of machines and equipment (1984-1986)
e) Automation of workflow and handling (1984-1985)

METHOD DEVELOPMENT (1978-1980)

The new method for manufacturing discs, introduced during 1979, recognised the impact on WIP of a reduction in the number of manufacturing operations. A computer model of the disc manufacturing facility quantified that a 24% reduction in WIP could be achieved for the proposed new manufacturing technique.

The new method involved the design and manufacture of a number of special four-axis CNC turning centres which could turn both sides of a disc diaphragm simultaneously. This technique ensured that machining stresses were in balance on each side of the disc, which effectively eliminated additional distortion removal operations. (See figure 3 - Heyligenstaedt machines).

FIG 3

Hevligenstaedt turning centres.

The effect of these new machines was to reduce, on average, the number of operations from 21 to 5 and the lead times on individual discs from 22.5 to eight weeks. (See figures 4 & 5).

1978 method (21 machining operations)

Turn	Turn	Turn	Turn	Turn	Hone	Turn
Grind	Grind	Turn	Turn	Turn	Turn	Turn
Turn	Turn	Mill	Broach	Turn	Turn	Turn

Today's method (5 machining operations)

CNC turn	CNC twin turn	CNC turn	Broach	CNC turn

FIG 4

Reduction in machining operations.

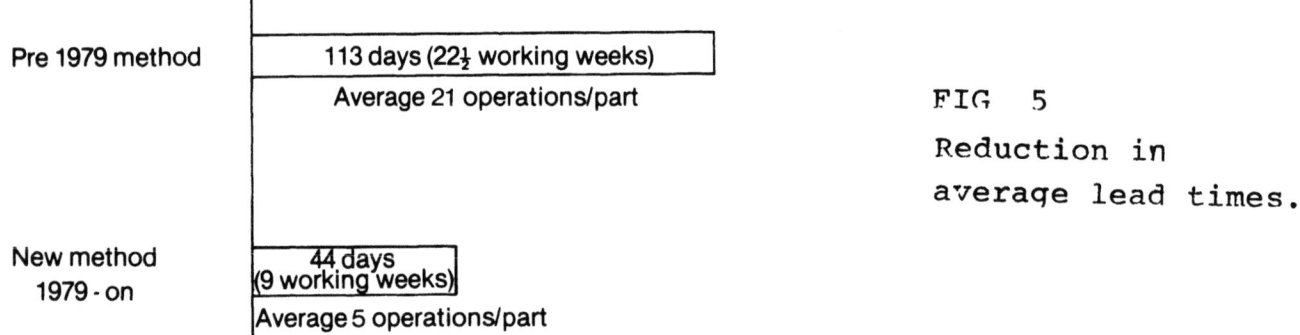

FIG 5 Reduction in average lead times.

Additional benefits were achieved in halving the machine tool population within the facility (from 62 to 31) and reducing the scrap rate by 40%.

By implementing this unique manufacturing method, the facility was in possession of:

1. A modern method of manufacture, installed and proved at an early stage in the project. This removed the pitfall of 'automating yesterday's method' and ensured that only proven technology was used in the new Flexible Manufacturing System.

2. A rationalised group of machines and ancillary equipment e.g CNC units, Auto-Tool-Setting-Units and cutting tools.

PURCHASED MATERIAL - CONDITION OF SUPPLY (1981)

Between 1975 and 1981, disc forgings were received, as today, in a rough machined, heat treated and tested condition. 60% of the material purchased at the condition of supply stage was swarfed during the manufacturing process. The material cost for these 'as received' forgings represented 75% of the component finished cost. The new machining technique (described earlier) obviated the need for excess material for repetitive distortion-removing operations. This permitted the adoption of slimmer condition of supply forgings from 1981 onwards, resulting in savings at 1985 load and value levels of £1m per year. (See figure 6-condition of supply).

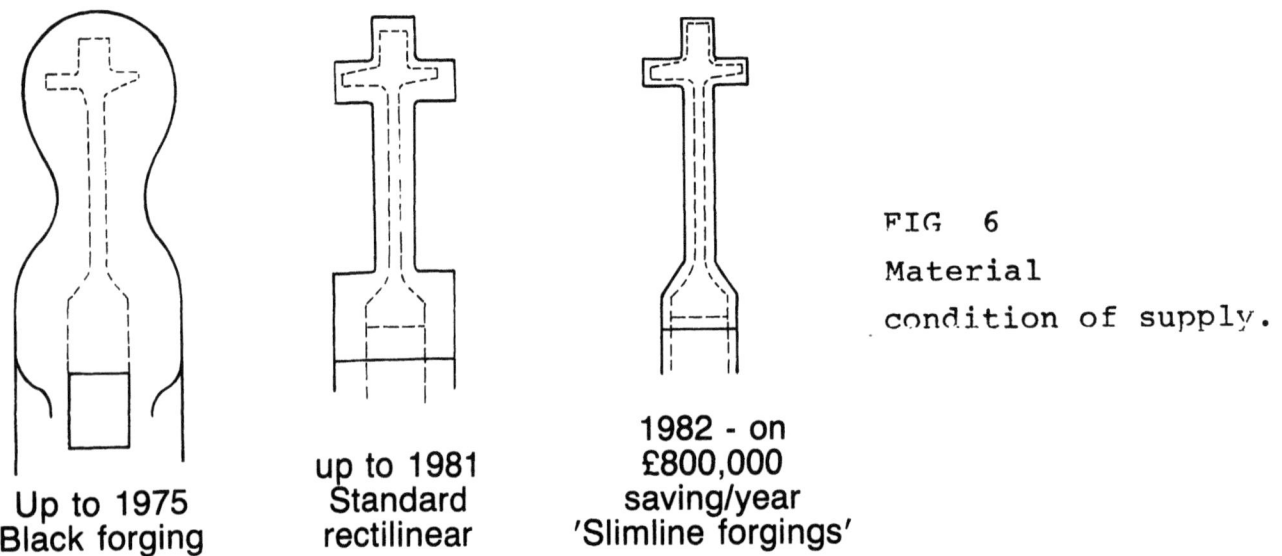

FIG 6 Material condition of supply.

REFINING THE METHOD (1981 - ONWARDS)

At this stage, attention was turned towards providing the means of achieving a 'batch of one' capability. This key target objective, which enables the manufacturing system to operate at its maximum flexibility, would ensure the shortest lead time and lowest WIP holding.

The 'unit-batch' can be linked, in its importance to manufacturing strategy, with the need for a successful policy of 'design for manufacture'. This particular element in the project therefore became one of the major challenges and drew a correspondingly massive resource from the Manufacturing Engineering organisation.

Before the launch of AIMS, batches were issued as units of 10, which meant that whilst one part was being machined or processed, nine were queuing. This extremely wasteful procedure was dictated by inherently high machine setting times and high movement costs and also by practical scheduling and shop control limitations.

As a rule of thumb, unit batches are possible if the operation setting time is no more than 5% to 10% of the machining or process time.

The techniques used during the project to reduce setting times were:

1. More refined Group Technology involving classification of specific features on similar discs.

2. Fixture standardisation and control.

3. Use of standardised 'Rings and Sleeves' - adaptors which effectively palletise the discs.

4. Standardisation of cutting tools - from 2000 to 100 standard tool sets.

5. Provision of more powerful software packages for the CNC machines (and hardware extensions in the processors to suit).

6. In-cycle inspection.

7. Introduction of quick change 'cassette' tooling.

(See figures 7 & 8).

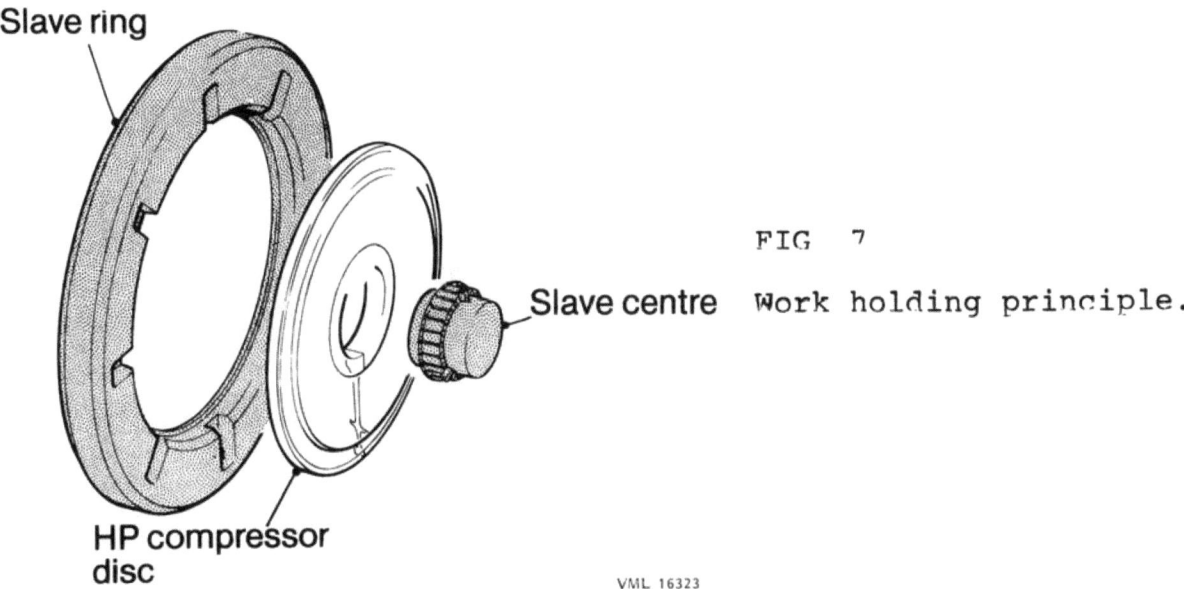

FIG 7 Work holding principle.

FIG 8

Mandelli 5-axis machining centre.

INTEGRATION OF MACHINES AND PROCESS EQUIPMENT (1984-1986)

This particular phase, together with the workflow and handling element, was launched in June 1984 and represented a large proportion of the total project spend (£2.45m).

The 'Integration Plan' for AIMS concerned re-arrangements of existing equipment into 27 cells which make up the total disc facility. Over 40 machines were deployed in 10 of the cells, the other 17 cells containing processes such as etch, descale, clean, X-ray and non-destructive testing. (See figure 9).

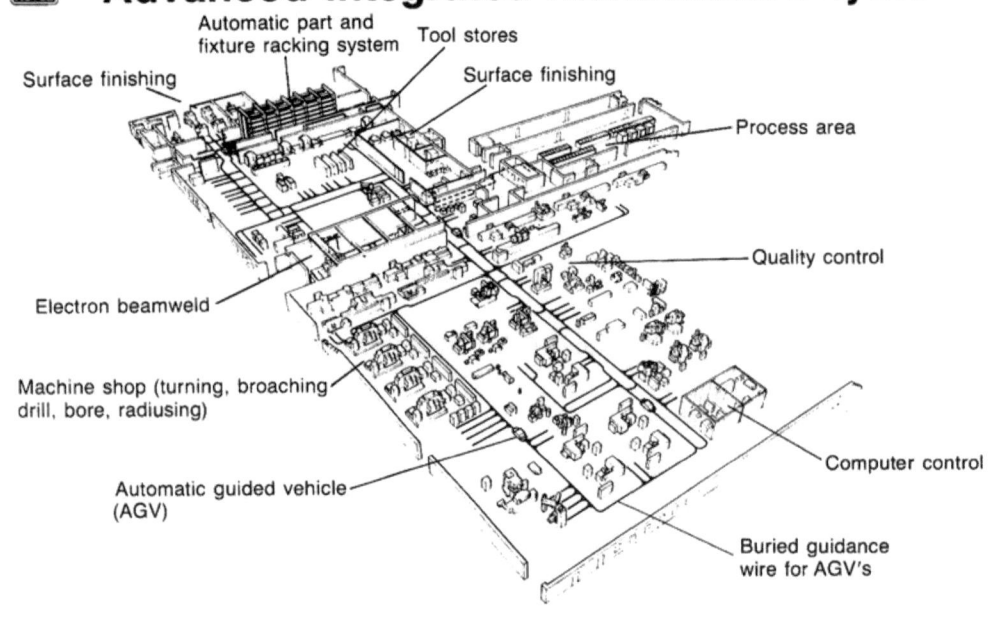

FIG 9

AIMS layout.

During the project study, in 1981, a scheme was considered for installing AIMS in a new 'greenfield site' factory at a cost of approximately £40m. This plan would have eliminated all problems inherent with the existing cluster of six ageing (circa 1907) machine shops which had to be linked together. After much deliberation, it was decided to take the cheaper option of utilising the existing factory buildings. Deficiencies in the old architecture included close-pitched stanchions, low roof trusses, varying roof heights and floor levels between the six shops and also inadequate services to new equipment. Not least of the problems was the discovery of four second World War large underground air-raid shelters, one of which, however, was converted into a settlement tank for a resited machine. The greatest challenge, during this project phase, was that of maintaining production output amid all the construction and re-arrangements. This challenge could not have been met without the full support of the workforce.

Costs for this integration phase, including associated civil works, accounted for £1m of the project budget. A further £1.45m was spent on new manufacturing equipment.

AUTOMATION OF WORKFLOW AND HANDLING (1984-1985)

The final step of ordering the workflow and handling system, at a cost of £1.6m, was taken at the same time that the Integration element was launched. In drafting the contract for this project element, there was a determination, expressed by penalty clauses, that the project would be completed on time, to budget and to specification. "Start-up" for AIMS was set at July 1985 - a 13 month project timescale.

This phase included a fully automatic racking area for components and fixtures, with two auto-stacker cranes, four conveyors, one profile gauge and eight Automatically Guided Vehicles (AGV's). Software for the Transport Control System (TCS), running on a DEC PDP 11/44 computer, was written by the handling system suppliers. The AGV's are guided by a total of 900 m of cable buried in the floor and linked to TCS through a network of 21 local traffic micro-computers. Receipt and despatch of pallets is via 45 Docking Stations. (See figure 10).

Thirteen potential suppliers of handling equipment were on the original vendor list which, after evaluation, was shortlisted to three. The contract was awarded following intensive discussions between Rolls-Royce technical and commercial staff and the consulting engineers. A considerable effort was spent drawing up and agreeing the technical and commercial aspects of this contract.

CENTRAL CONTROL

The key requirement of an integrated manufacturing system is control and this control was provided in AIMS by a computer system termed the "Central Control System" (CCS). CCS is aware of:

. What is happening
. What should be happening
. How to make things happen

It is a real-time shop scheduling system which sequences and loads work to the shop.

CCS runs on IBM 8100 series of computer, with links to the company mainframe, to a Singer 10 computer (which carries the Mechanised Work Booking system - MWB) and to the PDP 11/44. It communicates with cells by means of strategically placed terminals - 10 MWB terminals, 6 CCS VDU's and a printer at the tool store. (See figure 11 - System Layout)

FIG 10   workflow and handling system

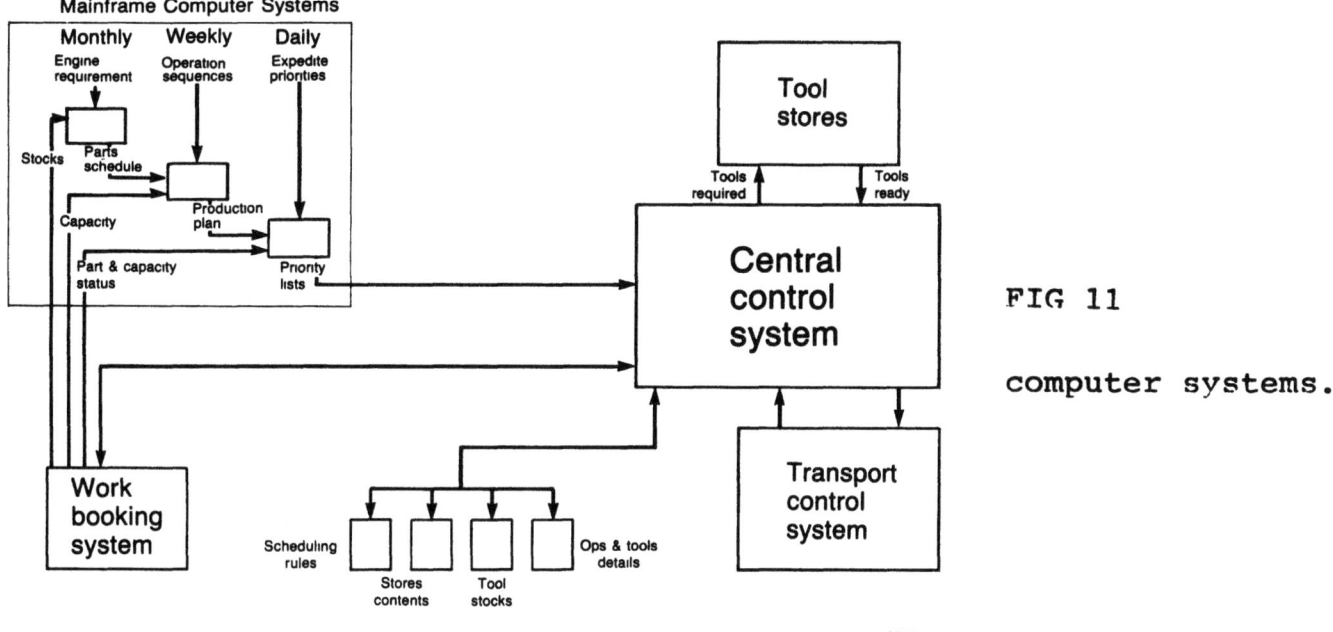

FIG 11 computer systems.

The functions of CCS are:

- recording progress of batches through their manufacturing sequence via the link to MWB
- maintaining shop status information - machines, WIP and docking stations
- managing and scheduling TCS and instructing required movements
- downloading priorities and operation sequences from company mainframe
- planning and instructing the loading of machines and instructing tool kits

The systems configuration and software content for CCS was of considerable size and complexity, requiring detailed company operating information. For this reason and not least for cost, the task was undertaken in house.

Up to the time AIMS was commissioned, 16 man-years of Rolls-Royce systems time was spent bringing CCS up to an operational standard. Further software enhancements were added in December 1985 and even more are planned for mid 1988.

FUTURE DEVELOPMENTS

Further steps in the evolution of the AIMS project are seen as:

1. Progressive increase in the level of automation within the cells, where this can be shown to be cost-effective.
2. Continuing Manufacturing Engineering activities directed at reducing setting times to improve unit batch capabilities and overall system flexibility.
3. Increasing the capacity of the AIMS facility by installing additional machines in certain key areas and by enlarging the Workflow and Handling capability.
4. Further integration of specialised processes.
5. Integration of AIMS with other existing and developing company computer systems.
6. Refinement of CCS.
7. Development of a Distributed NC system.

CONCLUSIONS AND OBSERVATIONS

The following points are considered relevant to other projects:

1. The decision not to build a new factory, but to install the facility in existing old buildings was, in this case, correct. Less than one tenth of the potential spend was needed and with few penalties.
2. It is essential to have a secure method of manufacture before entering into automation.
3. Never automate yesterday's method.
4. Planning in great detail is amply rewarded later - 40 man years of planning went into the final phase (1984-1985) of AIMS up to the day of commissioning.
5. Ensure that there is a well-defined suppliers specification and contract.
6. Maintain a company strategy to ensure that Quality is engineered into the manufacturing method and sealed after a match of process capability and specification has been proven.
7. Never underestimate the size and complexity of the software for an "intelligent" Flexible Manufacturing System. The quality of the CCS software determines, in the main, the quality and efficiency of the overall project.
8. The major software which controls the Flexible Manufacturing System should be planned, written and implemented in discrete stages.

Automated manufacture of small compressor blades
D. A. Glew and C. P. R. Hill
Rolls Royce plc, UK

Rolls-Royce has installed an automated manufacturing system to produce small high performance compressor blades. Rolled bar is delivered into the system in lengths between one to four metres. The bars are cut-up automatically into short blanks which are handled by robots selectively through three broach machines, a wash and an inspection station to produce the finished root fixings, and then into one of two 360° electro-chemical machining cells to shape the finished blade aerofoils. The automation has effectively combined the processes into an integrated manufacturing system yielding up to 75% reductions in shop-floor schedules and work-in-progress. Equipment designs are standardised to provide flexibility in coping with different families of blades and to minimise tool changeover times.

1.0 INTRODUCTION

Rolls-Royce produces gas turbine engines for the UK and world markets supplying to civil, military, industrial and marine applications.

The major engineering and manufacturing operations are based in the UK at Derby and Bristol. The engineering and manufacture of small engines for the helicopter market is at Leavesden near Watford. Industrial and marine requirements are handled from Ansty near Coventry. This particular system was installed in the Company's Bristol Manufacturing Facility and within existing factory space allocated to compressor blade manufacture.

The system was specified to initially produce compressor blades for two Rolls-Royce military engine programmes, the unique Pegasus vectored thrust engine for the Harrier and McDonnell Douglas AV8B jump-jets and the GEM-60 for the Westland 30 helicopter. These schedules required up to four thousand seven hundred blades allocated across nine different part numbers each four week period; the component load split between the two blade families being approximately 72% for Pegasus and 28% for GEM-60.

The Pegasus family contains five separate component part numbers with a batch size of approximately 700 per part number each four week scheduling period.

The GEM-60 family has four separate component part numbers with a batch size of approximately 350 per part number each four week scheduling period.

The blades for Pegasus are in a titanium based alloy and in a nickel based alloy for the GEM-60. These materials are expensive, contributing up to 30% to finished component costs.

These components are typical of the needs of the gas turbine engine business. The requirement for this automated system to accommodate small batch production over a range of high value components dictated the need to achieve the shortest permissible tool changeover times and the maximum flexibility of the handling and controls within the constraints presented by the broaching and electro-chemical machining processes.

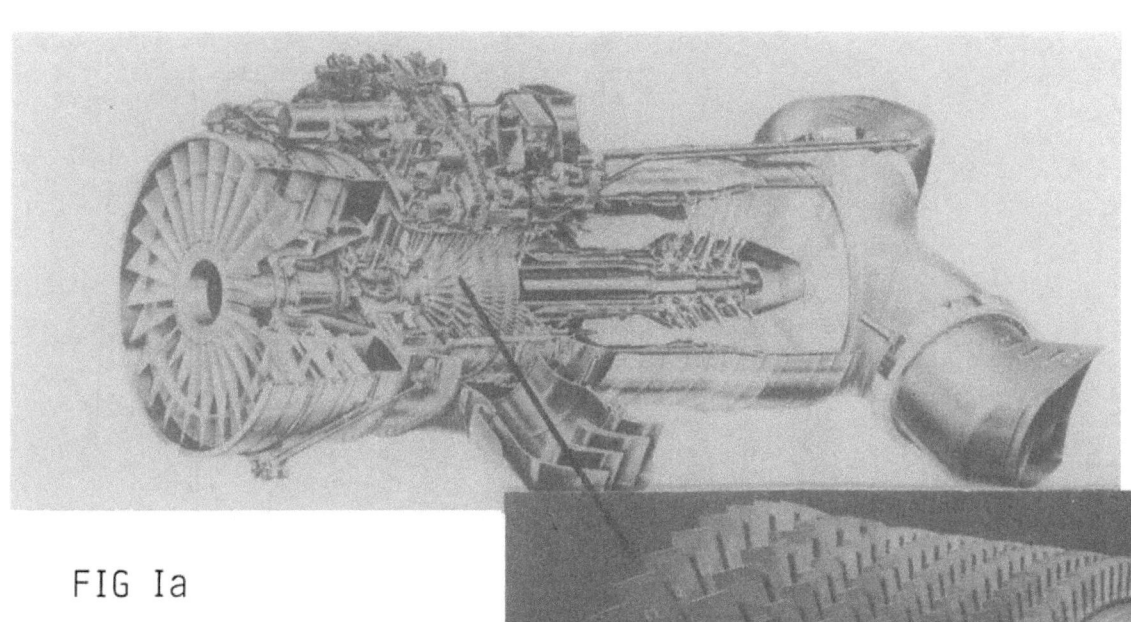

FIG Ia

THE PEGASUS
HIGH PRESSURE
COMPRESSOR BLADES

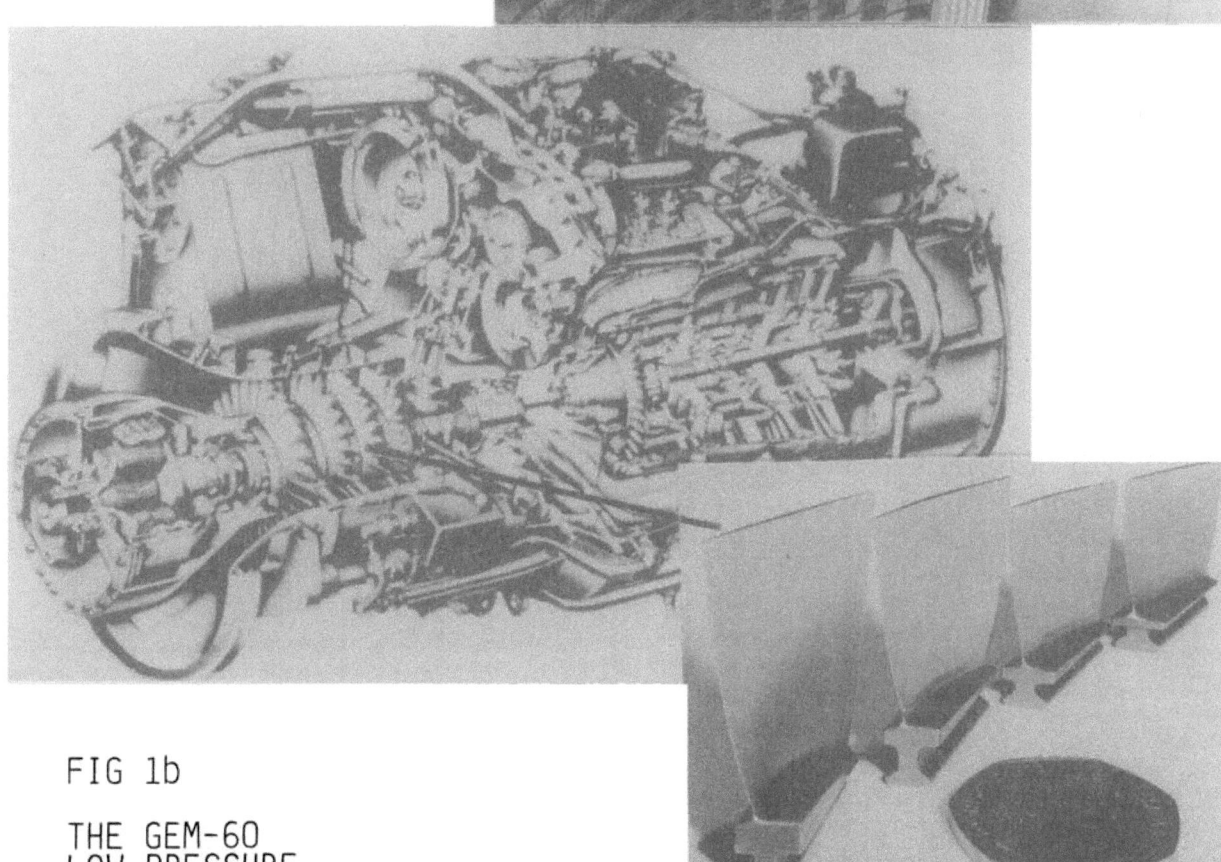

FIG Ib

THE GEM-60
LOW PRESSURE
COMPRESSOR BLADES

PEGASUS

GEM-60

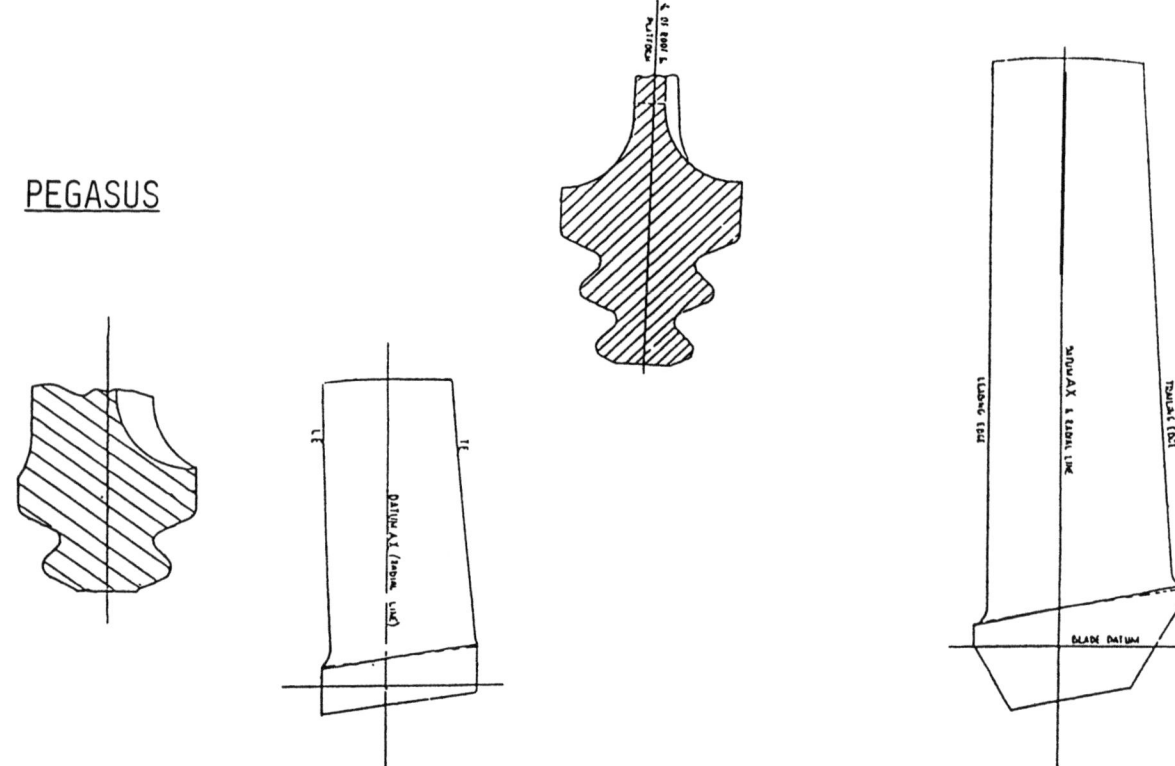

VARIATIONS IN ROOT SHAPES FIG.2

2.0 THE COMPONENTS

2.1 APPLICATIONS

The engines are shown in Fig.1. The high pressure compressor spool of the Pegasus features the components produced on the automated system. These comprise stages five to eight of the eight stage spool. The GEM-60 low pressure compressor blades are shown against the GEM engine section. The overall lengths of the blades range from 25mm to 47mm for the four GEM-60 stages and from 45mm to 80mm for the five Pegasus stages.

2.1 FEATURES

There are no dimensional features which are common to both blade families.

Significant features common within each blade family are:

- Material.

- The form of the root fixing.

    This is a multi-lobe "fir tree" for the Pegasus family and a single lobe dovetail for GEM-60. (See fig.2).

- Root form envelope tolerance typically +/-0.006mm (+/- 0.00025") for both Pegasus and for GEM-60.

- Aerofoil envelope tolerances

    0.125mm for Pegasus
    0.08mm for GEM-60.

- Axial and circumferential positions of aerofoils relative to the robot datums are:-

    +/- 0.2mm Pegasus
    +/- 0.125mm GEM-60.

- Final surface finish requirements for aerofoils:

    10 μ ins. CLA for GEM-60
    25 μ ins. CLA for Pegasus.

Significant features not common within each blade family: (fig 2).

Root fixings - Lengths front-to-rear
                End face geometries
                Platform widths
                Neck section depth (GEM-60 Stage 2)
                Single lobe fixing (Pegasus Stage 8).

Aerofoils   - Lengths, widths, twists.

3.0 THE PROCESS

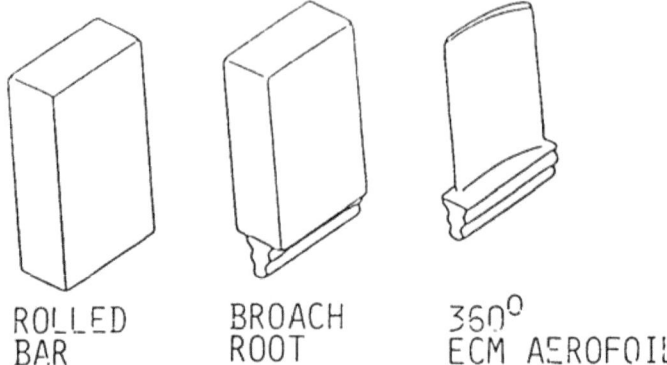

FIG.3

The overall process is simple in concept (fig.3). There are just three basic stages:

- The provision of unground as-rolled bar.

- Broaching the finished root fixings.

- Machining the finished aerofoil in a single operation by the 360° ECM (electro-chemical machining) process. (1)

The use of rolled bar minimises material costs. It is rectangular in cross-section with a permitted tolerance on side dimensions of 1.5mm and is solid and chunky to clamp during broaching (fig.4).

PEGASUS

GEM

FIG.4

(1) UK patent G.B. 2021645B Published 7th July 1982.

During broaching the same engine datum root fixings are created as are subsequently used to locate the piece during the machining of the aerofoil thereby ensuring precise positioning of that aerofoil in the engine.

The patented features of the 360° ECM tooling ensure that the aerofoil is completely encircled through the final part of the aerofoil shaping cycle (fig.5). The surfaces and edges of the aerofoil are produced accurately and repeatedly without handwork.

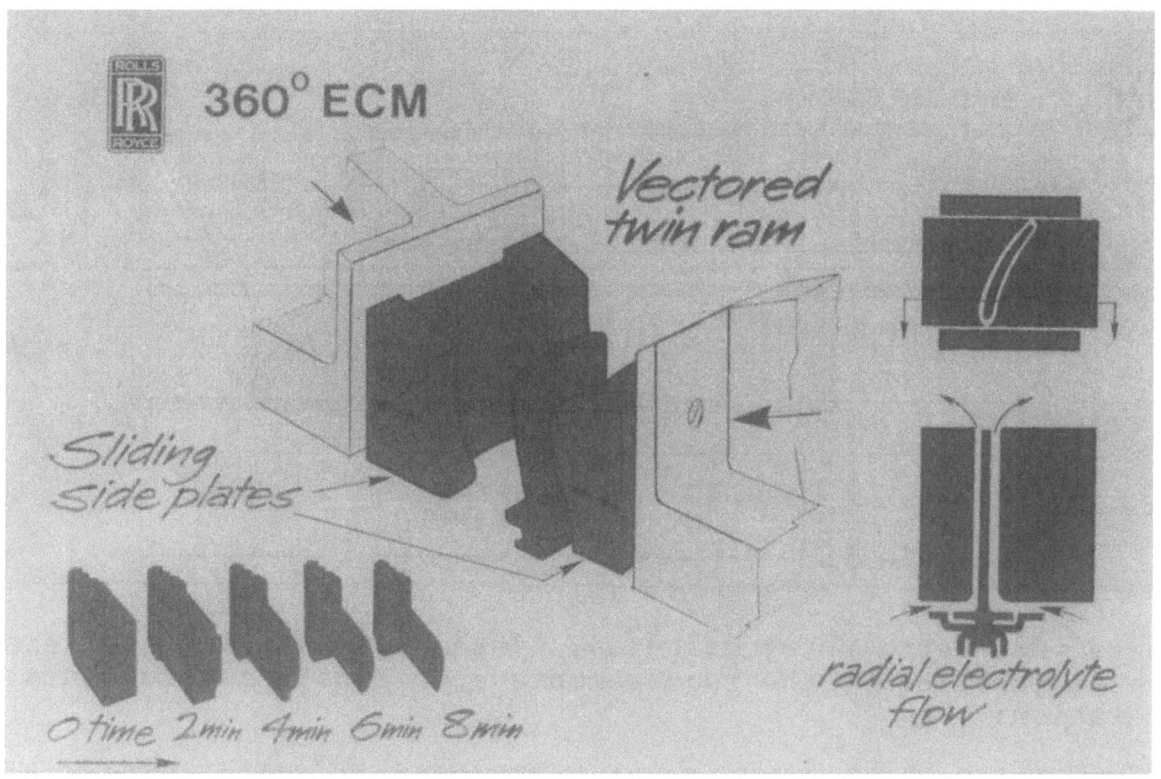

FIG.5

5.0 THE AUTOMATED SYSTEM

5.1 PLANNING

Process developments and feasibility studies were completed by mid-1982 enabling the system to be specified and costed. The management objective defined the Pegasus and GEM-60 programme commitment for the new facility with the in-house schedule lead times to be four weeks.

This last clause dictated an integrated solution designed throughout for short tool change times, and automation where appropriate to minimise inter-operation handling.

Process cycle time matching and systems reliability estimates identified a natural break point between the broaching of the root and the ECM of the aerofoil. The broach has the capacity to produce a fully machined blade root every minute. The average ECM cycle time is eight minutes thus the initial programmes required only two ECM cells. The reliability forecasts indicated that a minimum twelve-hour buffer should be maintained at each ECM cell. From the balancing of risks, reliability factors and costs emerged the extent to which automation and manning should be employed within the system (fig.6).

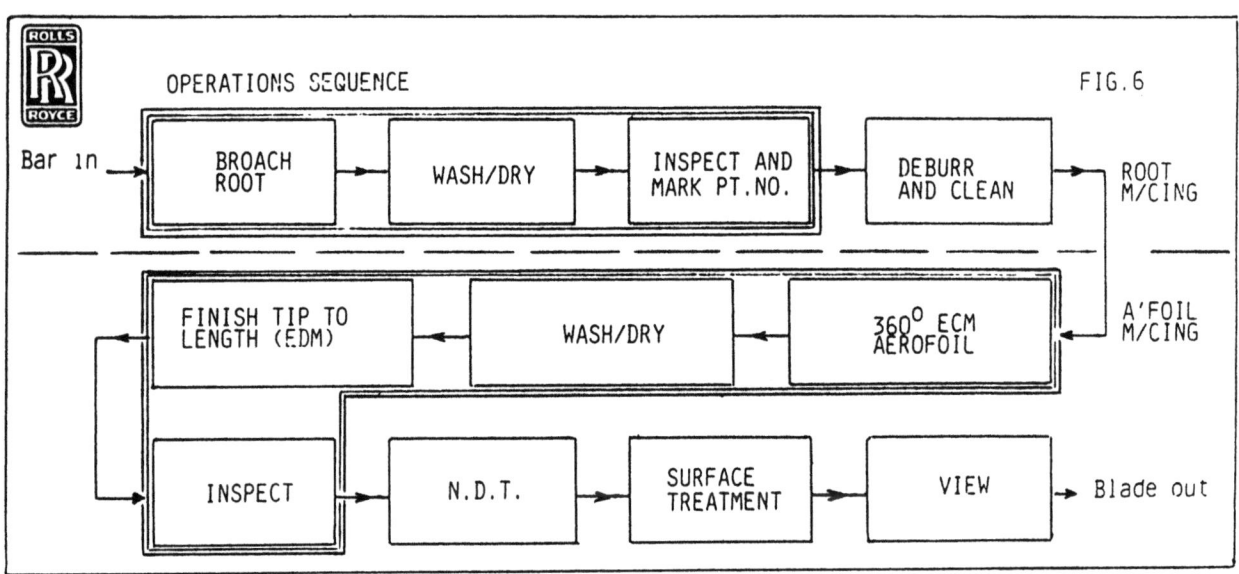

There were certain key criteria, listed below, which qualified the preparation of the requirements specification for economic operation:

- The total plant with associated personnel and services must be formed as an integrated manufacturing system designed to introduce a significant improvement in production efficiency and costs. It will do this by simplifying the production process, reducing inter-operation queuing and the associated production control problems and operating automatically with the minimum of manual intervention.

- To be economically viable the 360° ECM cells must achieve a utilisation in excess of 75%.

- The plant to be capable of continuous automatic operation between 7.30am Monday to 7.00pm Friday.

- The tooling to be capable of very rapid changeover using pre-set equipment and simple fixings. Four hours being the maximum allowable time for tool change stoppages. Setting to be automatic where cost effective. All sequences and controls to be pre-programmed in memory or cassette.

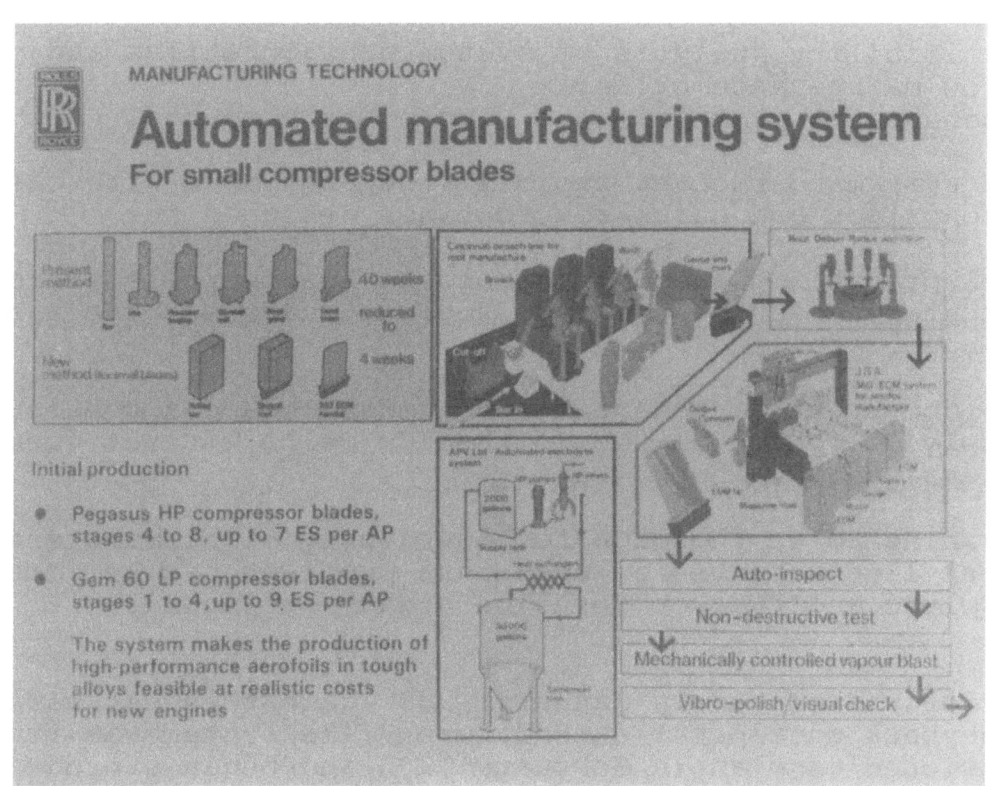

THE INSTALLED SYSTEM   FIG.7

- High plant availability to be achieved from:

    reliability analysis of system configurations and plant low maintenance criteria
    management commitment to planned maintenance.

- The system to recover rapidly from breakdown with clear fault diagnostics and quick-fix routines designed in.

- Tool presetting and maintenance to be controlled through the computer schedule to ensure tooling readiness for a part change or failure.

- The computerised schedule model must ensure a continuous supply of rooted blanks to the ECM cells without creating excessive work-in-progress.

- Training to be thorough for operators, plant maintenance and first line supervision such that problems can be immediately resolved and the workforce be highly motivated.

The order was placed in January 1983 against an estimated 2.2 year payback on capital expenditure. The system was finally commissioned to plan in February 1985 and began a progressive run-up to the production schedule (fig.7).

5.2 BROACHING THE ROOT FIXINGS

The equipment comprises a bar loader and cold saw, a blank identification system, three 30 x 66 Cincinnati Milacron vertical broaches, four Cincinnati Milacron T3-726 six axis robots, a wash station and a Vernon multi-probe inspection gauge, which also incorporates an Amchem dot peen marker to automatically identify each blade permanently on the broached section (fig.8).

The bars are automatically cut to length by the cold saw and the blanks of varying lengths between 30mm and 85mm, dependant on blade type, are fed into the stamping machine which imprints a temporary identification onto the surface. From here the first of the four T3 robots takes a blank to begin the machining sequence. A T3 robot is positioned directly in front of each broach machine to load each fixture and pass blanks down the line. A part station is positioned between each machine to receive a part from one robot and serve the next so enabling the blanks to be shuttled through the three machines in the required sequences (fig.9). The fourth robot transfers what has now become the fully broached preform into the wash and finally into the inspection and marking gauge. From here the part is automatically transferred-out ready for the next part of the system.

THE BROACH LINE   FIG.8

HANDLING THE BLANKS   FIG.9

The gauge system can show tolerance data on demand and will hold the complete line if the limits are exceeded.

The flexibility of the handling sequences built into the line by the use of the four T3 robots has permitted the set changes between GEM-60 and Pegasus blade families to be made without the need to move the substantial main table fixtures between machines. Each fixture allows for clamping in different axes to suit the various component datum face orientations required. The different handling sequence for each family is shown below (fig.10).

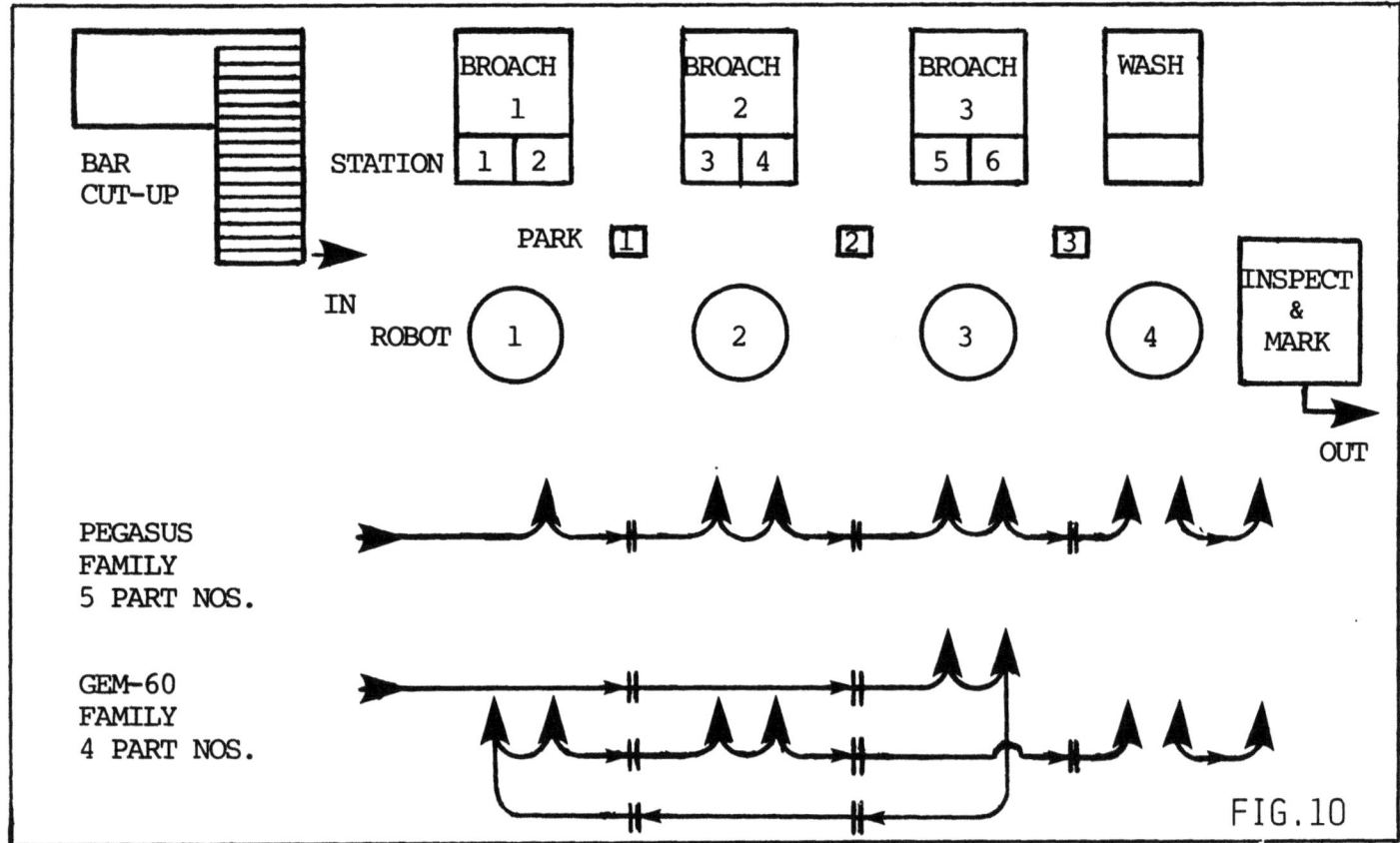

FIG.10

The tool changing is carried out manually between the machines and adjacent storage racks. Tooling is called up through the computerised production schedule which provides an average lead time horizon of 30 days for preparation. The economics did not justify automating the handling of the heavy broach bolsters, the fixture nests and clamps.

The fundamentally different shapes of the Pegasus and GEM-60 roots requires a complete tool change when moving from one family to the other. Changes within a blade family, the Pegasus for example, permits the broaches on machines one and two to remain whilst broach three is changed. The remaining tooling was analysed and designed for quick reliable changeovers.

OVERVIEW OF THE TWO 360° ECM CELLS   FIG.11

360° ECM CELL WITH SAFETY DOORS OPEN
LOAD MAGAZINE AT LHS, ECM AT RHS

FIG.12

Rapid positioning and clamping of the bolsters to the machine was achieved by the use of a dovetail location and a retractable clamp mechanism. Broach segments are pre-set in the bolsters off-line and then racked for easy access.

Programme changes are monitored through the keyboard control calling up the relevant inspection, marking, robot and sequencing programmes.

These features have enabled the four hour changeover time to be achieved with three operators employed during setting and one operator whilst machining.

Operating the one minute component cycle time for a batch of seven hundred parts represents a batch time of some twenty hours with approximately 20% allocated to setting.

5.3 MACHINING THE AEROFOIL

The two aerofoil machining cells are sited alongside the broaches (fig.11).

Each cell has a magazine capacity of approximately a hundred part-machined blanks - enough for fifteen hours worth of production (depending on the type of blade being machined).

Both cells are identical and comprise: a cartridge handling gantry robot, a blade preform loading system, a 360° ECM machine, a computerised gauge, a wash and dry station, a four-axis blade handling robot, an electro-discharge machining (EDM) unit and an output conveyor. The cells are controlled by supervisory and sequencing computers and are interfaced to provide limited management information in numerical form. Blade section repeatability is held to within 0.025mm by way of immediate post process gauging with corrective feedback to the 360° ECM process control computer. The system also identifies and records any faults, if necessary stopping the machine in fail-safe mode.

Blade pre-forms are automatically taken from the magazine, loaded and clamped into the cartridge. The finished broached root form provides accurate location for the internal gripping jaws and these are loosely torqued to hold the preform as the loading mechanism withdraws. Full torque is then applied to ensure that the blade is held securely and accurately in the cartridge. The overhead gantry robot then carries and loads the cartridge into the ECM machine.

The aerofoil with platform, leading and trailing edges and blend radii is machined in a single pass. Typical cycle times are from six minutes for a small blade up to thirteen minutes for the largest. The rams which hold the shaped electrodes are under full computer numerical control. They advance at variable rates preset to suit the workpiece geometry as it is shaped from bar to finished aerofoil.

The gantry robot then takes the cartridge to the adjacent wash station after which, to minimise down time, it immediately loads a fresh cartridge into the ECM unit. After washing the cartridge, the robot either takes it to the gauge to check chord-width and blade thickness or to the blade unload station where a floor mounted W500 robot removes the component and transfers it to the electro-discharge machine which sizes the aerofoil to length by "tipping" the end. Meanwhile the gantry robot has taken the cartridge for a thorough cleaning before returning it into the system.

From the tipping station the finished blades are handled through an output conveyor and to a programmable single contact-probe gauge which serves both cells and checks and prints out up to twenty four dimensions for each component. The gauge accepts both blade families, all nine part numbers, requiring only a programme instruction through the keyboard to identify the part to be inspected (fig.12).

Two operators currently oversee the ECM cells which both are run for twenty four hours per day Monday to Friday usually on different blade programmes. The weekend shifts are reserved for planned maintenance which is scheduled through the computer.

All tooling is preset and maintained off-line in a dedicated tool support area adjacent to the system. New tool sets are delivered on kitting trolleys in preparation for a tool changeover. Changes are achieved within four hours, involving cartridges, ECM electrodes and other quick change tooling. Programmes are changed by keyboard instruction. Setting time comprises less than 4% of total batch time.

7.0 LESSONS AND CONCLUSIONS

Several factors contributed to a successful implementation in this case:

- Extensive feasibility studies and management analysis gave confidence in the processes and the equipment specifications.

- The accounting rules were identified.

- The design engineering phase was thorough with both Rolls-Royce and its three main suppliers applying the required technical effort to meet the specifications before proceeding to build.

- A high level of commitment and support from sponsoring management.

- Extensive proving by the three main suppliers before delivery.

- A familiarisation programme for operators, maintenance personnel and supervision visiting the suppliers from the initial build onwards, and a monthly update of progress for the Rolls-Royce workforce.

- Time allocated to training before delivery and during build and commissioning.

- Adequate documentation before acceptance.

- "Core" teams of plant engineers and tooling engineers dedicated to the system as their first support priority within the framework of the normal shop disciplines.

- A management commitment to planned maintenance.

- Early acceptance of responsibility by shop personnel.

- Quick response given by supervision to operating problems.

- Key team members available throughout the run-up period.

The resultant manufacturing system has integrated several separate processes into a cost-effective and flexible production facility. Part of the management objective was to employ an extensive and yet appropriate degree of automation. The resultant working system has proved to be the correct approach for this need given the unusual mixture of processes, the large batch sizes, the special tooling, the component variations and the requirement to integrate with the Company mainframe scheduling system.

Increase of flexibility and productivity with computer integrated and automated manufacturing

E. Westkämper
Messerschmitt Bölkow Blohm GmbH, West Germany

Abstract

Computer Integrated and Automated Manufacturing ist the most important mode of manufacturing technology today.

In the aircraft production small batches, high variancy of individually optimized parts and new technologies are characteristics.
In order to industrialize the manufacturing areas computerized systems with integrated CAD/CAM, information flow and automated processes have been realized. Experiences are available and will be shown in this lecture together with technical concepts.

MBB — Definition of the System Components
COMPUTERIZED-INTEGRATED-AUTOMATED MANUFACTURING — ZFT

CIAM — Computerized Integrated Automated Manufacturing

- **CIAM–Design**
  - Preliminary Design
  - Analysis
  - Calculation
  - Drafting
  - Tool Design

  → Design Data →

- **CIAM–Planning**
  - Process Planning
  - Time/cost calculation
  - NC-Programming
  - Material Planning
  - Production Aids

  → Production Data →

- **CIAM–Production** (Integrated shop s)
  - Machining
  - Forming
  - Plastic Fabrication
  - Assembling

MBB — Objectives and Activities to Increase Effectivity in Aircraft Manufacturing — ZFT

Objectives	Activities
Design o Lead time product development o Cost of product development o Data quality	o Computerized project management o Integration of information flow o CAD o Computerized data management
Planning o Lead time process planning o Data quality o Cost of process planning o Flexibility	o Control of process planning o Computerized data management (administr., techn., geometrical data) o Dialog process planning o Integration of information flow
Production o Lead time production o Time/cost per unit o Material stock o Flexibility o Quality	o Process automat.(numerical control) o Automation of machine periphery o Automation of supply and removal o Integration of information and material flow

▷ Automated Manufacturing Computerized Integr

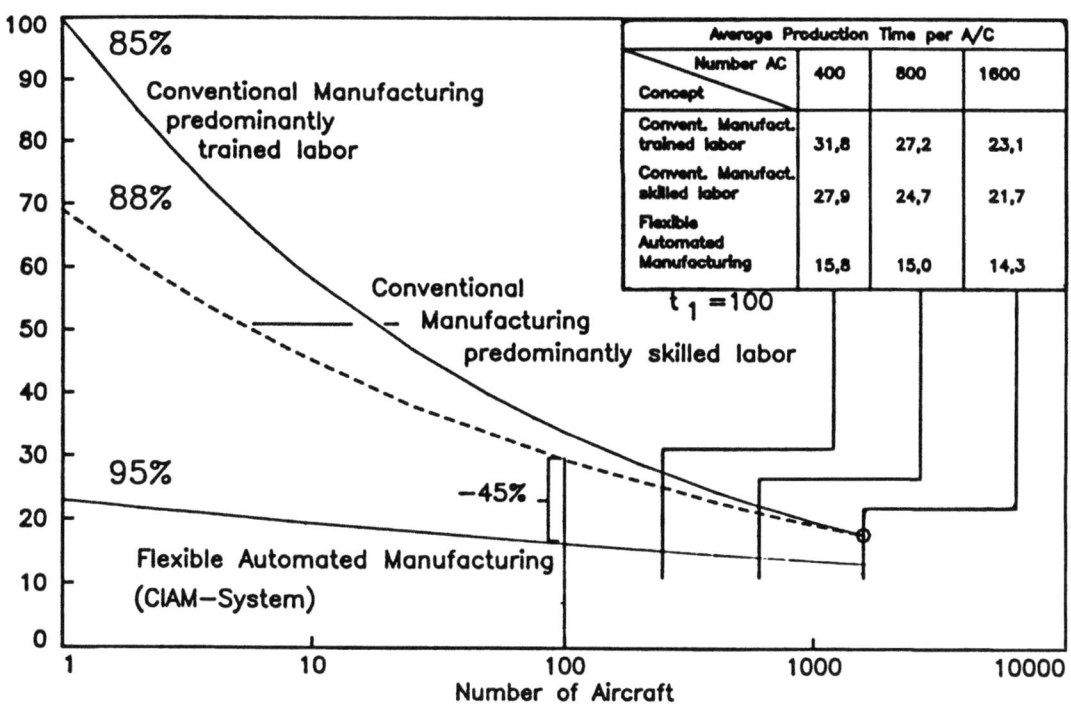

MBB — DEGRESSION OF PRODUCTION TIME IN AVIATION INDUSTRY IN DEPENDANCE OF MANUFACTURING CONCEPTS — ZFT

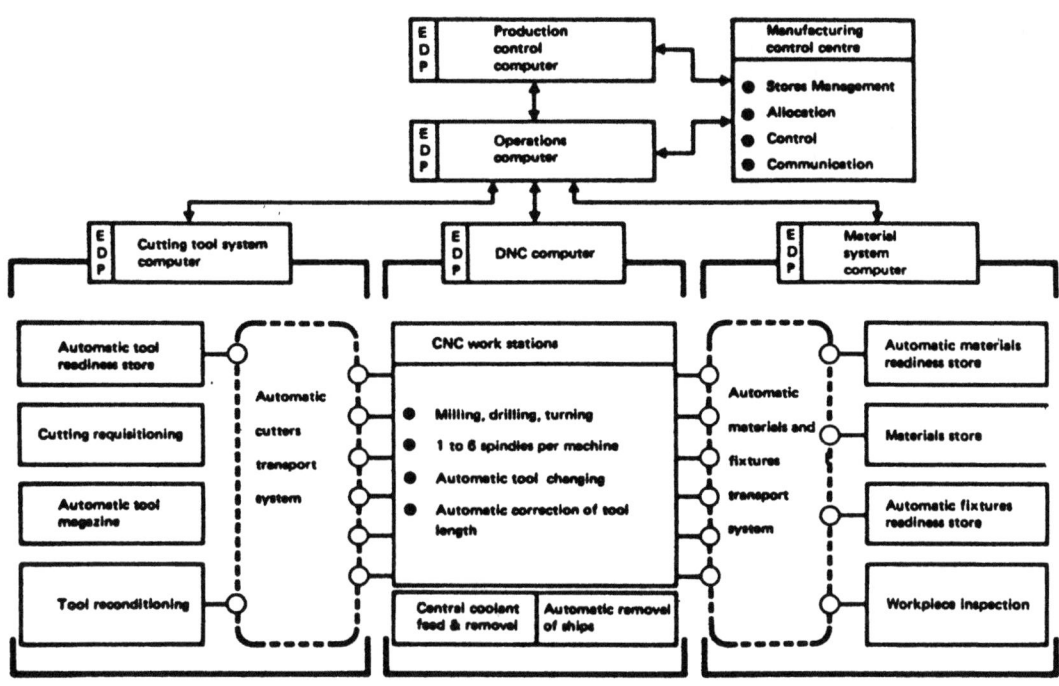

MBB — Structure of the CIAM Machining system for the machining of integral parts — ZF

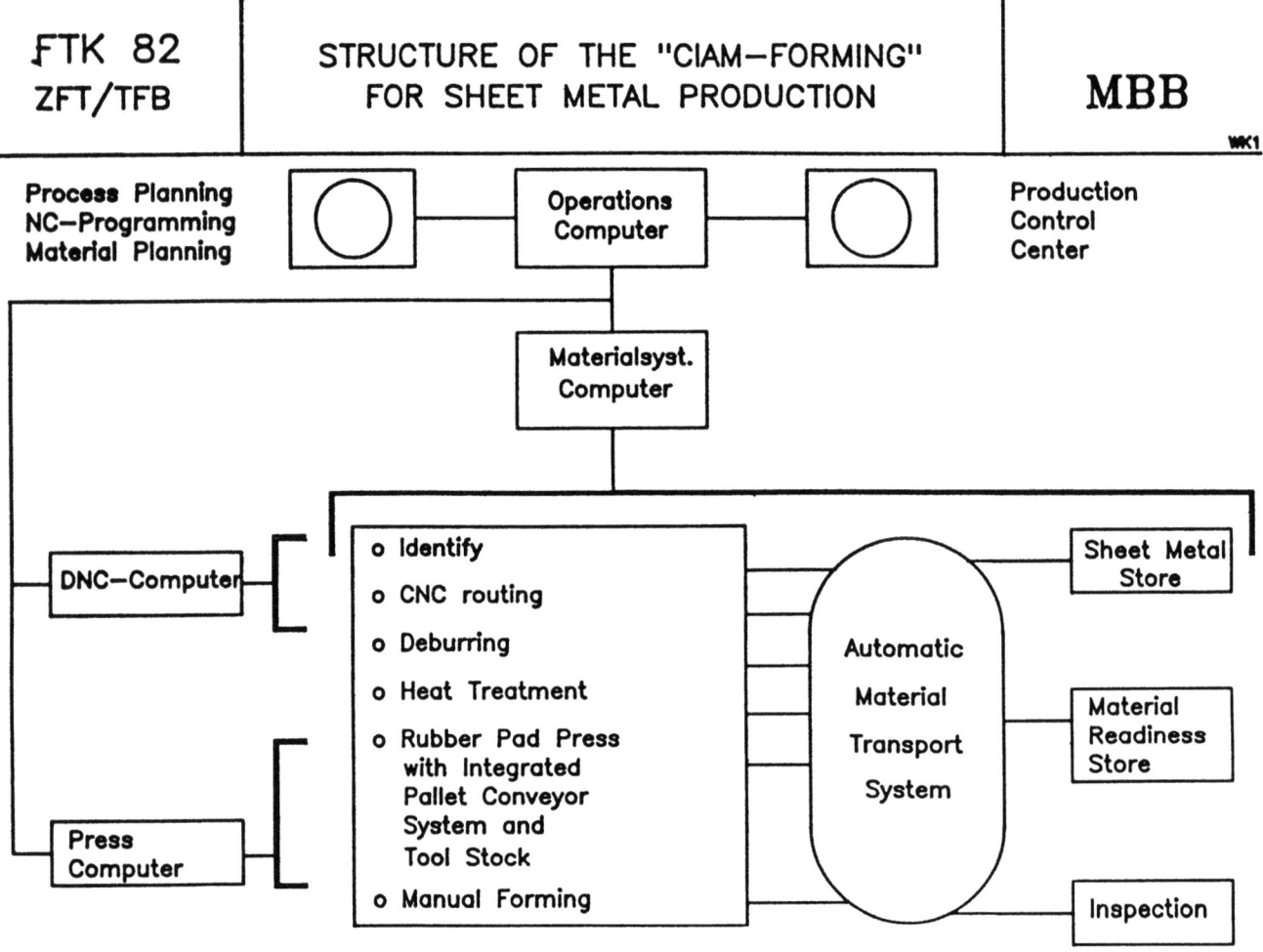

Flexible press in the CIAM-Forming Center

CNC milling machine with flexible tooling system

MBB | Effectiveness of the CIAM System in Comparsion to Conventional Measures Taken in Pursuing Productivity Augmentation | **ZFT**

General Measures to Increase Productivity		CIAM–Systematics contains	Remarks
Time Economics	– Alloting time – Time standards – 3 shift operation	– – o	} 95% processing time
Wage Systems	– Performance eval./Incentive wage system	–	
Work Structure	– Analyt. and methodical design of workplace – Arranging work procedure – Material flow optimization	o o o	
Planning Control Monitoring	– Production control center – Central commissioning – Stationary and rolling stock control for workpieces, cutters, jigs – BDE/BDA – DNC – Order control/monitoring	o o o o o o o	
Others	– Standardization of cutters, jigs – Manufacture of part families – Optimization of lot sizes	o o o	
Technology	– Optimization of processing values	o	
Automation	– Mashines, Periphery – Supply/Removal	o o	

MBB | Techno–Economical Advantages of CIAM Machining in Relation to Conventional NC Manufacturing | **ZFT**
WK1

o **Time Factor**
- Production time   –52,6%
- Lead time   –25,0%

o **Capacity**
- Personnel   –52,6%
- Floor space   –42,0%
- Number Machines   –52,6%

o **Costs Factors**
- Investment cost   –10,0%
- Tool costs   –30,0%
- Annual total costs   –24,0%

o **Quality**
- Higher accuracy
- Lower rejection rates
- Lower rework rates
- Lower inspection costs

o **Time schedules**
- Rigid adherence to schedules

o **Flexibility**
- Independent of batch sizes
- Independent of product type

o **Working Conditions**
- Independent of processing times
- Reduction in heavy physical work
- Lowered accident risk
- New, higher qualified tasks

Mandelli large scale FMS for Volvo car engines
P. Egalini
Mandelli SpA, Italy

ABSTRACT

A large sized flexible manufacturing system has been commissioned to Mandelli by Volvo, one of the major European car manufacturers. The machining of the 4 cylinder heads of the new engines, 16 valves of 2000/3000 cc. will be completely carried out in this plant which includes handling systems and automatic testing of the workpieces.

INTRODUCTION

Volvo has been present in the field of the advanced automation for several years, not only as users of flexible manufacturing cells and systems but also as an engineering company capable of developping its own projects and realizing flexible handling systems, including robot and robot cars. In this particular event, the case study submitted was that relevant to the manufacturing of cylinder heads in such quantities as to put as the main question whether it was more suitable to use rigid or semirigid systems (transfers or flexible transfers) instead of flexible solutions.

The choice of an FMS, which combines high productivity with basic flexibility, was taken in consideration of the following prime aspects: the quantities of the workpieces required (50.000 per year); the changeable elements already foreseen and those which will be foreseen in the future; the necessity of "adjusting" the production system in short temrs and with limited costs.

In particular, the eventual supplier of the plant was required to guarantee the productivity and the quality of the product. For this reason the choice of Volvo was oriented towards a "turn key" supply complete with:

- fixtures
- toolings
- machining programmes
- testing cycles

We can therefore understand how the eventual supplier ought to have had features of fundamental know how such as:

- management software with both numerical control and, particularly, central computer;
- technology in terms of specific experience in the machining of cylinder heads and of acquaintance with all the problems connected to a highly automated production with the integration of the relevant testing systems.

The choice of the supplier favoured Mandelli whose experience in FMSs designed for the automotive field have led to include several references such as:

Carraro
Case
Caterpillar
Ferrari Auto
Fiat Auto
Iveco
John Deere
Steyr Daimler Puch

The Machining Process

The "cylinder head", which also includes the camshaft housing cover, machined first separately and then assembled, is carried out on the machining centers, and subsequently on the Multispindle Units for the multiple and roughing operations, and again on the machining centers for the finishing operation, thus passing onto the washing machine and the control on the measuring units and finally to the unload operation.

The cylinder head undergoes N°4 different operations on the FMS, then exits from the System for the assembly of the guides and valve seats afterwhich it returns into the FMS to undergo the finishing operations of the two main faces.

After this, the head is assembled to the cover and returns to the FMS where two additional machining stages are executed.

Subsequently, the complete head is then executed on the specific machine by carrying out the finishing operation of the camshaft bearings, guide and valves seats, thus passing definitively to the final washing operation.

Before being assembled to the cylinder head body, the cover is completely machined on FMS in 3 machining stages which are carried out on the machining centers and on the multispindle units.

The Production

All the technical machinings with chip removal are carried out in the system, with the exception of finishing operations of camshaft bearings, guide and valve seat after the assembly and spring seats for valves.

The head is made in light alloy and obtained by means of the chilling method. The machinings to be carried out include borings according to precisions IT 6 and face finishing operations with roughness 1,6 microns Ra.

The component, which has N°3 different forms, is executed starting from the machining of raw material.

The characteristics of the plant are summarized in the following data:

```
-Overall annual production                    N.  50.000 pieces
-Available solar hours at 100%                N.   5.000
-Available time cycle for complete
 head at 100%                                 N.       6'
-Production per hour at a 100%                N.      10 pieces
-Average time cycle obtained                  N.    4,66'
-Average employment of the system                   78%
-Different machining stages                   N.      10
-Different types of fixtures                  N.       5
-Number of total set-ups on the fixtures      N.      38
-Quantity of fixtures on FMS                  N.      29
-Quantity of different tools employed         N.     140
-Quantity of different multiple heads
 for Roto                                     N.      16
-Standard pieces programs                     N.      12
-Total pieces programs
 (with alternatives)                          N.      17
```

If we concentrate our attention on the component-holding pallet, we can see :

- N° of tools to carry out one set-up: min.   1
                                         average. 6
                                         max.   8

129

- Machining time per set-up :
    min. 4,6 sec
    average. 12,8 sec
    max. 27,1 sec

- Average crossing time : 83 sec

The System Includes:
(see enclosed drawing)

N° 7 REGENT H 1001 machining centers equipped with :
- 40-pocket main tool magazine
- additional 36-pocket tool magazine complete with manipulating arm for automatic substitution in conceived time of broken and worn tools (tool monitor) and with a tool-insert station with bar-code reader for the identification of code on the tools loaded and un-loaded
- rotating pallet shuttle 800x800 mm.
- devices for un-manned functionings such as :-
  - electronic probe
  - monitoring of the spindle power
  - verification of tool integrity
  - control of tool life
- cooling system with :-
  - feeding through spindle
  - pre-washing equipment of component on board the machine
- CNM Plasma numeric control, based on PDP11/23 Digital Computer complete with 320 Kb, connection to Ethernet and dedicated to the local control of all the functions.

N° 3 rotating turret units with 6 faces "Roto" (dedicated to the multispindle operations) complete with :
- 2 controlled linear axes and one rotating axis
- automatic head changing system with 8-pocket magazine and exchanger for the total or partial reconfiguration of the 6 faces.
- tool integrity control system for the multiheads
- rotating pallet shuttle 800x800 mm
- CNM Plasma numerical control

N° 2 Washing machines with turnover of the component-holding fixture and automatic drying.

N° 2 co-ordinate measuring machines for the in-process control of the measurements of the components. The automatic and dynamic management of the data obtained is carried out by the computer which makes the relevant decisions.

N° 2 rail cars for the transportation of the component-holding pallets to the various modules of the FMS.

N° 2 chain rotating Multipallets with 10 pockets each for fixture buffer.

N° 5 Rototraversing stations for loading and unloading of components.

N° 1 CNM Plasma line controller for the management of the component movement modules such as robot cars, multipallet, load/unload stations. It is connected to the central computer from which it receives the sequence of production programs, messages, etc.; anyway it can make the system works without the central computer, except for the alternative decisions regarding component mesurements or other eventualities which can be verified (back-up at a low-level).

N° 1 central management computer, including:
. Microvax II computer connected by Ethernet to all the modules
  of the system.
It co-ordinates the system together with the tool room by means of a network of terminals.
-Computer tasks:
.Management of the procedures of intervention by signalling
 of the machining units.
.Management of verification of the machine load (simulation).
.Dynamic monitoring of the system.
.Control of the pieces being machined.
.Management of tools and tool wear with connection to the tool room
 for the relevant data.
.Statistics management of components produced and diagnostics of
 component control, etc.
.Diagnostics for assistance in trouble-shooting the plant.
.Management of the measuring machines by activating an automatic feed
 back procedure on the units, on the off-sets on the tools and by
 using eventual alternative strategies (degrading functioning)
.Sending the programs to the machining centers and measuring units
 (DNC)

Additional data of the system

. Occupied area	1.150 sq.m.
. Operators present per shift	N. 6 (7)
. Starting date of deliveries	October '86
. Starting date of production in automatic	November '87

Hardware Structure

The system is based on three levels of intelligence, developped on a network of computers of the Digital Equipment Corporation.

First level : Mandelli CNM Plasma Numeric Control based on a PDP 11/23.

Second level : line control based on PDP 11/23.

Third level : Central Computer based on Microvax II in the following configuration :-

. 9 Mbytes memory
. 456 Mbytes memory on Winchester disk (RA81)
. Magnetic back-up tape (TK50)
. 16 serial lines
. DEQNA interface to the Ethernet network

There is an Ethernet type network connection with a transmission velocity equal to 1 Mbit/Sec.

Software Structure

The various implemented packages, mentioned beforehand, have been developped by Mandelli by using the standard operative systems and the base Digital packages.

The various functions have been developped by carrying out and respecting the principals of modularity and a possible personalization, these being aspects of prime importance for the end-user who intends to have a software product made-to-fit his own personal requirements, thus giving him the possibility to make modifications even autonomously in the future.

CONCLUSIONS

The problem of elevated productivity, which only a few years ago was considered as synonimous of a transfer system, is brilliantly resolved here with a medium/low number of variants and therefore has the right to be placed within the flexible systems field.

This ulterior Mandelli creation constitutes an additional piece to the mosaic of experiences which the Company is putting together in order to achieve its strategical aim - the automated factory.

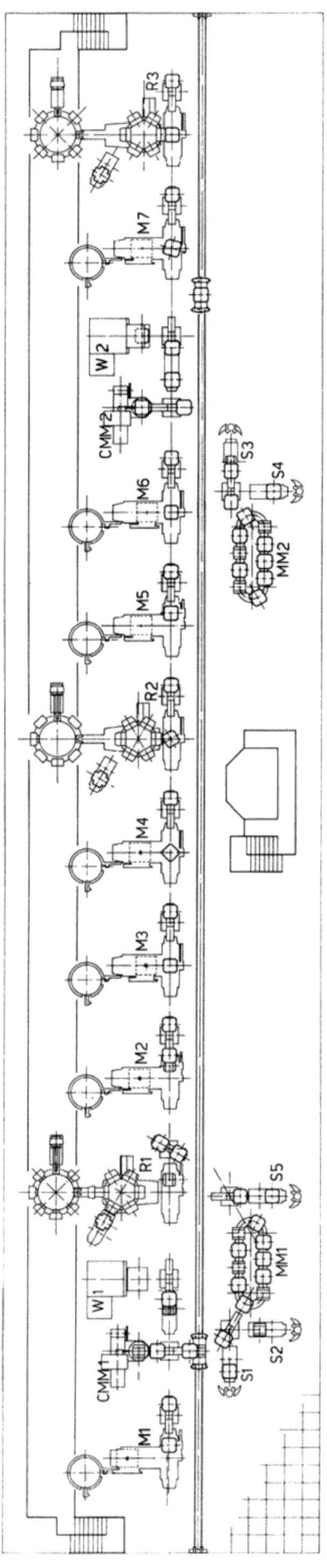

User project management for Holset FMS
S. Webb
Holset Engineering Co Ltd, UK

The paper discusses why and how a user designed, specified, and is installing a £4.165 million pounds FMS. It has been under development for the past two and a half years. Components being produced are the Shaft and Turbine Wheel sub-assemblies for turbochargers. The system includes a variety of machines grouped with gantry type robots into seven manufacturing cells. Automated guided vehicles move parts between the cells and the automatic stores.

1.0 INTRODUCTION

The main objective for most manufacturing companies is to manufacture products that the customer wants when they are wanted at the right price and quality.

Most successful manufacturing companies define the market requirements in very concise terms. From this definition a detailed manufacturing strategy illustrating company targets for improved productivity, cost, delivery and quality is produced. Because market trends dictate greater product diversity, manufacturers are required to improve their flexibility to enable them to respond to market changes and bring new products to the market place as quickly as possible. This implies being able to work with shorter lead times, smaller batches, lower inventory and still respond to new product introduction and fluctuations in volumes effectively. Hence there has been considerable interest in FMS for a number of years, but as yet it is still not clearly defined. Subsequently many possible users are confused about its application and potential.

This is a disappointing situation, Surely the users must take the initiative and "get to grips" with FMS. This is essential if users are ever to drive the suppliers to provide them with what they really require, rather than accepting a compromise for the nearest best-fit on the market. There is no doubt that in this country most of the potential users of FMS have taken the "soft" option and gone to consultants or turnkey suppliers, rather than getting to grips with solving the problems themselves. This has several disadvantages. Most importantly it leaves the user without the expertise that sent him to the turnkey supplier in the first place. It also leaves the knowledge gained during the development of the FMS with the supplier.

Bearing in mind that manufacturing technology is being developed outside the user areas of industry, that a large share of the talent is not working directly for the user, is it any wonder that many potential users of FMS are still unsure of its application and potential.

Upon completion of any project, the maintenance of the installation is an extremely important consideration. It is most likely and desirable that the maintenance should be performed by the user companies own employees. If they are involved in the development from the beginning, they are then in a better position to carry out the work more effectively. This must result in lower costs, a quicker response and should improve the running hours.

In the opinion of the author the users of FMS must become more determined and start the task of defining their own, specific requirements more effectively and subsequently improve their own skills in this area of manufacturing.

This paper discusses how Holset, identified their own requirements for FMS and subsequently decided to specify and supply their own detailed specifications to the suppliers.

2.0 COMPANY BACKGROUND AND GOVERNMENT GRANT

Holset is one of the world's leading manufacturers of turbochargers for diesel engines. The FMS is in its manufacturing plant situated in Huddersfield, England. It has a continued commitment to high levels of investment in order to remain competitive in the high technology automative components industry.

Holset originally approached the UK Government for discussion of a grant for robots to manufacture its Turbocharger Shaft and Turbine Wheel sub-assembly in June 1983. The Department of Trade and Industry(DTI) of the British Government in London, indicated that they might be more able to support a project if it were for an FMS. Suggesting that it should cover every process required, from raw material through to finished parts. Holset senior management agreed to the project provided that the DTI would also contribute to the funding.

Holset put the application together itself without using consultants. At the same time they also decided to carry out the project management themselves if government support was received. The reasoning behind this was that Holset considered that its Production Engineers knew more than any one single supplier of how to manufacture the parts. No supplier known to Holset was capable of supplying all the equipment for every processes together with the total material handling.

In December 1983 Holset went before the Technical Committee of the DTI and in March 1984 received an offer of £1,000,000 (24%) grant from the DTI towards the £4,165,000 costs of the project. Holset accepted the offer which included a start date of March 1984 and a completion date of March 1987. (3 year project)

3.0 THE COMPONENT

The shaft and turbine wheel is the most significant single element in a turbocharger both in terms of cost and difficulty to manufacture. It is a two piece welded sub-assembly. The "wheel" is a high precision investment casting made from Inconel 713C. The "shaft" is a forging from 8740 steel.

The shaft diameters are ground to close tolerances on size, and roundness together with good quality surface finishes and fine dynamic balancing tolerances. The high quality level must be attained due to the high operating speeds of the assembly. This being of the order of 120,000 r.p.m. A typical Holset shaft and wheel is illustrated in fig. 1.

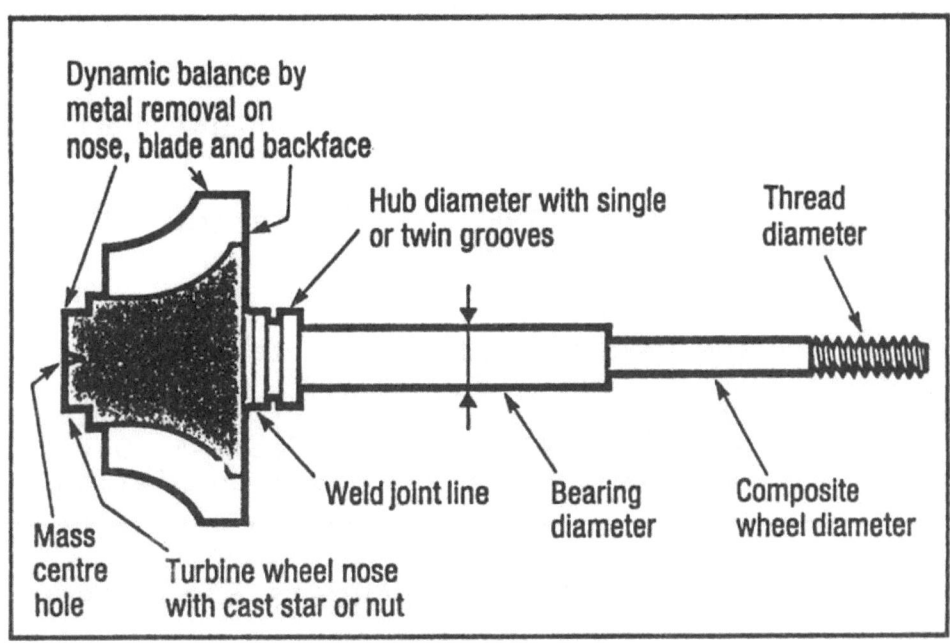

FIG. 1. TYPICAL HOLSET SHAFT AND TURBINE WHEEL SUB-ASSEMBLY

4.0 PROJECT AIMS AND TARGETS

The nature and sequence of the machining processes required in the manufacture of the shaft and wheel dictates an overall workflow, from raw material stores to finished parts stores, in a manner similar to that of a transfer line. Alternative routings need only be considered when there are machine breakdowns, preventative maintenance is being carried out or perhaps a machine or cell is being used to produce prototypes.

As is usual with components that have a large number of processes, the shaft and wheel component has a wide variety of cycle times from process to process. Hence it was considered crucial, at the initial design stage, that a balanced workflow for the manufacture of the shaft and wheel was required. The balanced line would constitute a unique flexible transfer line.

In September 1983 the project Aims was defined as. To manufacture the range of shaft and turbine wheel sub-assemblies for models H1, H2 and H3, at the same manufacturing cost on low batch sizes as if they were manufactured in high volume on dedicated plant. The specific areas to be concentrated upon, and targets were laid out. These together with the current position at the time of writing this paper in July 1986 are shown in fig. 2.

AREA	TARGET	CURRENT POSITION (July)
LOW LABOUR COST	NO DIRECT LABOUR	On target - only need to load Material and take off finished parts.
QUICK CHANGEOVER	FIVE MINUTES PER OPERATION MAXIMUM	4 Cells Automatic, 3 Cells part manual. All will achieve target.
SHORT LEAD TIME	TWO DAYS MAXIMUM [48hours]	Will better target - Looking to 36 hours.
BATCH SIZE	DAILY BUILD RATE	Will handle one offs.
ZERO DEFECT	NOUGHT PERCENT EXCEPT WELDING	Too soon to say.
MAXIMUM UP TIME	TWENTY HOURS/DAY	On target - will better.
HIGH THROUGHPUT	ONE POINT FIVE MINUTES = EIGHT HUNDRED PER DAY	On target.
LOWER INVENTORY	TEN DAYS MAXIMUM	On target.
FAST RESPONSE TO SCHEDULE CHANGE	CAPABLE OF REALLOCATING PART NUMBERS	Part number finally set at cell 5. Say 5 hours from completion.

FIG.2 HOLSET PROJECT AIMS - AREAS, TARGETS AND CURRENT POSITION

These were considered by Holset to be very tough targets at the outset. Perhaps they could or should have been even tougher?.

5.0 SYSTEM OVERVIEW

The system is designed to produce 50 different types of components at the rate of 800 per day. It involves 17 different sizes of investment cast Turbine Wheels ranging in size from 60 to 125mm, and eight sizes of forged Shafts.

Manufacture comprises 32 successive operations which take place in seven autonomous flexible manufacturing cells, layout as in fig. 3. These are served by three scissors-lift Automatic Guided Vehicles(AGV's) which carry work in pallets. At each of the cells, work is transferred between the pallets and the machines by 5-axis area gantry robot loaders, standard configuration of a gantry is shown at fig.4.

1
2
3
4
7
6
5
Stores

FIG.3. THE LAYOUT OF THE SEVEN CELLS - STORES AREA AND AGV ROUTES

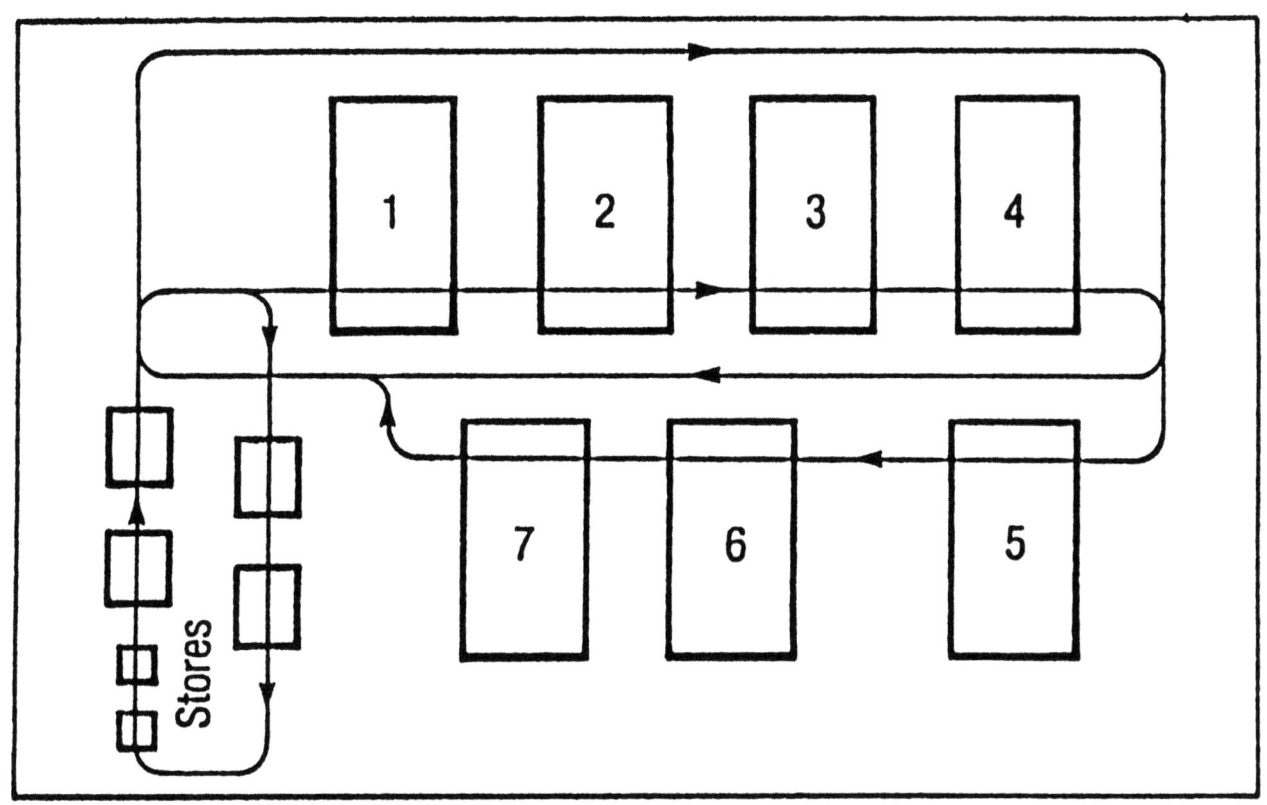

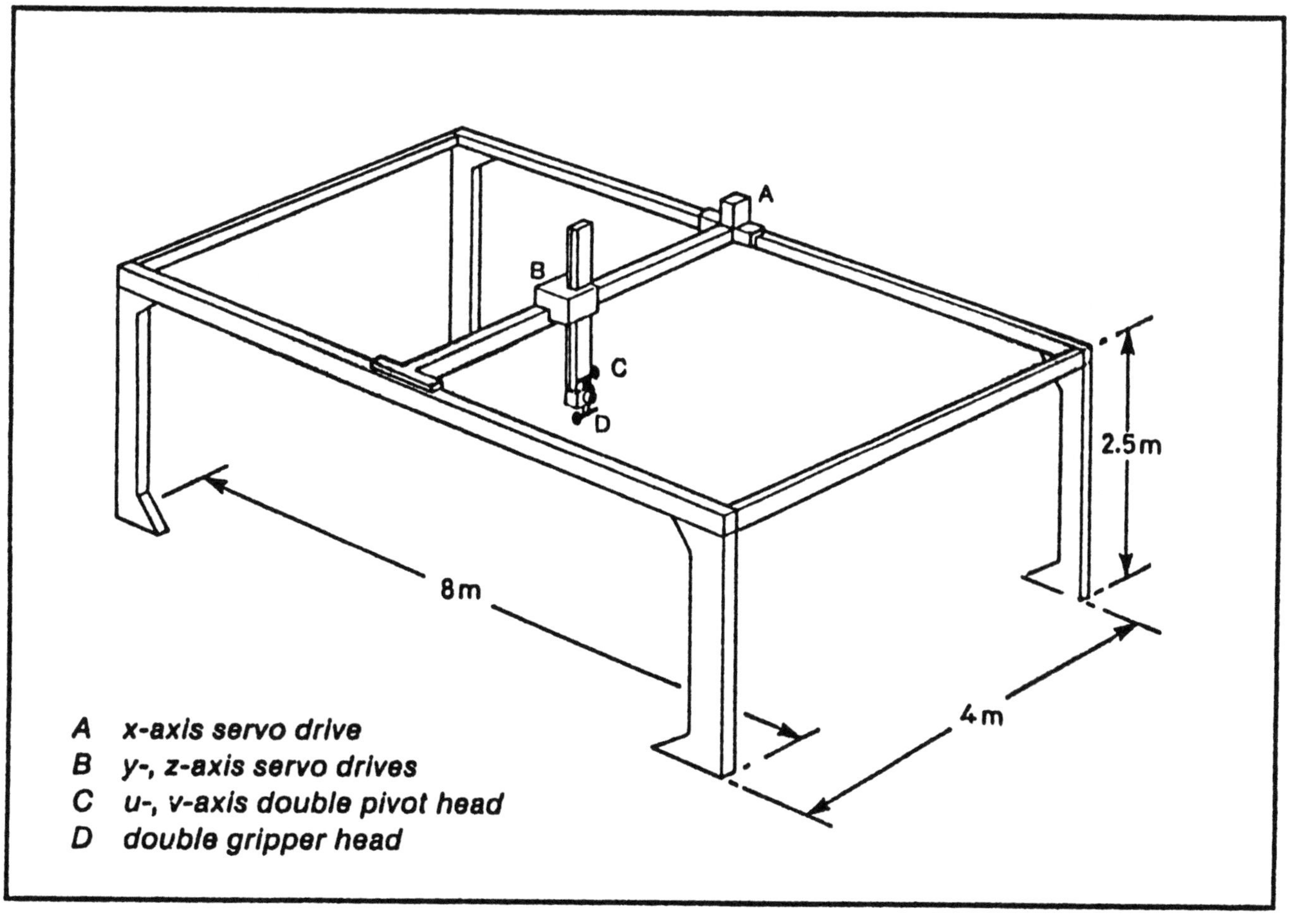

A
B
C
D
2.5m
8m
4m
A x-axis servo drive
B y-, z-axis servo drives
C u-, v-axis double pivot head
D double gripper head

FIG.4. STANDARD CONFIGURATION OF AREA GANTRY ROBOT

CELL 1. The two components are fed manually into the cell. The Wheel is centre drilled to its centre of mass. The Wheel and Shaft are then friction welded together and the weld is stress relieved in an induction coil. Empty pallets are called for by the cell controller after the components have been allowed to cool.

CELL 2. Face milling to length and shaft end centre drilling is performed on a special purpose NC Ending and Centring machine. The shaft diameters are then turned on one of the two, four axis CNC lathes.

CELL 3. Here the journal diameters are hardened and tempered by an induction coil under NC control. When the components have been allowed to cool, both centres are ground together on a special purpose machine.

CELL 4. The shaft diameters are all ground to finished size in one operation on one of the two, seven-axis CNC grinders equipped with in-process gauging. The quality levels of each grinding machine is however monitored 100% by an independent special purpose gauging station that is also gantry loaded.

CELL 5. A CNC straight approach grinder plunge grinds either one or two piston ring sealing grooves as required. Both the single and double forms are permanently crushed by a diamond roll on the front of the grinding wheel. The component is rotated by the robot at loading according to the groove configuration required for the component. Profiles on the blades of the Turbine Wheel are ground using a seven-axis CNC angular approach grinder.

CELL 6. A third, four axis CNC lathe is used to roll the thread on the shaft, and to deburr the shaft shoulders. The part is then loaded to an air-bearing balance checking machine for measuring the dynamic out of balance and angles in two planes. Rough balancing is achieved by removing metal from the turbine wheel on a special purpose CNC double headed grinding machine in two areas. The parts are then returned to pallet.

CELL 7. Deburring of the blade profiles and the area used for rough balancing is performed in a rotary transfer blasting machine. After cleaning the component is again measured for dynamic out of balance on an air-bearing balance checking machine. This measurement also monitors the rough balancing carried out in cell 6. Final balance is achieved by removing metal from the turbine wheel on an identical machine to that used for roughing in cell 6. The part is then rechecked on the same air-bearing machine prior to being returned to the pallet.

STORES. From cell 7 the completed pallets are passed into one of the two vertical, special purpose, random access pallet stores see fig.5. As parts are requested for delivery by either assembly or service the AGV's transfer the pallets to unload stands. The parts are removed by hand. Empty pallets are then moved, again by AGV to one of another two stores awaiting to be called out to cell 1. All the four stores are identical and can be used for either empty or full pallets containing either finished or part machined parts.

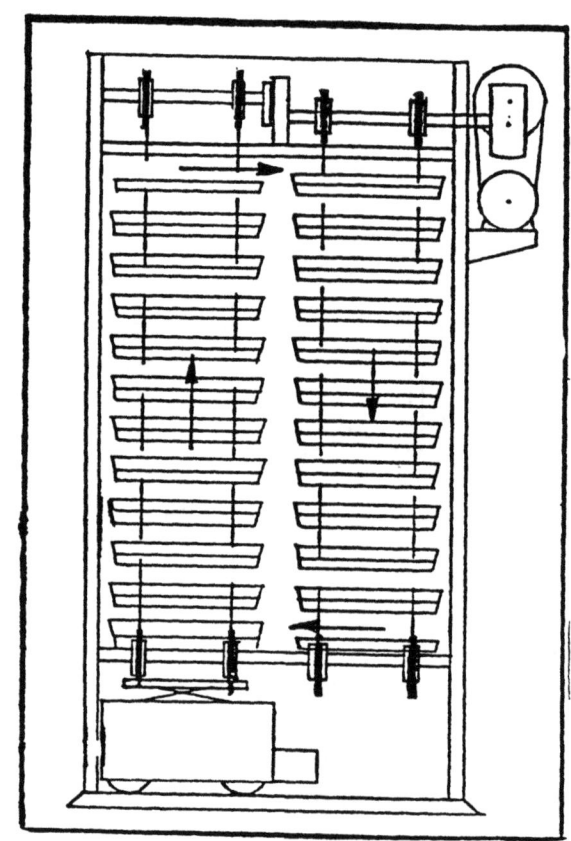

FIG.5 VERTICAL, SPECIAL MULTI - PURPOSE, RANDOM ACCESS, STORES

PALLETS & PALLET STANDS. Pallet stands are all identical. Five plastic Pallet types are used. Quantities per pallet range from 10 to 32. Accuracy is gained from the use of female cones on the underside of the Pallets and male cones on the AGV's and stands. Information containing Pallet Identity, Part Number, Batch Quantity, and operations completed etc. is carried on each individual pallet by means of a programmable identification plate attached to it.

6.0 PROJECT MANAGEMENT

Traditionally automation systems are justified with forecasted increases in customer demand. The increase in customer demand actually experienced is often disappointing, and this reduces the success of the total investment. Consequently when justifying FMS, if all feasible and realistic benefits are considered increased demand forecasts become significantly less important. It is often sufficient to consider that the current demand will be maintained and only by introducing the FMS as a cost reduction will the market share be increased by more effectively competing in the market place.

Holset had plans to improve the manufacture of the shaft and wheel component and the possibility of a financial grant accelerated this interest. It was realised from an early stage, that the design of the system would be innovative due to the unique characteristics of the component.

After the discussions with the DTI during the summer of 1983 a group of Holset production engineers were selected to investigate the feasibility of developing an FMS for the shaft and wheel component. However it was recognised that the teams limitations were in the areas of computer control, interfacing, software writing, and automated material handling. This knowledge it was thought, should be acquired and understood by the project team members themselves, rather than being hired from a consultancy or using a turnkey supplier. The results were favourable, it was found that when people were confronted with having to learn new skills, and hence take the responsibility for the success of a project area. They responded much more favourably than if they had been mere spectators. The responsibility for the success of the project must ultimately lie with people.

During 1983 Holset entered into a Teaching Company Scheme (TCS) with the Huddersfield Polytechnic. The scheme allowed for a full time graduate Teaching Company Associates (TCA). The associate started work with the project team in December 1983 for two years. He was given the responsibility of working as a Systems Engineer with the team. The Polytechnic had a high involvement with the development at this stage. They were of considerable assistance to the company. The critique by an unbiased source on the suitability/compatibility of the components of the "system" was very useful.

Holset appear by going from the "bottom up", to have taken an unorthodox approach to FMS according to the experts [1]. However to Holset the approach was quite logical. It knew the basic processes that it would require. It had been working with potential suppliers of actual process equipment for a number of years. Now it was seeking to put that latest process technology for producing shaft and wheels into practice. It wanted to avoid a syndrome that would make the system more important than the job of actually producing parts. Too often systems become the controlling factor with the business of supplying parts seeming to be of a lower order of priority. The skill in project management was, to obtain state of the art technology to process the parts, combine this with flexible handling equipment and provide effective control of the equipment and parts. At the same time meeting the completion date and remaining within the financial parameters. Perhaps the "key" to Holsets attitude was more "a way of thinking" [2]. The thought process was not that an FMS was being designed and installed, but that "the team" were and still are, having to find solutions in order to achieve the very tough targets laid down at the start. It must be emphasised that the majority of problems in a project of this nature have not been simply the interfacing of computers. Indeed they have been the normal, though often tough ones, that production engineers would normally be expected to resolve.

In order to carry out a project of this nature it was important to have both, a flexible approach to the system and flexibility from people themselves. An essential part of the project management was to establish a "team concept". With an environment that allowed "free thinking". Where ideas could be put forward, challenged, and worked through.

A major difference in how Holset interfaced its various pieces of equipment to that normally used, was in its approach to equipment suppliers. Users normally ask a robot supplier to interface to a number of machines. The price of the robot must therefore include interface design and development costs. In the case of Holset discussions between various suppliers was prohibited. The reasoning behind this was firstly private agreements might be made without their prior knowledge, Holset wanted to be able to hold to a standard through out the system. Secondly Holset wanted to be able to maintain and fault find on the interface themselves.

Holset approached the gantry robot suppliers and instructed them that all the robots would use exactly the same software and be wired in exactly the same way. To do this Holset had to specify, in detail, the interface to the machines and a cell controller.

It is important to consider that something of the order of twenty different types of pieces of equipment were required to be automatically controlled. The possibility of using the same control on each piece was remote without forcing some suppliers to use a control unfamiliar to them. Or worse, choosing equipment on the basis of its being more easy to interface with, and not being the best equipment for adding value to the components. Consequently the robot and machine suppliers were instructed how the equipment should be prepared. This ensured that when the equipment was connected together everything was compatible and that every machine had the same interface configuration. This decision ensured that Holset would have the flexibility to move machines from cell to cell as the order of machine processes changed during and after the project development. Indeed, if in the future, machines are required to be moved this will not present a major electrical or software problem.

By taking the responsibility themselves Holset were able to obtain (negotiate from strength) better prices for both machine tools and handling equipment at the same time getting state of the art technology.

Many machine cell suppliers build sections of equipment and then dismantle it, transport it and re-erect the equipment together as a cell before shipping the total system to the user. It was not considered to be cost effective to do this and consequently the first time that the robots saw the machines was on the shop floor at Holset. To be confident that this was possible facilities were required to prove that the interface had been provided correctly to specification before final acceptance. A comprehensive procedure had therefore to be planned. Each machine was checked with the use of a robot simulator and each robot tested with the use of a machine simulator.

At a very early stage a policy decision was made, that each machine in the system would be fully capable of being manually loaded, and capable of working as a stand-alone machine. This enables phased introduction and the process problems to be resolved more easily. They will also be able to continue producing parts in the event of a failure in either the system or material handling equipment. The later was a very important feature in getting acceptance of the FMS and building the required confidence level in other people within Holset.

The site of the capital equipment was not known until the first machines started to arrive in the summer of 1985. If it had not been for this flexible approach to the development of the system, major problems could have been encountered during the installation of both the cells and the total system. Indeed a total plant layout for the project was never fixed until after the first machines arrived. The only information that was crucial early in the project development was the total area that would be occupied by the FMS.

7.0 MANUFACTURING ENGINEERING PROJECT WORK

Fundamental changes in thinking were made during the first six months of the project, mostly related to the automated material handling field. Changes included a move from the use of conveyor systems towards AGV's, from one component per pallet towards several components per pallet, from free standing robots or pick and place units towards area gantry type robots, and from the use of no work-banks towards the use of strategically placed and sized work-banks.

Many configurations of gantry robots were considered before deciding that a standard size of 8 metres by 4 metres was the optimum size for the project. During the specification of the gantry robots Holset was not confident that the control usually used by the builder was the most suitable for the project. The robots were ordered without a controller to enable the builder to meet the required delivery date. The controllers were ordered some ten months later. Holset, with the assistance of a Huddersfield Polytechnic Systems lecturer, visited an alternative supplier previously used by the robot builder. After the visit it was recommended that a simulator control was purchased to continue with the investigation. This offered the advantage of educating the project team to the control and software possibilities and later educated assessments could be made regarding the future project requirements. Eventually the simulator control was used to specify the details required for the robot numerical and programmable control software. Later it was used to test the software produced by the supplier for compatibility with the "End Users" specification. In the future this control will be used as a maintenance tool. Importantly Holset will not have to rely on the builder for maintenance as the expertise is "in-house".

Many types of AGV's were considered before rejecting a fork lift type for the scissors-lift type. This configuration uses a vertical lift table mounted on the vehicle to raise and lower the pallets stored on the top of the vehicle. A unique idea considered worthwhile by Holset was to use the AGV's to load and unload pallets to and from automatic vertical, random access multi purpose pallet stores capable of holding several loaded or unloaded pallets. Due to the uncertainty of the layout during the project development it became apparent that total flexibility of AGV programming and guide wire routing was required. This was only possible if Holset was capable of programming and designing the route layouts for itself. The AGV's were ordered some nine months prior to a shop floor layout and some time before it was known which area of the factory they were to be installed in. AGV Layout, work tasks, task sequences and guide wire frequencies have all been done "inhouse". All the installation of wires and line drivers etc. has been carried out by Holset maintenance staff. The design of the AGV transport system is unconventional in the respect that there is not a central transport controller. Without exception every supplier considered that a system required a central controller. To specify the system their first requirement was always the plant layout. This is not always available as was the case with the Holset project. Therefore Holset ordered the AGV's without a transport control and agreed to specify with the suppliers a data communication link at a later date. Infra Red devices have now been developed as the means of communications. These have been fitted to the AGV's and accepted by testing on the floor at Holset. The distributed traffic control system is still being developed by the Holset project engineers.

An essential feature of any system is the identification of either the parts or the pallets. Programmable identification plates was chosen as the most flexible system. This system comprises of a small identification plate that is attached to the pallet and via inductive communication across free space the data held upon the plates can be read. The devices can be re-written at each communication station, updating quantities, operations completed etc. They are also key equipment for reliability, In the event of computer failure, then it will be possible to recover all the information relative to the location of the pallets and the parts in them, either automatically or manually according to the nature of the system failure.

When the system is complete Holset will have designed and written the total software for its FMS. The system is based on "cell" controllers. These coordinate the equipment within each of the cells see fig.6. They consist of two controls. The low level Programmable Logic Controller is used to monitor the safety equipment, interlocking, and movement of the AGV's through the Cells and adjacent areas. The higher level Micro Computer handles the Production Control, AGV task selection, Identification System, Machine Tool Programme Selection, Gantry Programme Selection, and will communicate with the other cell controllers on a local area network. The cell controllers have been built in house by Holsets own electronics engineers.

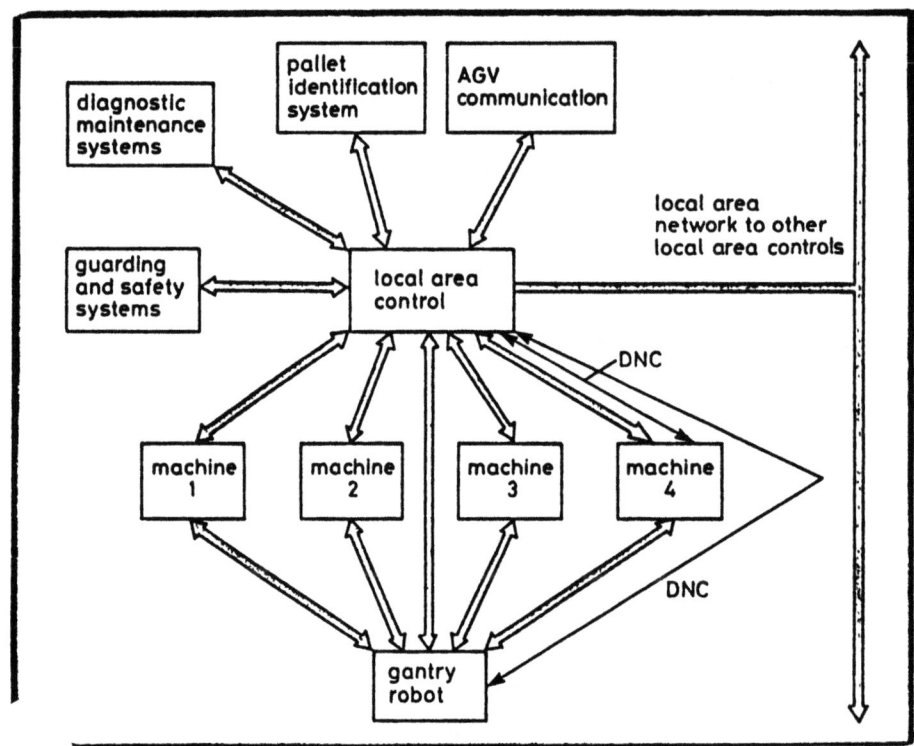

FIG.6 TYPICAL CELL CONTROL CONFIGURATION

8.0 CONCLUSION

The concept of flexibility at Holset has been developed amongst the project team over a period of time. The solutions to specific problems changed as the project advanced. This was and is being achieved by an iterative process of discussion and argument. The benefits derived from this process are significant but can only be reaped by the FMS user if he is prepared to invest the time and effort.

The author submits that there is a case for more users adopting this approach. It is the User himself who should decide the type and configuration of equipment that is best for his needs, and should in the future drive the equipment suppliers to meet those needs.

FMS should allow the phased introduction of plant and should be of major benefit in achieving Total, Cost, Quality and Delivery goals. With - Lower initial capital costs - Lower initial capacity - Without detriment to unit costs, quality or delivery. Then adding small capacity increments as and when sales volumes increase, or detracting capacity as volumes decrease towards the end of the product life cycles.

References
1 KOCHAN,A.: 'Holset takes unorthodox approach', FMS, 1986, July, pp. 133-135
2 WILDISH,M.: 'Flexibility is a way of thinking', The Engineer, 1983, 9th June, pp. 24-30

manufacturing cells and systems with computer intelligence
H. Hammer
Werner und Kolb Werkzeugmaschinen GmbH, West Germany

SYSTEM CONCEPT AND STATE OF REALIZATION

When in 1983 the first flexible manufacturing cells with computer control and a fully automatic supply of tools were presented at the EMO in Paris, a lot of people had not yet recognized that this introduced a trend-setting development (Fig. 1). In the meantime, this cell concept is generally viewed as the most expedient way to get into flexible automation for the boring, drilling and milling of prismatic workpieces. There are already several such computer-controlled manufacturing systems in practical use, and a large number are planned for the immediate future. This company alone, which the author is part of, has built more than 50 systems of different sizes but with the same basic design. Most of the users are medium-size plants whose main objective is to reduce production costs and achieve higher conversion flexibility.

What is distinctive about this cell and system concept is that a number of fully complementary machine tools of the same kind are linked together by a common supply of workpieces and tools to form an autonomous overall system. A higher-ranking computer control system, the so-called cell computer, controls and coordinates the extremely complex supply and removal of tools and workpieces, monitors the machining process and controls the entire sequence of operations. Moreover, all the organizational tasks such as job scheduling, checking the availability of tools, managing the NC programs and tool store as well as management of the entire data files for the same are seen to by the cell computer.

At first this system concept was presented, and also realized, in the form of duplex cells. But is was soon seen that efficiency grows in leaps and bounds when more than two machines are interlinked in this fashion. In the case of most of the orders placed so far four machines are interlinked to form a production unit, further expansion being possible and also intended. In the case of one large, computer-controlled cell now in the realization phase there are even 12 machines with interlinked tool and workpiece supply systems. They are used for the continuous manufacture of the same kind of workpieces in a large number of variations.

In the production of not only small and medium-size batches but also large series the use of flexible manufacturing systems - either in the form of autonomous individual cells or a self-contained interlinkage of cells - provided decisive advantages with regard to efficiency, service performance and availability. This is especially true in comparison with a system structure consisting of complementary machines and directional transport of workpieces. The following describes various design types and project stages of flexible manufacturing systems based on computer-controlled cell technology. The systems involved are exclusively ones already in operation or presently being installed.

DESIGN AND FUNCTION OF THE SYSTEM COMPONENTS

A computer-controlled flexible manufacturing cell consists of the following components (Fig. 2):
- machine tool with NC control system (e.g. machining centre),
- workpiece store and transport system,
- tool pool and handling device,
- system control (cell computer).

In the case in question the machine tools are powerful machining centres, size 500, 630, 800 and 1000 for pallet dimensions of 500 x 500 to 1000 x 1200 mm (Fig. 3). The workpieces are supplied and removed by a railborne, numerically controlled pallet vehicle (Fig. 4). It transports the workpiece pallets to and from the pallet-changing devices of the connected machines and the stationary set-up and/or storage stations. The vehicle is controlled by an on-board NC control system. The travel and control commands are transmitted directly from the cell computer via a contact line. It is also possible to integrate a washing station or also a measuring machine with direct data coupling to the cell computer and machine control systems.

Alternatively, workpieces can also be transported by inductively-controlled shuttles (Fig. 5). This is necessary whenever the machines cannot be set up in a row. Workpieces can also be loaded directly into the machine pallets by an NC-controlled gantry-type robot (Fig. 6). Preference is increasingly being given to this solution for large-series production and short piece times.

The computer-controlled tool supply system consists of the stationary storage shelves, loading and unloading station, robot carrier with dual gripper and the tool presetter with DNC connection (Fig. 7). There is practically no limit to the number of storage positions. When a tool is removed, a stick-on label is automatically printed out in bar code and clear text in order to preclude mix-ups outside the system. After being reconditioned and measured by the setting device the tool is deposited in a free storage position and identified there with a barcode reader. The respective setting data are already in the cell computer and are transmitted on-line to the NC control system as soon as the tool is introduced into one of the machines connected to the system. The robot carrier is controlled in three axes by an on-board NC control system. Here, too, the travel and control commands are transmitted from the cell computer via a contact line.

The automatic exchange of tools between the stationary chain magazines of the respective machines and the commonly used tool pool usually takes place during machining time. Accordingly, there is no need to stop the machine, thereby permitting continuous production flow with in-cycle changeover since the respective tools are "selectively" replaced prior to the job changeover and at the end of tool life, or when a tool breaks. This procedures result in considerable advantages:
- higher degree of machine utilization due to automatic tool change during machine time,
- lower tool costs due to full utilization of individual tool life,
- lower shop inventories due to production in small batch sizes,
- lower tool inventories due to supply of several machines from one common pool and
- lower fixture costs due to simultaneous production of several jobs.

SYSTEM CONTROL AND SERVICE PERFORMANCE

The automatic, timely provision of tools requires a complex computer control system. The cell computer developed for this purpose is responsible for managing not only tool inventories and storage positions but also all the NC programs as well as tool and data files derived therefrom. The handling device is controlled in accordance with a constantly up-dated "timetable", which is designed so that tools are always supplied and changed on schedule. This guarantees minimum down-times of the connected machine tools.

From the point of view of the hardware and software the cell computer forms a completely functioning, self-contained independent control unit (Fig. 8). Internally it is devided into a machine-related control section, the so-called machine computer level, and the information-processing section, the so-called organization computer level. The main task of the machine computer consists of co-ordinating and controlling all the activities of the cell components. This is done by an event-controlled timetable generator and constant monitoring of all flow and process functions. The main tasks of the organization computer consist of registering and evaluating operational data, communicating with the operator and processing the resulting data.

On the basis of the experience gained so far it is already possible to make concrete statements about service performance and system availability. Right after their start-up the systems are usually run on two shifts to begin with, a technical availability of 90 % being achievable after approximately 6 months. Most of the malfunctions that occur can be eliminated by the operating and maintenance personnel within a short time, failure of the entire system being relatively seldom (Fig. 9).

Unmanned and unsupervised operation is usually practiced for only a limited time. One of the reasons for this is to be found in the workpiece- and tool-monitoring systems, which do not work sufficiently universally and reliably. In many cases, however, the cost of making a number of sets of clamping fixtures is so high that unmanned operation for an extended period of time is not economical. For the time being, therefore, production is continued to only a limited extent in the third shift, mainly to bridge over capacity bottlenecks, and usually under supervision. But it is generally the case that the workpiece store runs out, which lengthens production time by 2 or 3 hours.

The number of machines to be operated by one person depends on the average pallet-machining time. With a machining time of some 20 minutes four machines can be seen to by one operator.

PRACTICAL EXAMPLES OF DIFFERENT FMS CONCEPTS

Several examples of flexible manufacturing systems that have already been realized or are in the introduction phase will be described in the following.

Standard machines with a pallet size of 630 x 630 mm are used in the case of the FMS 630-6 flexible manufacturing system shown in Fig. 10 and 11. The workpieces are supplied by an NC-controlled, rail-borne transport vehicle. In the event of a job changeover or tool wear the tools are automatically changed by a NC-robot. System control co-ordinates and steers all activities within the overall system.

The flexible manufacturing system was realized in 3 stages. First, 2 machining centres were set up to test the machining technology. In the second stage, and parallel thereto, a complete manufacturing system with 2 machining centres and an interlinked supply of workpieces and tools was put into opertion. Thereafter 2 more machines have been integrated. The system will then be expanded to 6 machines in a time of about 4 months. The system works in 3-shift operation. Several variants of 20 different workpieces (parts for axial piston pumps and motors) are continuously produced.

All the workpieces are clamped into "tower fixtures". 8 to 16 workpieces are accomodated on each respective pallet in manually operated clamping fixtures (Fig. 12). This results in pallet machining times of up to 40 minutes. In this case there is always only one clamping fixture of the same type in circulation. This is possible because the machining time is long enough for the tools to be replaced in the meantime so that an other workpiece can be machined, the workpiece being manufactured "in changeover".

As an alternative to a rail-borne transport vehicle for pallets it is also possible to use an inductively-controlled shuttle. This is especially advantageous when the pallets have to be transported to a number of stations such as measuring machines, washers, deburring stations or storage racks that cannot be set up in line with the machining centres. A practical example of this is shown in Fig. 13 and 14. The FMS 800-6 manufacturing system consists of 6 machining centres for a pallet size of 800 x 800 mm. The workpiece pallets are transported to the machining centres either directly from the set-up stations or from the pallet pool. After machining, the workpiece pallets can be transported alternatively to the integrated measuring machine, washer or the set-up stations.

The individual machining centres are automatically supplied with replacement or wearing tools by means of a rail-borne handling system. At the moment 18 different workpieces are machined in 30 set-ups. The parts involved are cylinder covers to be ready machined in mostly 5 set-ups. The average machining time per workpiece pallet amounts to approximately 45 minutes. Some 500 tools, which always remain in the system and are available to every machine for common usage until the end of wear, are required for machining of the intended workpieces. Due to the fact that the entire range of tools is constantly available for production, jobs can be planned in direct accordance with assembly demand so that a considerable reduction of process times and shop inventory is expected.

The complex manufacturing system FMS 500-12 shown in Fig. 15 and 16 with 12 machining centres with pallet size 500 x 500 mm is intended for the continuous manufacture of six different gearboxes for four-wheel drives. The workpieces are transported by four gantry-type robots. A gripper hand that can be controlled in five axes loads the workpieces directly into the hydraulic clamping fixtures on the pallets, where they are automatically clamped. Complete machining requires two set-ups each, which are carried out in immediate sequence on the same machine. The average machining time amounts to approximately 4 minutes for the first set-up and 8 minutes for the second.

The workpieces are transported on specially designed transport pallets (1000 x 1200 mm), each of which is designed to accomodate 30 properly positioned gearboxes. That corresponds to a work reserve of approximately 6 hours. After the first set-up is machined, the workpiece is first returned to the pallet until the machine pallet is free to be loaded with the second set-up. An inductively controlled pallet transport vehicle transports the pallets between the machine parking stations, the pick-up stations assigned to the workpieces and the central storage station for finished workpieces.

Tools are supplied, under the control of a computer, by an NC-controlled handling device with dual gripper. As many as 480 tools have to be replaced every day. As many as 1300 parts can be manufactured every day. This is the same output customary with transfer lines. In contrast, however, all the workpieces variants can be simultaneously manufactured here in accordance with the latest demand, all the machines always being put to optimum use.

Compared with a transfer line, the costs for clamping fixtures and workpiece pallets are much lower, since only one set per machine is required. From the point of view of personnel there is no great difference. In the case in question the flexible manufacturing system costs about 20 % more than the transfer line. But the much improved, more flexible operating characteristics, higher availability and much lower amount of capital tied up in circulating material should pay off in the short run. Due to all these advantages the user has ordered a second system with another 12 machines (Fig. 17). The machining centres are already installed.

EFFICIENCY CONSIDERATIONS AND OUTLOOK

The system integration of computer and robot technology typical of modern, flexible manufacturing systems has led to decisive improvements in machining conditions and efficiency. The experience gained hitherto shows that compared with non-interlinked, single-machine opertion the computer-controlled cell and system concept described makes it possible to cut production costs by 15 to 30 %. Individually this depends on the average machining time and tool time in the cut, batch sizes, number of interlinked machines, degree of system utilization and other influencing variables. The measurable cost savings for each individual case can be calculated in advance with the help of simulation programs, provided the necessary data are available.

In addition to the calculable savings there are advantages that cannot be precisely measured. Thus, "in-cycle" job changeover will lead to a considerable reduction in batch sizes, which will inevitably lower inventory in the shop and intermediate storage. "Selective" tool replacement and the automatic supply of tools from a central tool store lead to considerable cuts in tool costs. Simultaneous production of several orders ensures economic utilization of clamping fixtures. Above all, automation of the entire production and supply process leads to optimum utilization of machine and personnel capacity. This also has the positive effect of shortening process times and improving readiness to deliver.

But productivity and efficiency is primarily improved by extending effective plant time. To effectively cut manufacturing costs it is basically necessary for automatic manufacturing facilities to run during breaks and, to a large extent, during the 3rd shift as well.

The computer-controlled flexible manufacturing cells and systems based on complementary machine tools and standardized cell technology we have described are a new manufacturing alternative, not only to unlinked machines for one-off and small-batch production but also to flexible transfer lines for large-series and mass production. To a large extent the machine-tool industry is already in a position to manufacture universal automation modules and standardized, gradually expandable manufacturing systems so cheaply that they can also be used economically by small and medium-size plants for a wide range of applications. The intensive efforts to establish uniform interfaces will also surely make it possible in the near future to tailor systems to the customer and application on the basis of standardized modules and components made by different manufacturers. The optimum utilization of both system and personnel capacity, on the one hand, and the nearly unlimited flexibility of expansion, application and conversion, on the other, lead to a considerable increase in production and to long system life.

Fig. 1
WERNER FFS 630-2 flexible manufacturing system
with automatic workpiece and tool supply
by NC-controlled, rail-borne pallet vehicle
and 3-axis robot carrier

Fig. 2
Computer-controlled flexible manufacturing cell
(WERNER FFS 500-4) with 4 machining centres
and automatic supply of workpieces and tools
by rail-borne transport vehicles

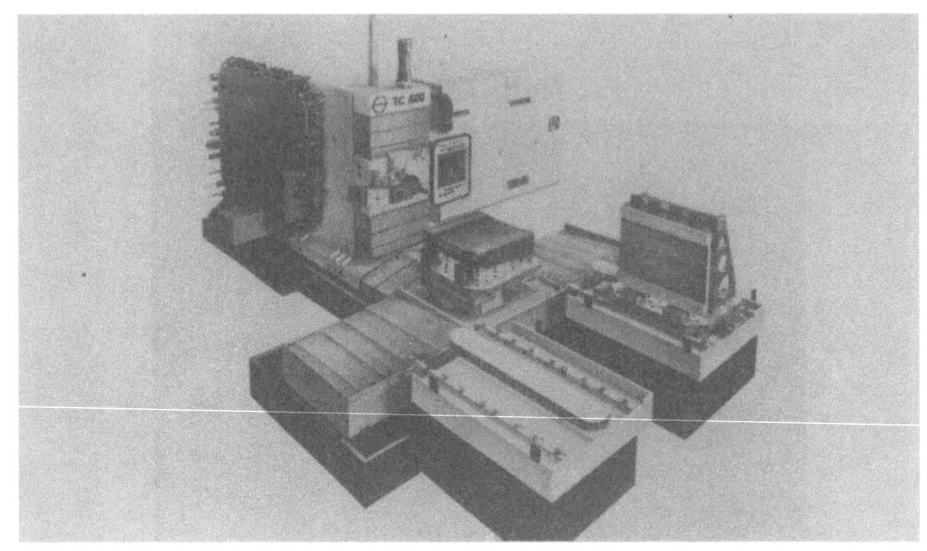

Fig. 3
Machining centre (WERNER TC 800)
with stationary tool magazine and pallet-changing device

Fig. 4
NC-controlled pallet vehicle
for automatic workpiece loading

Fig. 5
Inductively controlled pallet vehicle
for automatic workpiece loading

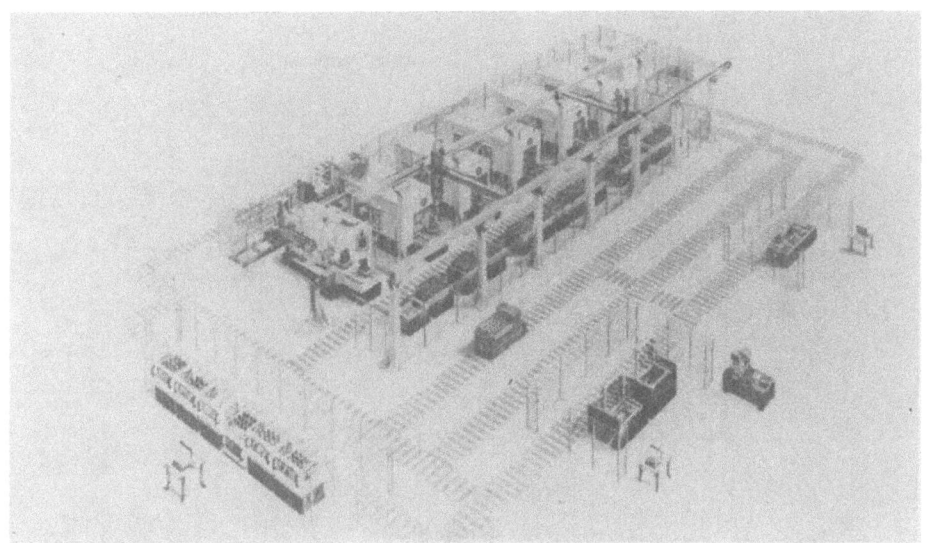

Fig. 6
5-axis, NC-controlled gantry-type gripper for automatic tool loading

Fig. 7
3-axis, NC-controlled robot carrier for automatic tool loading

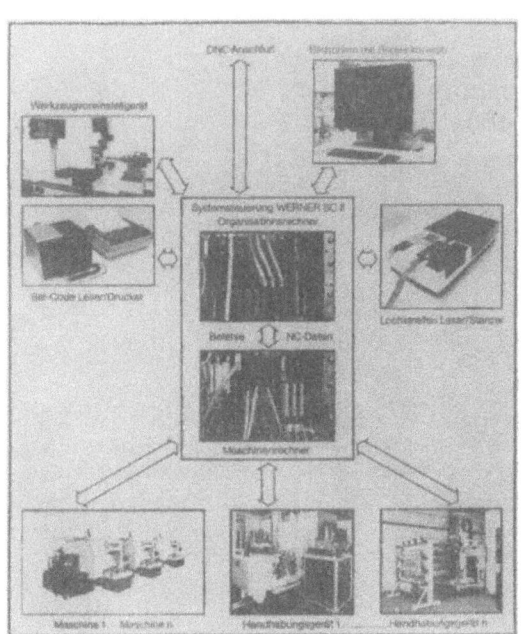

Fig. 8
Software and hardware structure of system control (cell computer) for flexible manufacturing systems

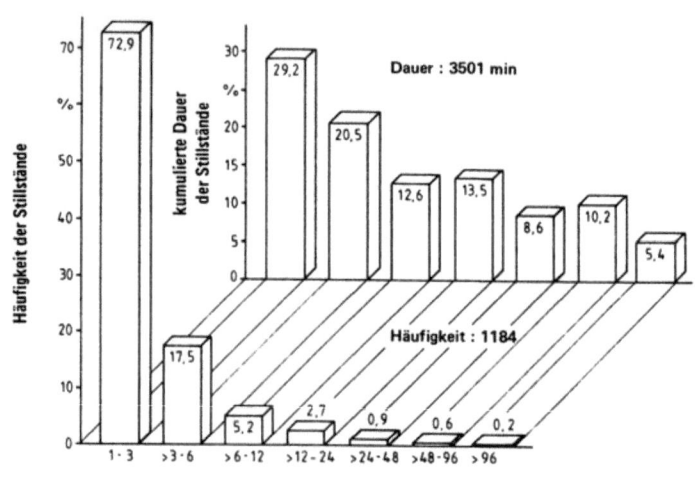

Fig. 9
Downtimes of a flexible manufacturing system
(K. Mertins, IPK/Berlin)

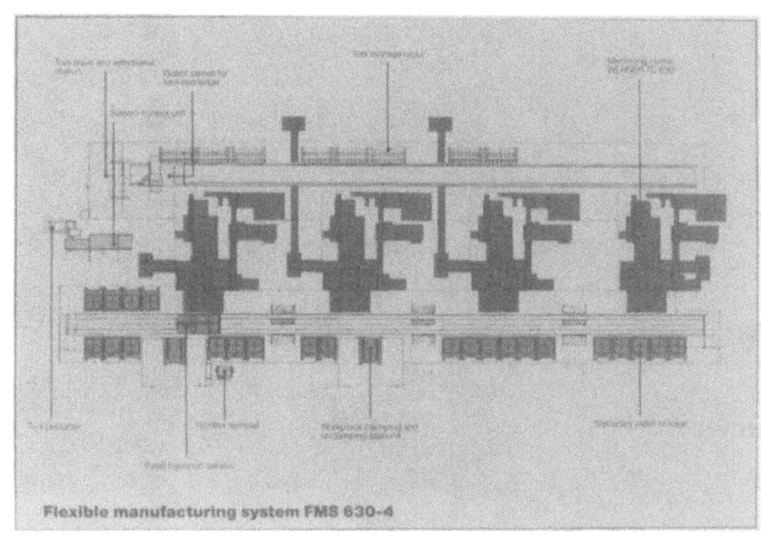

Fig. 10
FFS 630 flexible manufacturing system with
4 TC 630 machining centres and central supply of
workpieces and tools by pallet vehicle and robot carrier

Fig. 11
FFS 630 flexible manufacturing system
Picture of the system shown in Fig. 10

Fig. 12
Workpiece set-up station and clamping fixture
for gear parts
First and second set-up on one fixture

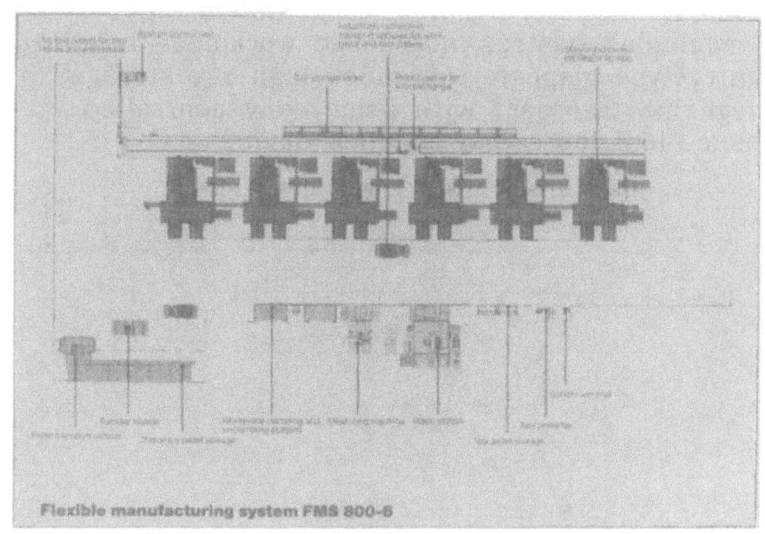

Fig. 13
FFS 800-6 flexible manufacturing system with
6 TC 800 machining centres with inductively controlled
shuttle for the transport workpiece pallets,
tool storage and handling device,
high-bay store, washing machine and
and integral measuring machine

Fig. 14
TC 800 machining centre with pallets supplied
by inductively controlled shuttle

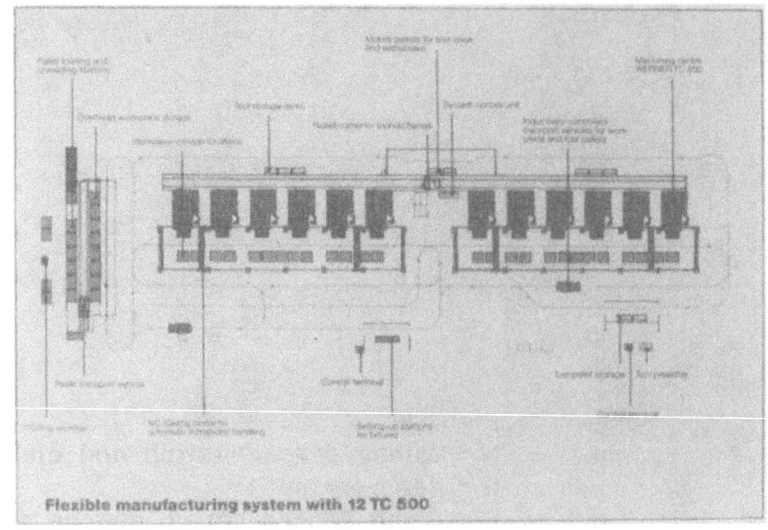

Fig. 15
FFS 500-12 flexible manufacturing system
with 12 TC 500 machining centres, automatic workpiece handling
with 5-axis gantry-type gripper, workpiece high-bay storage,
workpiece magazine transport with inductively controlled
vehicle, tool store and handling device

Fig. 16
FFS 500-12 flexible manufacturing system
Picture of system shown in Fig. 16

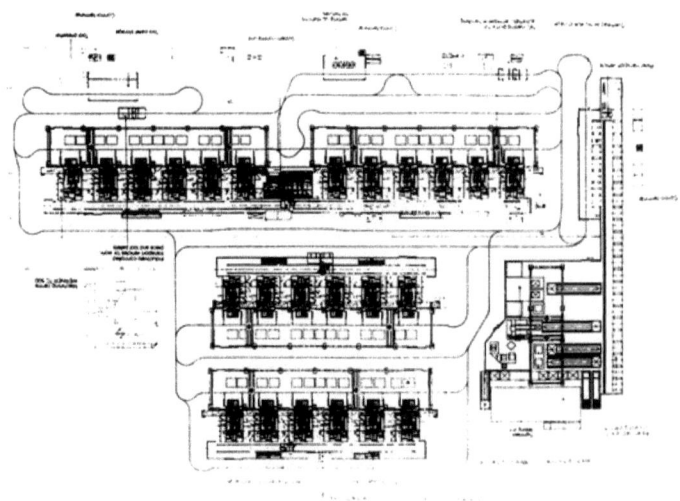

Fig. 17
Flexible manufacturing system with 2 single systems FMS 500-12 (see also Fig. 15)
and with higher-level workpiece transport system and
integrated parts washing, tightness-testing and measuring

CIM — THE INTEGRATED SOLUTION

Factors affecting the realiability of real time distributed control systems

P. J. Cornwell
Renishaw Controls Ltd, UK

ABSTRACT

The MAP initiative from General Motors Corporation has resulted in highly focussed user-vendor collaboration, original engineering development and International standards activity unprecedented in the manufacturing industries.

The emergence of completely specified communication standards and conformance testing guidelines to guarantee multi-vendor connectivity are significant steps towards an off-the-shelf approach to communications systems for CIM. This means that there is wide acceptance of a single type of physical network media and specific protocols are defined for all the layers of MAPs OSI communications model. It is notable that there now exists a MAP reference specification to "facilitate user companies in selecting a common set of communication protocols that will result in ... open multi-vendor communications"

Notwithstanding, MAP has wider implications for industrial automation than the provision of universal communications. Its new application protocols are having a profound effect on both the design of machine controls and system architectures. This paper briefly reviews the development history of these protocols and then examines some of the issues that arise in their use for remote and active control of manufacturing processes. The goals of the process industry lobbies for high integrity communication systems and their relationship to MAP are discussed.

INTRODUCTION

MAP was conceived by General Motors in 1981 to address the difficulties arising from multiple, proprietary communications systems installed in its factories. GM formed the MAP task force to define a generic, device-independent local area network for use within GM. The task force soon identified the International Standards Organisation OSI reference model as the network structure to be adopted.

Although it was possible to select suitable physical media and transport protocols from existing work, ISO had been concerned mainly with linking general-purpose computers, while application protocols to support factory automation did not exist. Consequently, the task force appointed the MAP Programmable Devices Committee (PDC) to specify protocols to provide 'interworking' services for industrial control equipment. Interworking services concentrate on the meaning, structure and message formation of application data, being concerned primarily with the interface to real applications. Accordingly much of the work of the PDC has been in development of application layer protocols.

Standard Message Format

A Standard Message Format (SMF) was defined to address program transfer and variable data access for Programmable Logic Controllers (PLC) and subsequently demonstrated running on a token bus network at the 1984 NCC conference. SMF provided effective generic support for programmable controller operations. It did not attempt to emulate protocols already supported by GMs PLC vendors, but it did address in a generic way the fundamental data typing and addressing strategies which these vendors employed. This made it possible for the vendors to implement a MAP compatible interface to their different PLC architectures in a straightforward manner.

SMF was not based on any existing PLC network because, apart from the obvious difficulties of vendor competition and media differences, many vendor networks were proprietary closed systems. Where they did use ISO protocols, the interworking services were tailored to a proprietary data and program representation. The responsibility was therefore put with the vendors to support communication using an application protocol which was designed, in relative isolation, to be vendor independent and to provide access to generalised PLC services. Such a sweeping request of vendors was made possible by the purchasing power of General Motors (in a similar way to a more recent call for all potential software vendors to be UNIX compatible (FinTech 4 56/1)).

Manufacturing Message Format Standard

The charter of the programmable devices committee was now extended in order to provide the same generic interworking services for computer numerically controlled machine tools and industrial robots. This extension was much more an application level development since much of the earlier work in the lower level protocol definitions could be re-used. However, the scope of the task was broadened because of the sophistication of these new device types and moreover, the far reaching consequences of specifying real time machine control operations over a network. The implication is that a network command alone can alter a large-scale machine activity. (This fundamental safety aspect has been broached by the process industries PROWAY effort.)

The EIA Numerical Controls Committee (IE31) had been working on a generic access and control protocol for machine tools since 1980 and its Working Group 1393 effort was combined with that of the MAP PDC. This joint team published the Manufacturing Message Format Standard (MMFS) in late 1984 and MMFS was subsequently adopted in the MAP 2.0 specification and incorporated in the Autofact '85 demonstration.

An important announcement from General Motors followed on from this work. GM decided to consolidate the MAP specification and freeze the MMFS implementation as MAP 2.1 for a period of two years for the purposes of GM purchasing requirements. This satisfied vendor pressure for a specification against which to supply equipment and resulted in real MAP installations operating in a manufacturing environment. This consolidation was beneficial in allowing many of the interworking techniques to be tested in a production environment but, it has had some far reaching consequences for the upgrade and migration of systems installed at that time.

Manufacturing Messaging Service - RS511

Meanwhile, it had been recognised that the specification of MMFS was not sufficiently concise as a permanent standard to be implemented identically by different vendors. Moreover the MMFS encoding method was not compatible with the ISO standard. This incompatibility had an impact on the MMFS networks installed in the field, since it required individual collaboration in order to allow different vendors equipment to successfully communicate.

Consequently MMFS was re-specified using the Abstract Syntax Notation, ASN1, and a new formal specification called the Manufacturing Messaging Service (MMS) was finally presented to EIA in late 1985. MMS-RS511 contains sophisticated mechanisms for accessing data structures in a wide variety of different equipment types. It allows both comprehensive data typing and the specifying of one variable in terms of others, so that application programs can work independently of the internal data representation of a particular control or machine. A semaphore system is also provided which can allow machine resource allocation for dedicated operations, so that real-time-critical tasks can be undertaken over the network without interference from other users.

The provision of device-specific functions for direct control of each of the three main equipment classes (PLC, CNC machine tool and robot) has particularly far reaching consequences. RS511 allows, for example, the direct setting of PLC outputs, remote manipulation of fixturing, tool changers and the tailstock of CNC machines.

Thus the MMS application protocol which forms much of the new material in the MAP 3.0 specification provides much more than a communication channel. It has considerable implications in the design of machine controls because many functions which were hitherto controlled only by a local operator can now be accessed over the network by a remote computer. Probably more important are the issues that this raises in system integration and the use of such computer control and machines in real manufacturing systems.

On one hand MMS, and to a greater degree FTAM (1), provide very comprehensive services built into the communications sub-system itself, which perform most, if not all, of the tasks currently expected of supervisory plant control and data acquisition systems. It is, therefore, possible to conceive of a plant system which implements most of the MMS/FTAM services and is controlled by a computer program describing only very concise, high level manufacturing strategies, while all the work of synchronisation, data manipulation and transfer between actual equipment is carried out on the network itself.

On the other hand the predominance of shop floor equipment and controls of earlier design, and the additional reliability requirement now placed on the communication system and its many programs and computers may continue to slow the uptake of this technology.

Other factors such as the current MAP 2.1 specification which was frozen in 1985 and against which suppliers are installing equipment, are causing an industry movement for continued compatibility with supply of the older implementation. This should be an issue of software upgrade, even of revision of downloaded software, but already at this year's International Machine Tool Show, vendors are talking of hardware updates to accommodate MAP 3.0.

Aside from these commercial issues, a review of basic computer communications is probably timely, especially to put in context some of the issues of remote real-time control.

PROWAY-LAN

The earlier mention of PROWAY-LAN defines an Industrial Local Area Network as distinct "from other on-line real-time computer networks in that control systems outputs cause material or energy to move". PROWAY is an independent development concerned with the integrity requirements of communication in the process industries although implementation of its services is also being considered in the MAP/EPA (Enhanced Performance Architecture) effort (see below). It is concerned with developing more quantitative standards for:

1. Event driven communication allowing real-time response to events

2. Very high availability of the communication service

3. Very high data integrity

4. Proper operation in the presence of electromagnetic interference and differences in earth potentials

5. Dedicated real-time sub-networks

As such, the PROWAY requirements could be more relevant to the real time control of machine tools, logic controllers and robots than support by individual machine controls of interworking protocols. It is notable that the main features of PROWAY are quite a low number of station connections (<100), local geographical coverage, high performance with a guaranteed priority media access time and quantitative error rate requirements. It is "not intended to provide an optimised interface for high speed computer memories or peripherals". PROWAY-LAN accepts a data circuit bit error rate of 1×10^{-8} i.e. it is designed to run on an inherently unreliable network.

(1) File Transfer Access and Management is an alternative application layer protocol that provides much more sophisticated file manipulation but is unlikely to be published until the end of 1986.

However, it expects applications to see a residual error rate of 3×10^{-15} after the data link protocol layer has attempted to detect and correct errors in the physical layer. This is quite low and we should perhaps be content to have a probabilistically reliable communication channel. However, the manufacturing automation systems that we have been considering are effectively distributed computing systems, and it is perhaps also worth looking at the implications that these other computers have for the overall reliability of CIM control systems.

Each of the layers of the ISO protocol stack and all of the machine and equipment controls on the shop floor, as well as controlling cell and plant level computers run in real time as software processes. That is, they are programs or groups of communicating programs in a state of continual execution. Many of them physically reside on different computers (each with associated individual reliabilities) and therefore have their own timebase and communicate over one media or another to function as a system. The possible number of states and relationships that these communicating processes can enter is quite large, certainly larger than it is practical to simulate for normal operation. The situation becomes much more complex in the face of complete or, generally worse, partial failure of individual system elements whether due to hardware, software or communications media itself. Modern multidrop communications systems also tend to increase the interaction of a single failure with other system elements.

Where sophisticated software communication protocols are heavily integrated with distributed machine computer controls then, the issue of reliability of a sub-system depends on rather more than that of the communication channel. In general, a communication transaction can have one of three possible outcomes:

1. good: the data is received and accepted by the proper receiving station,

2. bad: the data is received and accepted by a receiving station but is received incorrectly,

3. null: either the data is not received at all, or if received is not accepted.

There are however, only really two possible outcomes of the transaction that can be deduced by the initiating process:

a) success: the transaction succeeded

b) failure: it was not possible to conclude that the transaction succeeded.

The initiator will always be unaware, in absolute terms, of the actual outcome of a transaction because of uncertainties of transmission, acknowledgement, real time state synchronisation, etc. But it should always be possible, at some level of confidence, to conclude Success or Failure.

It is clear that of the six possible outcome combinations anything with 'bad' is dangerous in control terms. But it should be noted that while the use of sophisticated data link protocols enable us to convert every bad outcome to null, there is a considerable cost associated with the execution of software communication protocols which impacts the timely delivery required for event driven communication. This is particularly important where 'burst' errors may tend to cause difficulties for several concurrent transactions when increased latency caused by multiple, remedial access itself becomes considerable.

The following network classifications, due to Richard Bornat, were developed for process communication using the Cambridge Ring, but are also valuable in this context:

1. Reliable: only s/good occurs
2. Truthful: only s/good and f/null occur
3. Transmission truthful: s/good, f/good and f/null occur
4. Untruthful: s/good, s/null, f/good and f/null all occur.

We can regard the PROWAY requirement in two ways: either as that of a Reliable connection with no concern for the consequences of probabilistic, albeit low, occasional failure; or we can provide further software protection to turn a basically Untruthful network into a Truthful one. Software protection can be provided by using a scheme of idempotent messages which can be multipally received with the desired effect (and no undesired side effects). This would yield a Truthful network in absolute terms, which is acceptable, but it would be very expensive in network latency and additionally difficult to implement.

In the context of this discussion, it is sufficient to note that we have at best non-reliable networks available to us and that in addition interaction of these with distributed real-time software is likely to reduce overall system reliability still further. Moreover, maintaining the theoretical channel reliability by the use of software protocols has a direct impact on timely delivery (ie real time response to events) and is not without side effects.

MAP ENHANCED PERFORMANCE ARCHITECTURE

Before finally looking at the implications that reliable communication may have for real time network control semantics, one other issue should be considered. Namely of cell control architectures, especially in relation to the real time sub network organisations variously referred to as EPA (Enhanced Performance Architecture), Mini-MAP, etc.

The current consideration of MAP/EPA for support of PROWAY functionality was mentioned briefly earlier. A sub-net is a localised network attached to the backbone via some relay, eg. a gateway or router and which provides high-performance communication for time-critical operations. In EPA both geographic scope and number of stations are limited and some protocol levels are removed to lighten the computing load of communications. Different physical layer proposals are under review at the time of writing, but a carrier-band implementation of the IEEE 802.4 token bus is currently in work by several different vendors. A key issue in this discussion is the way in which EPA could allow (logical) access to equipment connected to the sub-net by multiple controlling programs or computers. The application protocols which are developing for interworking between machine tools, PLC's and robots effectively support cell control and adaptive real-time peer-to-peer communication, and could be used by multiple controlling programs for conflicting operations. It is, however, possible that machine commands such as 'start cycle' and 'advance woodstock' could be invoked concurrently on the same machine by otherwise un-related programs. This would, hopefully, never be programmed as such and an RS511 semaphore mechanism can be used to gain exclusive resource access. Nevertheless a complex real-time control system could still contain many states where such operations are valid if no specific prevention scheme is adopted. Probably the most dangerous case is where a program, executing on a backbone computer (perhaps plant data trending or management information system) or possibly on another sub-net cell controller, interferes with a stable yet time-critical sub-net task. This could be caused by the introduction of an exceptional delay, by inadvertently modifying the state of a machine while making an otherwise passive inquiry for data acquisition, or simply by issuing a direct command. Any command which can cause a change of internal machine state may have unfortunate side effects when operated remotely by an un-related control program. Such an event may indicate incorrect programming of the sub-net control task, but primarily the relay architecture connecting the sub-net and backbone is unsafe. However, the current protocol and system elements allow such an organisation to be constructed, and formal sub-net architectures will have to be developed to manage the problem.

An obvious structure is for each sub-net to have a controller responsible for information exchange with the backbone network. A single real-time representation of both system and individual machine states can then be maintained. Any request from initiating stations connected to the backbone can thus be screened for proper use of sub-net resources, which the controller can then co-ordinate. This organisation may resemble a conventional cell controller architecture, but it is important that single point co-ordination of real-time communications is preserved. However, an extension of this argument would be for peer-to-peer communications, performing adaptive control and synchronisation within the cell, to use this controller for local access management as well.

CONCLUSION

In general the issues of remote, real-time machine control are very complex at the cell level. The satisfactory execution of manufacturing processes over a computer network will involve not only reliable and deterministic communication but also structured cell architectures. Reliability of the communication channel would not seem to be the only factor in judging the safety with which real time operations can be managed. The support of Reliable communication itself has direct impact on the network's capacity to provide real-time response to events. In turn, the software overhead both to support a reliable channel and timely response, as well as to run plant control programs on highly concurrent and distributed multi-vendor hardware makes reliability of individual system computing elements very important. The complex relationships which result make the overall reliability of systems which implement remote real-time control difficult to gauge.

The U.S. PROWAY technical advisory group (ISA SP72) has proposed a PROWAY network based on IEEE 802.4 token bus in common with the MAP physical layer, but the IEEE 802.5 token ring is also under review. This is significant because the token bus topology is difficult to implement using current fibre optic technology. On the other hand, optical ring implementations exist today and while modifications may be required to satisfy PROWAY, the greatly improved immunity of optic fibre to electromagnetic interference and differences in earth potentials is very important. By increasing the intrinsic channel reliability real network performance for event management can be improved without software overhead. If used with carefully organised cell architectures, optical fibre communications could provide acceptably reliable communications for real-time control. This is in contrast to the current use of an expensive technology to combat gross network faults.

BIBLIOGRAPHY

Bornat, R. The Computer Science Laboratory, Queen Mary College, London University, 12 March 1982.
THE 30 YEAR TRUTHFULNESS PROBLEM

Hagar, M.L.
Moon, D.F. Group Technical Staff, Industrial Systems Divisions, Texas Instruments Inc. 1985
HIGH LEVEL INDUSTRIAL COMMAND LANGUAGE SPECIFIES CONTROL COMMUNICATIONS

MAP User Group The Society of Manufacturing Engineers, 25 February 1985
MANUFACTURING AUTOMATION PROTOCOL (MAP) REFERENCE SPECIFICATION

MAP Development
Group General Motors, Warren Michigan
7 March 1986
MINIMAP DIRECTORY SYSTEM, REV 3

Instrument Society of American 1985
PROWAY-LAN ISA S72.01-1985

International Standards Organisation 1985
Open Systems Interconnection: Specification of
ABSTRACT SYNTAX NOTATION ONE (ASN.1) (ISO DIS 8824)

EIA/IE-31/1393
Editing Group Electrical Industries Association. April 1986
MANUFACTURING MESSAGE SERVICE FOR BIDIRECTIONAL TRANSFER OF DIGITALLY ENCODED INFORMATION
PART 1: SERVICE SPECIFICATION Draft 4 Revised

Economical computer integrated manufacturing
M. J. Henderson
Siemens Ltd, UK

ABSTRACT

Use of Personal Computers and Local Area Network technology to provide a low cost integrated solution for CIM applications. This paper ex amines the historical development of CIM and the current state of the art approach using Personal Computer hardware, a Local Area Network (LAN), and software packages to handle the diverse requirements of CAD, NC programming, and control of a Flexible Manufacturing Cell (FMC).

HISTORICAL CONSIDERATIONS

To date, the majority of FMS systems have been installed in branches of industry where levels of investment are traditionally very high, e.g the automobile industry. For the large amounts of expenditure involved, users have occasionally had to wait longer than anticipated for the return on their investment. The learning curve which both users and suppliers have had to negotiate during the implementation of the new technology is indeed a tortuous one.

While the large companies have at least been prepared to invest in FMS, the sums of money involved have been beyond the means of most small and medium-sized companies. They have been obliged to restrict their investment to replacement of old machines on a piece-meal basis, and to adhere to the traditional methods of stand-alone NC with all the associated overhead of paper tapes, shop worksheets, paper-bound drawings and part lists, not to mention disappointing machine utilisation figures and high production costs.

In addition, the use of computers in industry has previously been restricted to the processing of data and information within departments, which led to the development of stand-alone solutions, i.e hardware and software was developed and applied independently, with the design, production planning and manufacturing departments each pursuing their own individual activities. A typical organisation would see the design department producing drawings and bills of material, the production planning producing the work schedules, the production control office issuing shop orders and the NC programmers supplying the NC tapes. In this scenario, data would invariably be exchanged in printed form, which meant firstly that the data had to be entered into each computer system, and secondly that it might be stored several times, thus causing updating problems.

Recent advances in Local Area Network technology, Personal Computer performance and CNC capabilities now open up the vista of a truly integrated, paperless system which enable the drawing office, the production planning office and the manufacturing process to be linked together, using central database facilities, at a price attractive to all engineering companies.

THE SUGGESTED CONCEPT

The last few years have seen a dramatic improvement in the performance of Personal Computers. They have now matured from being a toy for the children into a sophisticated piece of Data Processing equipment, with main memory capacities and disc storage previously associated with minicomputers. The price/performance ratio of modern PCs compares favourably with that of minis. About 15 years ago, minicomputers started to move into those market areas which had previously been the reserve of mainframe machines. We are now experiencing the same phenomena - Personal Computers are taking over traditional minicomputer functions.

At the same time, a great deal of work has been done to standardise networking concepts. This effort has borne fruit in the form of open networks such as Ethernet, based on IEEE802.3, and Token Passing, based on IEEE802.4. Most suppliers are now able to provide equipment which conforms to these standards.

The final breakthrough which has spawned the development of Flexible Manufacturing Systems is shown in the most modern range of CNC machines now on the market. CNC machines also make use of VLSI technology to allow sophisticated communications software to be stored in their memories. Tool life monitoring, spindle power monitoring and facilities for graphical simulation and diagnostics are also common features of these machines.

Figure 1 shows how the various units are being linked together in todays factories. This paper concentrates primarily on the CAD, FMC and NC programming components.

CIM COMPONENTS

Computer Aided Design (CAD)

CAD has been around for 25 years. The problem to date has been that these systems have only been executable on expensive mainframes. As mentioned above, powerful Personal Computers now make these systems available at a fraction of the price.

A typical low cost CAD system might use the following components:
- Intel 80286 processor and 80287 numeric processor
- 512 kB main memory
- 20 mB Winchester hard disc
- 1.2 mB diskette drive
- a colour graphics screen and an alphanumeric monitor
- a mouse
- interfaces for a plotter, printer, and access to the LAN
- remote data link for communication with other computers.

The alphanumeric screen would be used for the operator interface while the colour screen would be reserved for drafting. Such a configuration is commonly offered as a Work Station.

Despite this relatively modest configuration, such systems are capable of performing a wide range of CAD functions such as the drafting, scaling, and labelling of layout drawings, single-part components and circuit diagrams. Geometrical elements can be drawn in a variety of ways, and facilities for rounding and chamfering, cross-hatching, selective erasure are generally available. Automatic rescaling and cross-hatching is performed when the geometry is amended. Most systems also provide features for defining several component layers, which enable different components to be superimposed on each other.

Such systems demand no DP knowledge on the part of users, though a program interface would normally be available for merging other user programs.

Geometric data generated by the system can be passed directly to the NC programming system on another workstation for the development of a workpiece program for an NC machine tool. This not only reduces the overall workload by dispensing with the need for redefining the geometry, but also eliminates possible source of error as the data no longer has to be processed from drawings.

Module functions are available. These consist of basic geometric elements which are assigned names and stored on disc, from where they can be retrieved at will. Modules can be assembled from other modules, so it is an easy matter to compile your own library of symbols and drawings (figure 2).

NC Programming

We have already seen that the use of a CAD system facilitates the passing on in a paperless manner of the production planning and geometric data held in the system. The advantages for the NC programmer are less work, as the geometry created by the CAD system is already presented to him, and the elimination of errors arising from the incorrect conversion of drawing data.

This revolution in NC programming techniques is still going on. In fact, it is fair to say that the paper drawing is still required, but will gradually be discarded as the transfer of specified data from CAD increases. Prospective purchasers of a new programming system must ensure that their chosen system fulfills three requirements if it is to be integrated into a CIM environment:

a) facilities to handle both present and future data input methods, e.g in graphic interactive mode for conversion of drawing data in the familiar manner, and the acceptance and processing of product data presented by a CAD system.

b) dispenses with punched paper tape. If the NC programmer takes his basic data in a paperless manner, it is also expedient to pass on the results of the NC programming in the same way. In the interim therefore, the NC system must be able to output punched paper tape as usual, but at the same time have the ability to pass the NC program on-line to the machine tool or flexible manufacturing cell.

c) the system must also be usable by non-specialists.

This last point is particularly important since to be economically justifiable in small and medium size operations, elaborate production planning activities must be avoided. This support for the layman is provided by using so-called WIMP features (Windows, Icons, Mouse, Pointers) and numerous 'Help' facilities, all of which simplify the man/machine interface.

System Structure. The NC processor is embedded in a number of additional program blocks:

- CAD linking program
- File management system
- Postprocessor
- Off-line storage system
- On-line NC program input

Figure 3 shows the interacation between these various components as well as the corresponding files. During processing, the NC system will maintain a data model in which all geometrical and technological objects are assigned attributes, and relationships are established between objects. This activity is of benefit to the user at a later stage when intervention in existing part programs is necessary. This information will be passed to the postprocessor via a file containing cutter location data (see figure 3, file CLDATA). Both the NC processor and the postprocessor will also have access to technological files on the database.

Among these files should be a tool file, material file, machine file, machining file, standard tool technology file, postprocessor file, NC macro file, and contour file. The system will take decisions based on the contents of some of these files, thereby reducing the input overhead and permitting higher degrees of automation.

Machining Simulation. The majority of modern NC programming systems will support simulation. The purpose of a simulation should be to make the workpiece visible with as many secondary machine conditions as possible. Thus the tools and workholding devices should be displayed to scale and the workpiece updated dynamically as the simulation progresses. The simulation will normally be in two stages: during programming each machine step is represented visually as a feedback to the input, thus allowing immediate correction. When programming has ended, the entire machining process can be simulated in its full context.

Adaptation to the Machine Tool. We have seen that the heart of an NC programming system is the processor and postprocessor. The former processes the inputs and presents the results, which at the time will not relate to the actual machine tool. It is the function of the postprocessor to adapt this neutral data to the specific machine tool. A typical NC postprocessor will output the following data:

- NC program
- Setting instructions (tool setting, tool corrections, data relating to the work piece, material and workholding)
- Statistics relating to tools used, machining operations, non-productive time etc.

Flexible Manufacturing Cell (FMC)

The CIM concept based on workstations extends beyond the bounds of the engineering office to include manufacturing. Manufacturing does not become more flexible solely by automating the machine tools. It is only possible to achieve high productivity for small batches if the NC machine tools are linked together as a cell, and the whole production process and material flow are controlled together using a cell control computer (figure 4).

The control computer described below is capable of controlling a maximum of four machine tools, a tool setting device and a transport system. Non DNC machines, such as washing and inspection machines, can also be incorporated into the cell. Such a cell represents a standardised solution which is neither dedicated to any one machine tool type nor, more significantly, to any one machine tool builder. The hardware is the same workstation configuration as used by the CAD and NC programming systems described earlier. The paperless interchange of data via the LAN is therefore a simple matter.

Workpiece NC programs are downloaded from the NC program library or NC programming system to the FMC, where they are then prepared for subsequent machining cycles and transferred, as required, to the CNC control of the respective machine tool. The software structure is modular (figure 5), which allows tailoring to individual requirements. The main components are described below.

Data. The system recognizes four different types of data: master data, control data, status data and general management data.

- Master data will normally only be entered once when the cell is put into operation. This comprises system specific data relating to the architecture of the system, and resources master data, such as tool master data (geometrical data, tool life) and workpiece carrier master data.

- Control data are product specific data such as NC programs, tool layouts and work schedules, and manufacturing orders.

- Status data describe the current situation with regard to resources, i.e plant status data, workpiece carrier data, tool data and workpiece data.

- General Management data covers all operational and machine data which may be required for later analysis. Such data includes machine specific messages (e.g NC start, NC end), alarms, tool specific messages (e.g tool break, end of tool life), NC messages and fault reason messages, input by means of a dialog.

Basic System. This is the only compulsory software module. It provides standard interfaces for the individual function modules and to any modules developed by the user or on behalf of the user. Future software developments can therefore be easily incorporated. The basic system handles communication between the function modules and also between these modules and any peripheral devices. The basic system also supports or performs the following functions:

- system generation and parameterisation
- system initialisation
- collection and display of error messages
- logbook functions

Work Planning. The object of the work plan is to identify the number of operations to be performed on a particular component, and the fixturing which will be required. The work plan controls the routing of the component through the cell and identifies the NC programs which will be required. The system allows alternative machine tools to be used when possible and takes into account the time required for each machining operation.

Job Scheduling. Orders are entered by means of a dialog with the operator. Appropriate tabs allow manufacturing orders to be linked in such a way that parts are finished in the correct order for assembly. The planning method endeavours to use as few workpiece carriers as possible, achieve maximum machine utilisation and keep throughput times to a minimum.
During the scheduling, a check is performed to ensure that all necessary resources (NC programs, tooling, fixturing etc.) are available. If not the operator is informed. Should the operator decide that the required resources will be available in time, then the order is considered to be accepted and is scheduled. The availability check is repeated prior to manufacture.

Tool Requirements. Tool requirements planning is carried out in two phases: after job scheduling and prior to the start of manufacture. In the first planning phase, all relevant data, e.g. tool identification numbers and tool lives, are linked together to provide a list of complete tool requirements. These are presented to the operator who must confirm that the tools will be available.
The second phase comprises a requirements analysis to produce a net tool requirements list and tool loading and unloading lists for the machine magazine. If a tool setting device is included in the system, tool correction (TO) data determined by the device are transferred online to the cell computer.

Set-up Dialog for Workpiece Carrier. At the set-up location, the operator is informed of the set-up procedures to be carried out on that particular workpiece. Fixture zero offset (ZO) data are accepted by the cell computer and notified to the machine tool if necessary.

Clamping Dialog. At the clamping stations, of which more than one may be present, dialogs with the operator are performed for clamping, reclamping and unclamping. After each clamping operation has been completed, the workpiece status data are updated. Quality data (e.g remachining necessary) may be associated with the workpiece.

Material Flow Control (MFC). This facility controls the transport of workpiece carriers and their clamped workpieces through the FMC. Transport requests are issued by station specific programs (e.g NC program, clamping dialog) and processed by MFC on the FIFO principle (oldest request first). During transport, the status data of the transport system, source and destination stations, workpiece carrier and workpiece are being constantly updated by MFC.

Machine Tool Program. This program carries out all tasks arising at the interface to the machine tool. As these are on-line functions, each machine tool is assigned its own Machine Tool Program. These programs typically carry out the following functions:
- maintaining list of NC programs available in the CNC
- program loading by preparing the CNC and activating transfer via the NC program supply module (see below)
- erasing CNC memory
- reading tool magazine and passing data to tool requirements module
- forwarding TO data when loading the tool magazine
- issuing alarm and status messages
- synchronisation in the event of a restart by resetting the PLC and notifying the relevant modules.

NC Program Supply. This module allows NC programs to be:
- transferred to the CNC
- retransferred from a CNC following update or optimisation
- read from floppy disk, punched tape or via a host computer interface into the FMC
- managed
- deleted from the NC program library
The operator is able to interrogate the NC library for information relating to all programs, all blocked programs, all updated programs etc.

Tool Flow Control. This module oversees the bi-directional transfer of tools between the cell magazines and the machine tool magazines. Before a workpiece arrives at a machining location, a check is carried out to establish whether all required tools are present in the machine tool magazine, or those tools not present are available from the cell magazine. If the necessary tools are not available the workpiece is rejected. If the tools are available from the cell magazine, a tool change is initiated.
Physical tool exchange is accompanied by magazine management update.

Cell Magazine Management. This module performs the following tasks in the cell magazine:
- location management of installed tools
- machine oriented reservation of tools for current jobs
- statistics relating to machine occupancy
Exchange of tools between the cell magazine and the machine tool magazine is performed by a handling device and processed by the tool flow control module.

System Visualisation. This is a graphical representation of the FMC layout, the current status of the individual stations and the operating mode. Visualisation is normally in full colour.

Local Area Network (LAN)

The three components just examined (CAD, NC, FMC) are all available as stand-alone products. CIM is made possible by the fact that they can now all be linked together using a LAN.
The LAN most commonly in use is an open Ethernet system based on the IEEE 802.3 standard. Ethernet is widely proven as an office LAN and the addition of additional screening to the coaxial cable means it is now suitable for the noisier industrial environment.
The bus cable can be upto 500m long and it can support as many as 100 stations to form a "trunk" segment. Each station is attached to this segment by means of a drop cable of maximum 50m length. Area coverage is thus 100m x 500m. Several segments may be coupled using repeaters, of which two may be interposed between any two communicating stations. "Branch" segments may be added to a "trunk" segment, as shown in figure 6, which covers an area of 1000m x 500m. Two additional segments of 500m x 100m can be added to each end of the "trunk" segment, giving a maximum point-to-point coverage of 1500m. Such a configuration will accomodate 1024 users.
An Ethernet LAN offers good area coverage, expandability, low cabling and interfacing overheads, a high data transfer rate (10Mbit/sec), freedom of communication among a large number of partners and high reliability.

SUMMARY

In order to implement this concept succesfully, it is essential that the requirements of each department are taken into account during the initial concept design phase. Investing in an individual NC machine for manufacturing, or a robot for the assembly line, or introducing a CAD system for the drawing office will never produce an optimal result if an overall 'grand design' has not been considered.

Increased integration of the various components within an automation system highlights the problems of hardware and software interfacing. In these circumstances, developments such as MAP are most welcome, although it will probably not be until 1988 that a stable standard, based on version 3.0, will be widely available. In the meantime, any supplier who is able to provide all the necessary components in a well-proven integrated system will obviously have a tremendous advantage in the market place.

The more complex CIM/FMS installations are characterised by high capital investment. In many cases it is difficult to economically justify the installation of such a system. The difficulty lies in the fact that these systems are necessary in order to remove uncertainties about future planning. In other words they represent an insurance policy for the future. The economic justification calculation must not concern itself with the immediate present. As these systems represent a reduction in future risk, the calculation must also be future oriented.
The systems which have been described in this paper are of less complexity. It is hoped that they will offer the smaller companies the opportunity to invest in the new technology and at the same time keep the risk involved at a quantifiable level.

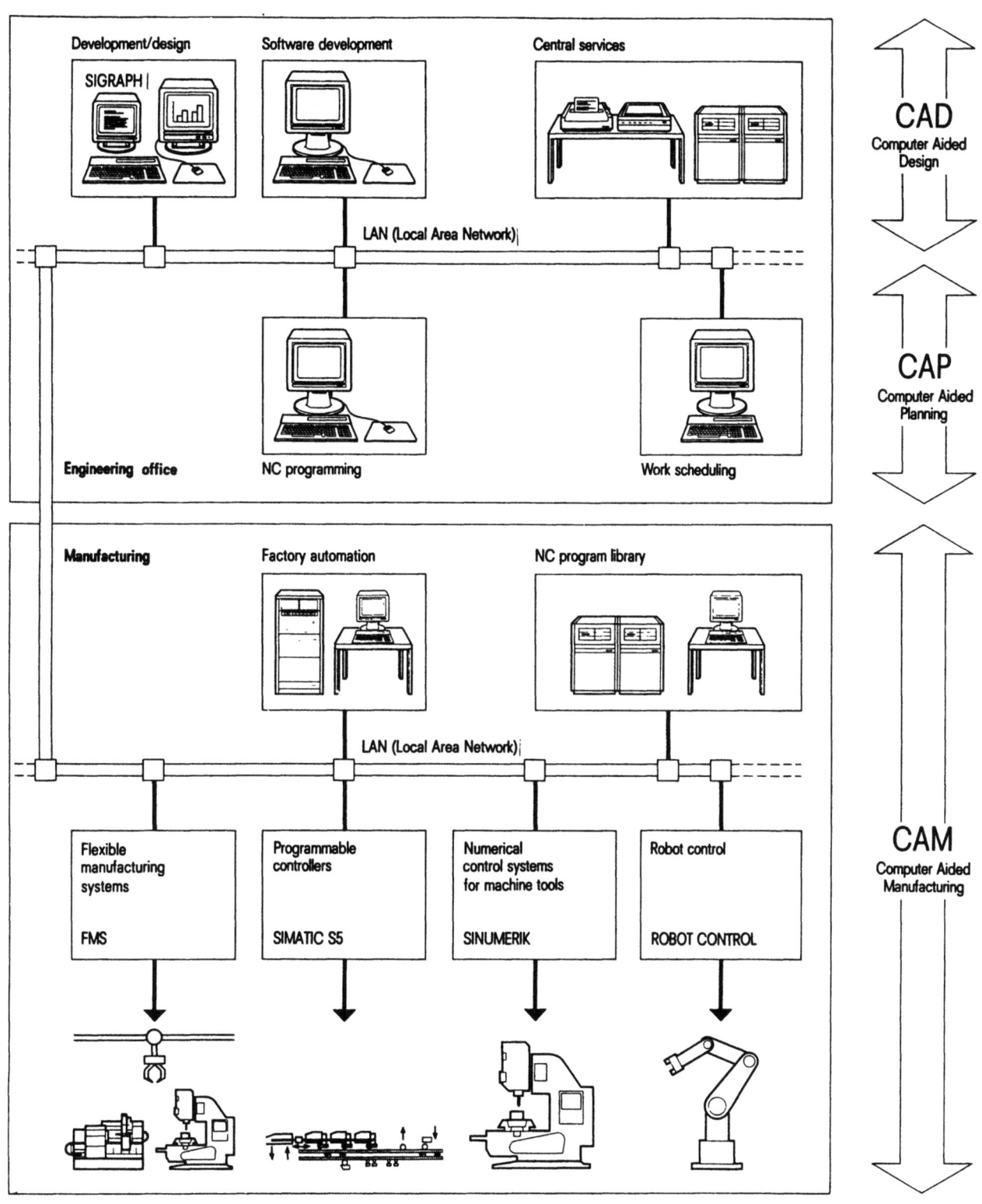

Figure 1: CIM Components

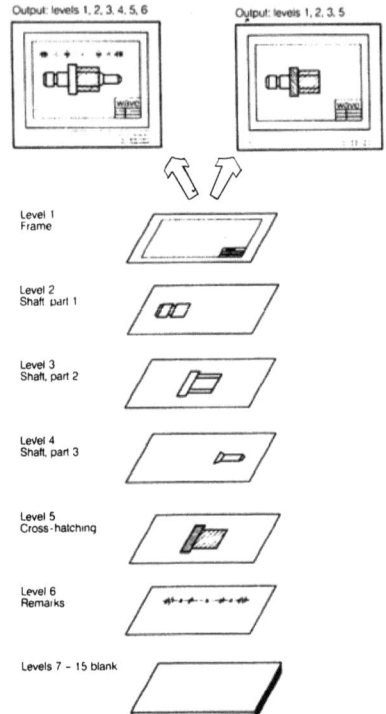

Figure 2 : Symbol Levels

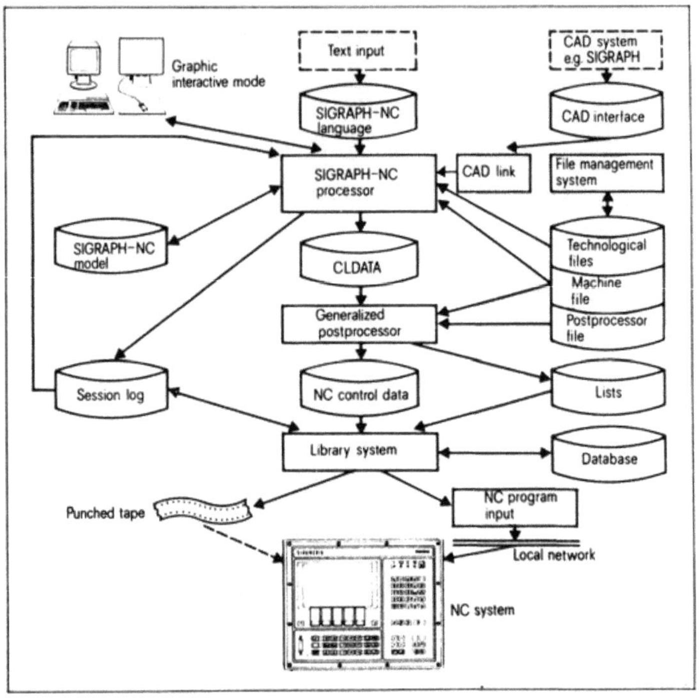

Figure 3: NC Programming System Structure

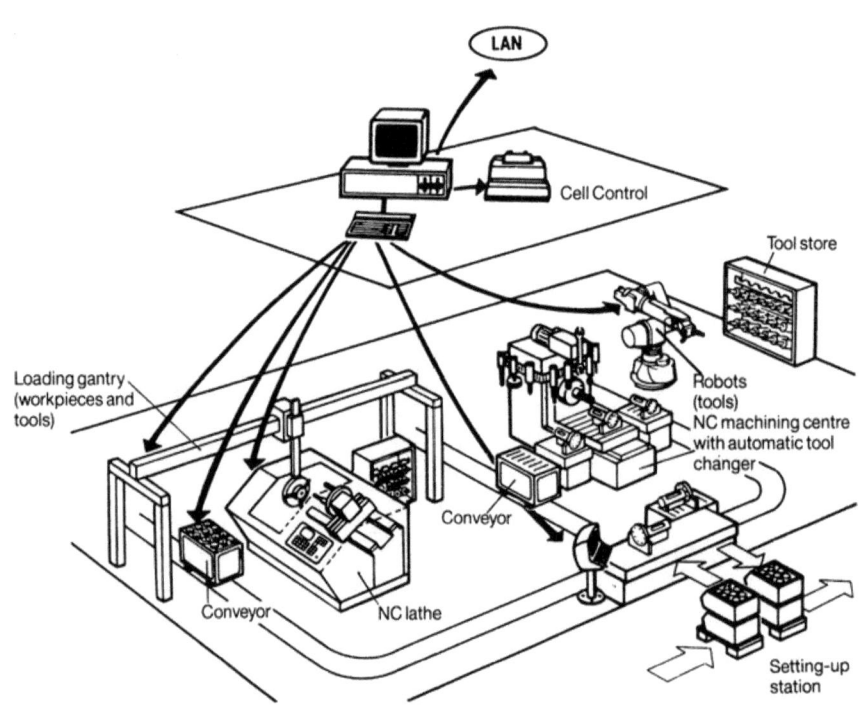

Figure 4: Flexible Manufacturing Cell

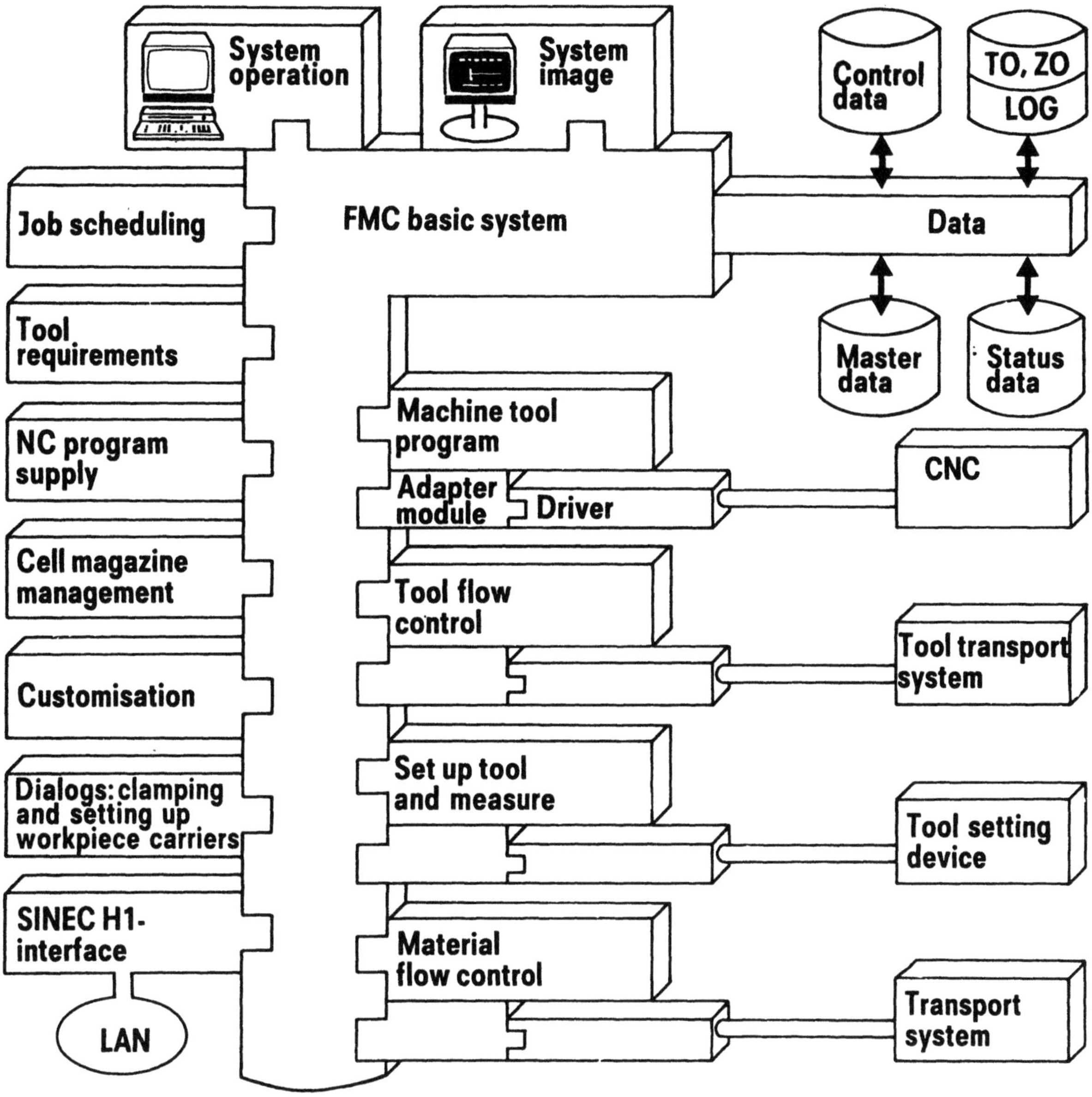

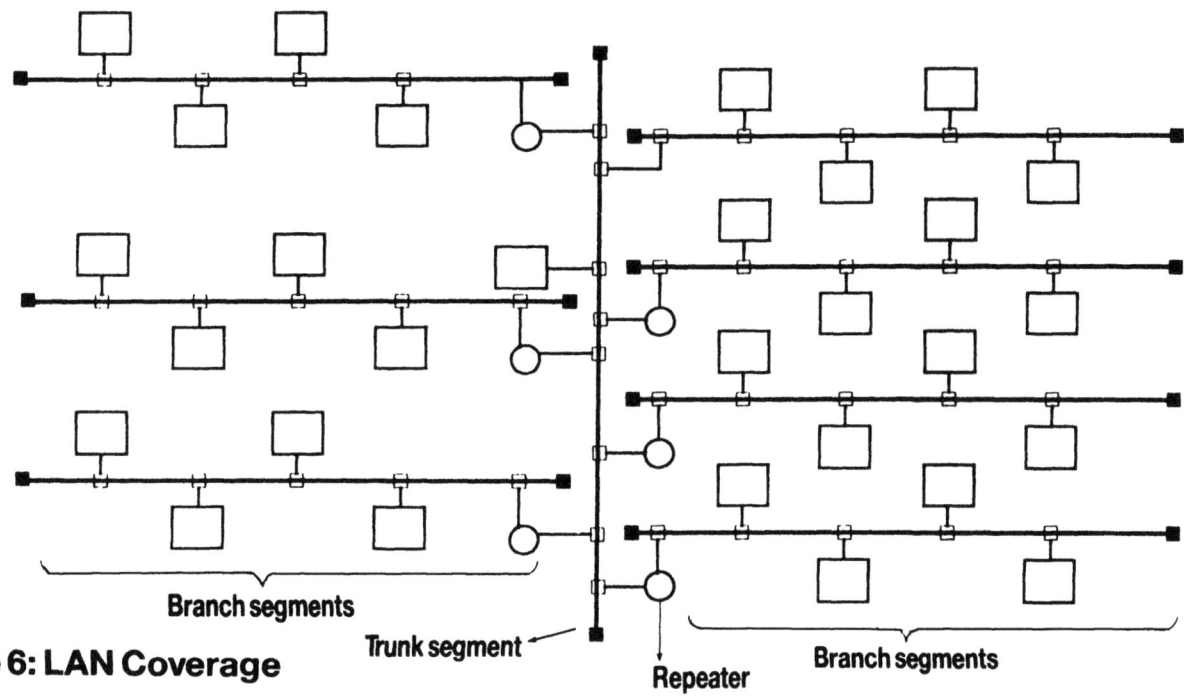

Figure 5: FMC Module overview

Figure 6: LAN Coverage

The integration of FMS, Just-in-Time and LAN technology into a common manufacturing system

D. A. Hearn and M. A. Donnellan
Project Managers, Handley Walker Co Ltd, UK

Abstract

The concepts of Flexible Manufacturing and Just-In-Time Manufacturing have been with us for some time. In many cases, although not all, they can only become practical solutions by harnessing the use of powerful computer control systems such as those based upon Local Area Network Technology.

This paper defines how these three techniques have been integrated to form a common manufacturing system for the production of hand-held portable radios. The paper also discusses how these principles may be extended for use in other diverse manufacturing environments.

The Product and The Problems

At the outset, fundamental problems were evident in both the product and the methods used to produce it. The product was innovative and attractive representing a significant step forward in communication technology with techniques such as microcircuitry, surfaced mount devices and VLSI all synonymous with the product. However a whole new set of manufacturing problems had been created by the technological leap forward, which were being compounded by surging market demand, and the associated need to rapidly increase the output of finished product.

A combined effect of premature market launch and the initial use of outdated batch manufacturing methods quickly lead to a serious deterioration in the quality of the finished goods and erosion of customer confidence. Consistent field failures were clearly attributable to poor design and inadequate manufacturing and quality control techniques. The situation was being worsened by the continual drive for increased output.

The Manufacturing System (Physical)

A new manufacturing system was needed to address the key problems already outlined however, these problems seemed small compared to the need to produce more than 2,000 product variants at a rate of over 1,000 units per week.

The added complication of raising quality in such a high variety, variable volume environment necessitated the use of a radically different manufacturing approach.

The solution adopted was based on the combined principles of Just-In-Time and Flexible Manufacturing. The production facility subsequently created is shown in figure 1.

The facility shown above contains 200 direct operatives configured around six production lines. The activities carried out on each of the lines differ widely depending on volume and mix requirements. With volumes maintained at a constant rate, the major focus of attention is product mix. Wide variations in product mix are normally incompatible with Just-In-Time manufacturing techniques. This constraint is overcome however by launching kits of variant conscious material to the lines with each kit uniquely identified via bar coding. With the bar code used as the work tracking mechanism it is possible to sequentially process each unique job through a series of common manufacturing resources.

This solution only satisfied some of the requirements though, and further flexibility was needed in the facility. The concept of "peer" groups was thus created in order to provide a means of coping with the constant changes in work content associated with each unique job. Unless overcome, severe line imbalance would quickly become apparent.

A "peer" group is basically a series of workstations which perform identical work to each other. They are used in cases where the assembly task takes longer to perform than the standard workstation cycle time set for attaining desired line output. A typical "peer" group is shown in Figure 2.

Using this technique it is thus possible to match differing time values for various assembly tasks by creating the appropriate number of "peer" groups needed to achieve a balance flow along the line and physically locate them in the position where the work task is to be carried out. The creation of such a group has to be carefully controlled in order to maintain the JIT discipline of "pulling" work from one workstation to the next. If work is not physically "pulled" from workstation to workstation whether it is between different assembly operatives or within a "peer" group, the primary principle of Just-In-Time breaks down.

To prevent this from occurring the computer control system was used to inform any upstream operator within a "peer" group to allow the next job to pass to a downstream operator who was available for work at the same time. Before this was allowed to occur the control system validated that the downstream operator was free to accept the next job (thereby satisfying the major "pull" principle of Just-In-Time). The downstream operator then "pulls" the job from the upstream operative and carries out the desired task. Once all the downstream workstations in the "peer" group are busy the control gate or first "peer" group operator is allowed to accept the next subsequent job passing down the line. In this way it is possible to process products which have differing variations in work content without causing bottlenecks, stoppages or line imbalances normally associated with traditional flowline techniques.

The complexity of producing a high variety product using JIT methods was compounded further by the need to manufacture an almost infinite number of accessory options which existed in the form of frequency signalling modules. (Functionally, these units customise the frequency signature of each radio such that the operational characteristic of each product is unique). It is in this area of the product where most variety occurred. The modules were designed as independent sub-assemblies which are subsequently incorporated into the finished product during the final stages of assembly.

A secondary ancillary production line was therefore created to produce the signalling modules. These finished items were then kitted ahead of mainstream production along with the other variant conscious material and uniquely identified via the bar coding mechanism in order to maintain its customised identity.

To achieve the desired level of flexibility operators working in this area of the facility possessed the highest skill rating and were used to build each unit right through to completion in direct contrast to the flowline techniques used to sequentially build the core elements of the product.

The manufacturing system as a whole was totally flexible (flexibility is not necessarily a technological function) by virtue of its ability to cope with diverse product mix and by its ability to manufacture other different products. Given that the PCB line is already producing an immense range of circuitry further physical expansion is possible. The requirements of increased output and the tolerance of further manufacturing diversity can thus be achieved in line with the requirements of different product types. Since the main assembly lines also exhibit common characteristics of flexibility it is not difficult to envisage these areas producing more than one type of product.

The Manufacturing System (Control)

It has already been mentioned that a computer based information system was used as the basis for controlling the <u>physical manufacture</u> of the product. The four primary functions of this system were as follows :-

- Control the flow of work through the facility

- Issue manufacturing instructions to each workstation based on the specific bar code attached to each job.

- Monitor the quality performance during manufacture and close 'Quality Gates' when necessary to stop the production of any product having an unacceptable quality level.

    Interface with the mainframe MRP system.

Three different solutions were examined before a final choice was selected. These varied in complexity and will be described later in this document.

There were a number of operational and strategic requirements which had to be satisfied, namely :-

- A practical solution was required rather than one which was technically elegant.

- A resilient system was needed. In the event of the failure major elements of the control system it must be possible for production to be maintained, (albeit at a lower level of control) until the problem was been rectified.

- An expansion capability must exist to cater for the application of more automation, or to cater for other products which were still under development at the time that the system was specified.

- The system must be supplied from a 'substantial' vendor. It was anticipated that a significant system such as this would be applied to other areas of the Company, perhaps to other Companies in the Group. It was therefore vital to source the system from a Company that could support international installations at some time in the future.

- Response time was not to be excessive. With 200 workstations on-line this was an important consideration. At first it was thought that the demands placed on the system by the high number of workstations would average out, effectively dampening excessive demand which would have slowed the system down. However, dynamic computer simulation of the facility revealed that any bottleneck, (there are always bottlenecks in a flowline with many of them randomly occurring, caused demand to be more sporadic. Hence the design of the system had to take this factor into account therefore.

The Data Gathering and Distribution System

The data gathering and distribution system consisted of two elements :-

- the input and output terminals used at each of the workstations,

- the connection of these terminals to the main control system processors.

VDU type terminals were one option considered for the display of information but these were dismissed for a number of reasons, such as acceptability by the operatives, cost, and because space on the workstation benches was at a premium and VDU's have quite a large footprint.

Simple single line alpha/numeric display devices were chosen instead. The type chosen were those based around plasma discharge technology similar to the type used in cash tills.

Bar code readers were chosen as the means for entering data into the system. These were selected because of their low cost and low bench top footprint. In addition, it was important to standardise on messages, particularly those associated with quality reporting. A communication medium based around bar codes allowed the number of messages to be standardised and constrained within set limits.

It was quickly determined that the barcode labels which were to be used needed to be of the highest quality, since experience has shown that it is the label, rather than the reading device, which maximises the integrity of data capture.

Having selected the terminal device a data transportation medium was sought to support the 200 workstations and give the desired response time under normal conditions. At the same time account needed to be made of expansion requirements.

A Local Area Network was therefore chosen as the distribution medium although its precise form was not initially known. This choice was initially the more expensive solution but in the longer term it ideally served the requirements of flexibility and expandibility.

Following evaluation of a number of different types of LAN technology Ethernet was eventually determined to be the most appropriate network technology for a number of reasons which were as follows :-

> The density of data traffic could be contained within an acceptable level. This allowed the standard Ethernet mechanism for handling data collision to be used to a reasonable level of efficiency. The problem of high numbers of collision detections occurring and the corresponding number of re-tries would not, therefore impede the efficiency of smooth data flow.

At the time, the General Motors backed Manufacturing Applications Protocol, (MAP), was being heralded as the panacea for standardisation. Ethernet, however, was an established system and many proven implementations are in existence. MAP, on the other hand, was new and nobody could anticipate what impact the impending IBM system, also utilising Token Ring technology, would have on the success of MAP.

A more general concern was the loading the network would place on the response times of the bar code readers. The technique of multiplexing was initially thought to be the most appropriate means of ensuring good response times. This solution also minimised the large number of input/output nodes on the LAN. A configuration was therefore envisaged, consisting of two levels of multiplexing, to service the large number of terminals required. However, this solution was eventually rejected on the grounds that the delays caused by the multiplexors and propagation on the network itself equated to the total response time specified for the full system.

The solution to this problem was to utilise one of the new barcode reader network systems which were emerging at that time. These systems incorporate intelligent decoder units in the reading device itself which could be used to relieve the main control computer of some of the burden of verification and local data logging. A major benefit also was the fact that these units incorporated a local one or two line display unit supported from the barcode network system.

The System Solutions

As previously mentioned three levels of design complexity were considered for the architecture of the system. A diagram of the simplest solution considered is shown in Figure 3.

This solution uses a single processor to act as the interface with the mainframe computer and act as the local control processor. In this case the barcode reading and display devices would connect directly into the system via a small number of network nodes.

This solution was ultimately rejected as it did not conform to the set objectives. The reasons for this are as follows :-

There is no resilience in the system. If the processor fails the system goes out of action and the manufacturing facility would come to a standstill.

The expansion capability of the system is limited. The requirement to expand the facility for the addition of further levels of automation or for the manufacture of new products on a common facility could not be satisfied.

A more complex solution which was also rejected is shown diagrammatically in Figure 4. This solution represented a complex Computer Integrated Manufacturing facility incorporating most of the major elements of factory automation including the direct link-up with an existing MEDUSA CAD system.

This proved to be too far to complex for immediate requirements. Extensive integration would have been necessary right across the company which would have lead an unacceptable level of disruption to the company at that time.

The solution eventually adopted is shown in Figure 5.

This shows four identical processors connected together via an Ethernet Local Area Network. Each processor serves a specific function in the overall control concept.

The Shop Floor Processor mainly serves as the connection to the mainframe computer, but also undertakes the role of data logging and report generation.

The Three Line processors each take responsibility for a specific portion of the manufacturing facility, one for PCB/Accessory Option build and assembly, one for Product Assembly and one for Store Keeping and Kitting.

None of these processors would be loaded to a particularly high level. This satisfied the desired response time criteria. However, in the event of a single processor failure, the load could easily be redistributed between the remaining computers and continue operating, albeit at a reduced capability.

An alternative means of satisfying the necessary resilience criteria was to hold a back-up processor which could replace a failed component as and when necessary. A recovery procedure based upon this approach is made feasible because each processor was of relatively low cost. Holding a spare processor therefore became a cost effective back-up solution

Expansion of the system can easily be accomplished through, either the addition of more processors to cater for a more complex situation or by the incorporation of another Shop Floor Host Processor which would provide the interface to alternative sets of Line Processors. All of these could, if necessary, be supported from either an extension of the existing Ethernet Network, or by the installation of a separate network system. This is not desirable however, since it can lead to large volumes of data travelling around the network. Automation could be added to the system as required, since standard network interfacing would be used and also because the transfer of messages between different workstations can be simplified.

Further expansion of the system is already envisaged in the area of material supply. The facility uses a "two-bin" supply method for the provision of bulk material to lines. It is planned to identify these material containers via bar codes and to initiate replenishment through the input of a "material required" messages direct from the workstation to the stores area.

In addition it is planned to integrate manufacture with an automatic stores system so that the request for material can be automatically handled. With the manufacturing system integrated to the MRP system via the stores area a high level of material control will be possible. The MRP system (in this case it was COPICS) can therefore automatically downdate its stock records as material is consumed on the shop floor. In essence this represents a closed loop material supply system.

Considerations for Other Manufacturing Environments

The manufacturing system already described is a novel one incorporating many of the popular elements of advanced manufacturing technology available today. However its fundamental design criteria is based on the application of Just-In-Time techniques which have been applied to a classic flowline environment. In many industries particularly those involved with metalworking, flowline manufacturing techniques do not initially appear to be applicable and more particularly the concept of JIT. This is because the process routings are complex with parts and products criss-crossing between various machine tools, processes and workstations in a highly variable manner. The application of Just-in-Time methodology and the systems needed to control it are, therefore, straightforward.

The problem of controlling the production of parts and products in "non-flowline" environments using JIT principles is complicated by the fact that many of the manufacturing resources cannot be located in close proximity to each other. In addition further difficulties are imposed by numerous jobs of different types simultaneously competing for the same manufacturing resource. This latter problem is particularly acute since the constantly changing priority that parts or products to be processed in at a given workstation can be infinitely variable.

Conventionally, this is overcome by the combined use of production scheduling and batch manufacturing methods. The penalty for this approach is the incurrence of high work-in-process levels and all the associated problems and costs that this brings. Many companies <u>only attack the symptoms</u> of the real problem by investing large sums of money on unwieldy MRP systems which <u>do nothing</u> for the root cause of the disease. Systems such as these fail to live up to their expectations due to the complicated nature of understanding what is really happening on the shop floor. Installing ever increasing levels of systems and hard automation to already problematical situations only adds to the basic difficulty of understanding what is really going on, hence little impact is made, and in many cases only make matters worse.

The solution lies in simplifying the total manufacturing process before any form of either shop floor automation or office automation is applied. One such simplification technique is that associated with the flexible use of Just-In-Time. The question is, how can it be made to work in the complex environments already described ?

A solution has been developed which centres on the following three factors :-

- Simplify the process routings. This seems an easy statement to make but it is a practical one. It is surprising to find the level of commonality that actually exists between apparently diverse parts and products.

- Physically impose the "pull" discipline by positioning controlling Kan-Ban squares between the various processes. In this situation a Kan-Ban exists immediately <u>after</u> each workstation for <u>each</u> workstation which is immediate<u>ly</u> downstream of it. As with all Kan-Ban processes, the upstream workstation must not operate on any job which is destined for a filled Kan-Ban square. This implies that if all the Kan-Ban squares for downstream workstation are occupied, or if there is no work available for an empty Kan-Ban, the upstream workstation must stop.

  - Define job priorities in terms of their business requirements, and maintain absolute control over these priorities and clearly display what they are to the appropriate personnel. The movement of parts from one process to the next will be controlled by a priority rating assigned to each job. A downstream process will accept work from an upstream process on the basis of interrogating the relative priorities of the jobs waiting in the upstream Kan-Bans. The job with the highest priority job will then be "pulled" to the next operation providing that it is available to receive work. Since the priority of a job can alter during manufacture consideration must be given to any changing requirements.

Incrementally changing the priority rating of each job as it progresses through the manufacturing system must be carried out according to a pre-defined set of rules. The priority rating assigned at each workstation can be determined by using a number of different methods. Some of these are discussed below :-

As a job passes through the production sequence its priority value will be increased. This could be a linear increase, i.e. a constant would be added to the priority rating.

Under certain conditions it is possible that a job could be continually overtaken by jobs with a higher priority. In order to overcome this problem the priority rating of every job will increase periodically, whilst it remains in an unfinished condition. This increase can be proportional, for example a percentage of the existing priority rating. The increase may either be time dependant, i.e. increasing in proportion to elapsed time, or event dependant, increasing on each occasion that the component is ignored in preference to another. Once the job eventually begins to progress again the added 'waiting' priority value can be removed.

The priority of a job will be raised if the job cannot travel to the downstream Kan-Ban because the downstream process is in turn in a waiting situation. A typical example of this would be where a group of jobs are collected together at an oven for batch processing and where the batch operation is held up whilst the remaining job that allows the operation to commence travels from the upstream workstation or Kan-Ban. In this situation the jobs held up at the oven can be released from the "wait" situation by allowing the oven process to proceed by increasing the priority rating of the jobs which have already been collected together.

An alternative method of removing the "wait" situation would be to completely re-assign a new top-level priority value midway through the manufacturing sequence. This could be based on a lead-time parameter in the form of an arithmetical quotient similar to the critical ratio used in many MRP systems. The value assigned to the quotient would obviously increase as the planned time remaining for the job to be completed elapses. The priority needed to satisfactorily progress the job through the system would be automatically flagged up as the time remaining to complete the job by the delivery date diminishes.

A control system similar to that described earlier in this document could easily track each individual job in terms of a priority value rating or leadtime quotient. Applicable information can then be supplied to the shop floor operatives such as the sequence that jobs must be moved from workstation to workstation in and indicate the relative priorities that these jobs are subsequently processed in. In this way the complex factory can be *simply* controlled. In this way significant benefits accrue via improved manufacturing performance.

In summary, it is not very difficult to envisage the benefits of the combined use of Just-In-Time, Flexible Manufacturing and Local Area Network based control systems. It is highly likely that they will be used on an ever increasing scale in many diverse manufacturing environments previously considered incompatible with the concepts, techniques and solutions outlined in this document.

Perhaps the most striking message to convey is that technology in its own right is not the panacea to all problems and that practical simplification of any problem or situation is a prerequisite to the introduction of any automation whether it is on the shop floor, or in the office.

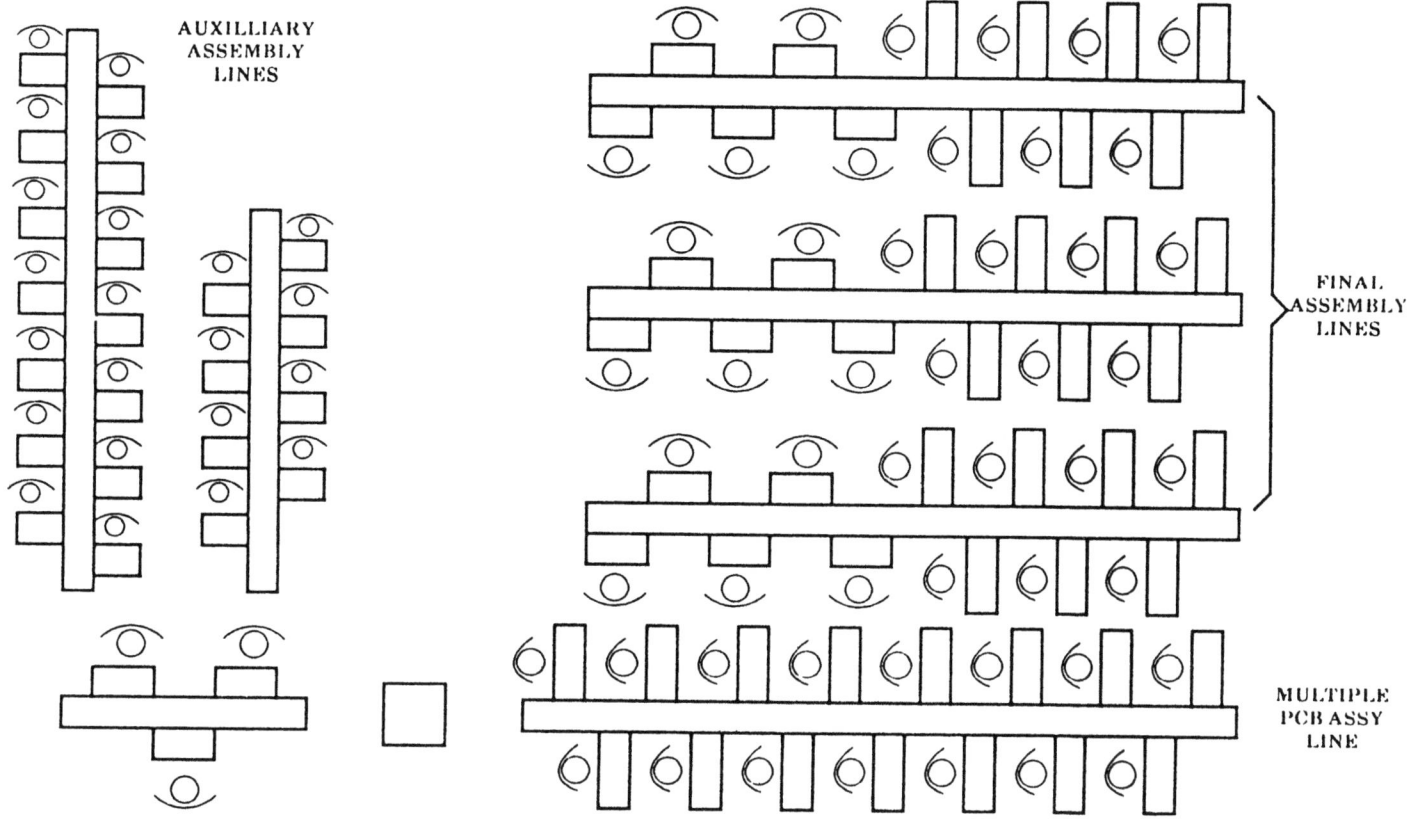

FIG. 1 - GENERAL LAYOUT OF MANUFACTURING FACILITY

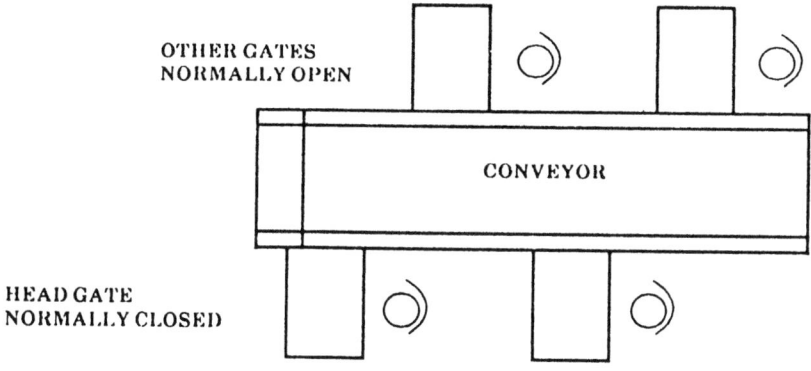

FIG. 2. - LAYOUT OF A FOUR STATION 'PEER' GROUP

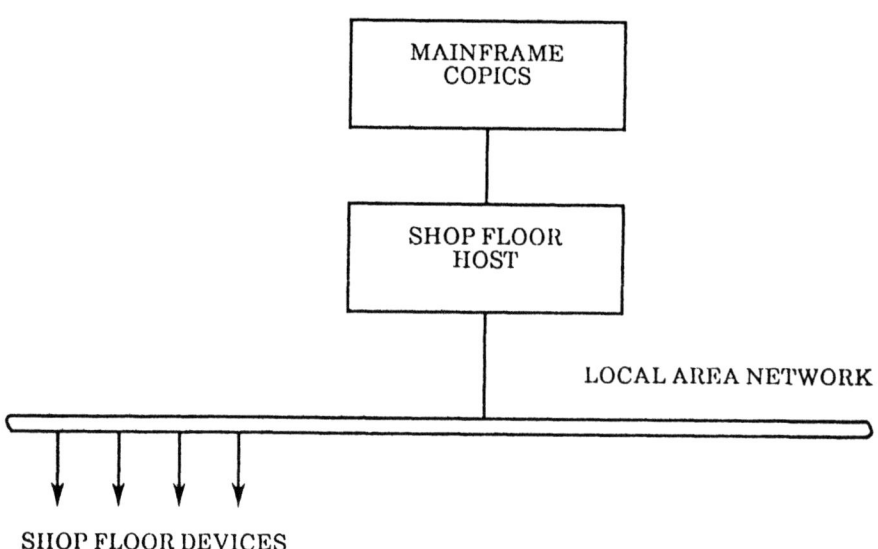

FIG. 3 - THE SIMPLE SOLUTION

FIG 4. - THE COMPLEX SOLUTION

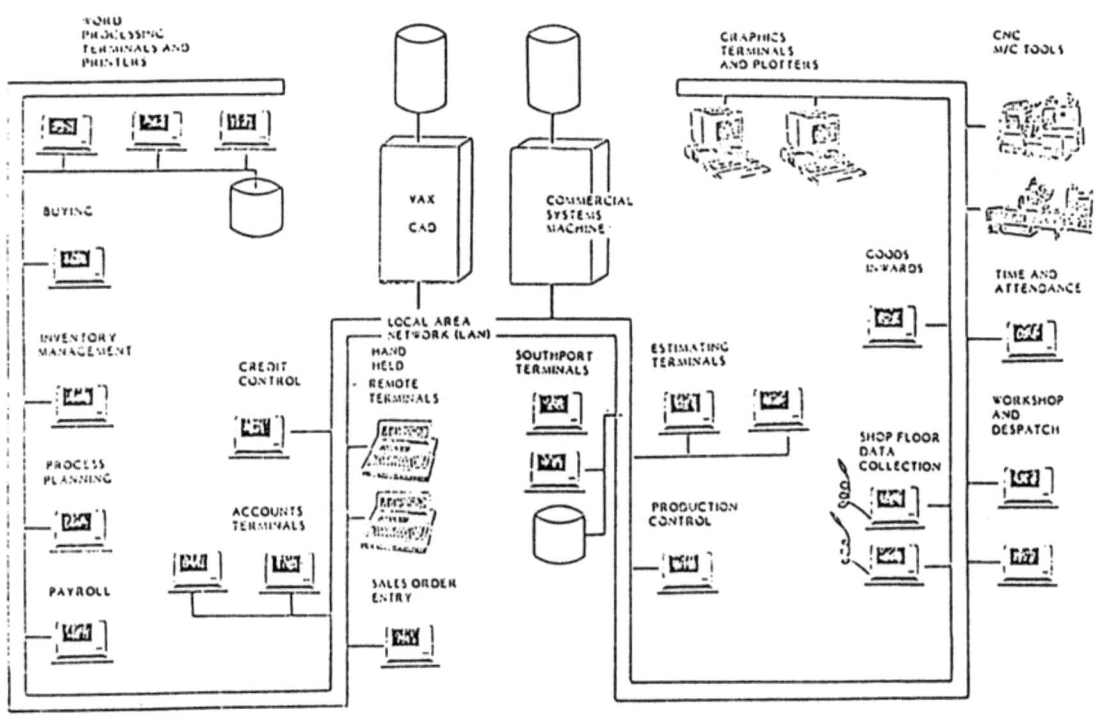

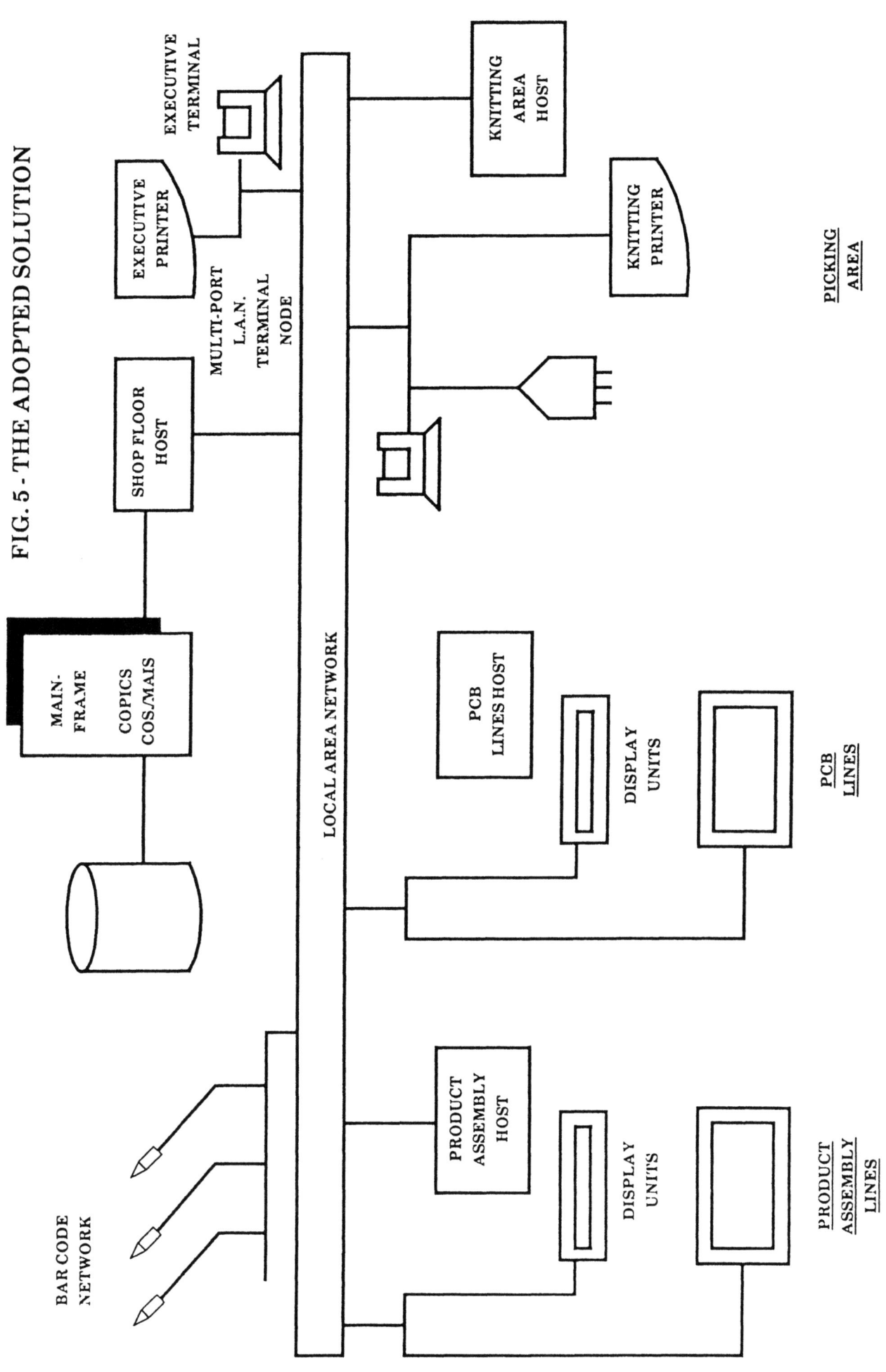

FIG. 5 - THE ADOPTED SOLUTION

The importance of information technology
M. W. Grant
Istel Automation Ltd, UK

THE IMPORTANCE OF INFORMATION TECHNOLOGY

INTRODUCTION

Advanced Manufacturing technology (AMT) is built around the computer, and involves one or more systems such as computer-aided design, robotics, group technology, and automated materials handling. The technology is available now. But the gap between actual results and potential is incredible.

To succeed in an increasingly competitive world, it becomes essential to improve the factors which play a major part in achieving success:-

- Market Share
- Profitability
- Response to Market Demands
- Product Quality

Factory automation is a critical strategic issue, and the automation "Winners" will be those companies that include factory modernisation planning within the strategic planning process. Marginal manufacturing operations cannot be tolerated. It's not the technology, but the way you manage the technology that will make the difference between success and failure. The "create, make and market" functions must be well interfaced and co-ordinated - it is vital to recognise the links between design and manufacture. Quality (reliability) is no longer an option but has to be second to none.

Implementation of AMT is not without pitfalls, some of the most common being inadequate detailed planning and 'piecemeal' introduction of automation. To avoid the resultant problems of decreased productivity, increased complexity of the overall manufacturing process and the consequent handicaps to further expansion, the application of AMT must be within the context of a planned Manufacturing strategy. Since this must inevitably impact on the other strategies within the overall corporate plan, the driving force must come from the top. The successful co-ordination of all individual strategies (R&D, Manufacturing, Finance, Sales and Marketing, Personnel) into a long-term master plan requires strong visionary leadership, backed by dedicated management commitment. A sustained commitment over three to five years by senior management and financial advisors is needed.

The best approach to the successful dovetailing of these strategies is via the Company Systems Strategy. Since this must address the long-term needs for company-wide data, communications and applications systems, it serves as an 'umbrella' under which to polarise the corporate objectives. Adopting AMT within this 'Company-wide Systems Approach', and viewing strategically over say five years hence, will enable the pursuit of a policy of sustained growth. Tangible benefits arising from such a policy can be:-

- Reductions in production and unit costs.

- Reductions in waste, stock levels and WI.

- Improvements in use of manpower and capital equipment.

- Improvements in throughput and quality.

- Improvements in volume and flexibility of manufacture.

- Improvements in operating profits.

AMT solutions are inherently complex and can cover some or all of the following:-

- Flexible Machining Systems (FMS)

- Computer Aided Design, & Draughting, Engineering & Manufacture (CAD, CAE, CAM)

- Automated Warehousing & Materials Handling

- Production & Process Engineering

- Specialised Plant & Equipment, including Robots, CNC Machine Tools and Automated Guided Vehicles (AGV's)

- Communications, including Local Area Networks

- Plant/Process Simulation/Modelling

DATA STRATEGY

The modern business environment is one in which management must be continually aware of what is happening, and in possession of accurate, up-to-date information to enable them to respond to new situations.

The availability of new, cheaper and more powerful technology has provided a means of achieving these objectives; but has also introduced the danger of proliferating large numbers of independent systems which can present a bewildering picture of unstructured, confusing and conflicting information to management if proper controls are not in place.

Such controls must break down functional barriers to allow full integration of all business activities. There must be full and appropriate information flows between all systems and functions, but with firm strategies in place to regulate the way in which these activities can take place.

Thus the key to successful information systems is their integration across the whole business, with appropriate controls of strategic areas such as data, processing and communications.

Most companies have experienced rapid growth in computer systems during the last decade, resulting in systems being developed in isolation from each other. Where an individual system required data for its own purposes, the data was usually created and maintained locally to the system.

Other companies have realised the need to address the problems associated with company data. Typically the situation they have been faced with is characterised by:

- A rapid growth over the last few years in the requirement to store data, with a substantial further increase foreseen in the future with the advent of more major systems.

- The same data being used by many different systems, by different company functions, and in different locations.

- The same (or similar) data being created and maintained by different company functions, resulting in a large number of data files.

- The growth of easily available information processing, resulting in a vast increase in the number of people who have access to information processed by computing facilities, coupled with a growing demand for "desk top" computing such as microcomputers.

- Increasing use of word processing and associated office automation systems.

As a result of this a number of problems were identified:

- Ostensibly people were using the same data but in reality it was inconsistent and inaccurate.

- The proliferation of data and files caused duplicated, and therefore costly, effort in data creation and maintenance activities.

- Security and privacy of data were difficult to enforce.

Overall, there was an unreliable base of corporate data to support management information requirements, and enable effective system integration to take place across the whole company.

By putting in place a data strategy to control the use and administration of data, a number of benefits have been identified and achieved. These include:

- Reduced effort in creating and maintaining data files, and a large reduction in the number of files.

- A major improvement in the accuracy and consistency of company data. Benefits in this area are difficult to quantify, but examples include production consistently to the correct specification, procuring correct parts from suppliers, and using correct costs and prices in accounting areas.

- Effective use of data management techniques (data dictionary) to improve the effectiveness of software development and maintenance activities, and offer opportunities for end user computing flexibility.

- A reduction in computer processing costs by reducing the number of data file maintenance and updating activities.

- Increased system security, and prevention of access to unauthorised data.

- The availability of a firm foundation for future systems developments, ensuring a quick and efficient evaluation and integration of bespoke or packaged systems into the existing company structure.

The ISTEL approach is to define, at the earliest possible stage, strategies in three main areas that will enable the necessary level of integration to be achieved. The strategies I refer to are strategies for <u>data</u>, <u>processing</u>, and <u>communications</u>. Alongside the data strategy complementary IT strategies for processing and communications are required. These are not just strategies for the factory floor but are strategies which apply to the whole company. A processing strategy is needed to define what should be processed where and what form of technology, be it micro, mini or mainframe computing; a communications strategy to enable the level of integration to physically take place.

INTEGRATION

True CIM requires the ability to integrate vertically in both directions, between the various levels from the planning level right down to the operational level, however and whenever you wish. An example of this might be the feed from a design system at the planning level through to the actual manufacture down at the shop floor level.

There is also increasingly a need for horizontal integration. This is the integration of all the operating activities to directly link one activity or operation with another. To achieve both vertical and horizontal integration requires the right information technology strategies to be adopted during that important planning stage.

The problem with integration however is not just simply being able to have strategies in the areas of data, processing and communications. Applications strategies must be overlaid on these IT strategies to form an application framework. This framework will consist of applications of manageable size that will link together. The glue that will link these applications together will be the communications network.

The development of flexible automation, particularly in areas of high complexity, has intensified the need for truly integrated CIM systems. The ability to generate, in digital format, geometrical, structural, performance and time related data for parts and assemblies; to automatically develop comprehensive manufacturing instructions; and to automatically control manufacturing operations is a necessity rather than a luxury for the manufacture of complex assemblies in a competitive world market.

Integration, with regard to manufacturing technologies can be defined as the characteristic whereby complex systems inherently interact with one another. Over the past decade, however, these complex systems have evolved as more or less discrete entities. Those systems can be functionally categorised as computer-aided design (CAD), computer-aided manufacturing (CAM), computer-aided production engineering (CAPE), computer-aided production planning (CAPP) and computer-aided storage and transportation (CAST).

Certain other enabling technologies, such as computing technology, database management, group technology concepts and communications network architectures, serve as potential catalysts or facilitators for integrating the discrete CAD, CAM, CAPE, CAPP and CAST functions. Incorporating these facilitators to achieve totally integrated system requires the application of a "systems approach" towards planning, development and implementation. The systems approach involves phased implementation and applying top down design with bottom up implementation.

How many companies consider simulating the business plan objectives from an operational viewpoint to see whether or not they can be achieved over the plan period? Simulation is a very powerful means of checking and testing the plans and investments of the future before funds are committed and those plans are put in place. This technique has been used extensively in Jaguar Cars where all new major facility upgrades or introductions are simulated before financial approval is given.

In support of the Manufacturing Strategy at Jaguar we have tested the company business plan using simulation to validate firstly whether the business plan can be achieved and secondly how best it can be achieved. This has been done by creating a model of Jaguar in the 1990's and defining the operating principles that need to be in place to achieve the required output, improvements in productivity, reducing product lead time, improvements in response to customer demands and improvement in inventory turn.

A great deal of information and knowledge has been obtained from the exercise as well as a clear understanding of the functional strategies that are necessary to support the plan, the facilities planning guidelines necessary to assist the Facility Engineer and the systems projects that are necessary to support the chosen direction.

This work has provided the basis for business requirements integration and is fundamental to the co-ordinated and coherent approach to planning. As plans change and they inevitably will, the change is reflected in the simulation model so that the effect of the change can be understood by the parties involved not only the direct result of the change but also the overall effect on the company as a whole. From this work a systems framework has been developed and has been prioritised at a project level.

A CIM approach requires a good deal of management vision and commitment and, it has to be said, a certain amount of faith. The benefits of CIM stem from the integration of all of the business functions, so that they operate effectively in accordance with the business goals and allow each part of the business to do a better job.

CIM enhances, rather than replaces, human capabilities and the synergy of the benefits of the whole, are greater than the sum of its parts. CIM provides the synergy to beat competition head on. It enables a company to maintain a viable business in the future by considering the whole enterprise as a system and not as a conglomeration of different functions.

IT SKILLS

To determine what skills are required, one has to look at the elements of an AMT project. It will start with consultancy in the areas of manufacturing, engineering and systems engineering. It will require specialist project management abilities that will involve the various suppliers of robots, plant and equipment, systems engineering skills, production and process engineering skills, a whole raft of different specialisations.

In this way, we believe successful AMT strategies can be developed as part of the overall company strategy for integration. It must be evident that the manufacturing engineering and systems engineering skills and disciplines must merge to achieve effective AMT implementations. I wonder how many of the so-called islands of automation are the product of the manufacturing engineers working in isolation from the systems engineers. One thing is for sure, if islands of automation continue to be the way forward then the unmanned factory is a long long way off.

There is no ignoring Information Technology and its ability to affect all aspects of business and industrial life. By using it effectively it can be used as a competitive weapon that can make an ordinary business become more profitable. The airlines and the banks use it as a fundamental and critical aspect of their day to day operation.

Manufacturing must use Information Technology not merely as a supportive mechanism but as competition increases much more as a strategic weapon.

A new concept in the control of manufacturing systems
P. Anstiss
BAeCAM British Aerospace plc, UK

British Aerospace have over the past ten years evolved a corporate strategy for the implementation of Advanced Manufacturing Technology. As a result of this strategy a new concept in the control of Manufacturing Systems has been established by British Aerospace. This paper outlines the strategy established within B.Ae and the factors involved in its implementation along with a description of this novel concept which is now being used by British Aerospace to provide control systems for its new manufacturing facilities.

A NEW CONCEPT IN THE CONTROL OF MANUFACTURING SYSTEMS

Introduction

The three sites of the Military Aircraft Division around Preston, Lancashire have a long and distinguished tradition of Military aircraft manufacture. Currently they are involved in the production and support of Tornado, Jaguar, Lightning and Canberra. New projects include major sub assemblies for the Airbus A320 civil aircraft and a technology demonstrator aircraft known as EAP - a forerunner to the European Fighter Aircraft.

Although each of these projects are large scale, multi million pound ventures, production rates by most manufacturing standards are low. Tornado, for example, is the largest European Military Aircraft programme for many years, yet only about 1000 aircraft are likely to be built in its first 15 years shared between 3 parent countries. As a result of this and the complexity of the airframe structure, British Aerospace works in a small batch environment making thousands of different parts at a production rate of less than 10 a month rather than a range of say 10 produced in thousands - a production environment more typical of the car industry for example.

The problems caused by low volume and small batch production are coupled with other complications such as:

o product complexity
o multiple variants
o frequent modifications
o sophisticated learning curves
o long makespans
o high work in progress costs

To ease these problems B.Ae has sought more cost effective production methods by introducing a wide range of Advanced Manufacturing Technologies. These have been made possible by the widespread availability of increasingly efficient computer systems.

A corporate plan was evolved several years ago to minimise duplication of development effort and expenditure and to co-ordinate investment in such a large organisation. This plan apportions lead and support responsibilities to sites of the company and has resulted in a balanced distribution of advanced production facilities.

This paper considers just one aspect of these technologies, the development and control of a Flexible Manufacturing System for Small Machined Parts, This system, capable of producing over 900 component types, includes the following advanced features:

o   Automated stores, transport and machine tools
o   Local Area Network Communications

o   Advanced computer control system for all automatic and manual functions
o   Comprehensive tool storage, handling and preparation facilities

Before describing the system some of the steps which were taken moving from an engineers concept (Figure 1) to a cost effective highly automated production system (Figure 2) are discussed and followed by a review of experience gained with earlier systems.

Feasibility

The two fundamental considerations at the very beginning of the project were:

o   Is it technically feasible?
o   Is it commercially viable and can we afford it?

Less obvious but nevertheless important questions were:

o   Do we have the appropriate skilled resource for successful implementation?

o   Can we cope with the step in technology?

Perhaps not surprisingly, the technical questions tended to receive the most attention although in practise they were often the ones most easily resolved. A combination of advice from in-house experts, suppliers or consultants, coupled with the availability of Government grants for feasibility studies can readily confirm the technical soundness of a project.

Far less straightforward than the technical questions was the commercial justification of a large factory automation project in particular, calculating the project's viability and adjusting to payback periods which were invariably much longer than the 2 to 3 year guidelines traditionally set.

The costs and savings categories in a typical automated machining facility are outlined in Table 1. The items do not equate with each other and of course the cost or savings are spread over individual timescales. Calculations of the overall cost and saving situations is correspondingly complicated.

Other benefits which are significant but even harder to quantify include increased sales from faster response times, better component quality and an improved image arising from a visible investment in the company's future.

Costs

o	Capital	(e.g. hardware & software)
o	Project Management	(perhaps several people for several years)
o	Development	(eg. non proprietary equipment, new processes etc)
o	Disruption	(eg. lost production during transition period)
o	Increased operating costs	(eg. depreciation, maintenance, etc)

Savings

o	Manpower	(Direct and Indirect)
o	Material	(Improved utilisation, reduced scrap/rework)
o	Once off savings in work, material and stock in progress	(from reduced production times)
o	Finance costs	(from reduced working capital)
o	Maintenance	(replacement of existing plant, not forgetting the sale of existing equipment if appropriate)

Table 1 Costs and Savings

Once these and other commercial factors were categorised careful presentation of the key facts was the next important activity. Of course the principal key facts were cash flow and money. As the project was defined the technique used varied from a simple payback which traded off savings of assessment, through discounted payback to internal rate of return (I.R.R.).

By recognising that money has a value which reduces with time and applying a factor to both savings and costs which represents this, (say 10% reduction per annum) then a discounted payback may be calculated.

IRR calculates the exact rate of return which the project is expected to achieve ie. the rate at which the net present value is zero. This is a measure of profitability at todays conditions and takes into account the timing of the returns on investment. If the rate of return is greater than company targets then the project may be considered commercially acceptable.

To add credibility and increase confidence levels in commercial justifications, consider optimistic, pessimistic and realistic values of key factors such as costs, future work load and system availability.

Figure 3 shows the effect of two expenditure levels (eg. with or without a 20% grant) compared with two savings levels to indicate the effect on payback in a large scale FMS.

When the engineers' concept had been demonstrated to be technically feasible and commercially viable and the financial resources were available, what are the other requirements were there for successful implementation?

Consider some of the different skills required:-

o   Project Management
o   Production, mechanical, electrical and control system engineering
o   Specification writing
o   Plant appraisal and maintenance
o   NC programming
o   Industrial relations diplomacy
o   Layout and method study
o   Cost monitoring
o   Progress reporting

The list of skills is extensive. Typically many of these responsibilities, a budget and some timescales were given to the production development department or were vested in an outside supplier such as a machine tool manufacturer.

Within the Warton unit of British Aerospace, the importance of a co-located multidisciplined team of project managers, engineers, accountants and system analysts/programmers was recognised early. In close conjunction with specialists from other departments such teams were created and given prime responsibility for the in-house co-ordination of all major factory automation projects.

The technology step is another important factor that was considered in preparing for the FMS project. Todays integrated manufacturing systems represent a very significant leap forward in production technology. Although the individual elements such as plant, the computer control system with its links to other emerging company systems, its inherent decision making and management skills, are far removed from the labour intensive, paperwork based control systems we were familiar with. It is in this area in particular therefore that the technology step was most noticeable. (Figure 4)

Fortunately, within British Aerospace, the steps towards todays technology were fairly progressive and the Small Machined Parts FMS implementation was made easier as a result.

Background (Previous Steps)

In the early seventies a Molins Systems 24 machining cell was introduced for small aluminium alloy part production. This was effectively an FMS conceived in the mid sixties incorporated:

o   Semi automatic workpiece and cutter handling
o   Magnetic tape control
o   Multi axis, multi spindle machining

In its fully developed form it could also embrace protective treatment facilities. It introduced new concepts such as component "nesting" in billets (a concept introduced by British Aerospace) is still in use 15 years after its introduction although reliability and spares availability is now very poor.

This experience was developed further in the late seventies when ten Mitsui Seiki machining centres were purchase for the production of small steel and titanium parts.

The main features of these machines are:

o   Automated tool changing from cutter carousel
o   Workpiece buffering on automatically indexing cubes
o   Computer Numerical Control CNC

They were further enhanced by British Aerospace to incorporate adaptive control, tool and workpiece probing and support facilities for minimally manned operations.

In parallel to these developments, a large scale Direct Numerical Control (DNC) system was introduced to large multi axis, multi spindle machine tools. These ranged from old devices incorporating ageing analogue controllers to those with the latest digital CNC. This system was based on DEC PDP-11 mini computers and has been progressively extended to embrace over 40 machine tools with an interface to the factory mainframe.

In 1984 the next major advance in DNC was proven when an Ethernet Local Area Network was successfully demonstrated in a "noisy" industrial environment. This provided DNC communications between a VAX computer and the Fanuc Control Units of the 10 Mitsui Seiki machining centres.

One of the many benefits of the LAN technology is to simplify the communications hardware required for data distribution when compared with the conventional alternative of individual serial links.

Providing these links to each of the devices suitable for DNC in one large machine shop would create a picture such as Figure 5. This may be compared with a representation of an equivalent L.A.N. installation shown in Figure 6.

The LAN may be likened to a domestic ring main with appliances attached around the house via sockets.

This background demonstrates the considerable experience base established in the company in both productive and control system,s technology before undertaking the large scale FMS implementation in house.

The Flexible Manufacturing Systems

The overall concept of the Flexible Manufacturing System was to integrate all primary and secondary machining operations, quality assurance stages, stores transport and tool/material management functions in a common area using the highest degree of automation that was cost effective.

This is best described by reference to figure 7 which shows the optimised layout with the various elements marked.

However, the FMS did not come into existence as the overall concept, it grew by various phases.

The initial phase known as the Automax Day 1 Cell, was to prove the integration of two new Automax machine tools (produced by Marwin Production Machines to British Aerospace) requirements for a twin spindle light alloy machining centre with automatic tool and workpiece loading), a two vehicle automated guided vehicle AGV system and vertical paternoster stores. These are shown in figures 8 and 9 and one also shown in Figure 7.

In these stores, prepared billets, pallet mounted workpieces with their associated cutting tools and a local buffer of cutters are stored. These are dispensed from the lower access part of the stores and delivered to the machine tools by the AGV system which also returns used cutters and delivers machined parts to the co-ordinate measuring machines. Use has been made of several simulation packages to help develop the cell configuration and to understand the stores loading requirements, for example.

A survey of proprietary cell control systems in 1982 indicated that a satisfactory package did not exist which was capable of meeting the ultimate requirements of the FMS mentioned at the beginning of this section. The company therefore took the decision to develop a generalised control system capable of satisfying not only the requirements of this project but also, with minimal modification, those of other emerging projects. This was made possible by building on the DNC experience already acquired and the extensive knowledge of digital computing obtained from Aircraft projects such as the Active Control Technology Jaguar programme or "Fly-by-Wire" aircraft as it is more commonly known.

The result of this work was proven in late 1984 and early 1985 during the commissioning and early production phases of the Day 1 Automax Cell. This interim control system permitted unmanned operation of the Automax Machine Tools and AGV's (on the ground floor of the two storey cell) with the kitting of cutters, assembly of billets onto pallets, and loading of stores (a manual function initially) being performed on the mezzanine floor above. Once the Automax Cell concept had been proven, work proceeded in a number of areas to further develop the FMS. Six machines had been specified to increase the aluminium part production capacity. The well established Mitsui Seiki machines were integrated into the system with automated cutter handling and improved workpiece loading facilities being provided. Co-ordinate measuring machines were installed for inspection purposes and the AGV system encompassed four vehicles.

A major development is the introduction in late 1986 of a fully automated billet preparation facility which stores, machines, drills and taps and then delivers prepared billets to the mezzanine floor, where a combination of robots and operators perform the billet to pallet assembly operations.

Robots will also undertake cutter kitting and stores loading functions making the total production cycle from raw material input to inspected parts with all primary machining operations completed, a highly automated sequence of events. Secondary operations, part marking, deburring etc. continue to be carried out conventionally but under the control of the advanced computer system.

Another important element of the project was the provision of a Bulk Cutter Store and Cutter Preparation Area which is located over a hundred metres from the FMS and serviced by an AGV link (Figure 10). This area incorporates four, eight axis, CNC cutter grinders, a large twin aisle store capable of accommodating 85000 cutters and a comprehensive computer tracking and management system based on bar code labelling and light pens. Integration of this to the cell tool life monitoring and other management systems is underway.

Control and Communication

Clearly, a very advanced and reliable control system is required to achieve the successful integration of the FMS elements described. In addition, it must be capable of communications with existing factory mainframe based systems for the supply of scheduling and part programme information for example as well as for the return of management feedback data for subsequent analysis.

The interim architecture of the Day 1 Automax Cell was based on the classical hierarchial structure with each processor performing specific functions (Figure 11).

This configuration, which utilises serial line communication links, provided an easy implementation for a set of specific requirements. It did however impose limitations on expansion, re-configuration, reliability and redundancy of the system.

The limitations were recognised when the control system architecture for the total FMS was considered, and a different concept in control system design was established. This concept provides a flexible, open ended yet very robust design capable of a accommodating a wide range of manufacturing requirements.

From a functional stand point this concept is formulated via a number of application modules. The definition of these modules such as DNC data supply, transport management, tool management and scheduling has been achieved by considering the requirements of the total small machined parts FMS plus a wide range of other manufacturing processes ranging from labour intensive to highly automated facilities. These application modules co-operate with one another to provide the total functionality required of the control system and they can be variably configured to satisfy a broad range of particular requirements.

In order to construct a total system from a series of such modules it is necessary that each module is capable of communicating with each other. In addition, to insulate each module from change in others it is important that this communication mechanism is standardised across all modules.

The control system design established for the FMS uses an inter module communication architecture known as BCA (British Aerospace Control Architecture) (figure 12).

The modules are buffered from each other by this communication environment, the module function being clearly separated from the communication media. From the logical viewpoint all modules are peers; that is they all share the same view of the communication media and hence of each other. This opens the door to a truly distributed system where individual modules need no longer reside on the same processor but can be distributed throughout a control system provided adequate communication links are used.

It is in this area that the use of a Local Area Network becomes a powerful and effective medium for the communication links between the co-ordinating processors and the equipment on the shop floor. The application of this technology is new within the manufacturing environment and it provides the perfect media for employing the application architecture described above. It allows the distribution of functionality to achieve an inherent reliability which is not possible from a hierarchical system structure.

This can be illustrated by viewing the hierarchical control system design originally considered for the Small Machined Parts FMS (Figure 11) in comparison to the rationalised architecture which a LAN based control system allows (Figure 13).

Further Developments

Recognising the commercial potential of such a generalised and modular control system (which may be configured to satisfy a range of control requirements from basic DNC through to sophisticated FMS) British Aerospace joined forces with GEC in 1984 to further develop, manufacture and market this technology for a range of industries.

Experience from other projects including PCB assembly and Sheet Metal FMS facilities is being incorporated as the product development programme continues, enabling widely different requirements to be satisfied by a proven product.

LAN and Serial line based DNC systems have already been sold for use in the production environment, and commissioning of the first full application of the FMS product will take place in the final quarter of 1986.

Turnkey packages are now being offered by a new organisation called BAeCAM from initial design studies through to the installation, commissioning and production handover phases, with coordination provided through both new and established project management and support organisations.

Conclusion

For British Aerospace, a major AMT investment programme of over 250 million pounds is already paying dividends and is helping to secure aircraft projects to take the company profitably into the next century.

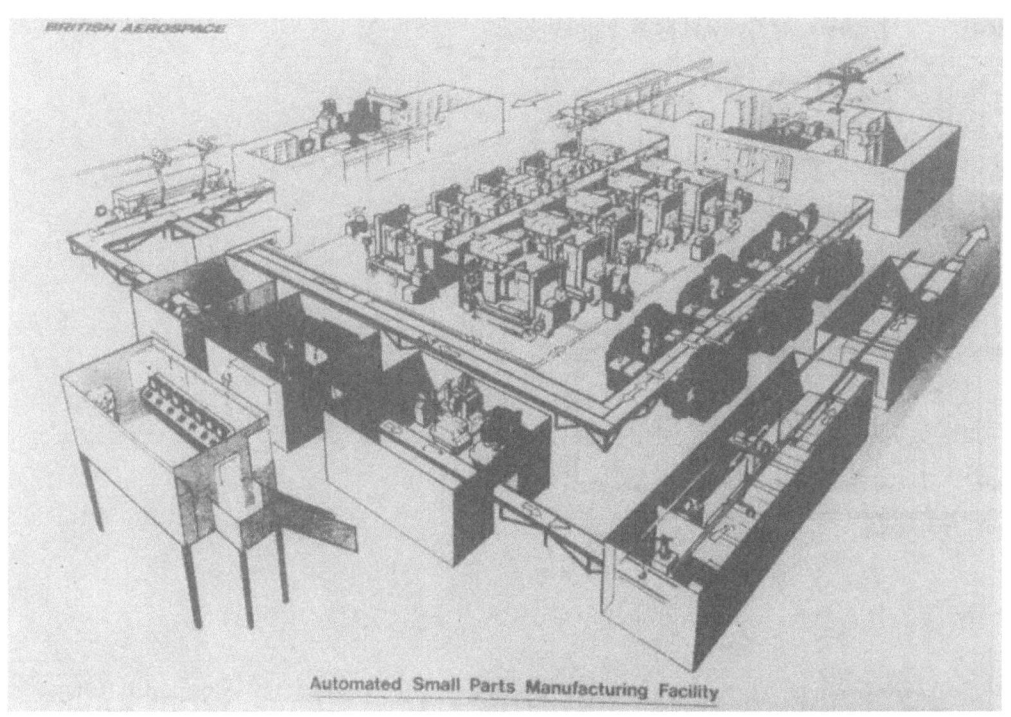

Automated Small Parts Manufacturing Facility

BRITISH AEROSPACE Advanced Manufacturing Technology

Alternative Paybacks

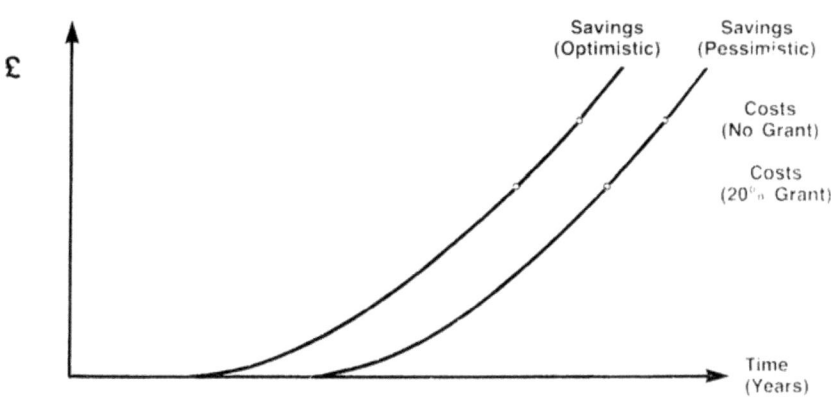

Evolution of Manufacturing Control   Overview of Control Environment

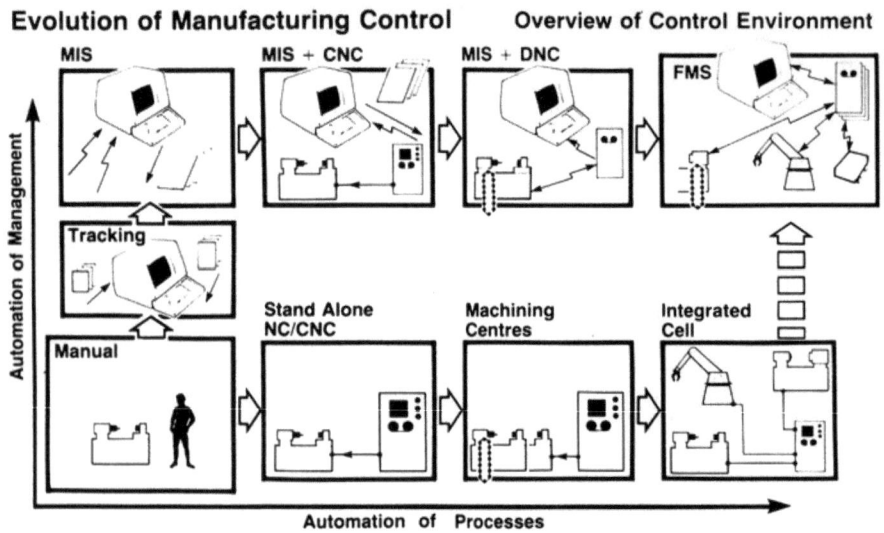

DNC Links Using Serial Lines

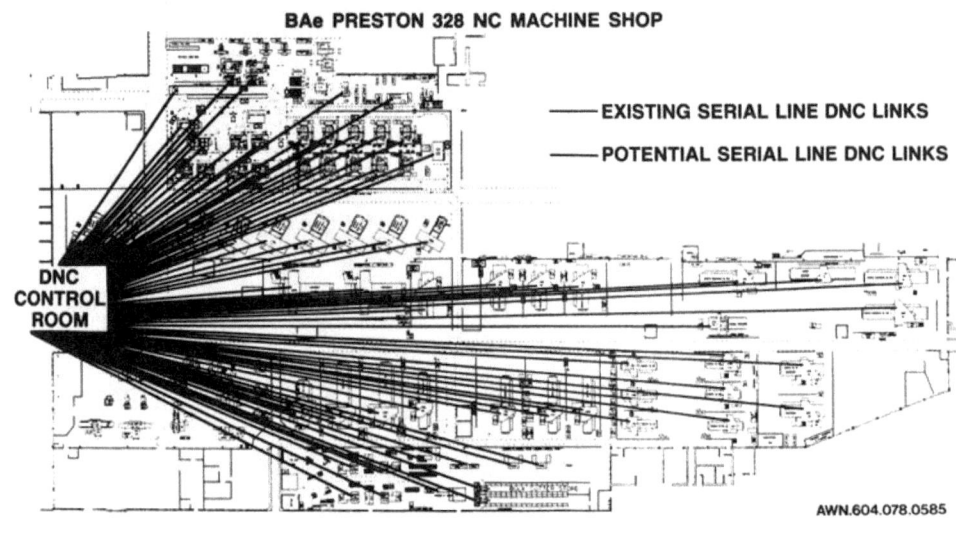

BAe PRESTON 328 NC MACHINE SHOP

— EXISTING SERIAL LINE DNC LINKS
— POTENTIAL SERIAL LINE DNC LINKS

AWN.604.078.0585

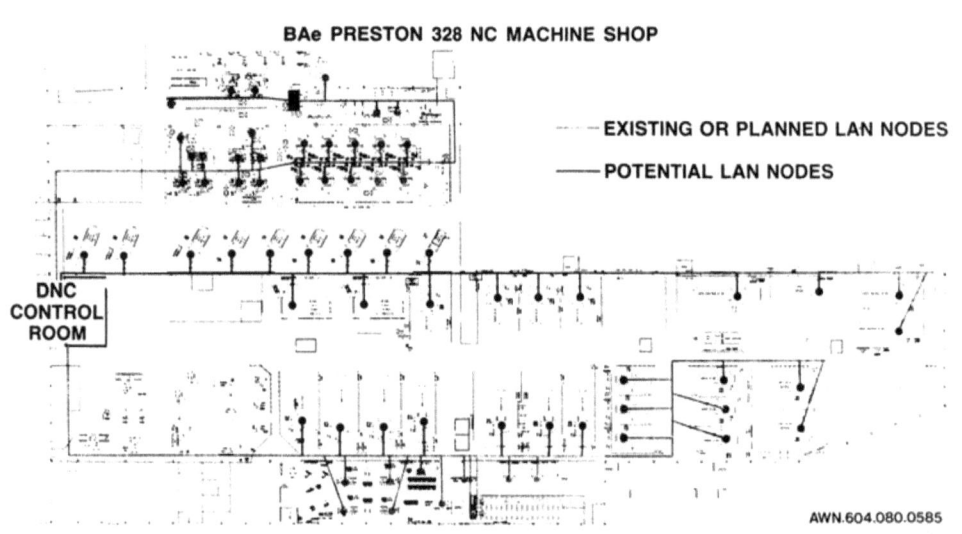

BAe PRESTON 328 NC MACHINE SHOP

--- EXISTING OR PLANNED LAN NODES
— POTENTIAL LAN NODES

AWN.604.080.0585

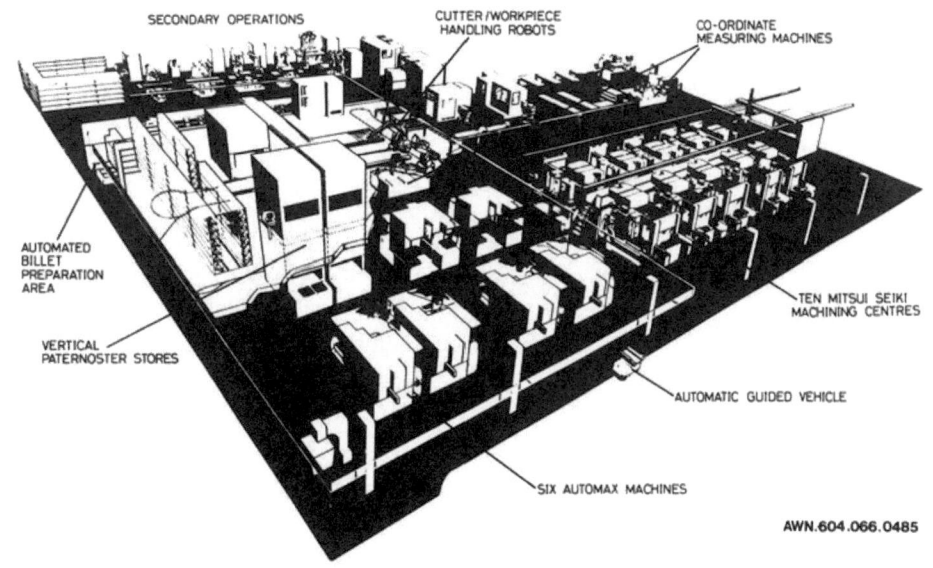

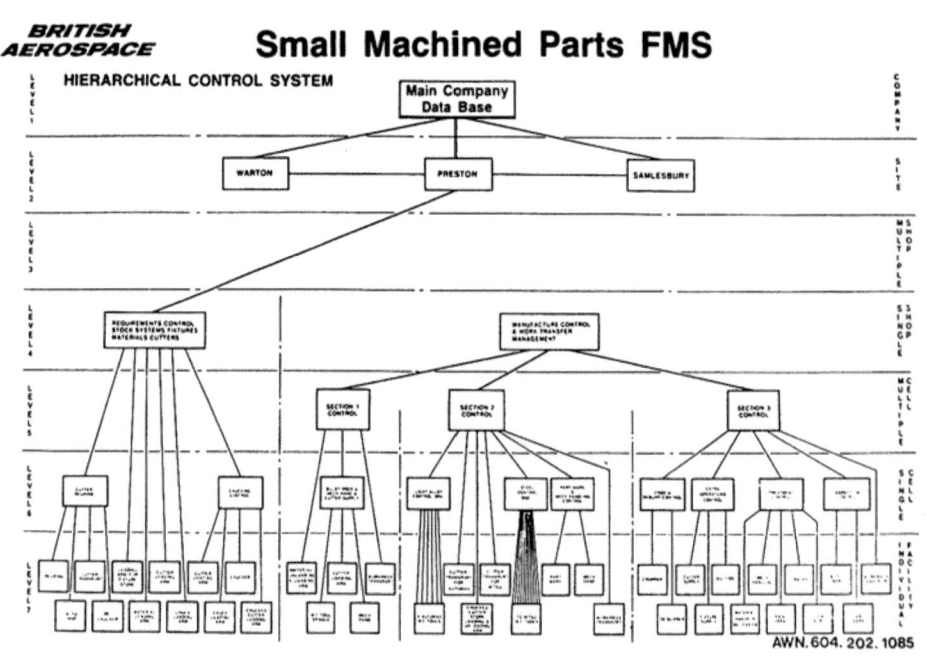

BRITISH AEROSPACE — Cutter Preparation Facility

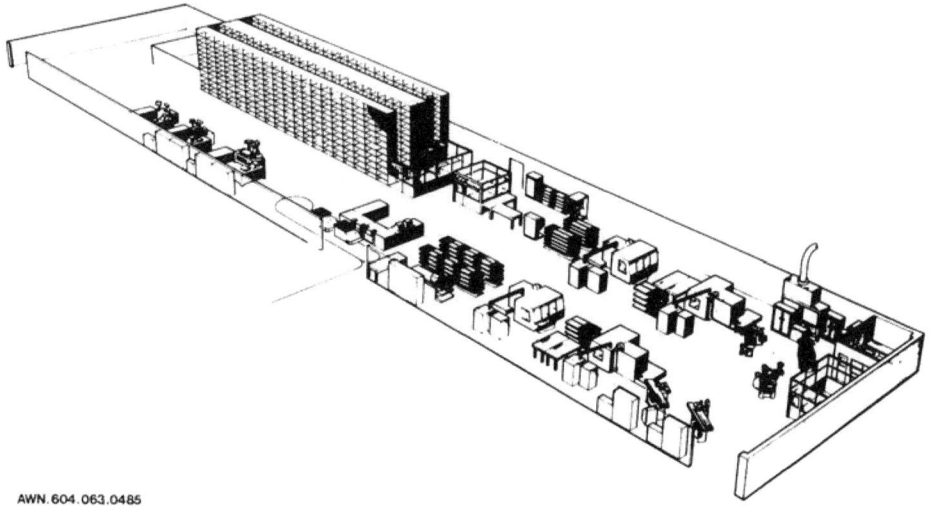

AWN.604.063.0485

BRITISH AEROSPACE

Logical View of Inter-Module Communication

Functional Aspects – Solution

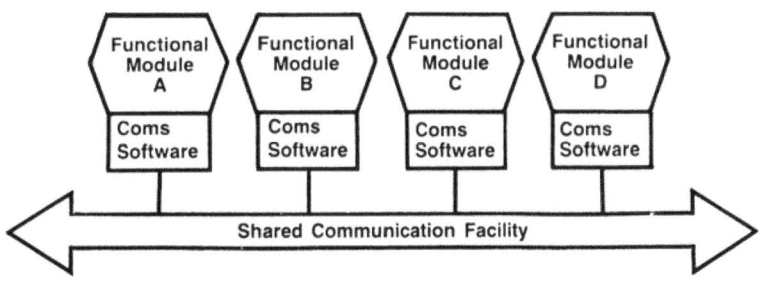

AWN 604 013 0285

BRITISH AEROSPACE — Small Machined Parts FMS

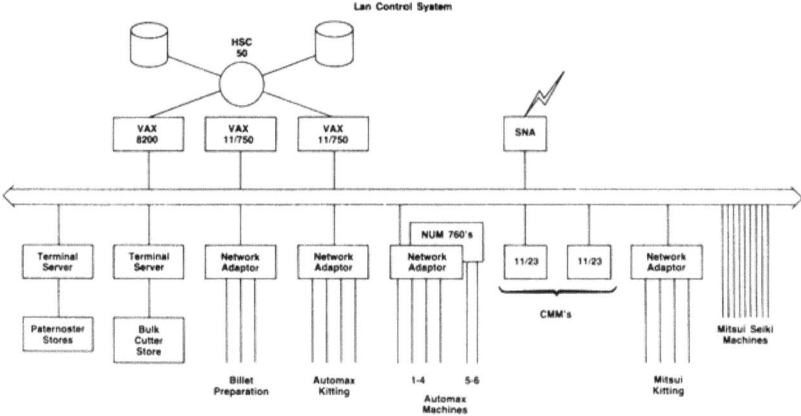

AWN.604.203.0486

CIM-OSA, a computer integrated manufacturing based on the open system approach

J. Huysentruyt
Cap Gemini Belgium, Belgium

ABSTRACT

The ESPRIT 5.1 P688 is aiming at the definition, the development and the validation of a CIM (Computer Integrated Manufacturing) architecture.

This project is carried out by a Consortium composed of 19 European companies and organisations.

The architecture will be open in order to cope with heterogeneous systems and to allow the evolutive integration of multi-vendor system components. Current research is leading to the definition of an Enterprise Reference Model, an Implementation Reference Model and a mapping mechanism which will be based on generic information structures.

FOREWORD

This paper intends to provide an overview of the ESPRIT 5.1 P688 project, which aims to define a CIM (Computer Integrated Manufacturing) Architecture based on the Open System Approach. This project is carried out by a Consortium of major European Companies regrouping aerospace, automotive and electronic industries, as well as major Information Technology suppliers (hardware and software), consulting companies and research laboratories.

The objective is to define a Reference Architecture which will allow adequate use of information technologies for satisfying user needs whilst ensuring the possibility of evolutive integration of systems in such a way that organisations can actually plan and control the development and further evolution of systems. This will lead to higher flexibility, lower cost and higher fitness, and hence, will support the competitiveness of enterprises in a turbulent environment.

The paper is based on the results gained so far and on current directions of research.

INTRODUCTION

Since the beginning of the 70's, individual CAD (Computer Aided Design) and CAM (Computer Aided Manufacturing) systems have been developed to satisfy the needs of particular areas of manufacturing industry. In many cases, these developments occurred to take advantage of particular breakthroughs in technology, such as developments in machine controls or interactive graphics workstations.

Because of their very nature and the fact that they were developed in response to the specific needs, these systems grew as 'islands of automation' and difficulties have since then been encountered in trying to link them together.

More recently, the advances in computing, particularly with 'computers on a chip' have meant that almost all areas of business are capable of computer assistance.

Fortunately, the techniques of analysing systems requirements have been developed to the point where a company's overall activities can now be broken down into a set of consistent but semi-independent businesss functions, according to a top-down approach. Similarly, the computing functions and communications methods have been analysed down to defined 'building-blocks'. This allows the development and implementation of systems to occur within one compatible master-plan. It also allows flexibility of implementation and subsequent enhancement of those systems.

The CIM concept

Although no unique definition exists for CIM, it is regarded here as a strategic organisational concept for manufacturing industries where maximum use is made of information, supported by a cost effective implementation of the possibilities of IT (Information Technologies).

Scope for an architecture

The major objective for this project can be formulated as the definition of an overall framework, i.e. models, rules, guidelines, to build CIM-system(s), guaranteeing through the use of standards that partial solutions and applications can be integrated into multi-vendor systems. Investment protection for the user companies is clearly one of the basic concerns.

However, in order to design this architecture, it is required to define its scope. This can be considered at different levels: first the manufacturing enterprise using one or more CIM-systems; second, the CIM-systems themselves. In particular, when considering the discrete manufacturing enterprise, it is clear that not only future enterprises or enterprise structures have to be described but also the existing ones, in order to allow progressive evolution from one situation to another.

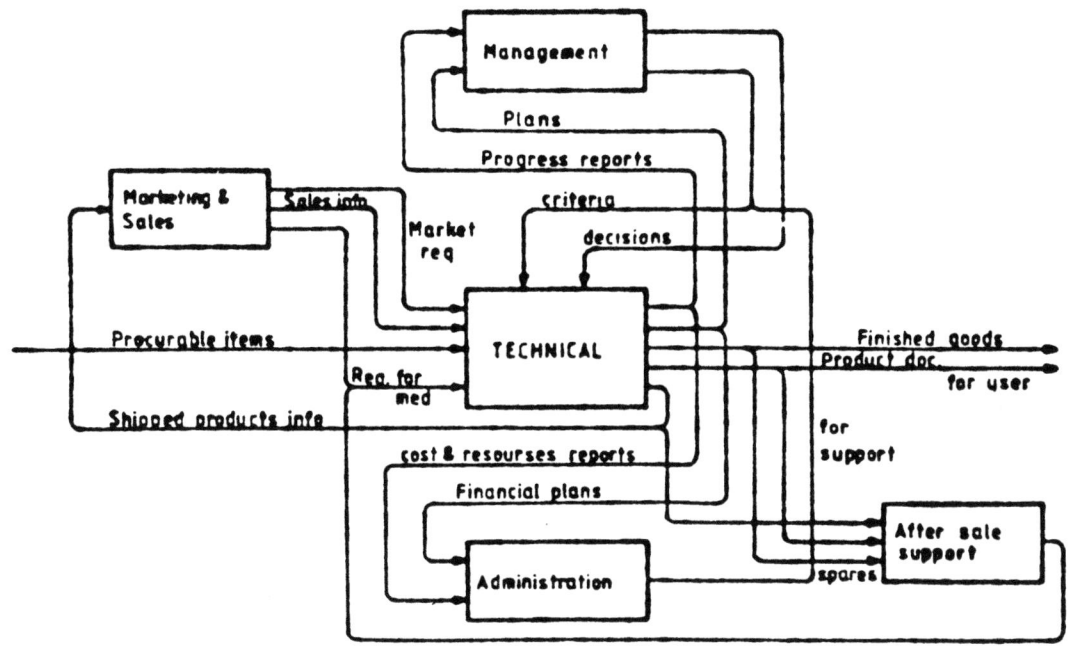

Fig.1: Enterprise functions in CIM

As in most cases, complete CIM-systems are too complex to be designed and built according to a pure top-down approach, it is mandatory to ensure bottom-up integration, by preference in an open-ended manner while partial solutions or sub-systems are designed in a top-down way.

Integration

Integration (see Computer _Integrated_ Manufacturing) relates to communication between system components so that the entities integrated in a system of higher order behave like one system. Communication happens not only at the lexical (form) level but also at the semantical level (content, meaning, for the applications involved) and even at the pragmatic level (in due time) so that control and coordination of action can really occur.

Therefore, when the acronym CIM is mentioned one should understand communication in the broades sense. The same applies for 'computer' which stands for 'Information Technologies'.

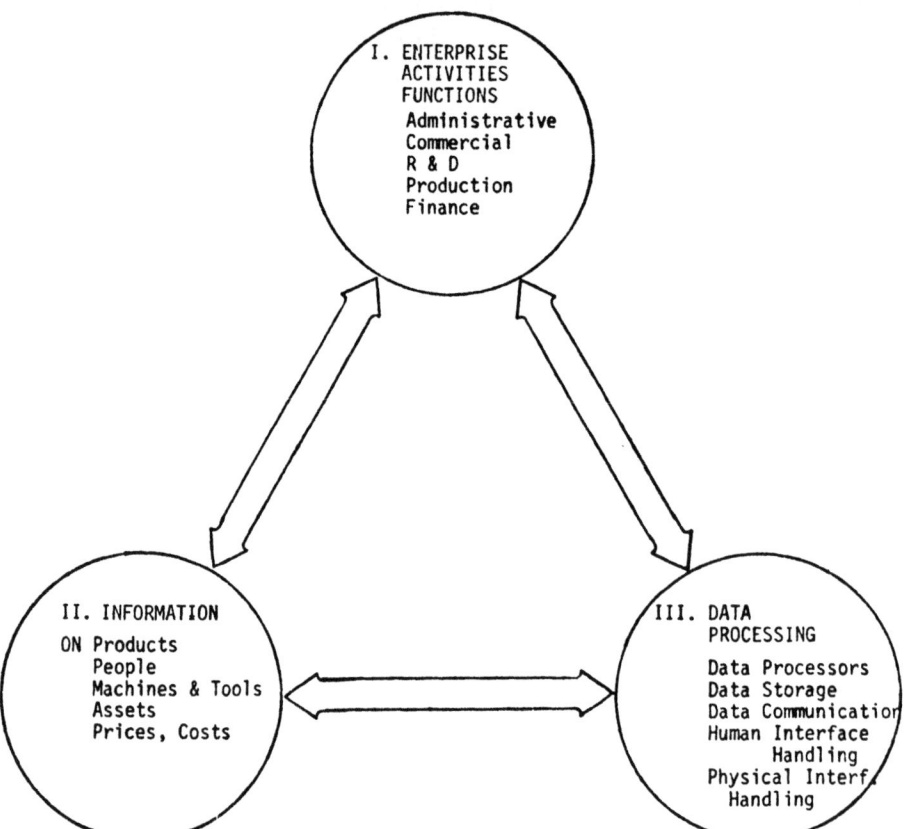

Fig.2: Aspects of integration

Architecture

The idea of an architecture covers two essential and complementary aspects, one being 'structural' i.e. the 'structure of what is built' or the common structural characteristics of a set of products or systems.

The other aspect is architecture as the 'art of building'; this means a set of methods, tools, guidelines, formalised expertise.

Both aspects are to be consistent such that the structural dimension provides the framework to the system to be built while the 'methodical' dimension generates the framework for the project producing this system.

THE PROJECT

The ESPRIT programme

The project 688 pertains to the well-known ESPRIT (European Strategic Programme for Research and Development in Information Technologies) programme.
Five areas heve been defined: Micro-electronics, Software Engineering, Advanced Information Processing, Office Automation and Computer Integrated Manufacturing.
The project belongs to subarea 5.1, addressing system architectures for CIM.

Project start and status

After an initial definition study carried out between November 15th, 1984 and March 29th, 1985, the project was restarted on October 1st, 1985. It aims to define CIM-OSA i.e. an Open Systems Architecture for Computer Integrated Manufacturing and is being carried out by the AMICE Consortium constituted by 19 major companies: CAP GEMINI SOGETI, Prime Contractor and AEG AKTIENGESELLSCHAFT, AEROSPATIALE, ALCATEL, AT&T EN PHILIPS TELE-COMMUNICATIEBEDRIJVEN, BRITISH AEROSPACE, BULL, COMPUTER RESOURCES IN-

TERNATIONAL A/S, DIGITAL EQUIPMENT G.m.b.H., DORNIER, GEC, IBM DEUTSCHLAND G.m.b.H., ICL, ITALSIEL, PHILIPS AND MBLE ASSOCIATED, SELENIA-AUTOTROL, SIEMENS, VOLKSWAGEN AG, WZL-AACHEN UNIVERSITY.

The project plan

The operational objective of the project is to publish the first version of the Architecture 40 months after project start.
As stated above, the project activities resumed on October 1st with a short Consolidation Phase carried out by a kernel team, in order to consolidate more thoroughly the results gained during the previous Definition Phase and to allow a gradual build-up of the project teams whilst maintaining a common conceptual framework.

Thereafter, the Specifications Phase was started and is currently under way with a certain number of work areas having been defined, such as:-
- the definition of CIM systems and CIM-OSA user profiles and corresponding requirements
- the definition of general properties and functional specifications of the CIM target systems
- the development of an initial draft architecture which is the starting point of future iterative development
- the investigation of scenario's, provided by partner companies, which will serve to illustrate and verify architectural properties, and support the formulation of further requirements
- the investigation of relevant State of the Art information.

In addition to the definition of the model of the architecture itself, four domains have been recognised to contribute to the architecture: they provide typical requirements and validation criteria:

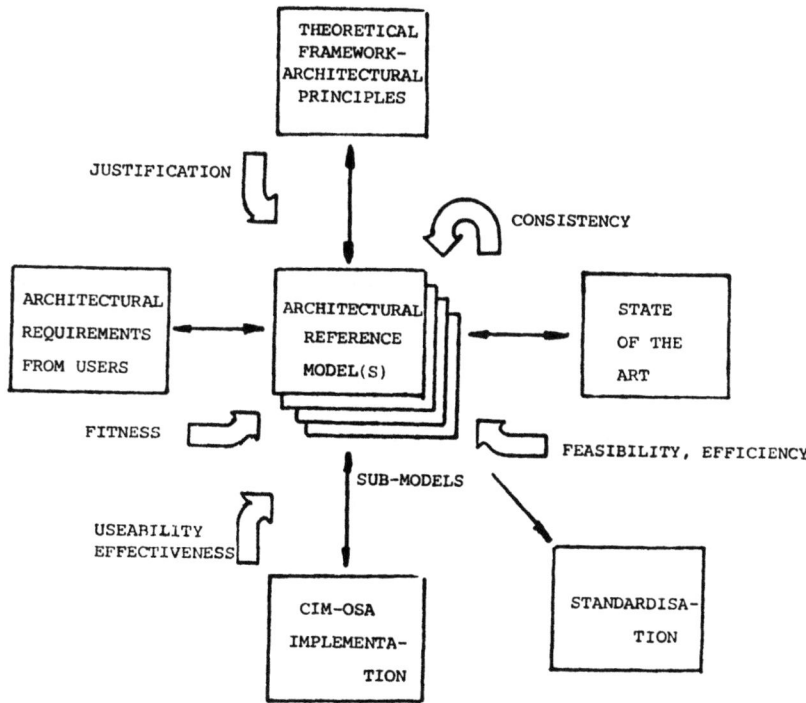

Fig.3: The architectural model(s) and related domains

The figure indicates that submodels are recognised within the main architectural model. This generates of course, additional requirements on cross-consistency between the models.

OSA, A CIM ARCHITECTURE

Users of the Architecture

In the definition and design of an architecture, the term 'user' is to be enlarged in order to cover users of the architecture. These are basically:
- the policy or decision maker in the company making plans and issueing requirements on the development and the characteristics of CIM-systems. He is particularly concerned with investment protection.
- the responsible for CIM-systems (system planner and manager) having very close views to the former and ensuring the translation of the business objectives and requirements into technical properties. There, supporting the enterprise flexibility and performance with appropriate systems is a key issue.
- the CIM (sub)system builder interested in a widely accepted architecture which enlarges the market for optimised system components.
- the CIM system integrator who is concerned by the problem of integrating possibly multi-vendor components into optimal solutions.
- the end-user of CIM systems. For him, the architecture should ensure transparency and a unified view on the information processing systems.

General properties of CIM-systems

CIM will provide to the industry opportunities to streamline production flows, to reduce lead times and to increase overall quality whilst adapting the enterprise at maximum to the needs of the market.

Flexibility in a turbulent environment is a key issue. Obviously the flexibility of the enterprise is depending on and supported by the flexibility of the information system.

This leads to a certain number of requirements to the CIM systems and the CIM-OSA Architecture:-
- evolutionary move towards mor integration
- compatibility with existing organisations and applications
- dynamic interconnecting of IT based processes should be possible instead of unflexible pre-programmed sequences of activities
- the user's view of the information he needs and requires, should be independent from implementation aspects such as actual distribution, storage and processing
- as integration of applications can restrict the enterprises flexibility, the level of integration should remain a managerial decision. Hence, one could find in some parts of a CIM enterprise, a set of tightly coupled systems and elsewhere, a set of loosely coupled systems according to choices made by this particular enterprise.
- the multi-vendor characteristic of the Architecture, both in terms of hardware and software, relates to the needs for strategic flexibility
- re-configurability is required as a CIM system should necessarily be reconfigured to re-optimise its structure and mode of operation according to major changes in the organisation and in the technologies applied.
- open-endedness, in such a way that the integration cost (in effort and money) remains stable independently of the number of subsystems interconnected. The open-endedness principle leads to providing 'escape mechanisms' which are standard procedures for coping with non-standard situations. Another aspect of open-endedness is the possibility to allow multiple modelling of the same object with flexible constraints on consistency.
- compatibility with existing and emerging standards.
- other aspects such as performance, fault tolerance, ...

The Reference Model

The architecture for CIM (Computer Integrated Manufacturing) systems and applications is to be valid for numerous situations and contexts. Hence, the architecture is called a Reference Architecture and it is expressed in a Reference Model. It will act as a reference in systems description, design, implementation and integration. It will provide a framework wherein more specific sysem characteristics and standards will be defined.

Furthermore, it is expected that the Reference Model will be used in particular situations, i.e. a specific enterprise or a specific CIM (sub)-system, to generate a specific model. The purpose of the Reference Model is to define generic structures for:-
- the complete CIM enterprise as a system and its information processes, information flows and communication networks (the information side of the enterprise)
- 'neutral' data and related rules in order to describe and manage the data of the system
- the Information Processing support environment
- functional application modules.

To summarise, the Reference Model is about structuring a CIM system both in terms of processes, information and control and providing the infrastructure wherein specific i.e. problem-oriented applications can be defined, operated and coordinated.

Outline of the Architecture

The architectural Reference Model can be seen as follows:

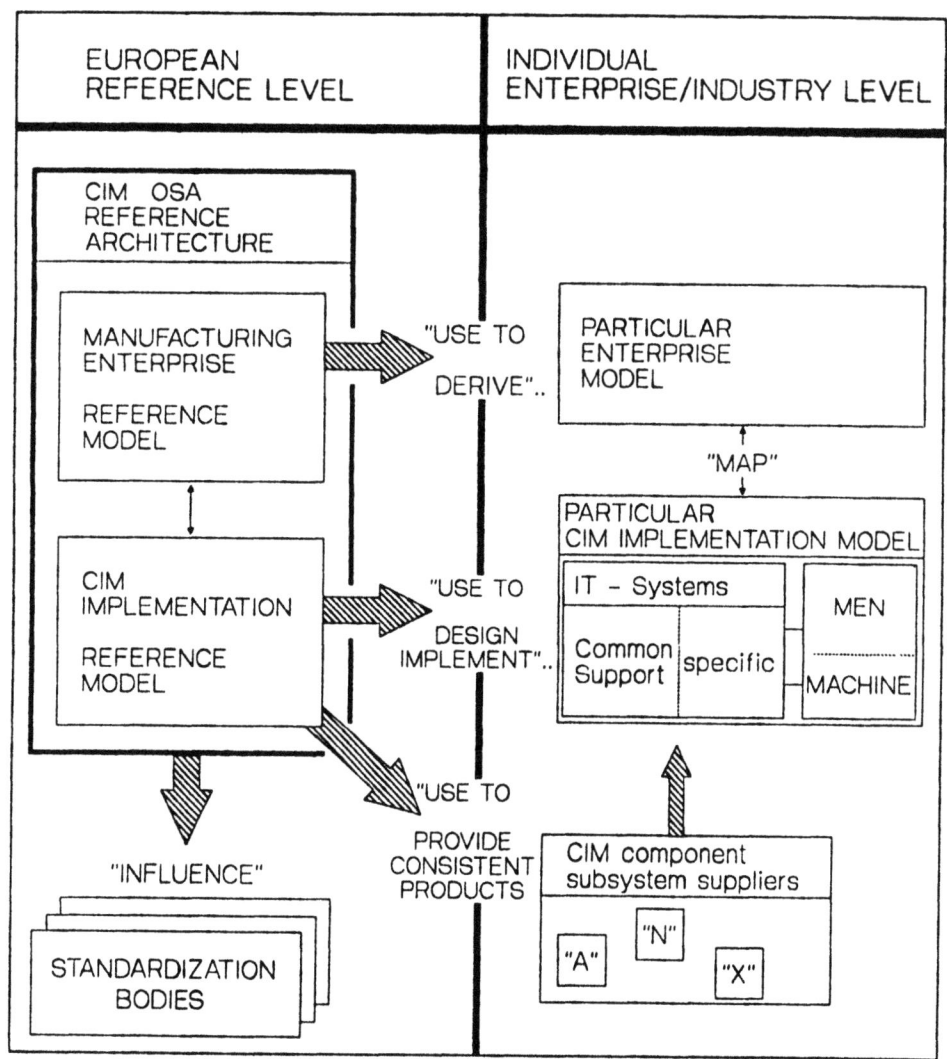

Fig.4: Outline of the Architecture

The left section in the diagram relates to the 'Reference Part' while the right section is addressing the implementation of the model in particular contexts.

The Reference Model encompasses three parts:-
- the Manufacturing Enterprise Reference Model
- the (IT) Implementation Reference Model
- a mapping mechanism between those models.

The manufacturing Enterprise Reference Model. This is aiming at providing a model of an enterprise in terms of activities interconnected through rules. The approach followed is to derive from enterprise strategies, plans, policies, the set of rules which interconnect enterprise activities in a logical and dynamic way, taking into account the specific constraints for each activity and for the interrelated activities.

The connection of activities through rules lead to processable 'project's' which are sets of sequential and/or parallel processes. Conditions are verified at given milestones and the follow-on activities are activated or, if necessary, exception handling projects are started, or even more, control is given back to the people in charge of monitoring the process.

The following remarks should be noted:-
- the model is concerned with control, information and activities which either are immediately executed or which are the exact image of those which are machine-operated, in such a way that execution can be initiated and controlled automatically
- the rule-based approach does not mean complete automation, as projects will be organised taking into account the possibilities of available technology, the operational requirements and constraints, and the adequacy of automated control
- in cases where the rules cannot be derived in a deterministic or programmable way, human and/or artificial intelligence will be required at checkpoints of a project to guide its execution.

The Implementation Reference Model. One of the basic problems raised in this area of the Architecture is to make an inventory of the IT related functions and to organise them in groups or clusters, according to architectural principles such as:-
- grouping functions by nature
- isolating the effects of change.

The model will be characterised by architectural boundaries. An illustration of such an approach is the seven layer OSI (Open System Interconnection Model) where the interfaces between adjacent layers and the protocols for peer-to-peer communication are to be considered as architectural boundaries.

The relevant Information Technologies involved can be categorised as follows:-
- data storage
- data communications
- applications control
- man-machine interface
- physical/spatial interface, for communication between the IT system and robots, numerically controlled machines, ...

Current thinking leads to a subdivision of the functions according to the particular technologies. This has a major advantage to cope with different life-cycles and technological evolutions. Of course, there is probably more than one subdivision.

Mapping

The mapping or at least the partial translation, in both ways of one sub-model into the other, is mandatory for several reasons:-
(a) cross-consistency between the models
(b) the Manufacturing Enterprise Model issues feasible requests i.e. which can be fulfilled by services provided by the IT Implementation Model
(c) choices made at each level can generate constraints for the other model. This is important when addressing real organisations and real systems, especially as it is recognised that the CIM-OSA based systems will have the capacity to cope with:-
  - existing organisations, which evolve progressively
  - existing subsystems (for obvious reasons of investment protection).

It should be noticed that the mapping provided by the Reference Model is generic in the sense that it provides the rules and consistency constraints between the Enterprise Reference Model and the Implementation Reference Model.
When applying the Reference Model, a particular mapping will occur between the specific Enterprise Model and the specific Implementation Model, where criteria depending on the context will be taken into account such as cost, cost-effectiveness, performance, for a given choice of technologies.

For the mapping provided by the Reference Model, it is expected that generic information structures based on concepts such as:-
- object, view, context
- type, version, instance
- other concepts, relating to projects and operations
- and the corresponding combination rules
will play a fundamental role.

PROJECT METHODOLOGY

The innovative character of CIM-OSA in terms of object of design and development implies that standard methodologies are not applicable as such and that a multiple viewpoint approach has to be followed. The development of CIM-OSA will require a certain number of iterations of definition, analysis, detailing and validation.

Therefore, a number of subsequent versions of CIM-OSA have been set forward in the project plan both as technical and project management milestones, with the intention to propose the results as objects of review and possibly, as early instruments for longer range planning.

The proposed workplan for the Development Phase is organised around a number of phases. During the early phases, a central team is working to develop a common view with a set of consistent concepts. When the reference model and 'kernel' concepts are sufficiently defined, parallel work starts in three areas:-
- enterprise activities modelling
- information structures
- IT support
while a central group is consolidating the contributions from the parallel areas and performing technical coordination, liaison with other ESPRIT and not ESPRIT projects, and awareness activities.

As this is in particular an integration project, essential design considerations are:-

- the intention 'not to reinvent the wheel' i.e. to take into account as much as possible valuable results from other research projects
- the active involvement of 'user' groups and user companies
- the intention to establish forecasting activities to validate the architecture against possible evolutions not only in technology but also in the environment.

SOME CONCLUSIONS

Defining, developing and validating a CIM Architecture is a huge effort. It will have very important benefits, both for users and IT-suppliers and even more for European industry. This CIM-OSA architecture defined at many levels, supported by industry and by standardization efforts, constitutes the crucial enabling integration technology for CIM.

This endeavour requires close collaboration and support not only from the partner companies in the AMICE Consortium, but also from future users and concerned European Organisations.

Therefore, it is considered that the close support and very active interest of the ESPRIT organisation was and remains a key factor for success of this project.

SIMULATION: PLANNING TOOL FOR FMS

Simulation for manufacturing systems – a critical review
R. I. Mills
Ingersoll Engineers, UK

ABSTRACT

The large number of simulation packages now available has given rise to confusion amongst prospective users. This paper outlines the types of packages available, describes the use of each type, and notes some commercially available packages. A strategy for using the various types of package is described and some of the difficulties of applying simulation techniques are pointed out. Overall, the conclusions are that simulation is an essential tool for planning certain classes of manufacturing systems, but that its difficulties are not well understood.

INTRODUCTION

There are now at least thirty different simulation tools available to assist the planning of everything from a post office layout to a complex manufacturing system. This variety of packages, ranging from a few hundred to many thousands of pounds in initial cost, has served mainly to confuse the prospective purchaser of either the actual software or the services using it. Each supplier, of course, maintains that his is the best package for your application.

The choice of simulation package is only the start of a lengthy process to a final stage of a usable and effective simulation capability. The actual cost of the hardware and software usually represents only a small portion of the total cost, as training and learning curve related costs are high. This investment must be used effectively within predicted timescales.

This paper attempts to make some form of order out of this chaos, by explaining the categories simulation packages can be divided into, and what each category is useful for. The paper will not choose a package for you, but it should allow you to make an informed decision.

HISTORICAL BACKGROUND

The development of discrete-event simulation for the analysis of manufacturing systems has taken place in various parts of the world. Probably the two most commercially significant areas are those of the three-phase (or "English") languages and the US developed network (or "process") packages. There is a major conceptual difference in the method of operation of these two families of simulation language. Different approaches have also been taken during the development of these packages. These differences are summarised in Figure 1.

Let us consider these two families in greater detail. The approach originally developed by Tocher at United Steel [1] has three basic elements (the three phases) at each time advance; first the clock is updated, next the activities that can finish are finished, and finally, in a defined order, all activities that could start are tested, and if appropriate, they are started. This approach is shown in Figure 2 (a). The technique has the merits of being conceptually simple, easy to code in a computer program and very flexible. The best known languages available in the UK today are based upon this approach (see Figure 3).

In an effort to make models easier to build, code generators were developed. These are basically question-and-answer style computer programs which allow a model to be described to the computer. The actual computer program is then written by the code generator. The most popular target simulation languages are the three-phase ones. Usually code generators are not 'full-featured', in that they cannot access the full set of facilities provided by their target language. This reduces the complexity of models that can be handled. Also, the final machine-generated code, usually FORTRAN, is not easy to understand or to modify.

Interactive colour graphics were also developed initially on the basis of the three-phase framework, giving the capability to view the operation of a model, rather than just obtain printed output. [2] This graphical output can provide excellent communication benefits for model builders, as well as for dissemination of the results.

The family of packages developed in the United States uses a two-phase approach. This technique advances the clock and then starts or finishes activities in the order that they are listed on an event calendar, as shown in Figure 2 (b). This implies that greater consideration of activity ordering and resolution of possible conflicts is necessary at the model design stage. Several of these languages are now available in the UK (see Figure 4). As well as these, many others based on the same techniques are available in the USA.

Development of two-phase languages has focussed upon the use of queuing networks. These are easy-to-use flow networks using processes and queues. A typical network is shown in Figure 5. The network can represent the flow of information, as well as of physical items. Sub-routines in a computing language, such as FORTRAN, can be patched in where required, e.g. for detailed scheduling algorithms.

Graphics were originally applied to this type of language on a 'playback' principle, where program execution details are dumped to a file during a simulation run. A separate program is then used to give animation of the results. Playback graphics do not allow any "what if" alternatives to be made whilst the graphics are displayed. They are thus less useful as a de-bugging and communication aid. Interactive graphic facilities are now becoming available for this type of language.

Overall, the two families of packages offer broadly the same level of facilities, but in different styles. For ease of use in complex model building, the US network packages are better, but UK-based support and training is obviously less impressive.

A REVIEW OF CURRENTLY AVAILABLE SIMULATION TOOLS

This section will review the major types of simulation tool currently available for the analysis of manufacturing systems. These tools will be detailed in approximately ascending order of complexity. Brief mention will be made of some of the major products available within each section.

Other Techniques

Before starting to consider actual simulation packages, it is worth mentioning conventional static capacity planning and the use of queuing network theory based models.

Static capacity planning is now usually computerised using a spreadsheet package. The output includes lower limits for both the number of processing units of each type and system lead times. These figures are based simply on processing unit utilisation and the number of productive hours required, and the total processing and travelling times for each component type. The use of this technique should be the first stage in any analysis approach - if only because data errors will be found quickly. Obviously this analysis is static and has no ability to predict dynamic system measures such as the level of work-in-progress.

Queuing network theory models have become very popular because of their ease of use. Most models available are based upon CAN/Q, originally developed at Purdue University by Solberg. [3] This program models the workflow of components through a manufacturing system as a network of queues. Overall, this class of model is useful for identifying bottlenecks and obtaining "ballpark" information on production output. However, as lead time is directly linked to the work-in-progress level, predictions of both these parameters are poor. This latter point is especially true for smaller, tightly scheduled systems. CAN/Q itself also performs the essential functions of static capacity planning.

Generalised Manufacturing Systems Simulators

Generalised simulators models consist of a validated model of a particular type of system, which the user adapts to his particular requirements by selecting the input data. This is in marked contrast to other methods where some programming is necessary. In this approach, the user provides only numerical data that is usually available in a data base or from a feasibility study. These simulators are often referred to as "data driven". More advanced users can also incorporate patches of code into the model to allow special features to be handled. This latter option needs to be handled with extreme care, and a thorough understanding of the model and its underlying assumptions is required in order to avoid mis-use of the concept used within the original program.

Several commercially available packages are worthy of mention. Modelmaster by General Electric Corporation, and MAST from CMS Research, are fully fledged packages complete with playback animation facilities. MAST also has a static analysis package which pre-processes data ready for MAST itself.

Slightly more unusual is the EXPRESS 1 simulator from ISTEL. This package actually produces a SEE-WHY program, which can then be run in the usual way. The full interactive features of SEE-WHY are available. This is the first commercially available generalised simulator built on a three-phase approach.

The advantages and disadvantages of generalised simulators are common to all types. They give a good representation of a system for very little effort in data preparation and model definition. However, the original generalised simulation model incorporates many basic assumptions and it is crucial that these are understood when a model is being set up.

Three - Phase General Purpose Simulation Languages

As can be seen from Figure 3, there are five packages that are currently available and supported in the UK. These all incoporate interactive graphics capabilities, and most have code generators. They are all based upon FORTRAN and can be used on many types of computer from micros to mainframes.

These packages are all genuinely general-purpose, and can be used for a wide variety of purposes, not merely manufacturing systems. Accordingly, systems can be modelled to whatever degree of detail is necessary, although this power and flexibility requires experience and skill. The full benefits of simulation for communication purposes become available with these packages, owing to their interactive graphics capabilities.

There are numerous examples of complex models for manufacturing systems being built using these packages. The general results are similar, the system performance difficulties observed stemmed from organisational factors and operating rules. Frequently, the level of pallets and/or tooling estimated could be reduced by improved scheduling and control. Another common feature is the difficulty of collecting and collating the large quantities of data needed.

Network-Based General Purpose Simulation Languages

Generally the comments upon three-phase based languages also apply to network-based ones. However, network-based packages tend to be easier to modify, allowing a simple model to be extended as required. The use of networks as the source for the system logic is beneficial, as it can be easily communicated to technical personnel who are unfamiliar with the simulation package in use.

Simulation and Control Packages

The use of simulation methods as an on-line scheduling or advanced planning tool has been discussed for several years. Some commercially available tools have been developed in this area, mainly aimed at small and/or simple systems. Their use is still limited by the need to maintain up-to-date information on many parameters if a genuine on-line tool is needed. Other drawbacks are the need to have a dynamic real-time scheduler involved and the need to balance the long-term view that simulation takes of breakdowns and other interruptions with short-term scheduling needs.

WHICH TOOLS ARE MOST USEFUL?

We now review the use of the various simulation tools available, pointing out that these are our own opinions, based on our experience of them in our planning role.

Static workload analysis is not a particularly useful option for smaller, automated manufacturing systems, although it is almost essential for larger systems. The queuing network model CAN/Q performs a static analysis, as well as providing other useful information. Accordingly, CAN/Q can be used to handle both static analysis and a queuing network approach with one set of data. The results from CAN/Q give a good overview of system bottlenecks, relative utilisation and have proved capable of predicting production figures accurately. However, this model's ability to predict lead times and work-in-progress levels is not so credible.

The usefulness of generalised simulators is a debatable issue. They tend to be expensive and not well documented in terms of their underlying assumptions. Special features are extremely difficult to model and frequently require undesirable simplifications to be made. It is usually almost as quick to produce a network model, which has the great advantage of being capable of expansion in detail at a later stage. However, at a feasibility study stage or for in-house verification of a concept, this class of model is extremely useful. They lend themselves to short timescales and so to the comparison of many alternatives configurations.

Generalised simulators are possibly the worst example of 'technology hype' in the simulation area, as their sophisticated graphics and easy-to-build models of simple (unrealistic?) systems seem to provide the answer to the classic requirement of "accurate models quickly". In practice, they require highly experienced model-builders and careful interpretation of their output. Given these, this type of model can give good quality information within very tight timescales.

General purpose simulation languages have one major failing. The learning curve to use them effectively is lengthy, typically six to nine months. The efforts to reduce this period by using code generators or building network models have given mixed results, although network models tend to be a more flexible and understandable option. Unfortunately, when more than a certain level of detail is required, there is currently no alternative to these languages because of their power and flexibility.

Until recently, network-based packages have tended to be less good than three-phase packages for communication aspects. However, this is rapidly changing as dynamic graphics enhancements become available for network-based packages.

To summarise, the type(s) of model to be used for a specific situation are dependent upon overall requirements and the stage of the project. However, for a normal analysis of an automated manufacturing system, the use of a queuing network model, typically taking 2-3 days, is the first step. The next stage is to build a pure network model and then enhance it as required, using the more powerful and flexible discrete - event features built into these languages. At later stages, interactive graphics are essential both as a de-bugging aid and for communication purposes. This steps will run concurrently with the project as it matures.

HOW TO USE SIMULATION EFFECTIVELY

The first step is to define exactly why simulation is needed. It is an expensive and time-consuming method of analysis, and thus if any other technique can be used, it should be. However, for many systems the only realistic analysis method is simulation and although it is expensive, it will still be cheaper than a manufacturing systems that does not meet its design parameters.

It is also necessary to decide whether an in-house capability is required after completion of the project. If the requirement is for only one project then no such need exists. The opposite situation is when an initial project is to be used as a vehicle to build up in-house expertise.

Having considered exactly what is required, then there are several options to follow. Briefly, these options are:

a. To use the supplier of the manufacturing system. This has the disadvantage that only the supplier's preferred option will be simulated. Standard models configured to suit a given system are usually used, and the level of detail is low. This approach simply confirms the basic design of one system.

b. To use a university or other academic establishment. This is a low-cost option where the quality and level of detail are extremely good. The main disadvantage is that timescales tend to become prolonged.

c. To 'do it yourself'. This approach appears attractive, but the costs (both direct and indirect) tend to be underestimated. Also, learning curves take in the order of six to twelve months before satisfactory results are obtained.

d. To use an independent consultant. This is the most expensive option, but a rapid service of high quality is obtained. One disadvantage is that, when the project is completed, no transfer of skill has occurred, and for subsequent projects the whole process has to be repeated. Final project definition at an early stage is crucial to the success of this approach.

e. To form a joint project team between yourself and an academic establishment. This is an excellent long term approach, but does not meet immdiate needs especially for a single project.

f. To form a joint project team composed of an independent consultant and yourself. This is the best compromise alternative, combining the advantages of using an independent consultant and at the same time reducing the real costs of starting to use simulation in-house. However, it will cost more and take longer than a pure consultancy route, although significant knowledge transfer to your personnel will occur. This route can have significant long-term benefits.

HOW NOT TO USE SIMULATION

Simulation is an extremely powerful tool for the analysis of manufacturing systems and, as such, there are certain drawbacks to its use. These are inevitable in the light of the complex nature of simulation concepts.

First, the trend towards simpler and more user-friendly packages could mean that the personnel using them do not understand their implicit assumptions. This means that the package could easily be used outside its design scope with unpredictable results. Any user of these generalised packages needs both a thorough understanding of simulation concepts in general and of the basis of the package.

Secondly, to expand the previous point to a more general level, training in simulation concepts and theory is poor. Training is usually directed at one language and does not address the statistical background, the use of variance reduction, etc. This is in contrast to the United States where some form of simulation teaching is available within most engineering degree courses.

Overall, simulation should be regarded as a "last resort" technique. If any other method can provide, with confidence, the level of detail needed then it should be used. Simulation is usually more expensive and more time-consuming than originally anticipated. Paradoxically, the experience of knowing when not to choose simulation is crucial to its successful use.

CONCLUSIONS

Simulation has become an accepted tool for the planning of complex advanced manufacturing systems. This acceptance has resulted in a large number of both simulation packages and methods of using them, and has led to confusion among many prospective users.

There are many levels of simulation, each suited to differing needs. Care taken in the selection of a package and in the route to using it, assessed in the light of the needs of the user, will pay for itself many times over. An ambitious in-house programme is not justified if only one simulation model is needed.

Looking to the future, development of simulation and associated packages as decision support aids is inevitable. The development of these aids as packages based on generalised simulation models increases their utility. It would also permit the application of computer-aided decision-making to otherwise low-technology manufacturing units, such as manual job-shops.

Finally, the use of simulation methodology in conjunction with decision analysis based on Artificial Intelligence is an area with almost limitless potential. This is one stage advanced from decision support, and is definitely a long-term concept.

REFERENCES

1. Tocher, KD, "Review of simulation languages", Opl. Res. Q., 16 (2), 1965.
2. Hurrion, R.D., "An investigation of visual interactive simulation methods using the job shop scheduling problem", J. Opl. Res. Soc., 29 (11), 1978.
3. Solberg J.J., "CAN-Q Users Guide", The Optimal Planning of Computerised Manufacturing Systems, NSF Grant No. APR 74 15256, Report No. 9 (revised), Purdue Univeristy, West Lafayette, Indiana, USA, July, 1980.

FEATURE	US NETWORK OR PROCESS APPROACH	'ENGLISH' THREE-PHASE
TIME ADVANCE	TWO-PHASE	THREE-PHASE
CLOCK ADVANCE	QUASI-REAL TIME	INTEGER
SIMPLIFIED MODEL BUILDING	NETWORK DIAGRAMS DIRECTLY INPUT	ACTIVITY CYCLE DIAGRAMS INPUT TO CODE GENERATORS
GRAPHICS CAPABILITY	PLAYBACK FROM DATA FILE	SIMULTANEOUS AND INTERACTIVE

Figure 1. Major differences between Simulation Language types

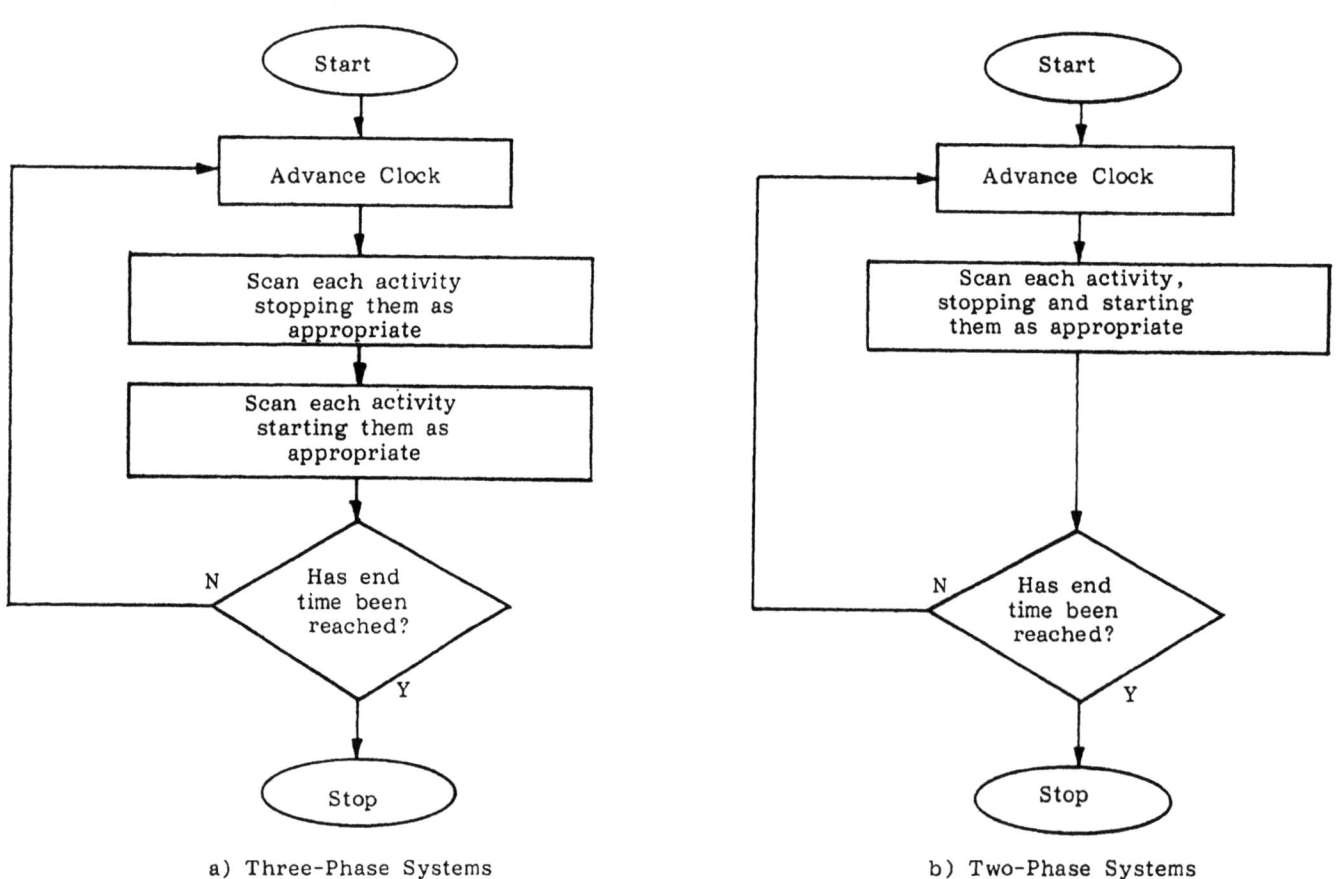

a) Three-Phase Systems   b) Two-Phase Systems

Figure 2. Simplified flow charts - two- & three-phase simulation systems

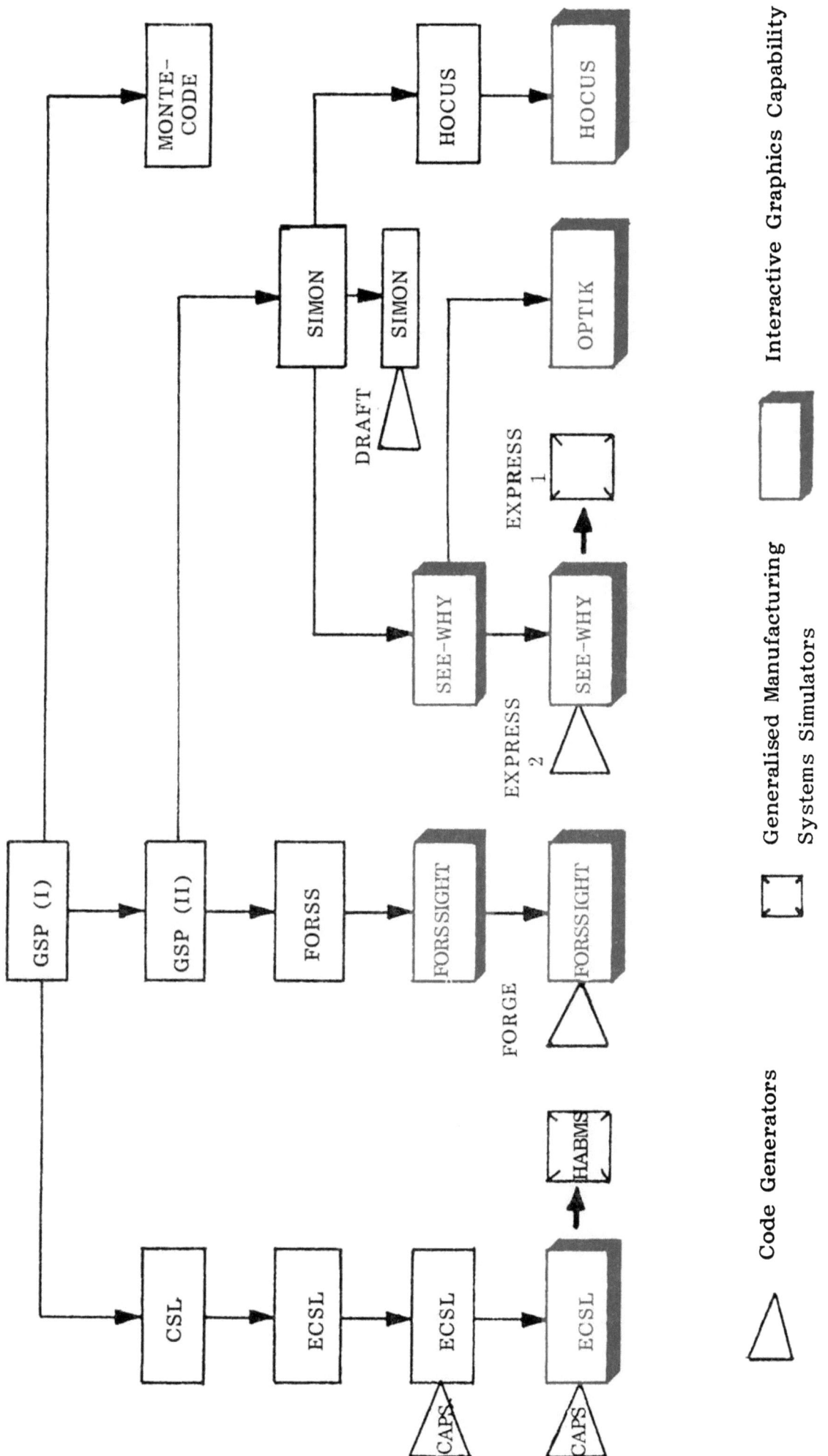

Figure 3. Simplified Development of Three - Phase Simulation Languages

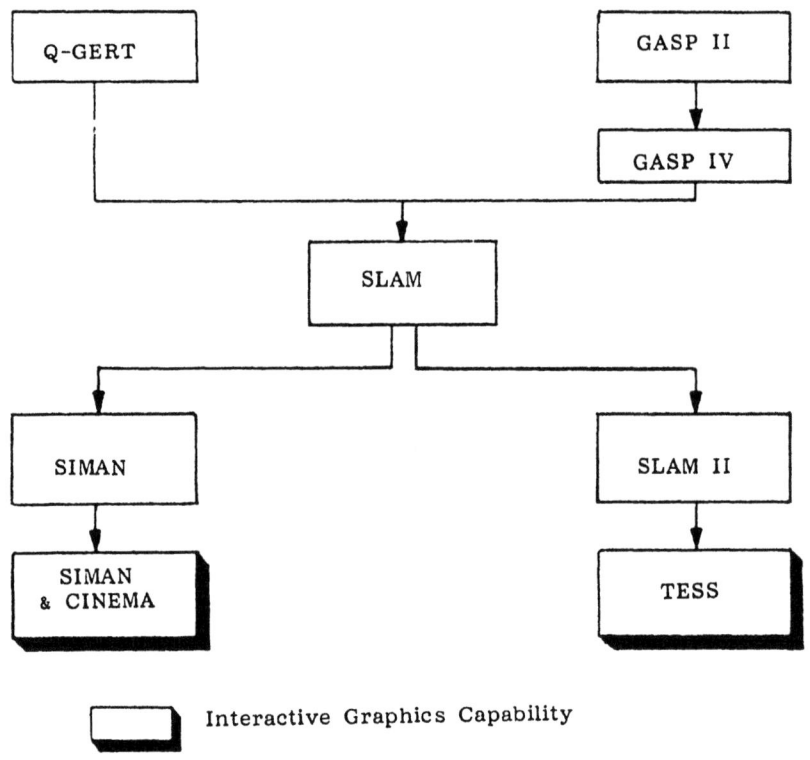

Figure 4. Simplified Development of Certain Two-Phase Simulation Languages

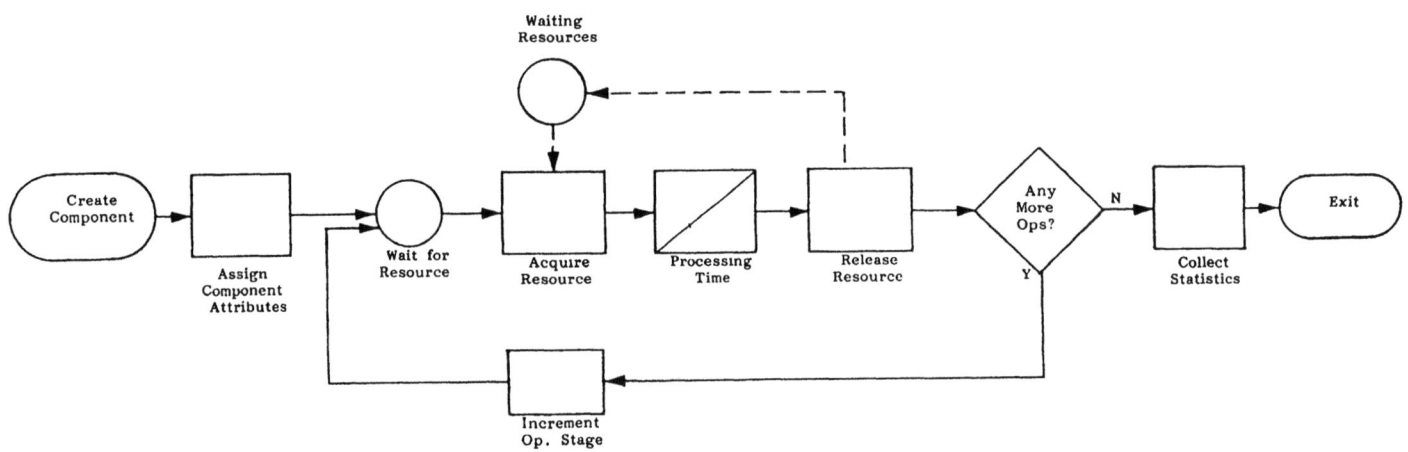

Figure 5. A Typical Network Diagram

Strategies for material transportation in a computer integrated manufacturing environment

C. L. Moodie
Purdue University

and

D. BenArieh
AT & T Bell Laboratories, USA

ABSTRACT

Within any discrete part manufacturing environment there will always exist the three basic elements: processing, transportation, and storage. Processing might be subdivided into component processing, and assembly. Also, in a contemporary FMS environment one would attempt to minimize the storage component. Upon closer observation of the transportation component one can discern transportation within a cell and transportation between cells. The former can consist of the synchronized interaction of conveyors, robots, carousels, turntables, etc., and the machine tools. The between-cell transportation would probably consist of the flexible movement of such programmable devices as automatic guided vehicles. One can view the between-cell transportation problem as a delivery problem and the within-cell transportation problem as a sequencing control problem. Both employ elements of Artificial Intelligence. The between-cell delivery task is work-in-process part oriented and utilizes a knowledge based routing system (KBRS) which is coded in PROLOG. The within-cell strategy centers on control of the interacting materials handling equipment (conveyor, robot, etc.). A pattern recognition methodology is utilized to match real time FMS state with predefined table states.

1. THE FMS CONCEPT AND MATERIAL FLOW

One hesitates to offer a definition of a flexible manufacturing system (FMS) since there are a number of them out there. One basically thinks of such a system as a collection of NC-type machine tools with a captive material handling system, all integrated under the same computer control. T. Ohmi, of the Mechanical Engineering Laboratory (MEL) in Japan, offers a three level definition of such systems [1]. He defines the flexible manufacturing cell (FMC) as a computer-controlled system consisting of a single machine tool, a workpiece loading system, software and hardware interface to allow a large number of machining functions on a number of different workpiece configurations, and unmanned operation. Mr. Ohmi says that a smaller system could not be classified as flexible and anything larger is an FMS. He says that a very large system could be classified as a factory-scale FMS.

In this paper we will consider material flow control in a large FMS environment. The FMS consists of a number of flexible cells which achieve computer integration in a manner similar to that defined by McLean [2], for the National Bureau of Standards Automated Manufacturing Research Facility. He proposes a five-level hierarchy in which the cell level is approximately equivalent to a small FMS by the above definition. A group of these cells would be equivalent to a factory size FMS. The levels of the hierarchy are shown in Figure 1. In general, the Facility level could represent the factory and the Equipment level could be the machine tools. The Workstation level is defined as a small integrated grouping of shop floor equipment. A Cell would consist of a number of workstations. Control at the cell level is "..... responsible for sequencing batch jobs of similar parts through workstations and supervision of various other support services, such as material handling or calibration [2]."

2. MATERIAL TRANSPORT IN AN FMS ENVIRONMENT

Within any discrete part manufacturing environment there will always exist the three basic elements: processing, transportation, and storage. Processing might be subdivided into component processing, and assembly. Also, in a contemporary FMS environment one would attempt to minimize the storage component. Upon closer observation of the transportation component one can discern transportation within a cell and transportation between cells. The former can consist of the synchronized interaction of conveyors, robots, carousels, turntables, etc., and the machine tools. The between-cells transportation would probably consist of the flexible movement of such programmable devices as automatic guided vehicles. One can view the within-cell transportation problem as a sequencing control problem; the between-cell movement can be seen as a delivery problem. Solution to both problems must be coordinated in order to facilitate the total material movement within a FMS.

A component of the research being carried out at the Engineering Research Center for Intelligent Manufacturing Systems, at Purdue University, is concerned with the FMS transport problem. The on-going investigations directed at intelligent control of between-cell and within-cell material movement will be described in this paper.

3. A WITHIN-CELL STRATEGY

As soon as material or WIP have been delivered to a specific cell for processing they will attempt to follow the machine sequence defined by the appropriate process plan. If there are other jobs in process, or waiting for processing at that cell, then the cell will need some planning intelligence in order to resolve machine demand conflicts. Additionally, the cell will require the ability to sense environmental variables (i.e. status of WIP) so that physical equipment (i.e. conveyor, robot, etc.) interaction can be controlled. Of these two cell-based problems, the former is best handled by a rule-based system, while the latter can be accomplished with a state table approach. The decisions to be made will require much real time data concerning the status of the system; it is assumed that contemporary sensor and communication network technology can capture the required data in a timely manner. The state table system will be described here.

The basic idea of this control strategy is to make on-line decisions to control the system by retrieving, and applying intelligently, appropriate operating information which has been previously stored regarding the overall system. Unlike must artificial intelligence systems, the approach here uses a supervised learning process for the computer rather than the method of extracting the experience from human experts. Therefore, the operating information is created by solving many possible control problems in advance, using a computer simulation procedure. The information is then stored and grouped. The retrieval is based on the similarities of this stored information with the current system behavior. The developed control strategy will perform in the following way:

A. Recognize the current status of the system,
B. Retrieve operational decisions automatically, and
C. After a certain period adjust the control parameters by repeating the procedures A and B.

This control procedure is repeated periodically to maintain a desired system performance. Based on various criteria such as maximum throughput rate, minimum work-in-process, or others, the system can be controlled to react adaptively to various circumstances. The adaptive decisions will be made to consider the interaction, the synchronization, and the operation of the material handling equipment. This control procedure is diagramed in figure 2.

A laboratory FMS to which this control procedure was applied is shown in Figure 3. This system contains two robots, a carousel, a conveyor and a turntable. The conveyor has two sides, each moving in opposite directions, so that parts can be delivered to, and transported from, the carousel, where assembly operations are performed on the parts. The three different part types which flow through this cell are initially placed on the turntable to be picked up by the first robot for transferring to the outbound side of the conveyor. the capacities of this material handling equipment, in parts, are: carousel (10), turntable (10), and conveyor (20) on each side. Parts arrive at the turntable in a random order, and the time between arrivals is not a constant.

Considering only work in process (WIP) queues as state variables, there are (11X11X21X21) different states possible here, if the empty state is considered; this is probably too many for a practicable state table. Even if it were possible to contain this number of states in one table it would be a formidable task to determine the proper, corresponding values for the decision variables for each set of status variables. The methodology which will be outlined below will reduce to a manageable quantity the number of state variable sets (vectors) for the situation where a single state table system is to be used for controlling the material flow in a computer controlled, manufacturing cell.

3.2 Determining Values for the State Table

Returning again to Figure 2 it can be seen that the state table resides in the database and that the values for the state table were generated by a simulation-optimization procedure. The steps involved in setting up this control hierarchy are, generally, as follows:

Develop a simulation model which will simulate the real cell system with interacting material handling equipment in a dynamically changing environment.

Combine the simulation model with an optimization model (SAMOPT) [3]. The combined simulation-optimization model is abbreviated as SIMOPT. SIMOPT will be used to find the optimal control parameters for the cell system operation with certain criteria. The result of a SIMOPT search will be saved to construct an operational support system. The stored information can be used for controlling the system with various objectives and constraints.

Integrate the SIMOPT model with a database management system. For this investigation, a relational database, SDL (Simulation Data Language) [4] was used. The results from SIMOPT will be stored on-line in different relations/tables of SDL. These data may be used for many operational support purposes. The information can be retrieved by using either a supported query package or user defined subroutines.

Develop an operational support system (OSS) which uses the results from the integrated SIMOPT and SDL models to support the system operation. OSS will provide different decisions for the cell system operation by considering the current system status, possible future inputs, and the desired output. OSS will include some artificial intelligence concepts such as clustering analysis, pattern recognition, and learning control.

Implement the developed control strategy in the modeled real cell environment and perform a validation.

Figure 4 is a diagram of the defined state table generation procedure. A more detailed description is given in Chu [5] and Chu and Moodie [6]. The proposed intelligent, transportation equipment, control strategy for an FMS cell has been subjected to some experimentation to evaluate its worth and determine some of its weaknesses. A SLAM [7] simulation model was used to represent the real system. The control strategy was found to be very effective where there was much variation in the work input to the cell. For situations where a steady state type environment existed, the proposed control strategy was not found to be superior to more conventional strategies. Work is underway on the used of hierarchical state tables, and state tables which combine expert system production rules as well as state data.

4. A SCHEME FOR BETWEEN-CELL MATERIAL FLOW CONTROL

A within-cell material flow control method has been presented in the preceding sections. Of equal importance in a manufacturing cell oriented environment is the between-cell material flow control problem. In a sense this is a delivery problem. Raw material, tools, and or WIP must be delivered to different cells for different processing operations. In this situation we have resources (delivery devices) which do not belong to any one cell (as, say, a conveyor or robot might) and they must be effectively "managed" to properly serve all cells. Also we have the so called "just-in-time" delivery problem which must be considered if work in process inventory is to be minimized.

The previously mentioned Center for Intelligent Manufacturing Systems at Purdue University is investigating the concept of a small, free-roaming, computer-controlled vehicle which could deliver materials to cells and not be constrained to wire-in-floor fixed paths. The potential flexibility of these vehicles requires an intelligent control methodology.

Ben Arieh [8] devised a knowledge-based procedure for delivering parts to different work cells which appears to be superior to non-knowledge based (traditional) scheduling methodologies. One assumption of this work is that some of the cells can perform similar work; this will allow for alternate routing as a material flow strategy. This method is different from that which was described above for the within-cell control problem. Here we consider the solving of material flow problems in real time, based on the current, status information and previously stored, static data which defines the system. The within cell method relies on use of stored solutions to previously solved problems.

4.1 Description of the KBRS

The knowledge based routing system (KBRS) to be described below is designed to introduce context (or environmental) knowledge into the decision making process. The objective of the proposed system is to utilize all the data available in a computerized manufacturing environment, to create a good control mechanism to supervise the system, and generate real time answers to material delivery situations which occur while the system is in operation.

The control system interacts with the process controllers, gathers data from the shop floor and makes simple decisions. This system also contains procedural knowledge. For control purposes two types of knowledge exist: production knowledge (rules), and procedural knowledge. The representation of the production type of knowledge is done in predicate form, using PROLOG.

The data base is the part of the expert system which stores system states, properties of the system components, and interacts with the system. The data base for this system is composed of a static data base, and a dynamic one. The static data base contains information about the required processes, the part structures, and cells available and their capabilities. The dynamic data base contains data about system queues, parts' current process, time a part is required and assembly unit states.

During run time of the system, decisions and hypotheses are made based upon the system state at that time. In order to have the desired information this system utilizes the query capability of the PROLOG language. The query process is based upon the predicate form of PROLOG and imitates the required predicate with a variable name for the required data item. This implies that the predicate form of the data base must be known in order to retrieve any desired item.

4.2 The Production Rules

There are two components of knowledge in this system. The first one consults with the data base and the system behavior knowledge and decides upon the system response to basic changes in its states. The behavioral knowledge of the system is represented by a combination of predicate logic and production rules, which is given by the PROLOG language.

The rules in this part can be divided into two main groups:

A. Finding to finding rules. These rules relate to events that occur in the system as the antecedent, and the hypothesis as the consequence. In this system the hypotheses are stated in action form; they are the actions that are assumed to be the right answers to the findings discovered.
B. Hypothesis to hypothesis rules. These are the more common rules in this system and they connect every hypothesis assumed to be true with all the other hypotheses that need to be checked (by performing action).

The second part of the knowledge is the algorithmic knowledge. This part solves the routing decision according to the current system states. The algorithm is dynamic and dependent upon the system current states as reflected in the dynamic data base, and not on mean off-line data.

4.3 Control Strategy

The main technique for rule selection is a matching technique. In PROLOG the matching is similar to unification in predicate logic with some variations. Sometimes several rules have the same L.H.S. (therefore, the same name), and all of them can be triggered. In order to choose a specific rule to fire, the system (through PROLOG) selects rules according to their order in the data base. In order to make a rule more favorable, it is possible to move the rule upwards in the list (or downwards for a lower likelihood rule).

Another strategy partially used is a "context limiting" strategy. This approach checks the context of the rule, and only rules with the right context can be triggered. In this system the context is mentioned immediately in the R.H.S. (consequent) so the rule is first chosen by matching its name, and then by its order, and only then by the context. In this system any rule that fires leads to an action. This action is a change in the system state as is reflected in the data base. A part that moves from a queue to a process, or enters a different queue, start assembly or finishes assembly, has all of these changes reflected in the data base. Changes in the data base utilized through predicates; these are specified rules whose task is to create the desired change in the data base. Figure 5 defines the components of KBRS and the direction of information flow.

4.4 Evaluation of KBRS

In order to evaluate the KBRS a simulation model was tested. This model simulated production of fourteen different parts, and assembly of nine of them. The production environment consisted of five machining cells and one assembly cell. The machining cells were flexible enough to be able to duplicate, to a certain extent, each others work, with varying efficiency.

An assumption taken in designing the system is that assembly times are not negligible in comparison with the production times. This assumption forces the production system to consider the state of the assembly cell.

In general, the KBRS was found to be significantly superior to traditional scheduling procedures for system configurations tested. One must assume that the availability and use of much system information contributed to the good results. One example of this is seen in Figure 6 which shows that the KBRS controlled system assembled at least four times as many products as the system controlled by more conventional means. The same order of magnitude improvement applied to number of parts completed. Note that the cross hatched portion of the graph applies to the KBRS.

5. HARDWARE WITHIN WHICH TO IMPLEMENT THE CONTROL STRATEGIES

The discussion up to now has been concerned with intelligent control strategies for material transportation in an FMS environment. An important question which is only mentioned here concerns the material handling equipment within which this control will be embedded. Certain material handling equipment was mentioned in Section 2, but it is computer controlled equipment of a type which is now available. One must assume that there will exist different material handling equipment configurations in the future. Currently, automatic guided vehicles, which depend on a wire embedded in the floor, are an acceptable mode of transportation for the between-cell delivery. However, free roving transport vehicles, currently under research consideration, might be better for implementing the intelligent control strategy.

6. REFERENCES

[1] Kidd, J.M., "FMS: Lets get the definitions straight first", *American Metal Market Metalworking News*, (June 16, 1968).

[2] McLean, C., Mitchell, M., and Barkmeyer, E., "A computing architecture for small batch manufacturing", *IEEE Spectrum*, (May 1983).

[3] Azadivar, F. and Talavage, J., "Optimization of stochastic simulation models", *Mathematics and Computers in Simulation XXII*, North Holland Publishing Co., p. 231-241 (1980).

[4] Standridge, C. and Wortman, D., "The simulation data language (SDL): A database management system for modelers", *Simulation Today*, SCS, LaJolla, CA.

[5] Chu, C.C., "An Adaptive Decision Making Methodology for Material Handling Equipment in a Computer Integrated Manufacturing System", Unpublished Ph.D. Thesis, Purdue University (December 1984).

[6] Chu, C.C. and Moodie, C.L., "Experimentation with an adaptive control methodology for material handling systems containing robots and conveyors", *Material Flow*, Vol. 3, Nos. 1-3, pp. 141-151 (February 1986).

[7] Pritsker, A.A.B., *Introduction to Simulation and SLAM II*, Halsted Press, (1984).

[8] BenArieh, D., "Knowledge Based Control System for Automated Production and Assembly", Unpublished Ph.D. Thesis, Purdue University (August 1985).

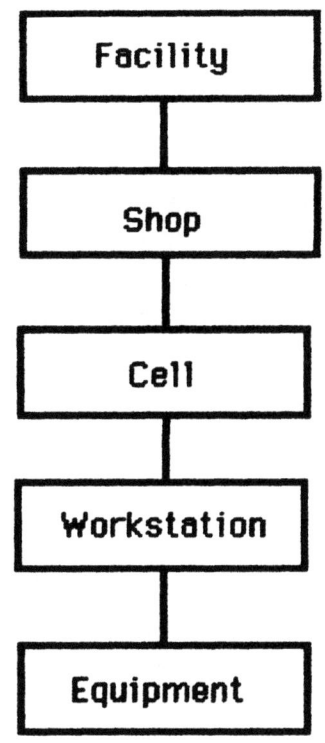

Fig. 1 AMRF Control Levels (From McLean, et al.)

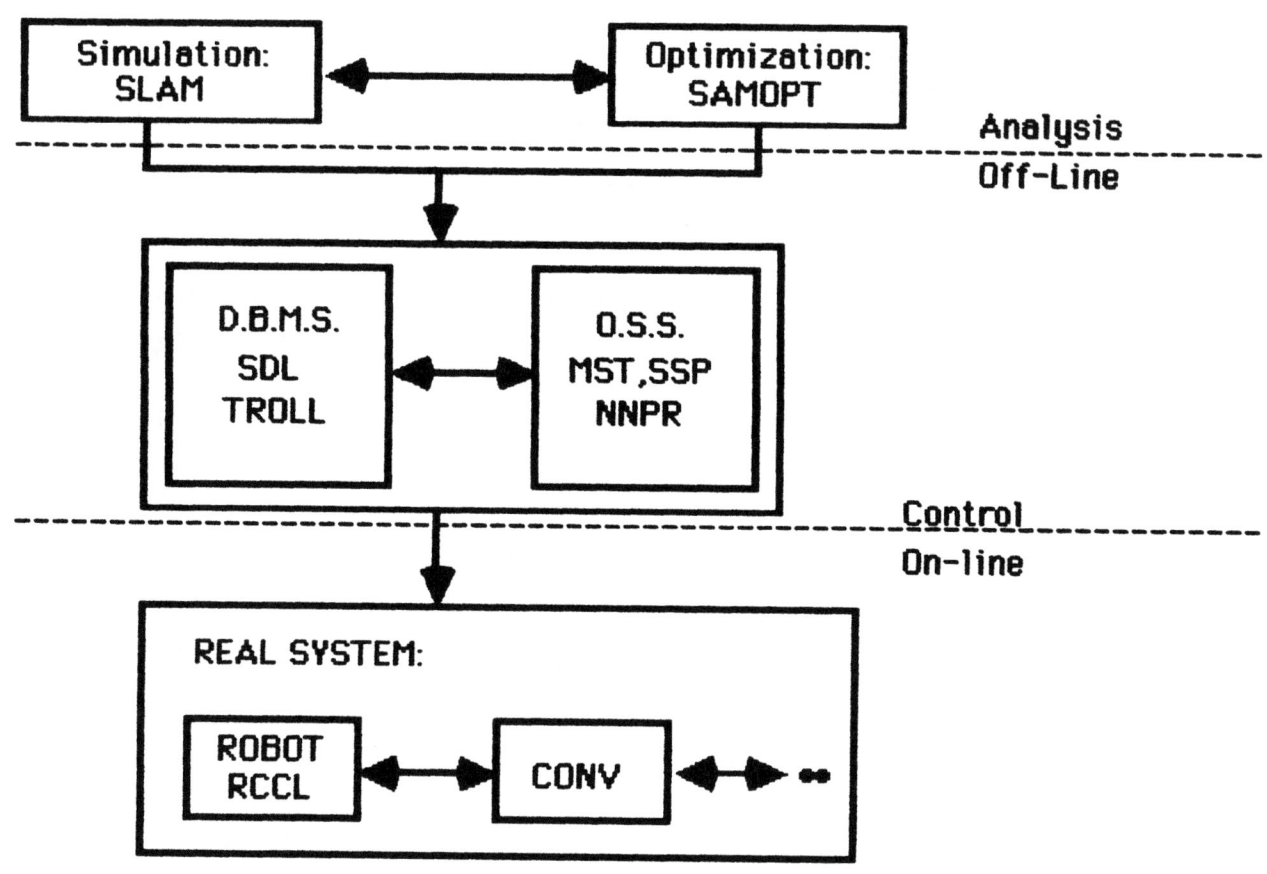

FIGURE 2

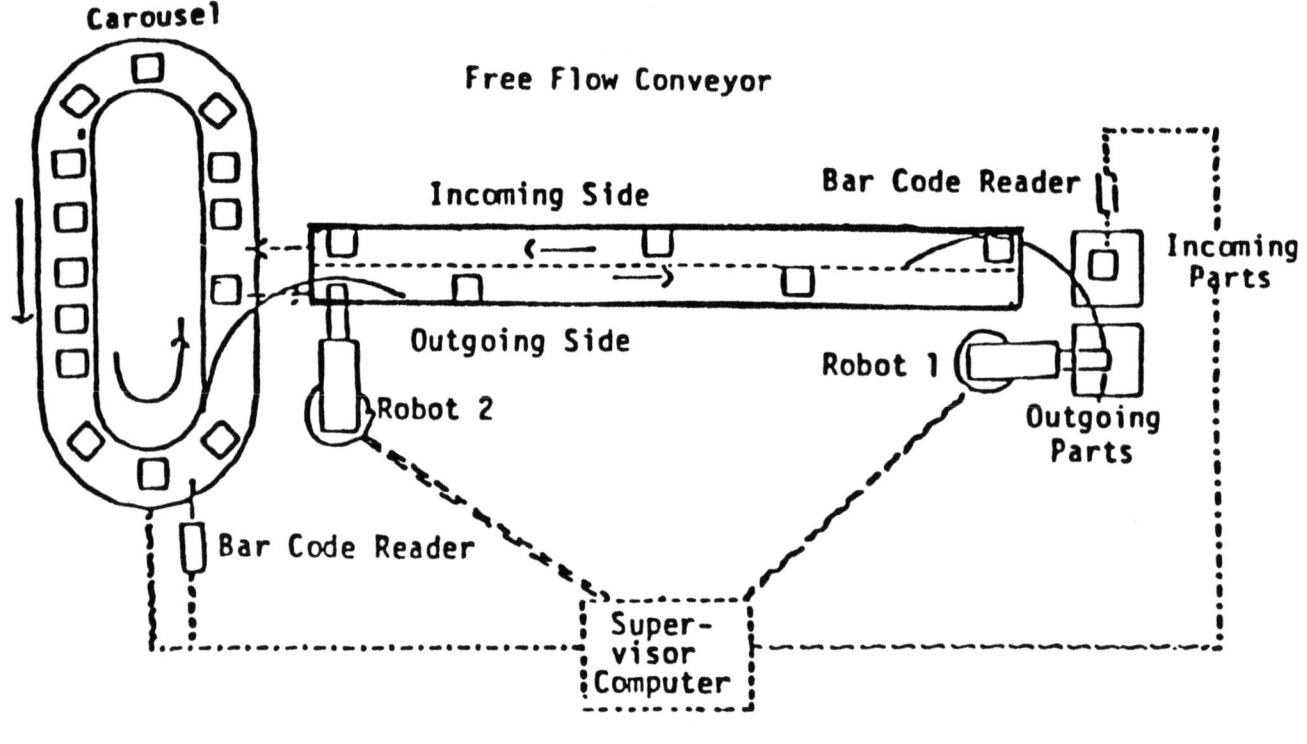

FIGURE 3

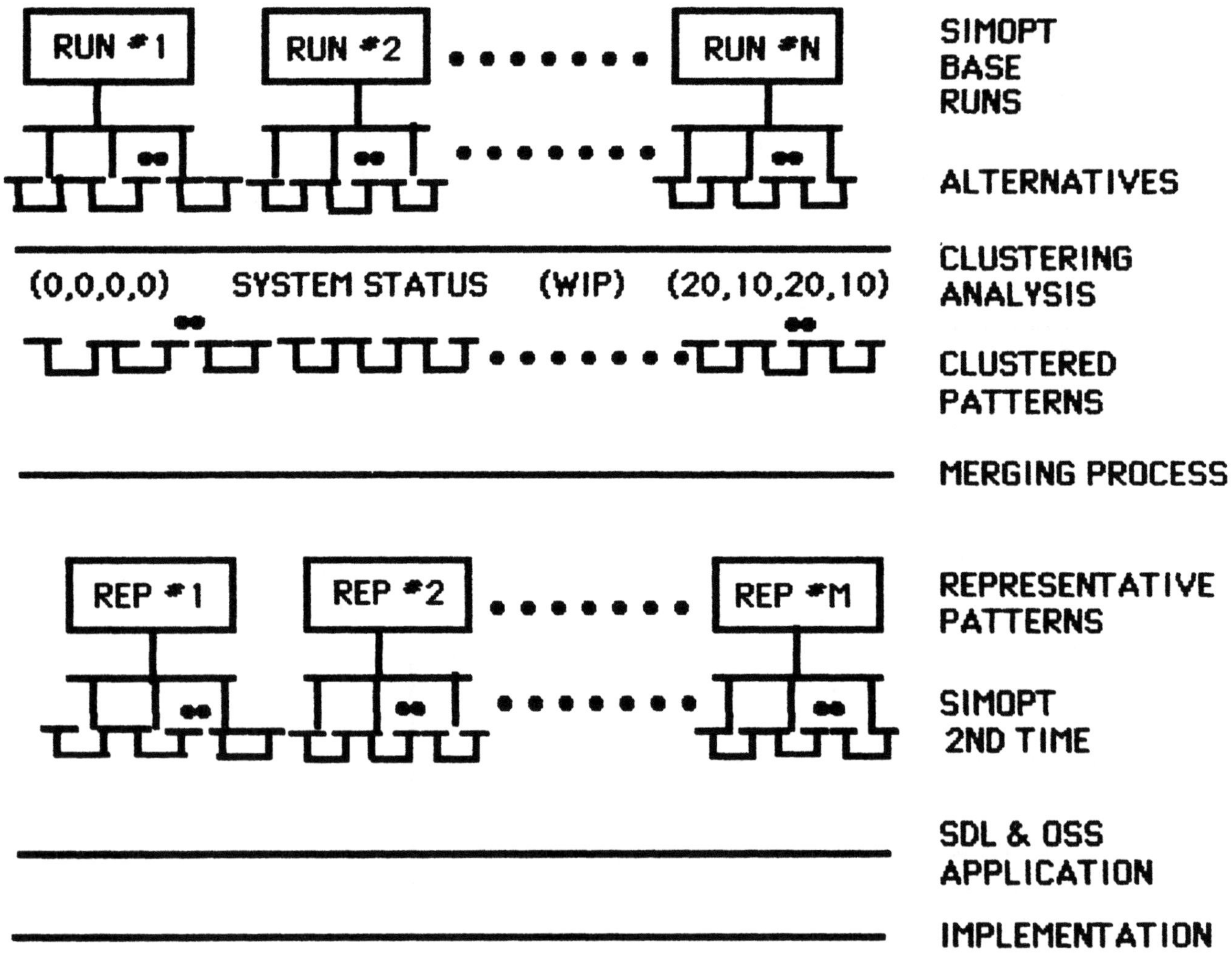

FIGURE 4: The Development Procedure of the OSS Model

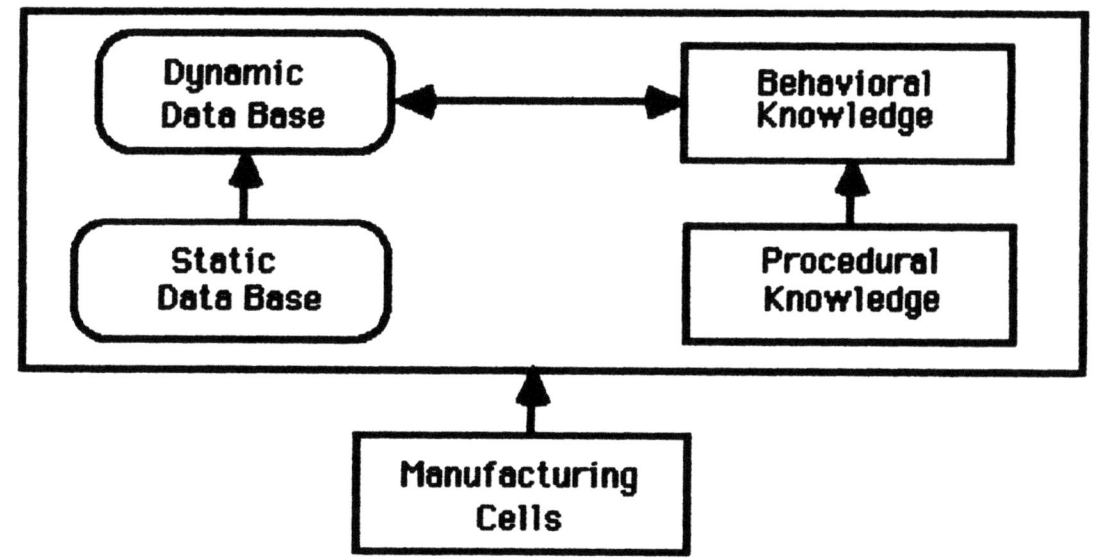

FIGURE 5: KBRS Architecture

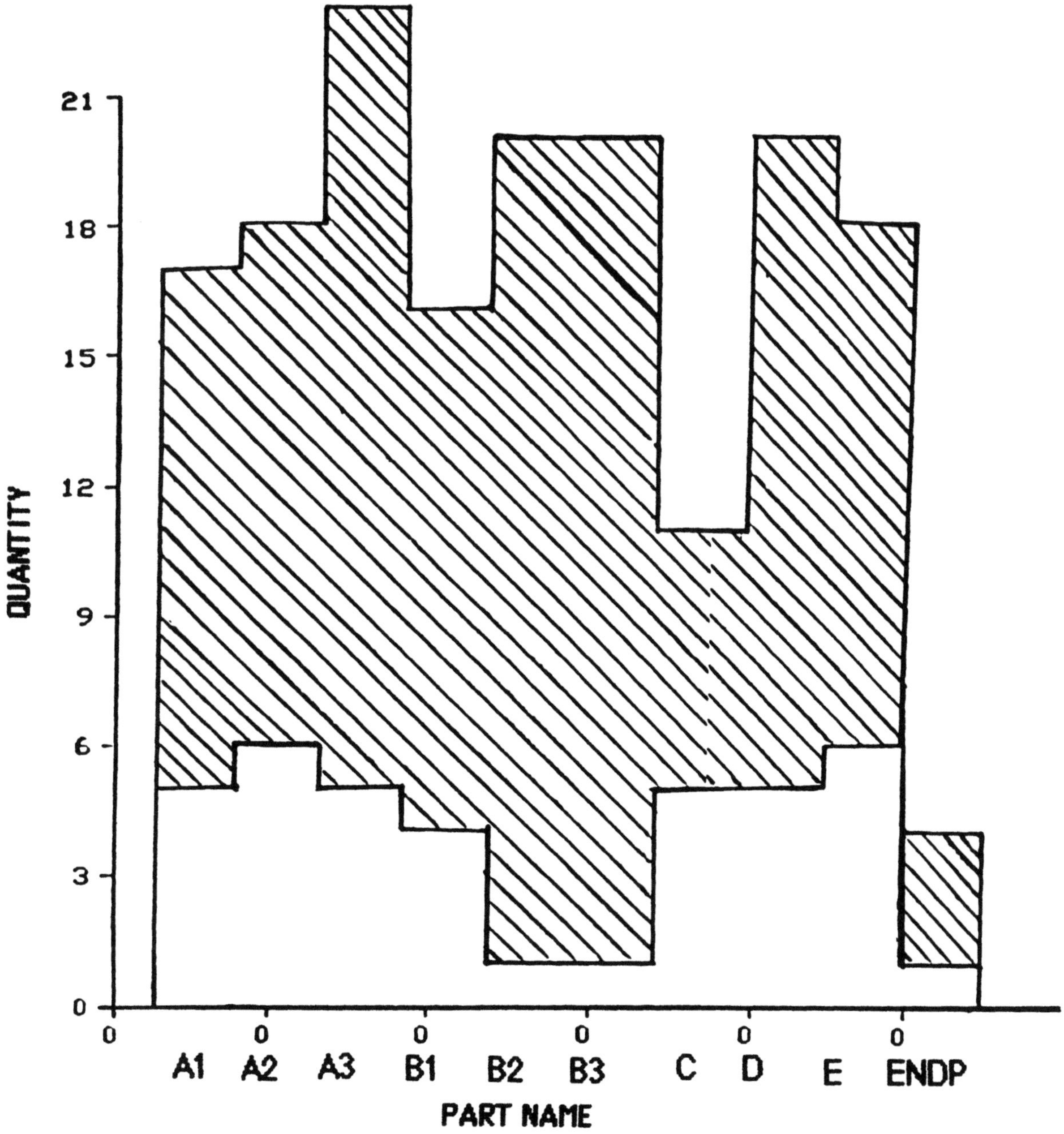

FIGURE 6: Expert System Routing Algorithm

TDL, a task description language for programming automated robotic workcells

A. Adler
Tecnomatix GmbH, West Germany

ABSTRACT

A design and simulation system for automated workcells, which consist of robots, process machines, and transport systems, provides tools to create geometric and kinematic models of components. It enables to simulate and optimize the workcell layout and to program robots and other machines off-line.

In order to allow graphic animation of the workcell a new language for the simulation of concurrent tasks (Task Description Language, TDL) was developed. It is a high-level structured language that includes instructions for robot manipulation, and provides an interface to most robot controllers but is yet independent of the robot type.

INTRODUCTION

The versatility and flexibility of a robot system depends to a large extent on its programming system. Existing and proposed robot programming systems fall into one of the three categories: **guiding systems** in which the user leads a robot through the points or motions to be performed, **robot-level** programming systems in which the user writes a computer program specifying motion and sensing, and **task-level** programming systems in which the user specifies operations by their desired effect on objects.

Robot guiding [1] is a programming method which is simple to use and to implement. Because no general purpose computer is needed, this method was mainly used before it was cost effective to incorporate computers into the programming. During guiding, the programmer specifies a single execution sequence for the robot; there are no loops, conditionals, or computations. But for certain applications, such as mechanical assembly and inspection, one needs to specify the desired action of the robot in response to sensory input, data retrival, or computation. In these cases, robot programming requires the capability of a general-purpose computer programming language.

Explicit or robot level languages enable the data from external sensors, such as vision, to be used in modifying the robots motions; hereby the robot can cope with a greater degree of uncertainty in his environment. The disadvantage of this approach is the need to be experienced in computer programming to use effectively this method [2].

Another approach, **task level** programming, requires specifying goals for the positions of objects, rather than the motions of the robot needed to achieve those goals. In particular, a task level specification is robot-independent. Task level programming systems require complete geometric modelling of the robots environment. This approach was selected in developing TDL.

The Task Description Language (TDL) is a structured high level programming languge developed to enable accurate simulation of robotic workcells and the generation of codes to be loaded directly to equipment controllers on the factory floor. TDL is workcell oriented and includes motion instructions for robots and other devices; object handling instructions to deal with workpieces, components, and other workcell objects, and intertaskt communications, which simulate signals between parallel tasks and interrupts in the workcell.

The TDL language, was developed as part of ROBCAD, a computer aided system for the design, simulation, and off line programming of robotic workcells. This system adresses two important issues that have received little attention in other robot programming systems like AL [3], VAL [4] and AML [5]:
- The execution of multiple tasks with multiple devices in parallel.
- Defining robot independent programs.

WORKCELL SIMULATION

Before simulating a workcell on a ROBCAD system, several steps must be taken:

Definition of the geometric and kinematic models of the workcell components.
The first step in building a new workcell model is the retrieval of components from the libraries, and the mechanical design of additional components, e.g. robots, lathes, presses.

Design of the workcell layout.
This second step includes:
- Placement of the elements in the workcell.
- Test robot reach and detect collisions.
- Verification of cycle times of specific motions.

Interactive robot motions can be recorded and used later in the simulation phase, instead of writing motion instructions in TDL. This graphic programming technique is similar to teach mode in programming at the robot controller.

Writing TDL tasks that describe the sequence of actions of the robots, machines and other active elements in the cell.
Each task file describes the sequence of events executed by a machine. All tasks have to be compiled and linked to create the simulation file.

THE STRUCTURE OF TDL

TDL is a structered, block-oriented language based on modern robot languages and on general purpose high-level programming languages like PASCAL. It enables discrete parallel event simulation for components having a mechanism.

Each active component in the cell is described by a separate task. Each task has a block structure containing three main blocks.

>   task of <component>
>   * Declatation Block
>
>   begin
>   * Program Block
>   end;
>
>   * Procedures and Functions Block
>   end task;

The DECLARATIONS BLOCK includes selection of modes, units, declaration of variables, and definition of interrupts; the MAIN BLOCK is the main task; the PROCEDURES AND FUNCTIONS BLOCK contains subprograms which may be invoked from within the main block, or from within another procedure or function.

Several tasks, each describing the events sequence of one active machine, together with a connections file that defines the connections between the communication variables of the tasks, constitute together a cell operation sequence. Each task is compiled separately and linked to other tasks to create a cell operation simulation file.

DATA TYPES

TDL supports the following basic types of variables: integer, real, boolean; and the following ones which are specific to robotics/CAD/CAM: location, orientation, point, vector, pose, compname. For all those data types, the array constraction is available.

TDL INSTRUCTION SET

The TDL instructions include **control statements** which support program loops, conditions testing, and transfer of control.

The **motion commands** move the joints of mechanisms in the workcell. In TDL there are two classes of motion commands:
- Commands which move joints to a given joints values; these commands are used with devices or robots:
  * DRIVE <JOINT>
  * MOVE <POSE>
- Commands which move the robot so that the TCPF of the robot is superimposed upon a given location; these commands are used only with robots (they employ an inverse-kinematic solution):
  * MOVE <LOCATION> / BY <OFFSET>
  * APPROACH, DEPART

There is a number of options which may be employed with motion commands. So for example the **straight-line-motion** ensures that motion of the TCPF between the current-location and the destination-location(s) will be on a straight line; the **fixorient** ensures that the orientation of the TCPF along the movement will remain costant.
The motion commands include options for defining the speed of the motion or its duration.

The **object handling** commands are used to deal with workpieces, components, or other workcell objects, and with families of them, and to define relations among them. These commands are simulating real-world events. For example, in a real workcell a new workpiece may be brought into the workcell manually; in simulation, we have to "generate" a new workpiece.

There are two kinds of object handling commands:
- Commands that alter the object: GENERATE, MODIFY, REMOVE.
- Commands that affect the current attachment between objects: ATTACH, DETACH, TAKE, LEAVE, RELATE.

Object handling commands do not change the location of the handled object; the transfers are effected by the motion commands.

The **object selection feature** permits selection of a group of objects, without naming their specific identities. The selection is based upon three factors:
- The component model or family.
- The proximity criterion.
- The attachment criterion.

In order to simulate parallel sequences, much attention was given to **intertask communication** in a workcell. Two forms of communication are possible:
- Synchronous communication.
- Asynchronous communication (interrupt).

Synchronous communication are effected by creating a unique signal or pulse in one task and having a WAIT instruction which expects this signal or pulse in another task. The WAIT instruction suspends task execution until the signal is received.

Asynchronous communications are effected by a special WHEN procedure that expects an input signal or pulse:

```
when <condition_on_input> do
   <statement_list>
end when;
```

The WHEN statement block, as previously explained, is located in the declaration block of a task and not in the program sequence. Whenever a signal is received from another task, during task execution, the normal flow of the task is interrupted and the list of instructions in the WHEN block is executed. The WHEN statements block is the method for interrupt handling.

EXAMPLE

As an example, the following TDL task is written in TDL:

A S-360 robot is part of a cell that includes the following components:
* a robot: name = S360
* a rotating index table: name = index_table
* an assembly table: name = assembly_table
* 4 pins (workpieces): model type = pin
* 4 boxes (workpieces): model type = box

The task of the robot is to put each pin into one of the boxes.

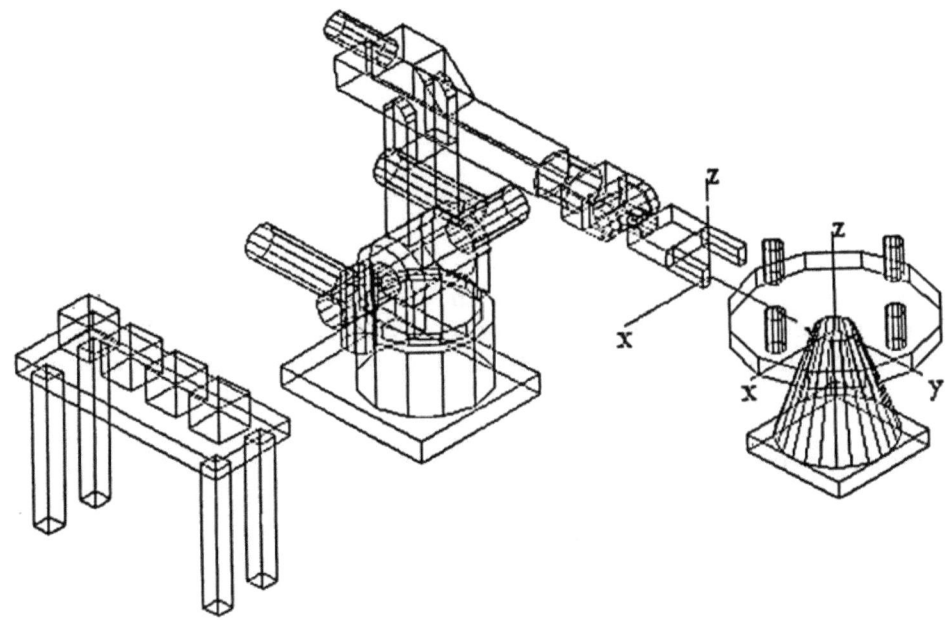

The TDL task for the the robot would be:

```
task of s360;
var
        i : integer;
        s360_start_point : location;
        box_location[4] : location;
        box, pin : compmodel;
begin
    s360_start_point = [0, 1880, 1620, 0, 0, 0];
    box_location[1] = b1;
    box_location[2] = b2;
    box_location[3] = b3;
    box_location[4] = b4;
    for i = 1 to 4 do
      move [0, 0, 450] rel index_point;        -- move above pin
      move index_point;                        -- move to pin
      take pin;                                -- attach the pin
      move by [0, 0, 450];                     -- move above pin
      pulse rotate_table;                      -- pulse to index_table
      move [0, 0, 450] rel box_location[i];    -- move with pin
      leave pin onto box;                      -- leave pin on box
      move by [0, 0, 450];                     -- move above box
      wait ready;                              -- wait from a signal
    end for;
    move s360_start_point;
end;
end task;
```

The first part of the task, preceding "begin", is the declarations block which defines the type of each variable. The second part of the task, starting with "begin", is the program. The task contains various elements such as COMMUNICATIONS (ie pulse, wait), OBJECT HANDLING COMMANDS (ie take, leave), and MOTION COMMANDS (move statements).

Two signals are involved in COMMUNICATIONS: The "pulse rotate_table;" signal is sent by the robot to cause the index table to rotate; the "wait ready" signal informs the robot that the index table has finished rotating.

Two OBJECT HANDLING COMMANDS are used in the task: The "take" instruction disconnects the object from its current attachment, and attaches it to the robot's gripper. The "leave" instruction places an object (the pin) held by the robot on to an object (the box) and attaches it accordingly. The two instructions do not change the location of the handeled object, but they determine its attachment.

The line "S360_start_point" identifies a location. It is a sextuplet, posessing both position (x, y, and z) and orientation (a, b, and c). The variables "index_point" and "b1" to "b4" are also locations (working points) which were created and defined in the workcell layout phase.

The MOTION commands can take a variety of forms in TDL: the "move index_point" instruction moves the robot TCPF to a location named index_point; the instruction "move (0,0,450) rel index_point" moves the TCPF to a location which is 450 mm above the location index_point in the z-direction.

The index_table task consists only of a "when procedure":

```
task of index_table;
when rotate_table do                    -- pulse from s360
    signal ready = off;                 -- signal (ready = off) to s360
    drive idjoint by 90;                -- rotate index_table by 90
    signal ready = on;                  -- signal (ready = on) to s360
end when;
end task;
```

When the pulse "rotate_table" in the robot task is activated, the commands in the "when procedure" are executed.

OFFLINE PROGRAMMING

One of the most important features of ROBCAD, which greatly facilitates the application of automatic workcells, is the provision for offline programming of the robot and other controllers in the workcell. Two basic methods for offline programming are implemented:

- Appending native robot instructions within a TDL program.
- Activating built-in translators which automatically translate the TDL into various robot languages.

The first method is more efficient and accurate. The user writes the program on the ROBCAD system in the robot native language eg KAREL, and simulates the sequence of events. The robot program can be downloaded directly into the robot controller.

With the second method, the user works with TDL only; the translation is accomplished automatically upon request. If many robot types are implemented in the same factory, this method is superior to the first (since TDL is robot independent). Yet a higher level of accuracy and efficiency in the code generated for the robots is easier to achieve with the first method.

SUMMARY

TDL is a simulation language for automated workcells. The TDL is used to verify the design of a workcell, by simulation of its operation, and is used as an offline programming language for robots and other mechanical devices.

The Task Description Language is robot-independent, and so different robots can be programmed and simulated for a given task without the need to rewrite the program.

REFERENCES

[1] Grossman, D. D., "Programming a Computer Controlled Manipulator by Guiding through the Motions", IBM Watson Research Center, Research Report RC6393, 1977.
[2] Lozano-Perez, T., "Robot Programming", MIT Artificial Intelligence Laboratory, Memo No 698.
[3] Mujtaba, S., and Goldman, R., "AL user's manual", Stanford Artificial Intelligence Laboratory, AIM 323, January 1979.
[4] Unimation Inc., "User's Guide to VAL: A Robot Programming and Control System", Unimation Inc.,Danbury, Conn., Vers12, 1980.
[5] Taylor, R. H., Summers, P. D., and Meyer, J. M., "AML: A Manufactoring Language", Robotics Research 1, 3 (Fall 1982).

General theories of flexible integration
J. E. Lenz
CMS Research Inc, USA

Abstract

The three basic categories of manufacturing are the job shop, the transfer line and the flexible manufacturing system (FMS). Each can be considered the result of evolving manufacturing requirements. The FMS, first introduced in the early 1970s, is the most recent solution to manufacturing problems and it is constantly being improved as the technology evolves. The FMS allows production volume to be increased or decreased and allows the product to be varied within specific limits.

The FMS has often been treated as a hybrid of the job shop and transfer line systems. Because of this approach, many traditional design and evaluation tools, ranging from job shop priority scheduling rules to transfer line balancing techniques, have been applied to FMS development. All of these tools, however, treat the FMS as either a job shop or transfer line system. They use algorithms developed from relationships that do not exist in the FMS. The result is an FMS that is both unpredictable and inefficient.

Thus, there is a need to establish the FMS as an independent system of manufacturing with its own tools for design and evaluation. These tools are based on the lessons learned in the job shop and transfer line systems, yet they also take into account observed or predicted relationships within the FMS.

"GENERAL THEORIES OF FLEXIBLE INTEGRATION"

Introduction

There are many different views of manufacturing and each one promotes some theories which assist in explaining and understanding the strategy. For example, the job shop has scheduling rules, printing rules and product versus process layouts. With each of these views are algorithms which can be applied. These range from shortest processing time scheduling to due date priority schemes. In all of these algorithms, the objective is to obtain efficient use of all stations. However, not one of these rules deals directly with integration effects. Integration effects are defined in Table 1.

In a sense, all of these job shop algorithms tend to assume there are no integration effects and the problem is to constantly prioritize the work. However, when running in a low inventory environment, there is no large buffer of work to prioritize and more importantly, the service of one station will directly affect the usage of another. In this environment, there is a need to identify and measure these integration effects. This is the subject of the remainder of this paper.

Table 1

INTEGRATION EFFECTS

1. STATION IS AVAILABLE WITH NO PARTS

2. STATION IS BLOCKED BY PART WHICH HAS NO PLACE TO GO

Production versus Inventory

The relationship between inventory and production has been learned from many job shop applications. In the job shop, inventory is raised until the operation of one station is isolated from what might happen to any other station. That is, each station is surrounded by inventory so that its operation is independent of all other stations. This creates a situation where production is tied to inventory level. If the inventory level were to drop, stations would become less independent so their dependency will cause some interruption in service. These are termed integration effects. In the job shop, integration effects are eliminated by maintaining longer inventories around each station. The penalty for increasing these inventories is in extending the flow time of parts. But since inventory can accumulate without creating serious trafficking problems, the flow time has no effect upon station usage (production). However, flow time does have a critical relationship to production in transfer line manufacturing.

Production versus Flow Time

The objective of the transfer line is to obtain perfect balance. In this situation, no inventory is needed to smooth over integration effects because all parts spend exactly the same amount of time at each station. As a result, changes to the flow time have direct relation to production and as well, station usage. For example, if the flow time decreases (either by shortening the line or decreasing the cycle), production-station usage must increase. The converse is true as well; as the flow time increases, production will fall.

Table 2

MANUFACTURING STRATEGY

JOB SHOP   -   INVENTORY

TRANSFER LINE   -   BALANCED OPERATIONS

FMS

LOW INVENTORY MANUFACTURING
WITH UNBALANCED OPERATIONS

Triangle of Integration

Table 2 contains the three major manufacturing strategies: Job Shop, Transfer Line and FMS. The job shop can eliminate integration effects by raising inventory levels. The transfer line eliminates integration effects by obtaining balance cycle time for all operations. The FMS becomes the hybrid of the two previous strategies and can take on the following definition.

Flexible Manufacturing Systems are any manufacturing environments where low inventory exists with unbalanced operations. The degree of automation in this environment has very little input on these characteristics. Therefore, the characteristics of FMS can be applied to any job shop with low inventory, unbalanced transfer line or any other type of integrated manufacturing system. Identifying the integration effects within the FMS is the topic of the next section.

Raising inventory levels within the FMS will have a positive impact upon station usage. This basic relationship is understood from job shops. As mentioned above, a side effect of larger inventories is increased flow time. In the closed FMS, this change to flow time will have effects upon station usage similar to that seen in transfer lines. Raising inventories will increase flow times which will have a negative impact upon station usage. The residual impact upon station usage due to an increase in inventory will be equal to the positive effects of larger inventory minus the negative effects due to increased flow time. This relationship is termed the "Triangle of Integration" and is graphically shown in Figure 1.

For station usage below 80%, an increase in flow time will have a small impact upon production and an increase in inventory leads to predictable increases in station usage. However, at 90% usage or above, an increase in flow time will contribute greater effects upon station usage. At some point, the effect upon station usage due to an increase in flow time can be greater that the effect due to increased inventory. In this case, production will actually fall by increasing inventory levels. Many more examples can be made, but each one enforces the fact that integration effects within the FMS are due to both inventory levels and flow time. Also, these two are always contributing offsetting effects. If one is positive, the other will be negative. But to use this knowledge, actual values must be measured for each. Expertise is needed to quantify the inventory effects and flow time effects upon production.

Expertise in Measuring the Triangle of Integration

One useful tool in quantifying the integration effects of an FMS is the WIPAC Curve. WIPAC stands for Work In-Process Against Capacity and shows a graphical relationship between inventory and FMS capacity. The curve itself represents the minimum inventory level needed to fully utilize all available capacity of the FMS. An example of the WIPAC Curve is shown in Figure 2.

GENERAL THEORY OF INTEGRATED MANUFACTURING (FMS)

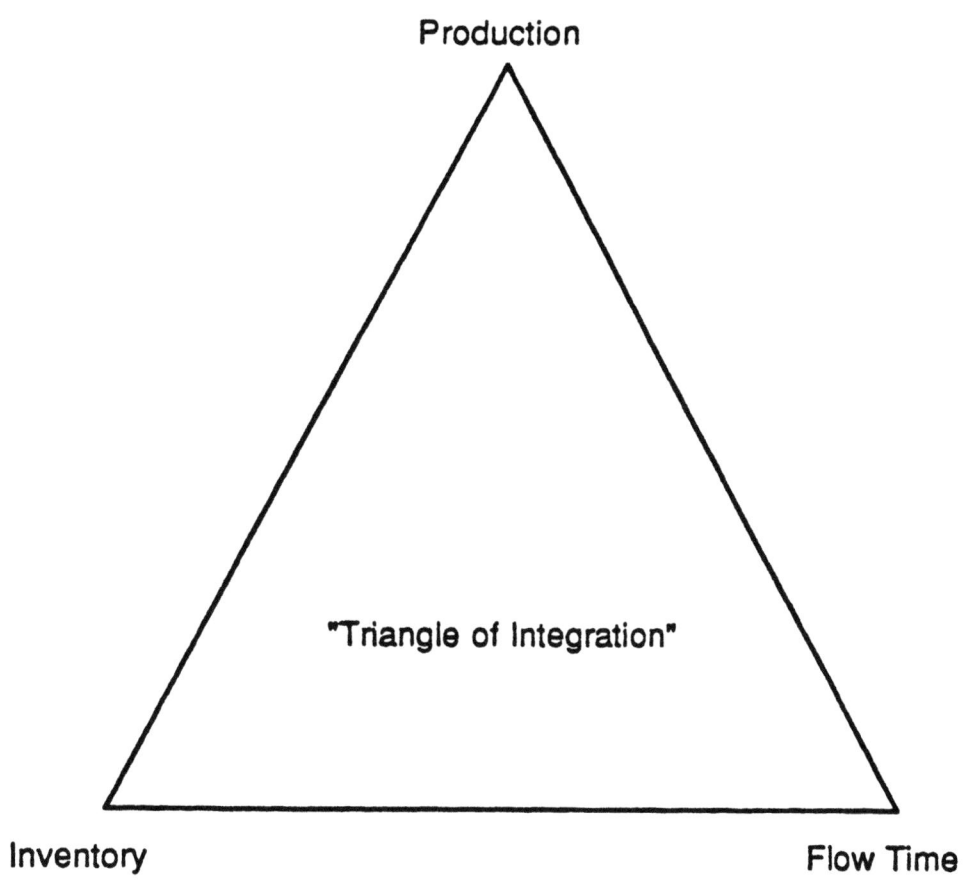

Figure 1

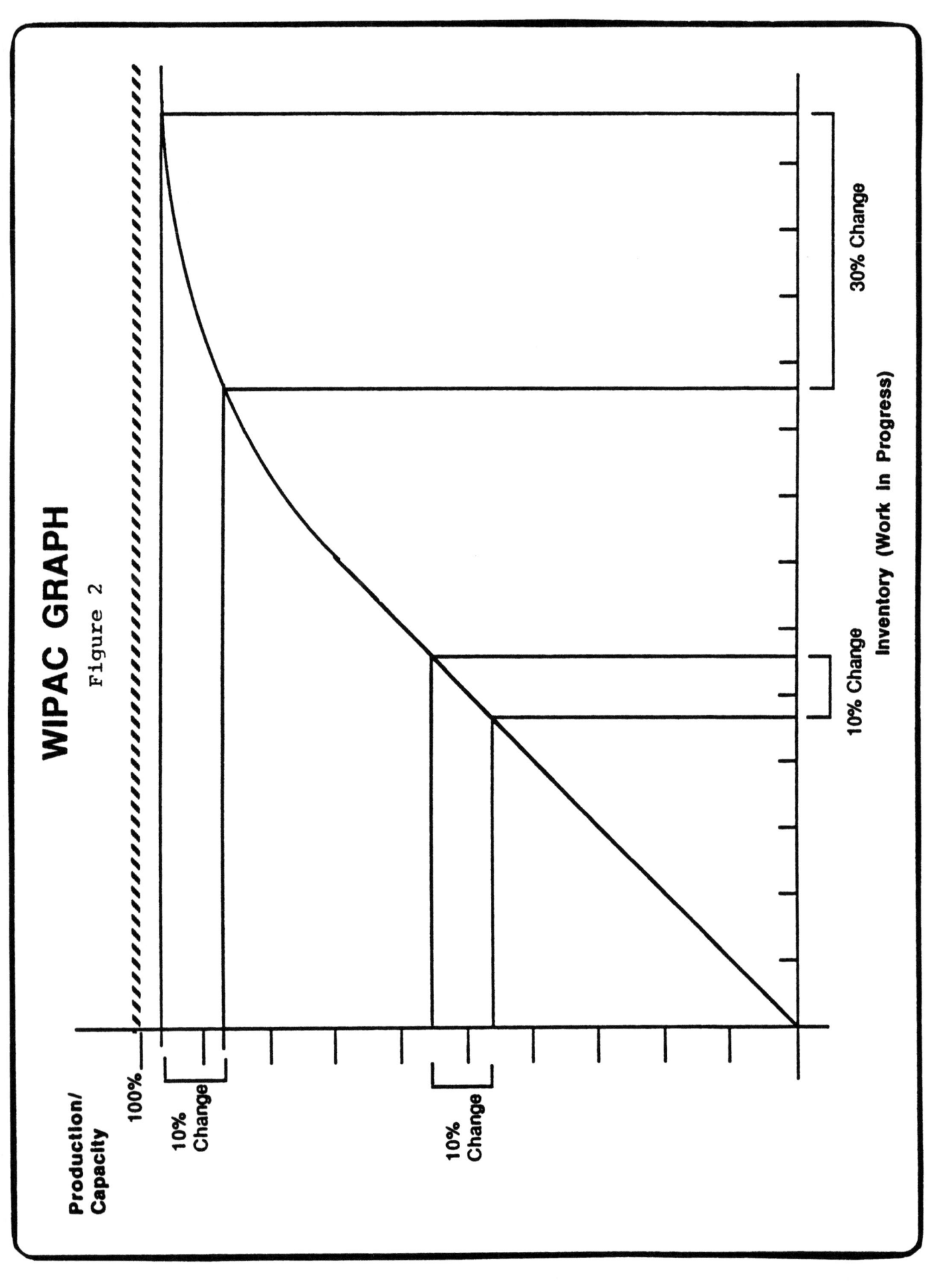

The curve flattens out due to increases in flow time resulting from higher inventory levels. This has negative effects which offset some of the positive effects upon station usage due to the addition of pallets. Following is a discussion of the characteristics of this curve.

The WIPAC Curve has two corollaries associated with it. The first is shown in Figure 3. In this characteristic, the flow time increases as you move along the curve. The increasing time is what causes the curve to bend.

The second corollary to the WIPAC Graph is shown in Figure 4. In this characteristic, the production rate or station usage will be equal to station availability at any point along the curve. The reason for this characteristic is that flow time, inventory and production are in balance along this curve and at any other point, they are not. This imbalance results in higher integration effects which lower production even though stations are available. The remainder of this paper introduces the concept of expertise and defines what expertise is needed in applying these General Theories of Flexible Integration.

The expertise which is needed in development of the curve is described. The first stage in generation of the WIPAC Curve is to establish lower and upper bounds as selected levels of system capacity. This can be accomplished by utilizing some features within the SPAR Planning Software for FMS. Specifically, the planning horizon is adjusted so that station usage does not exceed decreased levels and the corresponding number of pallets establishes a single point in the chart. SPAR also permits the ability to adjust the flow time for pallets and for each flow time, a bounding line can be established for the WIPAC Curve. These bounds are depicted in Table 3.

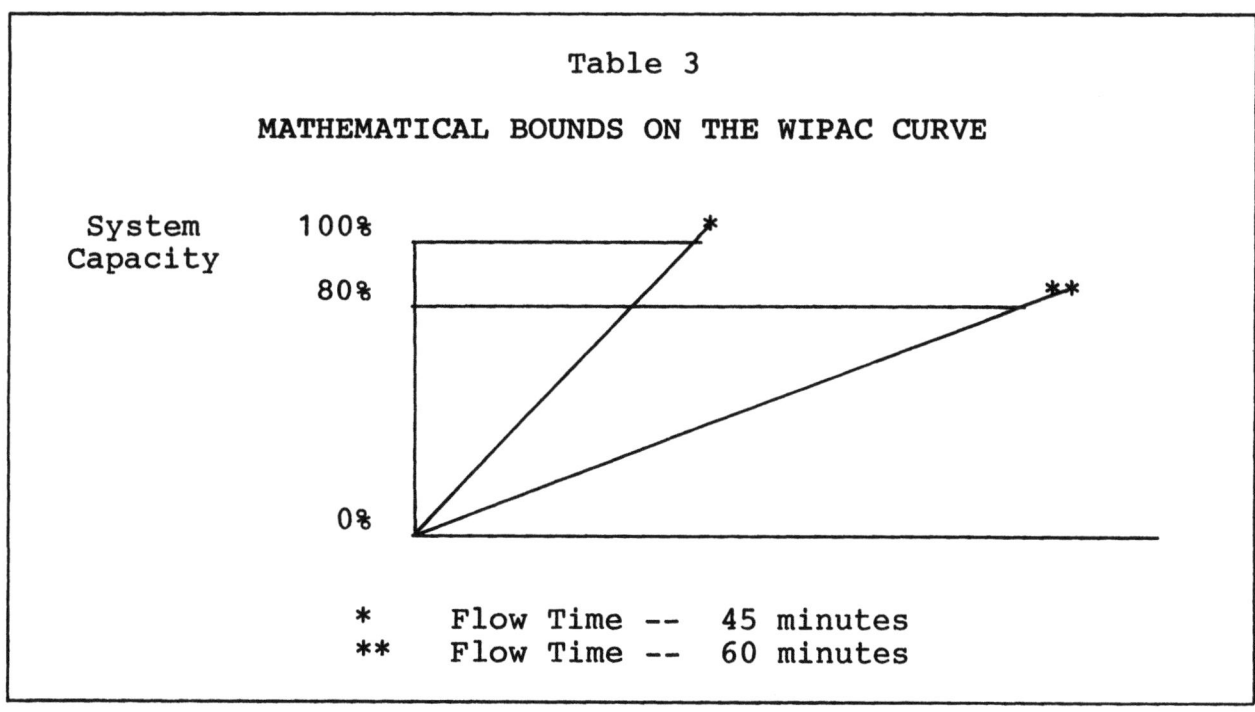

Table 3

MATHEMATICAL BOUNDS ON THE WIPAC CURVE

*   Flow Time -- 45 minutes
** Flow Time -- 60 minutes

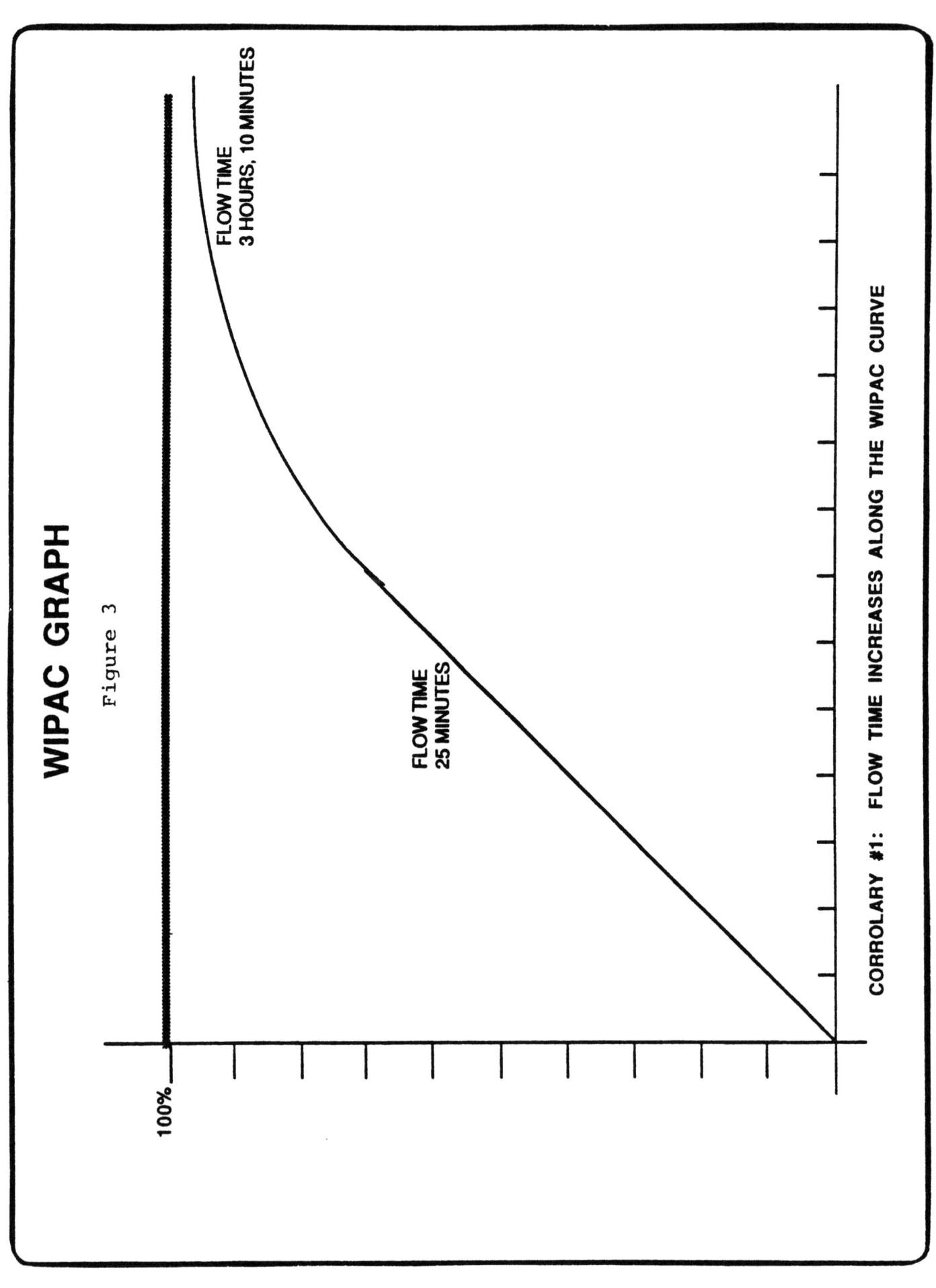

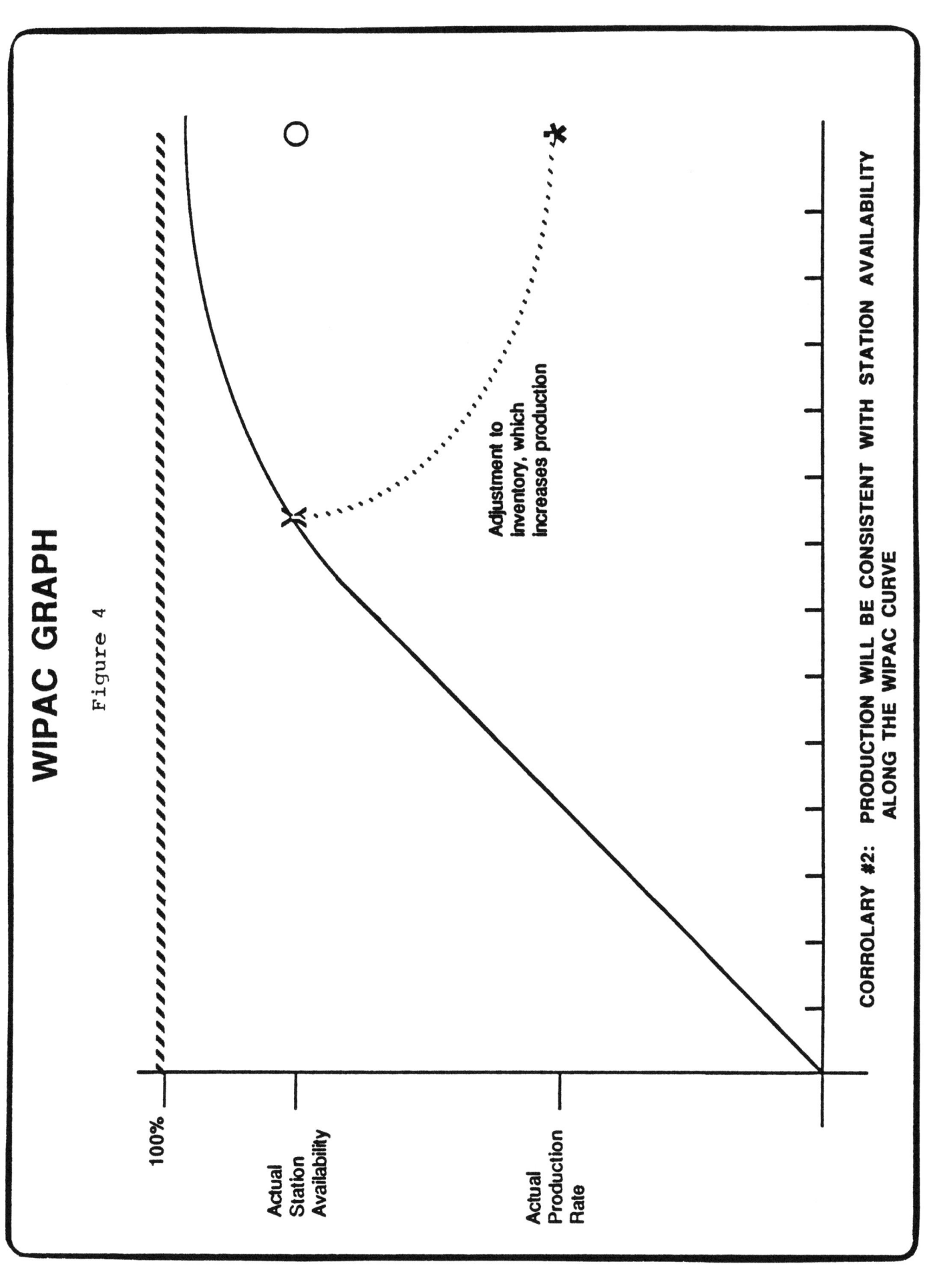

These curves will be straight lines because the flow time will not vary as the number of pallets increases. The next step is then to establish the actual WIPAC Curve using the bounds as starting points for evaluation. The relationship between flow time, inventory and station usage has not been defined mathematically, so computer simulation is used to measure this relationship. Each point on the WIPAC Curve requires a computer simulation run with judgment of controlling the number of active pallets in the system. Here again, expertise is needed as to what points should be selected and for selection of specific values for decision parameters. All of the expertise described above can be gained from training, education and repetitive procedures. The portion of the expertise which is not as easily gained (or transferred) is that for evaluation of performance.

Before a simulation result can be used as a point of the WIPAC Curve, the results must not contain any bias because of inefficient system utilization due to poor control algorithms, unrealistic constraints or errors in the model. This area of expert analysis of FMS performance is the single most difficult area in which to gain expertise. But tools such as expert system, knowledge base and inference engine offer potential solutions.

Table 4 contains a formal procedure for FMS Design and Evaluation. This procedure establishes a basis for expert system application to the problem of FMS.

Table 4

'STEPS TO FMS DESIGN AND EVALUATION

1. "BALL PARK" THE SYSTEM:

    USE OF MATHEMATICAL MODEL TO
    ESTABLISH TARGET REQUIREMENTS

2. QUANTIFY THE INTEGRATION EFFECTS:

    USE COMPUTER SIMULATION TO PRODUCE
    ACTUAL LEVELS OF USAGE

    COMPARE THESE TO THE TARGETS

3. ESTABLISH OPERATIONAL STRATEGIES:

    QUALITY ASSURANCE
    SCHEDULING versus "PULL-TYPE" SYSTEM
    DATA COMMUNICATIONS

management system of an FMS of prismatic workpieces

J. Pylkkänen
University of Oulu, Finland

ABSTRACT

The efficient production should have short and precise lead times, small batch sizes and zero defects. The parts and subassemblies of a product are produced directly to the assembly section by means of just on time. High flexibility, availability and capability are required for and FMS an its subsystems. One important subsystem of an FMS, tool management system, is discussed in this paper.

INTRODUCTION

The FMS of prismatic parts for diesel engines of medium size was designed by the means of just-on-time production. The parts and the subassemblies of a product are produced precisely according to the orders from the assembly section. The efficient production with short and precise lead times, with very small batches, with zero defects and with high utilization grade of the facilities was achieved by

- minimizing material handling,
- grouping together the parts belonging to the same subassembly,
- utilizing identical machine tools in parallel and
- utilizing permanent setups in parallel.

The evolution steps of the diesel engine production is shown in Figure 1.

The layout of the FMS discussed here is shown in Figure 2. The system consists of three identical machining centres with tool magazines of 300 tools, a robot station for deburring, a washing machine, an L/U and impection station, a tool presetting device, an automatic ware house of 244 places for machine and wooden pallets, and a subassembly section. The tool presetter will be integrated into the automatic system by a microcomputer in the near future. Further, the automatic subassembly is under investigation. The company has also other flexible manufacturing systems and cells designed according to the similar principles.

WORKPIECES AND TOOLS

The workpieces to be machined in the system are at the moment cylinder heads and connecting rods of medium sized diesel engines. The main accuracy requirements are as follows:

- in boring IT 6
- in circular milling 0,02 mm
- in surface finish R_a 1.6 μm
- in parallelity 0,01/200 mm

All workpieces can be complete machined at random in any of the machining centres because of permanent setups in parallel. 260 different tools including sister tools are required for the workpieces. Efficient combination tools like boring bars with five cutting edges, solid carbide step drills and special formed milling heads and end mills are widely used.

Number of different kinds of machine tools was reduced from four - two lathes and two machining centres - to one. Complex rotational forms of the cylinder heads are completed by milling and boring.

CAPABILITY

Zero setup times and zero defects in machining were realized by built-in quality. The sources of errors were investigated (Figure 3). The influence of the error sources on the repeatability was minimized by

- accurate and rigid machining centres, tools, fixtures and clamping devices,
- a precision tool presetter,
- keeping the facilities in good condition by efficient maintenance,
- keeping the environment clean and steady,
- maintaining raw materials of even quality.

The machining centres have temperature control and cooling devices for the spindle, the main bearings and the coolant, and symmetrical construction for elimination of temperature distortion. Further, the machining centres have an automatic tool shank cleaner, and precision tool clamping and pallet clamping devices.

The correlation between presetting and machining machine by machine and tool by tool, was investigated. It differs machine by machine with same tool. Thus, the finnishing tools are never changed from the MC to the other.

The tools are mainly special design for the workpieces. Number of different tools of permanent setups was reduced by the special combination tools. The number of fittings in a tool was reduced and the rigidity was increased by the special design.

The required accuracy in process can be reached unattended by preset tools. Unefficient and unreliable measuring and compensation in process is not necessary because of high repeatability of the automatic setting change.

Responsibility of quality and material flow cover also the maintenance people. The maintenance has a great influence on quality and on repeatability in automatic setting change. Waiting times in process are automatically used for preventive maintenance and for capability inspection of the facilities.

AVAILABILITY

The system availability is the probability that at time t the system will be available for mission. The FMS requires high availability because of high price and of 24 hours utilization. Failure rate will be the greater the more complex the system is.

To assure high availability of the FMS the machine tools including setups and other facilities are linked together in parallel. The machining centres are independent and identical provided with identical setups permanently. Further, the system is easy to control and to balance. The comparison of two applications based on data measured in the FMS discussed here is shown in Table 1.

TOOL MANAGEMENT SYSTEM

Every machining centre has a tool magazine of capacity of 300 tools (Figure 4). Tool shanks are cleaned by a rotating brush and press air in the tool changer automatically to assure high repeatability of preset tools.

Each machining centre has a tool control and management system of its own. The machining centres, the tool presetting and the NC programming will be linked together by a microcomputer in the near future. The tool management system discussed here is a new development of Yasda. The diagram of the tool management functions are illustrated in Figure 5.

Tooling control by tool number ensure the easy centralized control by utilizing tool numbers of a part program everywhere in the manufacturing system. The tool number and the magazine pot number can be different. The tooling control data can be input and output as a part of part program. Registeration is divided into three kinds as follows:

(1) Tool control data (Max. 300 groups)
    - tool number, magazine pot number, tool diameter/length compensation data, tool life setting data, spare alarm setting data for tool life, cutting power setting value (4 kinds), cutting thrust setting value for Z-axis (4 kinds)

(2) Spare tool control data (Max. 300 groups)
    - main tool number, spare tool number (7 kinds)

(3) Data of tooling content confirmation (Max. 400 groups)
    - part program numbers (Max. 400), main tool numbers (Max. 600)

Content of registered data can be corrected as well as confirmed on the CRT display panel. Some exambles of display are shown in Figures 6, 7 and 8.

Spare tool control system controls spare tools for main tool. If tool breakage occurs in machining, the workpiece is taken out from MC table as a half-machined workpiece, and a new workpiece is set on the table at the same time as a damaged or broken tool is returned to the magazine pot. The used tool is changed to spare tool automatically. Moreover, when returning a damaged tool to the magazine, the system permit to return the tool to a pot reserved for damaged/broken tools. Therefore, tool replacement is easy. Through CRT display the initial magazine pot number of a tool, returned to the speciality pot for damaged/broken tools, the tool number and causes for the damage can be confirmed (Figure 7).

All tools for machining are judged by the tooling content confirmation system just after a part program number for machining is read by address code sensor or obtained through input from host computer. Even though only one tool is impossible to be used, without starting to machine, the system informs the host computer that this kind of workpiece can not be machined, and requires to transmit other kind of workpiece. A display examble is shown in Figure 8. Further, the part program list of machinable and unmachinable workpieces, the part program list by a tool and the tool list by a program number can be displayed on CRT.

Only the principle of the tool management system developed by Yasda and its main functions are briefly discussed here. The content of all functions mentioned in Figure 5 can be displayed on the CRT of MC, printed out on a printer and transmitted to the host computer. Further, the Yasda PSC system has functions for production management and for self-diagnosis.

CONCLUSIONS

Tool management system is an important subsystem of an FMS. For achieving high utilization grade of an FMS in flexible environment the capability and the availability of all subsystems integrated together must be high. If the capability and the availability of machine tools and tools themselves are poor, it is very little to be done with an advanced tool management system, and vice versa.

REFERENCES

[1] Pylkkänen, J. (1984). Finnish FMS exceeds exceptation. The FMS Magazine, April 1984, 82 - 85.

[2] Pylkkänen, J. (1985). Wartsila. FMS in Action. Flexible Manufacturing Systems. Edited by H.-J. Warnecke and R. Steinhilper, IFS (Publications), UK and Springer-Verlag, Berlin, Heidelberg, New York, Tokyo, 237 - 244.

[3] Pylkkänen, J. (1986). Auslegung eines Flexiblen Fertigungssystems für Belange der Produktion auf Abruf. Vorträge, Internationaler Kongress Metalbearbeitung, März 1986, Karl-Marx-Stadt, DDR.

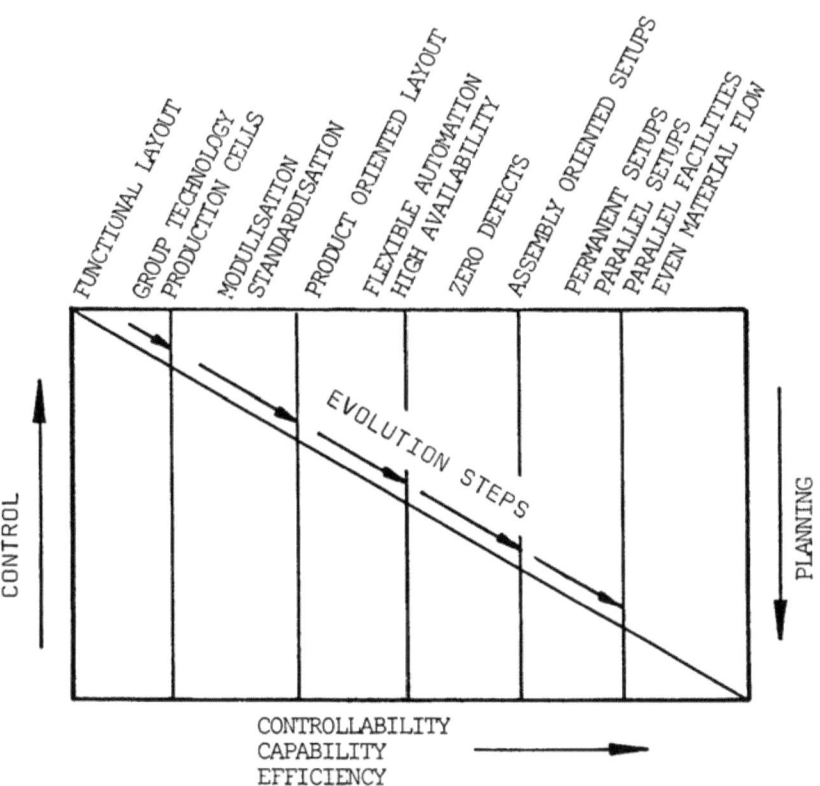

Fig. 1. Main evolution steps of the diesel engine production.

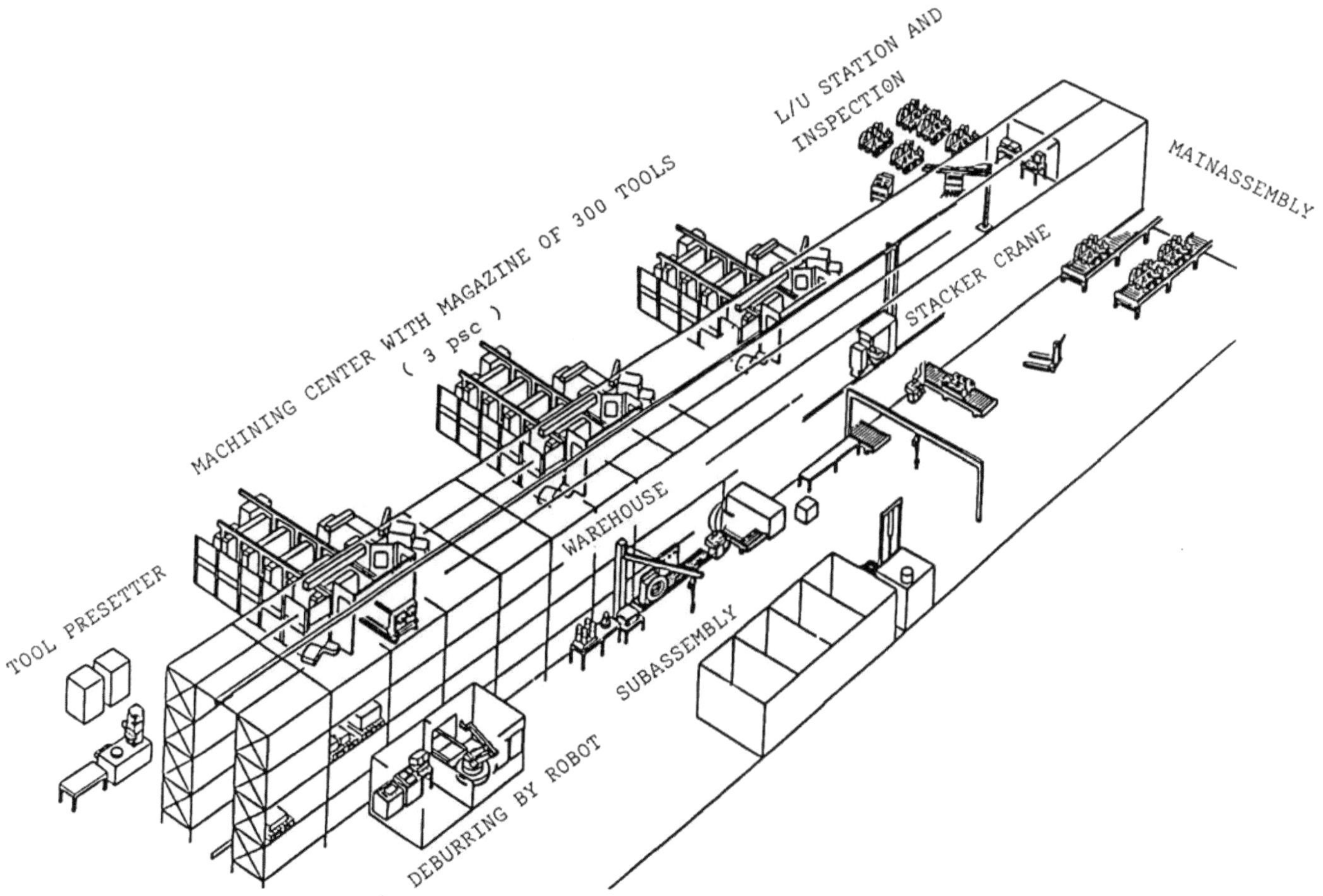

Fig. 2. FMS with three identical machining centres with identical permanent setups.

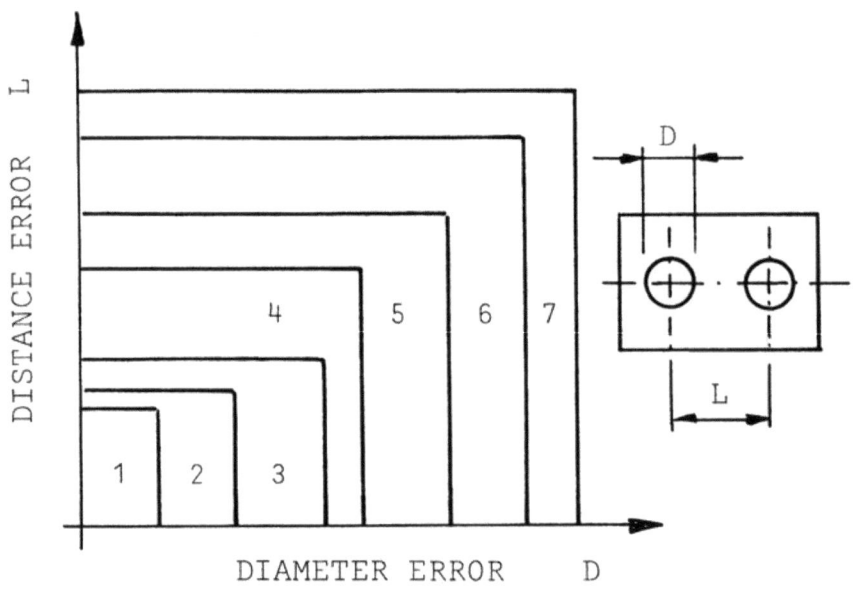

1) Basic machine tool with clamping devices
2) Tools, tool wear and presetter
3) Raw material
4) Maintenance
5) Uncleaness
6) Random variation of temperature caused by very small batch production
7) Ventilation

Fig. 3. Error sources in boring.

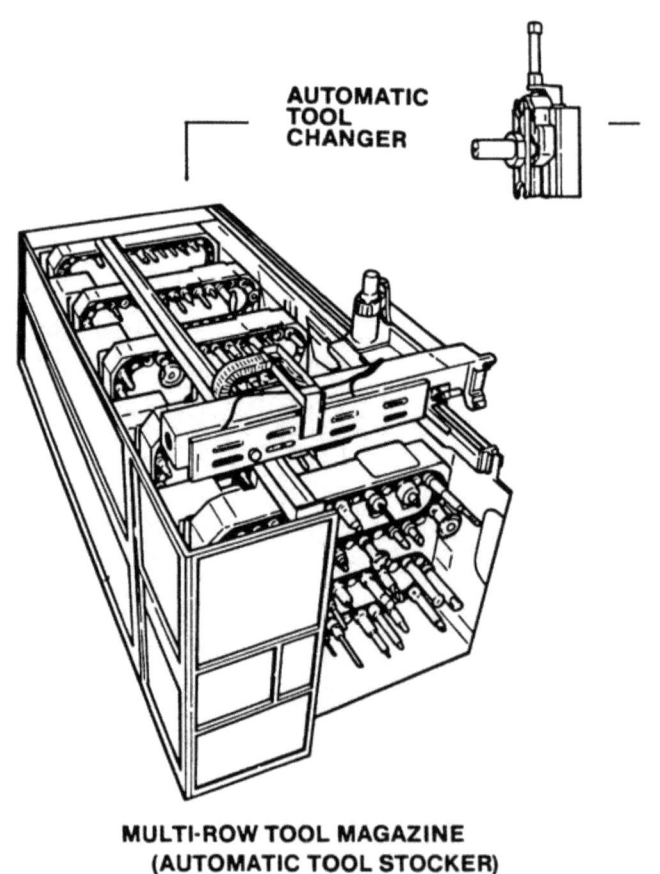

Fig. 4. Multi-row tool magazine with ATC and tool shank cleaner.

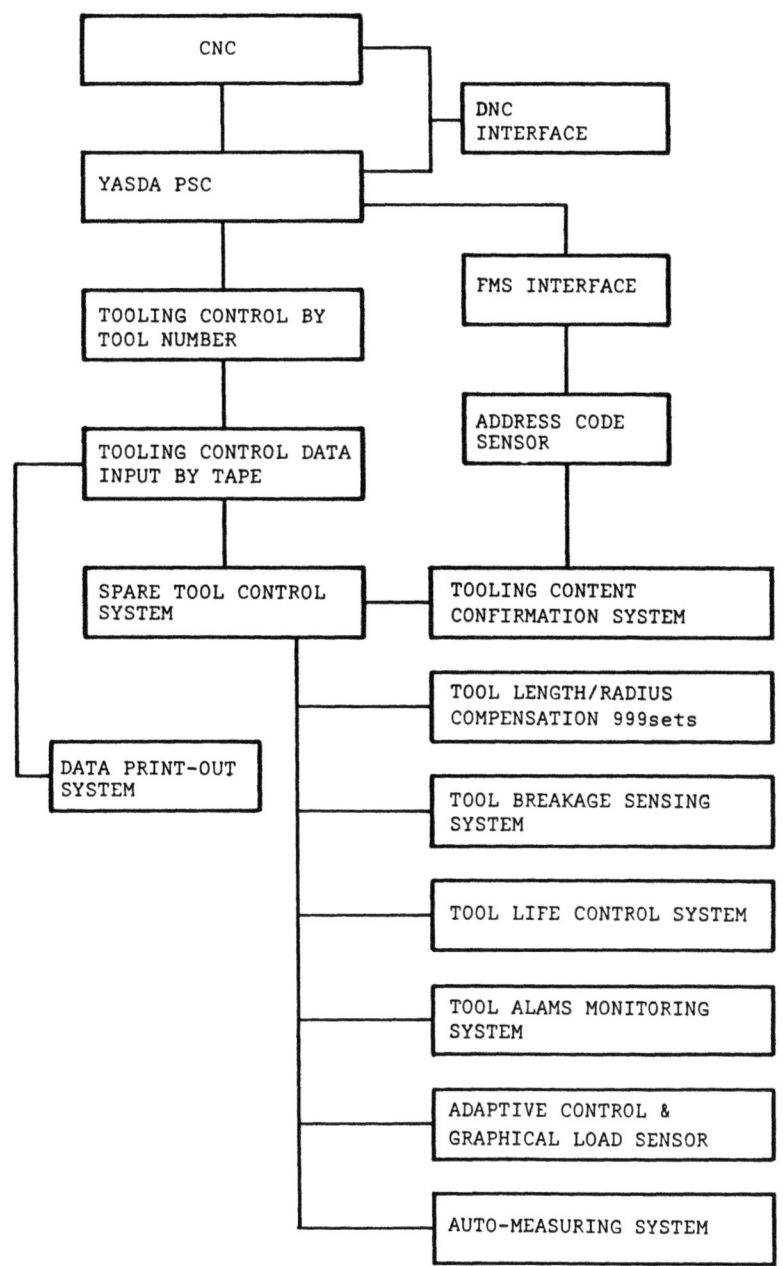

Fig. 5. Diagram of tool management system by Yasda.

Fig. 6. Display example of registered tool control data.

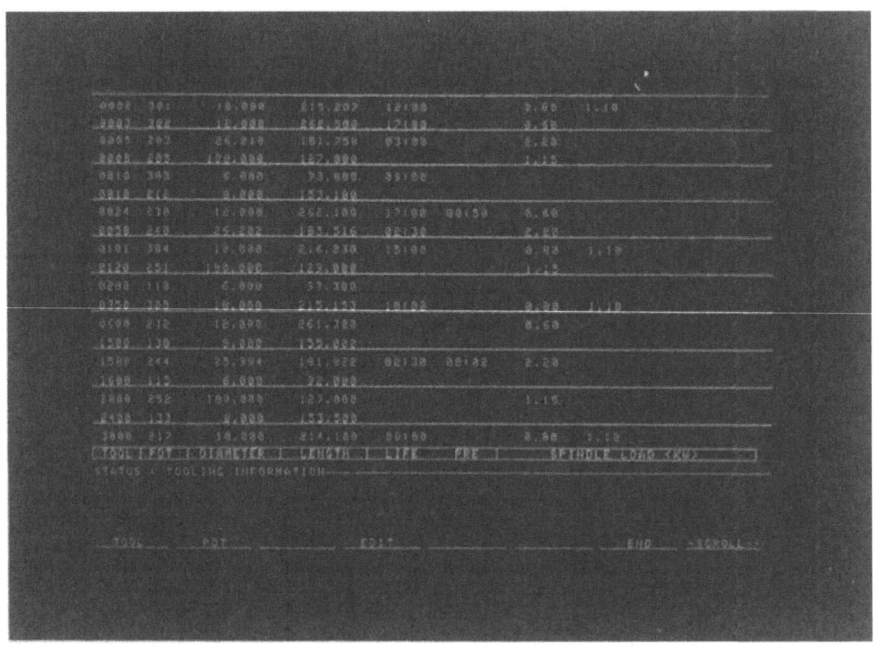

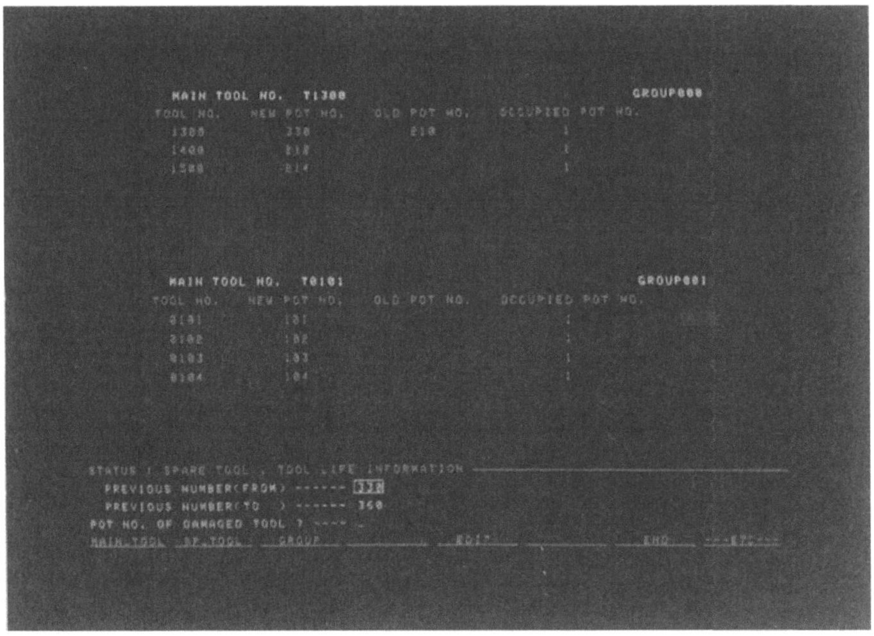

Fig. 7. Display example of spare tool control data.

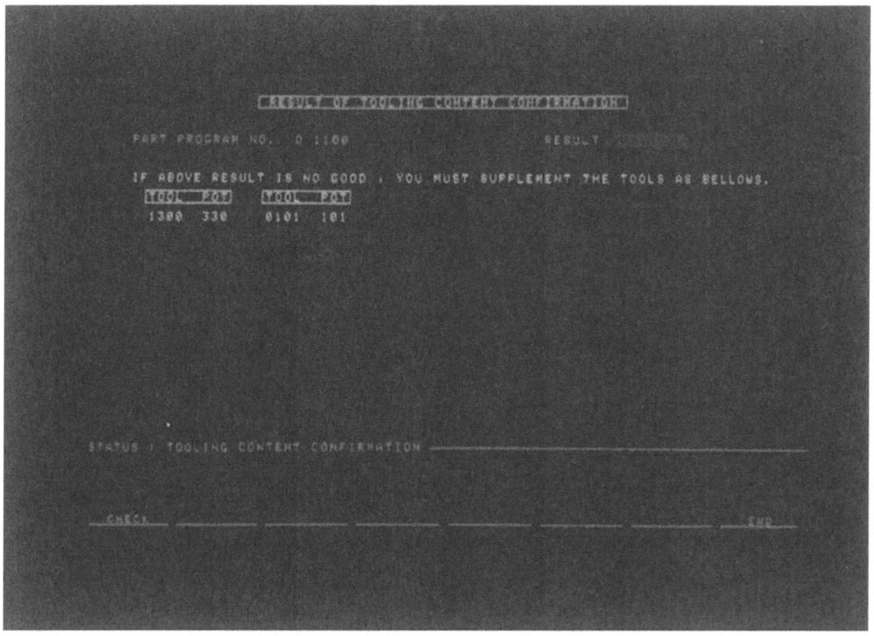

Fig. 8. Display examble of stored tooling content confirmation.

Interconnection technology on basis of automatic crane installations
H. J. Roos
Erwin Mehne GmbH & Co, West Germany

The crane is the oldest means of conveyance, and the one on which most research has been done on all of its constructional components. The available standards and regulations applicable to cranes are wide-ranging, and hoisting system classifications cover the entire service life. Conversions to changing applications are always possible. This enables clear definition of an important interface between the manufacturer and user. A hoisting system's favourable degree of space utilization (space required/operating area) provides the best possible prerequisites both for changes in use and also for extended utilization in existing buildings.

AUTOMATIC CRANE INSTALLATIONS IN PRODUCTION

The crane is the oldest means of conveyance, and the one on which most research has been done on all of its constructional components. Its classical function is to transfer goods and interlink components in the production plant. With reference to various practical examples, this paper illustrates the possibilities offered by automatic crane systems in production.

Free competition is forcing all enterprises to continually reduce costs incurred per product unit. Thanks to rapid developments in automation technology and computer-aided factory organization, it is now possible to adapt all planning procedures to the model of the automated factory (CIM concept).

In the concrete production plant, it is always possible to assume a combination of three systems, i.e.:

- Human mental and manual labor
- Mechanized labor
- Automated labor

In recent years, automatic systems have been incorporated into production and assembly lines as insular solutions and must now be integrated into overall industrial complexes for flexible production. These systems are only ever conceivable in conjunction with automatic warehousing systems (i.e. stores for raw material, finished parts and goods for shipping).

The means of conveyance used in the interconnection of plant systems has become an integral part of production and assembly. The consideration given to the conveyor in the planning phase is as extensive as the planning work devoted to the machine tool, the chucking device, the tool or workpiece pallet.

Investments are always planned for a specific product spectrum, and the return listed in the investment calculation always refers to the existing market situation. We know at the same time, however, that products' life cycles are becoming shorter and shorter and that capital goods frequently survive the current product. No information on future products is available and so a few general statements covering easy conversion and simple programming etc. have to take the place of an exact technical specification. Referred to the means of conveyance in use, this means:

- Simple replacement of the load suspension device by means of mechanical interlocking
- Simple conversion of the control to new load and distance curves

- Verifiable operational strength and high availability in the event of changing tasks.

PRINCIPLE FUNCTIONS OF DISCONTINUOUS CONVEYOR USED FOR LARGE AMOUNTS OF SMALL GOODS

At the start of our consideration we specify the load units (LU) to be handled and conveyed in the form of large amounts of small goods. Small goods are handled quasi-continuously with cycle-controlled continuous or discontinuous conveyors. The conveyor links the material flow Q (LU/h) between production or transportation processes.

There are various reasons for us to recommend a track conveyor for the solution of these conveying tasks (Figure 1).

a) The practical construction reasons are:

- Space savings (cost-relevant): up to 40 % in comparison with FTS operation

- Room height utilization (cost-relevant): new load bearing members e.g. tubular cranes with tangential crabs (Figure 2)

- Calculation of approach dimensions for warehousing design (cost-relevant), i.e.: new arrangements such as use of a telescopic crab or crane warehouse (German utility model)

- Direct mains connection of drives via a contact line or trailing cable (cost-relevant)

- Minimization of risk: the crane, including its main assemblies, is the one means of conveyance which has undergone the most research in engineering science.[1] This is reflected in the available technical and accident prevention regulations.

- The possibility of use in existing buildings without too many conversion measures detrimentally affecting production.

b) System reasons are:

- "Intelligent" load units with mobile data memories (Premid, Moby-M etc.), e.g. workpiece containers

- Development of a dispatcher, a new type of machine operating as a buffer and feed unit on the machine tool (Figure 3)

- Flexibility in handling by the replacement or automatic adjustment of a load-bearing strut bars.

- Development of the teach-in control for easy programming of automatic discontinuous conveyors

- Configuration of a process control in four levels: i.e. master production computer, field computer, field control and machine control (Figure 4)

c) Design reasons are:

- The use of laser systems enables optimum crane track alignment

- Inexpensive, industrially proven control hardware facilitates transmission of the automation cycles of retrieval conveyors or trolleys and their connection to the crane

- Positioning accuracy is improved by new drives and transmitters (by up to $\pm$ 0.5 mm)

- The calculation models based on structural mechanics provide a reliable forecast of the dynamic system behavior.

CONVEYING PROCESS

The crane picks up the individual loads (LU) at a starting point A (x_1, y_1, z_1), moves them over a defined and reliable spacial track curve to the destination point B (x_2, y_2, y_2), positions them in a destination field $\pm \delta(\delta_X, \delta_Y, \delta_Z)$ and lowers them. It then travels unloaded to a new starting point. The time needed for all necessary travels must ensure that the entire production process can be served appropriately. The operator therefore defines or estimates the following items, and thus specifies them:

- Size, nature and frequency of the load mass (LE) (= load collective)

- Number of operating cycles during the calculatory service life of the crane and its drive

- Frequency distribution of the starting and destination points, and also average distances travelled

- Size of the destination field $\pm \delta_i$

- Scope of the material flow data to be recorded for destination control

These data must be laid down in a specification by the manufacturer and operator, thus enabling the designer to plan the crane in accordance with requirements.

PRACTICAL EXAMPLES

Even if ordered for manual operation, modern cranes should be equipped with monitoring facilities and designed in such a way as to permit future automation.

The following examples illustrate automated crane operations over a wide range of automation levels.

AUTOMATED EMPTY RUN: TUBULAR CRANE WITH REMOTE CONTROL

The remote control system consists of portable, pocket-sized transmitters similar to those used for TV sets. Receiver circuits installed at optionally chosen intervals on the crane track can be activated with these transmitters. Automated empty travel of the crane to the selected operating field is achieved by means of a simplified positioning process. That is to say, only one of the crane's operating fields can ever be activated at one time. Once the crane has finished its operation in this operating field, it is released for automatic travel to other fields. This remote control system can also be used for crab motions and systems consisting of several cranes. There is thus no need for the time-consuming task of fetching the crane for travel to its respective location of use.

MASTER CONTROL LINE: LOAD AND DISTANCE CURVE OF A SHORE UNLOADER
(Figure 5)

The grab of a ship unloader is controlled on an automatic master control line which pulls the grab over the bunker edges after manually-actuated grab operation, and which opens the grap in the center of the bunker, after which it returns the crab to the jib. During this travel, the grab is again manually controlled by the crane operator.

HALL WITH A CRANE OPERATING AS A GOODS TRANSFER AND STORAGE MACHINE (3) (Figure 4)

The bridge crane has always been the optimum storage and transport machine in warehouse halls and workshops, and a large number of demands must be fulfilled simultaneously when developing an automatic crane.

These are:

- High starting accuracy

- Mechanical load-bearing devices such as nippers and grabs etc.

- Designs requiring little maintenance and allowing good accessibility by servicing technicians to maintenance points

- Comprehensive status displays and reports for the control center

- Precautions for the rapid detection and elimination of mechanical, electrical and electronic malfunctions

- Comprehensive safety devices in compliance with the accident prevention regulations, including safeguards against operating errors

- Simple light-weight construction and use of standard elements

- Easy, low-noise operation

If these basic functions are observed, an automated crane is capable of considerably simplifying goods transfer tasks in a hall or the operation of single machines.

AUTOMATIC BRIDGE CRANE FOR HANDLING STACKER RACKS (4)

Stacker racks can be loaded with an extremely wide range of goods. This example illustrates three rows comprising approximately 300 stacker racks in a storage hall. Using an automatic crane, up to 6 of these racks can be stacked on each other. A cross beam which holds the stacker racks on specially designed hooks serves the purpose of automatic load carrying. If necessary, restacking is also performed automatically. The double girder bridge crane has a double hoisting unit so that the cross bean is held on 4 x 2 strands of cable. In its highest load position, the cross beam is guided on the crab frame to prevent oscillation during travel.

In its initial capacity stage, the system is controlled from a stationary control console by means of punch cards. Connection to the warehouse administration computer is planned for the second capacity stage.

FULLY AUTOMATIC INTERLINKAGE OF INJECTION MOLDING MACHINES AND OF THE CORRESPONDING TOOL STORE BY MEANS OF AN AUTOMATIC CRANE (5)
(Figure 7)

A fully automatic crane system is used to considerably reduce set-up times for changing the tools and melting units of injection molding machines. The crane is equipped with a double grab unit to achieve the shortest of retooling times. The grabs are guided rigidly to achieve extreme positioning accuracies. The required data are provided by the production control computer to ensure that the crane's automatic connection, defined precisely in respect of time, between the tool store, preheating station and injection molding machine, and also for the realization of automatic retrieval tool stores.

FULLY AUTOMATIC CRANE WITH CRANE STORE IN A CRANKSHAFT PRODUCTION FACILITY

The computer-controlled bridge crane in use here was designed for fully automatic feeding and discharge of the production machines, and also performs the task of intermediate storage which is necessary due to the different operating times of the individual machines. Here also, the crane system is an essential component of the automated factory, and has a direct link to the master control computer.

FULLY AUTOMATIC CRANES WITH CRANE STORE AND LINKAGE TO A HIGH-BAY WAREHOUSE IN AN AUTOMATIC SHAFT PRODUCTION FACILITY [6]

Here, eight automatic cranes serve to feed and discharge around 130 machining units for shaft production. The machines are installed in sheds over an area of approximately 110 x 60 m and their layout and type do not depend on the transport system. The cranes run parallel to each other and, thanks to use of an auxiliary crab travelling on the main crab, are capable of transferring the containers to the next bay. The necessary storage areas serve as buffering space and intermediate stores. In addition, a single-lane high-bay warehouse is located on the face side of the crane tracks; this is integrated into the overall system and is linked to each crane by means of a chain conveyor system.

The purpose of the crane transport system is to convey the workpieces in containers from the workpiece infeed point via all machining stations up to their point of completion. The containers used for this purpose are "intelligent" containers capable of detecting the workpieces and of corresponding with the master control computer, that is to say the computer is informed at all times about the quantity and degree of machining of the individual workpieces circulating or held in the store.

Thanks to comprehensive protection and safety facilities, the transport system is capable of overhead travel above persons.

OUTLOOK

The components of the crane were the basis for development of the bay conveyor and trolley. They were automated in the storage process and interconnected in the material and information flow with other conveying machines. The resulting automation cycles are now able to convert the crane into an automatic interconnection machine.

Coupled with the wish to economize on industrial sites and make good use of available space, knowledge derived from modern workplace design is enabling hall configurations with automatic cranes operating flexibly, thus reducing the investment risk. These cranes are pointing the way towards an automated material flow in existing buildings without placing excessive demands on structural changes.

In view of the importance placed on such crane systems in the material flow concept, new, successful and individually adapted crane concepts can only be put into reality in a close cooperation between the user, after devoting sufficient thought to all advantages and disadvantages, and a cautious manufacturer who has proved his competence by a sufficient number of reference installations.

LITERATURE

(1) Neugebauer R.: "Zum praktischen Festigkeitsnachweis der Stahlbauteile mit Hilfe der EDV" Konstruktion, Jg. 35, H.5, S. 173 to 181 (1983)

(2) Roos H.J. Der automatische Kran als Teilsystem einer Fertigung, Dhf; 30, H.12, S.424 to 426 (1984)

(3) Copetti T.: "Fortschritte der Automation bei Spritzgießmaschinen", Kunststoffe, Jg.73, S.170 to 176 (1983)

(4) Rösel W.: "Mehr Intention statt Perfektion"; Industriebau, Jg.6, S. 366 to 368 (1982)

(5) Krebser R.: "Der vollautomatische Sprizgießbetrieb, eine Realität von heute", Plastverarbeiter", Jg.34 S. 307 to 308 (1983)

(6) "Letzter Handgriff an der ersten Maschine" Materialfluß, H.11., S. 18 to 24 (1985)

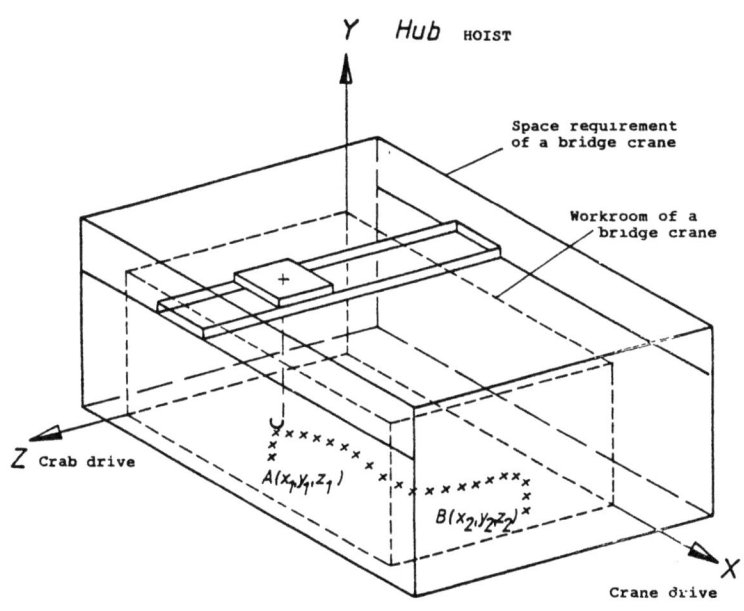

Figure 1. Layout of a bridge crane with a clear illustration of the crane's function

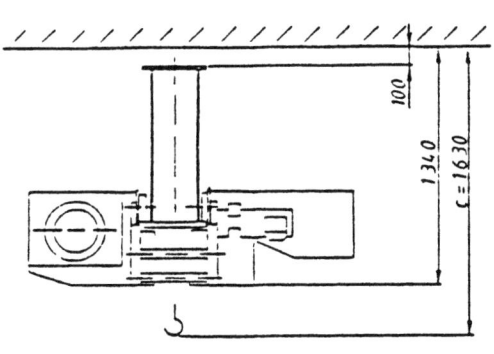 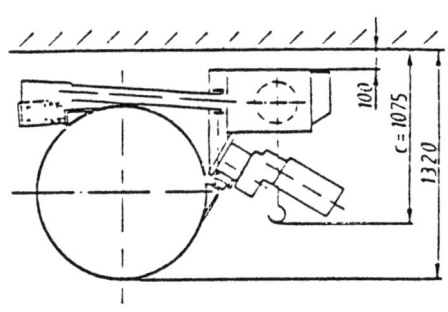

Figure 2. Comparison between a conventional crane cross section and the tubular crane with tangential crab (the c dimension of the tubular crane is 555 mm less)

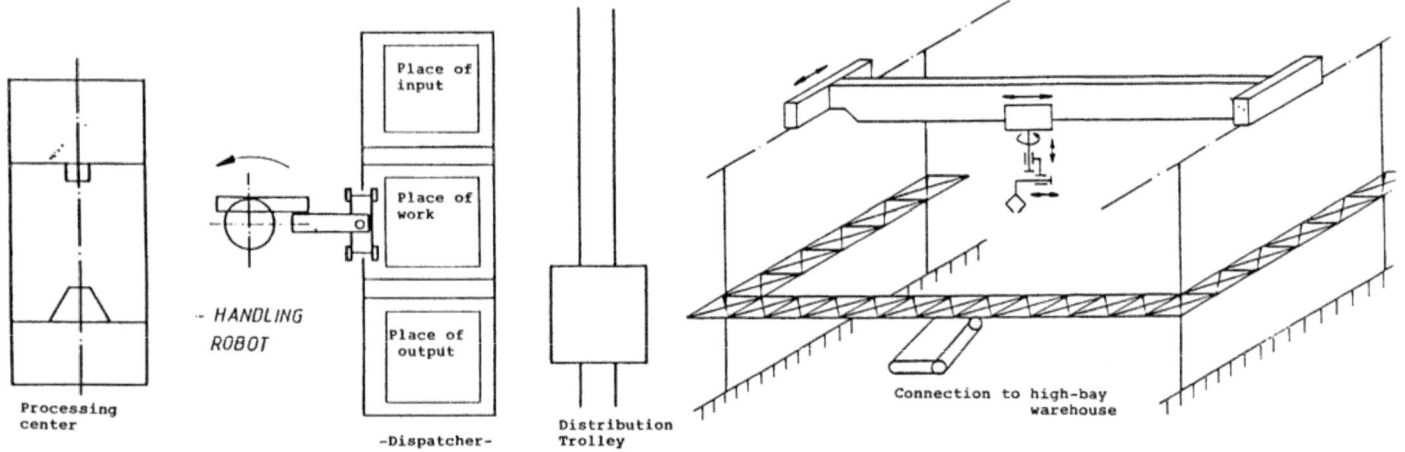

Figure 3. The function of the dispatcher is to refine conveying system's accuracy to adapt to the precision of a handling robot.

Figure 4. Combined crane-operated warehouse with telescopic crab linked by conveyors to a high-bay warehouse

Figure 5.   Ship unloader for cement clinker

Figure 6. Cross beam frame for sheet metal transport

Figure 7. Overall layout of an automatic factory with high-performance injection molding machines. The crane automatically changes the injection molding tool and melting unit.

(Photos: Erwin Mehne GmbH + Co., D-7100 Heilbronn)

CIM in sheet metal production
M. Bitz
Trumpf GmbH & Co, West Germany

ABSTRACT

The CIM-components and systems which can be bought at software houses or computer manufacturers always have some mistakes. Either they are only runable on a certain hardware, are only island solutions for particular problems or are not able to connect existing systems.

With the system introduced by this paper the machine tool company TRUMPF solves these problems by themselves.

COMPANY PROFILE

TRUMPF GmbH & Co. reached a turn-over of 345 million DM in 1985/85 with 1700 employees in 8 plants all over the world. The product range includes machine tools and working cells for nibbling, milling, laser and plasma cutting as well as NC-programming systems and laser resonators for cutting, welding and surface finishing.

INTRODUCTION

CIM (Computer Integrated Manufacturing) is no longer only theory in the production of flat sheet metal components, but is reality. Any manufacturer of sheet metal components has to withstand the increasing pressure of the market, which forces the industrial plant

- to produce in smaller batches
- to handle a widespread range of material in size, thickness and quality
- to reduce cycletime and material inventory as well as
- to react highly flexible to changes in demand.

These problems and possible solutions TRUMPF GmbH & Co. faced in 1983 after introducing studies in their sheet metal shop with the decision to develop - together with the software house SCS - the specifications and then the software for the Computer Assisted Sheet Metal Production (in German: "Rechnergestützte Blechteilefertigung" RBTF).

Since DP started to enter the machine tool factories, crossovers between most of these parts have been developed:

Files containing geometrical data, developed on the CAD-system are used to generate NC-programs;
production data, collected by manual data preparation was updating the order data in the PPC-system and in later accounting procedures.
To connect these islands based on organisational and informational criteria was the main aim of this project (Figure 1).
The sheet metal shop is now developed to a decentralized production unit controled by a Production Control Computer (PCC) embedded in a PPC-System. Beside this overall task the following commercial ideas founded the base of our planning:

- connection of the different automated sheet metal machines and working cells into an integrated material and information handling system,
- realisation of a demand/assembly orientated manufacturing,
- optimization and improvement of the manufacturing process at the DNC-working cells,
- shift working with a third shift on reduced manpower at the DNC-working cells,
- feedback and evaluation of production data,
- synchronisation of parts production with larger components,
- transfer of the software wholly or in components to the cutting shop in our plant and to the sheet metal shops of our customers.

STRUCTURE OF THE COMPUTER ASSISTED SHEET METAL PRODUCTION (RBTF)

1. Information System (Figure 2)

a) Production Control Computer (PCC)

   Hardware: HP1000/A700
   4 MByte memory
   2 x 132 MByte disc storage
   24 serial ports to connect terminals, factory terminals printers and different other computers (V24, 20mA, modems)
   Factory Data Link Bus (FDL) for DNC-download
   DS1000 to connect HP3000 (PPC-System) and another HP1000 (Tool Management) with the PCC.

   Software: SCS-Stuttgart (RBTF)

b) Connected Systems

   - NC-Program Development

     Hardware: PDP11/84 (DEC)
     Software: NC-program development system TCAPT
     Functions: Generates the runable NC-programs and a file which contains information about the machine tools, the resulting parts and scrap, as well as the raw material and the different characteristics of the program.
     These files are transferred by modem link to the PCC.
     The characteristic data is stored in the database on the PCC and the program on disc.
     The source program remains on the PDP11 and is updated there in case of changes.

   - TRUMPF Tool Management

     Hardware: HP1000/A600
     Software: TRUMPF Tool Management (HP Boeblingen)
     Functions: Together with Hewlett Packard TRUMPF is developing a tool management system for its whole plant in Ditzingen. It is intended not only to manage tools, tool parts lists, machine tool usage and inventories but also to connect tool assembly, paternoster storage and NC-program development.

     Until completion of the software, a more simple tool management system satisfying all demands of the sheet metal shop is in use on the PCC.
     The interfaces which are to be used in the final system have already been realized.

   - Production Planing and Control System (PPC)

     Hardware: HP3000/48
     Software: PS-System (PS Bremen)
     Functions: Management and maintenance of production schedules and parts lists of all TRUMPF's German plants. The dispatcher starts the PPC-System to generate out of the parts lists actual production orders originating in the assembly shop or in customer orders.

The starting time is arrived at by calculating backwards from scheduled completion. The durations of the different working steps are added to the workload of the production unit on the particular day. The availability of NC-programs and machinery material is first checked.

The process control department schedules the stock of orders for every week and starts allocating, dividing, changing schedules or contracting the work out if there are too many orders to be produced during a particular week.

The production documents are printed shortly before starting the first working step.

The progress of the work is supervised by reports from the PCC or manual input based on documents.

The time really needed for the number of parts produced is reported to the commercial DP for later calculation.

- DNC-Data-Concentrator

    Hardware: A&B (Aigner & Berger, Schw. Gmuend)
    Functions: Buffering of up to 2 DNC-programs of 1 MByte in total.
               Ring buffers store alarm and operation messages.
               They are read regularly by the PCC.
               To shorten the time required to download the NC-programs to the machines they are transferred line by line and executed at once. This means that only about 20 lines are in the control unit of the DNC-machine at any time.

- Factory Terminals

    Hardware: HP3081A
    Functions: 10 programable function keys (job report, working step report, control commands for the AGV, download of NC-programs, mounting tools).
               Alphanumerical keyboard for rough surroundings and a 32 character display.
               The barcode reader allows fast and correct input of the production data.

2. <u>Machines (Figure 3)</u>

    - Sheet metal working cell TRUMATIC LASERPRESS 180 LKR. Automatic supply of raw material and removal of finished parts and scrap by TRUMALIFT SB and SE (Figure 4).
      The industrial TRUMPF LASER with a power of 1500W is used here to cut sheet metal with a thickness of up to 10 mm.
      2 tool towers each containing 100 tools are used as storage for an industrial robot to mount the tools. Intersections between the control unit of the machine and the AGV ensures a troublefree flow of material.

    - Sheet metal working cell TRUMATIC PLASMAPRESS 300 PW.2G (Figure 5). Automatic supply of raw material and removal of finished parts and scrap by TRUMALIFT C.

    - Punching and nibbling machine TRUMATIC 240 equipped with TRUMAGRAPH-Control Unit for programming on job.

- Manual Jobs

    2 folding presses
    2 welding robots
    1 japanning belt device
    10 manual workstations (deburring, laying dressing, cleaning)

- Automated Storage and Retrieval System (AS/RS)

    Automated Material Handling System.
    72 pallets for raw material and semimanufactured flat products.

    Automated Guided Vehicle System (AGV)
    1 vehicle with 3 t maximum load.
    10 loading and unloading stations.

3. Characteristic Data

Production Orders
- 350 orders current in the sheet metal shop
- 4-5 working steps per order
- 3 weeks cycle time
- 1-1000 parts per order
- 30 parts average
- 3500 different actual parts

NC-programs
- 1500 active NC-programs running on the TC 180 LKR
- 1500 active NC-programs running on the TC300 PW.2G

Tools
- 800 tools administrated by the tools management system
- 200 tools directly accessible

Raw Material
- 1000 t per year
- 50 different kinds of material
- 0.5 - 12 mm thickness

4. Processing of a Production Order (Figure 6)

4.1 Selection of Orders

Production orders are generated and scheduled by breaking down of the parts lists of ordered machines in the PPC-system.
The selection program chooses all orders which have to be produced under the control of the PCC in the sheet metal shop about 5 days before they reach their starting date. Each night it generates a transfer file which contains all working steps which can be done in the sheet metal shop or which are already present and should be changed or deleted.
Urgent orders can however be transmitted during the day. Batch transfer runs during the night.
If necessary the orders can also be input at a terminal of the PCC. Thus if the PPC-system is not able to select and transfer order data it is still possible to control the sheet metal shop.

4.2 Storage of Orders

When taking the orders into its data base, the PCC checks them for validity:

- are the values of the data within defined limits
- do they have the right format
- do the items belong together

An easily understandable error protocol allows the process control to correct mistakes.
The availability of raw material, NC-programs and tools is also checked. Material and modification index in the NC-characteristics and in the production orders are compared amongst other things.
If there is not enough raw material in store a warning is printed in the failure protocol to initiate the loading of the relevant pallets before the planned start of production.

4.3 Sequence Optimization

Blocks of working steps with the same start time in which no mounting of the machine is necessary are compiled for the DNC-machines. The basis of this optimization is formed by the characteristics of the NC-programs and the information on the working steps.
Most significant are the raw material, claw positions and tools. The necessary tools and sheets are reserved for this working place.
Based on the optimated sequence the process controler is able to take measures like allocating, dividing, changing schedules or contracting the work out.
The suggested sequence of DNC-blocks can be evaluated if necessary. With the aid of the final list, the PCC generates mounting orders for tools and claws.
The workers at the manual working places get a list of working orders which they have to process sorted by urgency. These lists contain information about characteristics, raw material and schedules.

4.4 Processing an Order

By activating automatic processing, numerous actions running parallel are initiated in the PCC:

- The first task of the first DNC-block in the chain is taken by the Shop Floor Control from the Order Management.
- The suitable pallets for raw material and processed parts are choosen and the driving orders transmitted with high priority to the AGV.
- The machine operator tells the PCC via a factory terminal which special tools he has mounted on which places in the tool magazine. The final acknowledgement is given after the claws have been moved.
- The NC-program is prepared for transmission. The tool commands are updated and the unloading positions on the pallet are changed from default to their actual value.
  The data is packed together and down loaded at high speed for some seconds to the concentrator.

After the starting order from the PCC when all preparations have been finished the program is emitted sentence by sentence to the machine and executed at once.

Only a buffer of about 20 sentences is held in the machine control unit. During execution time the next NC-program is down loaded simultaneously into the DNC-adapter.

The messages of the machine are buffered in the concentrator, frequently checked and analyzed by the PCC. Errors like an unexpected end of the raw material are recognized early on and corrected automatically during production time. Only NC-programs using whole sheets of metal are allowed to run automatically. In this case loading and unloading of sheets can be supervised by the PCC.

The message "NC-program finished" before the last execution causes the PCC to check whether a new pallet of raw material is necessary for the next order. If it is, he changes the pallets at once.
After the last time of running, the items in the order are completed.
The upload report containing number of sheets, parts and used time for the PCC-system is generated.
The order is assigned to the job which the next working step is supposed to carry out.
In the meantime, the next NC-program has already started.

Besides automatic production two more methods of operation are intended:

- Parts which are not able to be loaded or unloaded automatically by the TRUMALIFT are manufactured during the day shifts by hand. Supervising the Material Handling System, mounting the magazins and calling the NC-program is done by the worker. He can use his factory terminal

    - to start the working step
    - to call the pallets to be changed
    - to mount the tools in dialog with the PCC
    - to load the updated NC-program
    - to keep a tally of sheets and semimanufactured parts on their pallets
    - to empty the scrap pallet
    - to finish the working step

- At demonstrations for customers it is sometimes necessary to switch the machine to punched tape mode. The other utilities of the PCC however are available.

The elapsed time is monitored up on the time account of the machines. Either automatically in auto-mode or manually via the keyboard the state of a job is set to "trouble", "production", "maintenance", "demonstration" or "standstill".

At manual working places the functions of the PCC are reduced to generating a chain of orders to be produced (sorted by schedule and raw material) and to recording manufacturing data quickly and exactly.

Via factory terminals the workers in the sheet metal shop first announce each working step and finally report the number of parts they have produced.

Input by barcode decreases the input time and the number of errors. Unreadable barcodes can be bypassed by the alphanumeric keyboard. The number of production documents is thus decreased by 50 %.

5. Tools for Process Control

The state of each order whether scheduled, waiting to be manufactured, in process or finished can be supervised via a terminal. All orders in the shop with a certain identification number are shown on demand with their actual status.

Graphical information shows the load of a particular working place for the next 5 days (longer periods are considered in the PPC-system):

The first 3 columns show the situation today (Figure 7):

left column: T - scheduled orders as a percentage of the capacity of this job. Either the start schedule has not been reached or the preceding step has not yet been finished.
E - Orders ready to be processed. These orders are able to be started immediately at this job.

Above you can see the sum total of times for all scheduled orders. Below you can find the sum in hours of all orders ready to be started.

middle column: R - Reserved capacity for maintenance or demonstrations in %
F - Useable capacity. This schedule can be updated regularly for one or multi shift working.

right column: T - Orders which have to be processed by day.
N - Orders which can be processed on a night shift with a reduced number of workers.

Listings of the succession of orders for each working place are distributed every day and used to control production.
Reallocating on alternative jobs, rescheduling or the decision to contract out are made by the aid of terminal dialogs.
For each working place a picture of the current conditions including capacities, reservations and added production data is kept on the PCC.
Information about tools, parts and NC-programs are available on terminals and listings. Maintenance and geometrical data of tools are administrated automatically.
Status and occupancy of all stations of the AGV are supervised frequently and shown on a layout mask (Figure 8).
The occupancy of the stations can easily be set after a failure by entering the allocation of each material system point into this mask.
The dialogs are kept easy to allow less qualified workers to deal with it during the night shift.

The status of the manual working places

F - out of work
B - ready to work
P - producing
S - fault
W - maintenance
V - demonstration

can be seen at one look. Whether a DNC-working cell is in automatic (A) or manual mode (H) is also shown.

Stocks of raw materials and the contents of the pallets with semi-manufactured parts, depending on variable limits, can be printed or displayed.
The layout of pallets containing automatically worked parts is shown graphically for easier identification (Figure 9).
The parts belonging to the displayed identification number are highlighted by stars. The surfaces of the other parts are marked by dots.

The progress of each order including required and reported actual data can be retrieved at once. Process control, foremen and workers can get an actual view of the order (Figure 10).

Data, corrupted by wrong or missing input can be corrected in a similar mask. With an adequate password level you can correct the required or the reported number of parts, times and schedules.

6. Production Report

A report sentence determined for the PPC-system is generated at once by any assignment for a working step or for material usage. The required dates can be compared with the reported numbers and times and if necessary the reason for the difference can be determined at once.

During the nightly batch process this file is transferred to the PPC-System on the HP3000. There the orders are updated and the significant parts of the information are sent to the commercial DP running on an IBM-machine. The production data is used to order sheet metal on a just in time basis, to validate the schedules for the production plans, to clear the scheduled capacities at the working places and to calculate the manufactured parts after production.

STRUCTURE OF THE SOFTWARE

The software of RBTF is modular in its design.

- Order Management
- Material Handling System
- Tool Management
- Shop Floor Control
- NC-Program Management
- Production Reporting

are stand alone parts which can run on separate computers or which could be omitted if not needed.
The interfaces are minimized to avoid difficulties in interconnection of different computers.
Changes in Storage Management, Order Acceptance or installed machines can be made by simply changing the relevant software module.
These points are important for our aim to transfer this package to other parts of our company. To achieve this without huge additional expense would be the greatest costreduction.

STATE OF PROJECT

Software development and tests were finished in July 1986 with the acceptance of the automatic process at the DNC-machines. Only the connection to the PPC-system remains to be done by October.

The introduction started in January 1986 with the interconnection of DNC-machines and AS/RS-system via PCC. This also required the dialogs on factory terminals. The NC-programs could then be loaded and driving orders could also be transmitted to the AGV.

Process control with real orders in the manual part of the sheet metal shop was introduced during February and March. Listings and dialogs are constantly improved due to feed back from the workers.
During the introduction of the system theoretical, but much more practical training of all involved persons, separated by their functions took place. First general information for everybody to understand the whole concept and sense of the system was followed by direct training of the different tasks like unloading a raw material pallet.
Each working place received a manual which is continuously improved. After initial sceptisicm about the computer, the workers willingly accepted the additional work they have to do.

The workers council had no objections, because no personal datas are stored. Payroll is carried out on a spearate system
The process times on DNC-machines cannot be influenced by the operators so piecework rates make no sense here. Much more important are highly skilled workers who are able to repair small electrical or mechanical faults by themselves to reduce the amount of downtime.

Process Control, Process Planning and NC-Programming are gradually benefiting from this installation.
Faults and weak points are shown clearly and can be eliminated. By the deliberate release of orders on the PCC, the shop is not overloaded with orders and it becomes possible to keep to delivery schedules. Work in process and cycle-time are gradually dropping.

One problem which remains is the higher complexity of the system in automated mode. This necessitates the establishment of easier set up modes after a fault has occured than first was intended.

We have to try much more than before to increase the availability of the single components within the system especially of the older parts such as AGV, material storage and the connecting stations to the DNC-cells. Certain functions of the TRUMATIC-cells were found nobody had thought to automize before the first tests in automatic mode.

These efforts have improved the availability of the system not only in automatic mode but also clearly in manual mode.

FINAL STATEMENT

With the development of an integral Process Control System which covers existing flexible Sheet-Metal Working Cells, an Automated Guided Vehicle and a Material Handling System as well as the manual working places TRUMPF has taken a further step into the future of manufacturing technology.

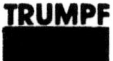

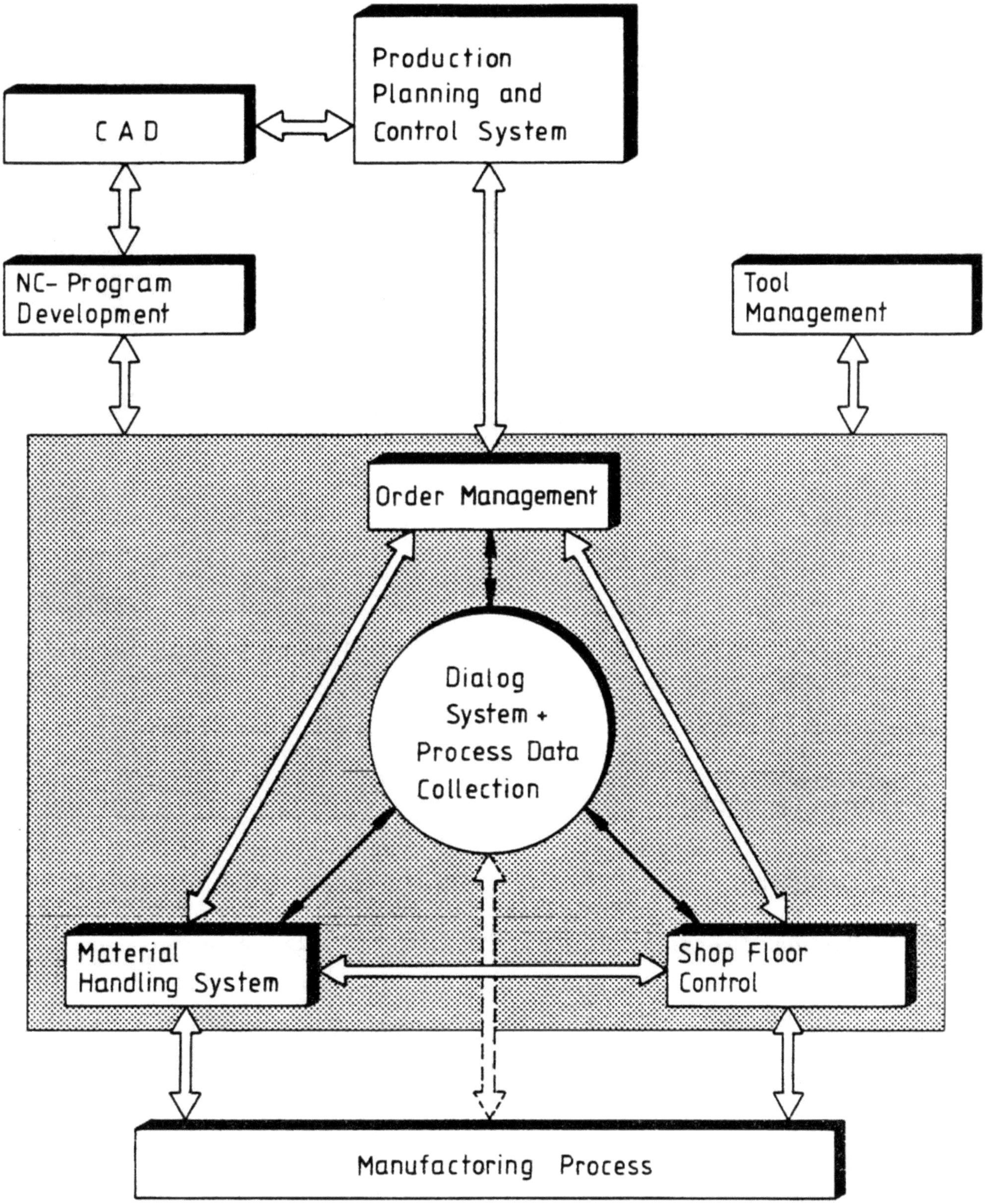

Production Planning and Control System
CAD
NC- Program Development
Tool Management
Order Management
Dialog System + Process Data Collection
Material Handling System
Shop Floor Control
Manufactoring Process

Figure 1

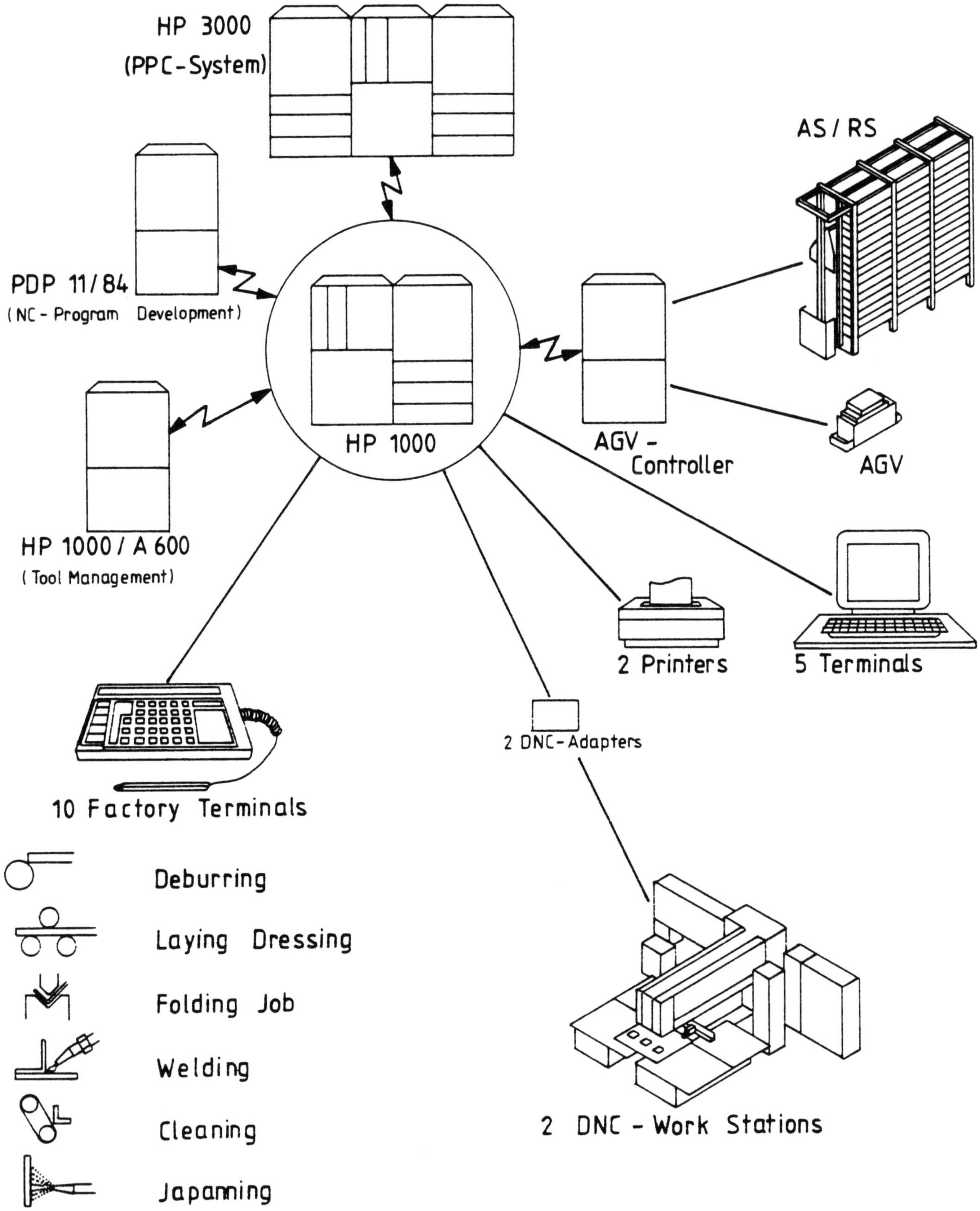

Figure 2

SHEET METAL SHOP   TRUMPF DITZINGEN

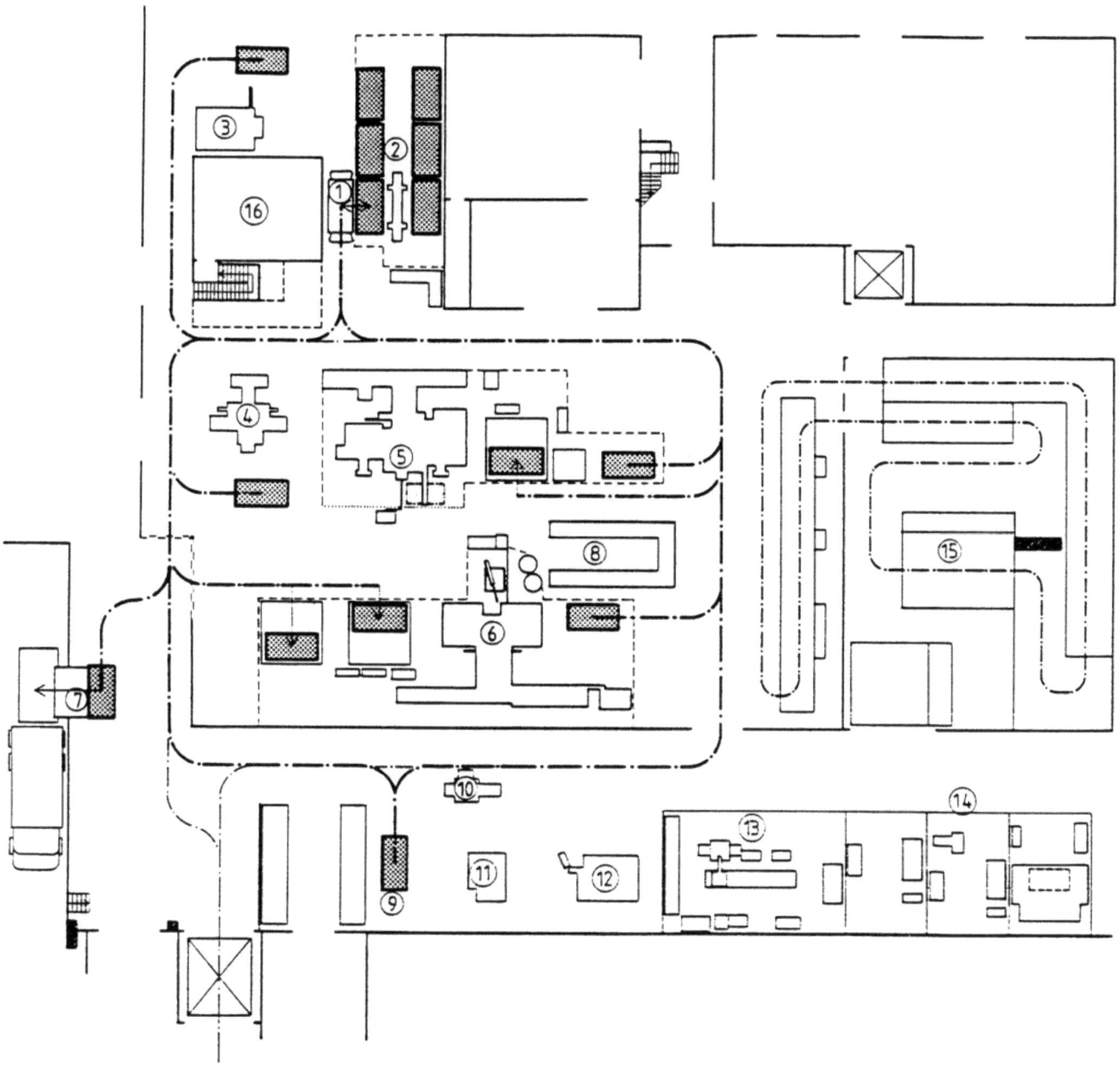

1   Automated Guided Vehicle System
2   Automated Storage and Retrieval System
3   Sheet Metal Shear
4   Sheet Metal Punch   TRUMATIC 240
5   Sheet Metal Working Cell TRUMATIC LASERPRESS 180 LKR
6   Sheet Metal Working Cell TRUMATIC PLASMAPRESS 300 PW2G
7   Scrap Output and Raw Material Input Station
8   Tool Assembly
9   Combination of Welding Parts
10   Laying Dressing
11   Deburring Station TRUMAFIN
12   NC-Folding Station
13   Welding Robot
14   Manual Welding Jobs
15   Automated Japanning Cell
16   Production Controling Center

Figure 3

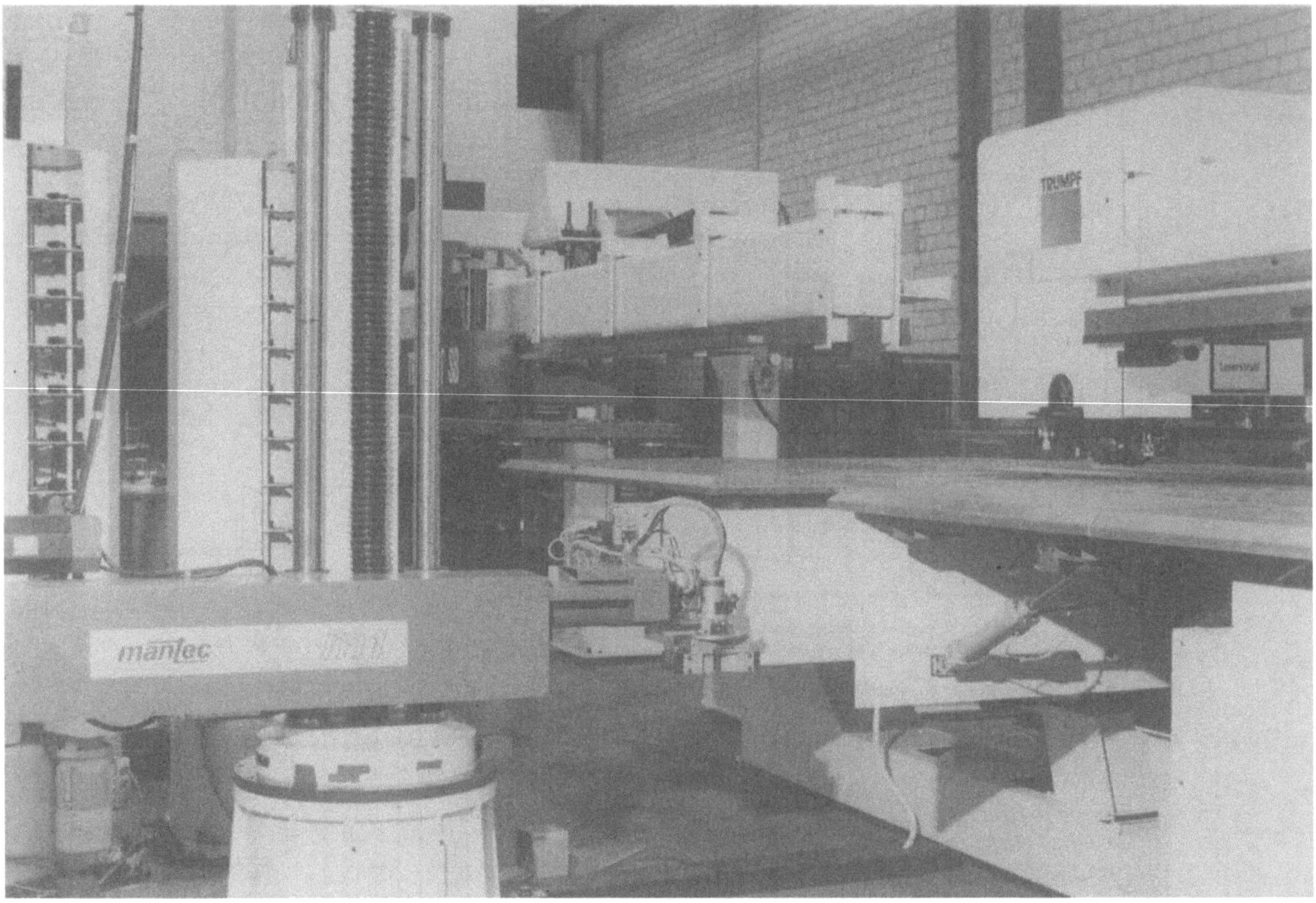

Figure 4 Sheet metal working cell TC 180 LKR

Figure 5 Sheet metal working cell TC 3oo PW.2G,
AS/RS - System and Process Control Cabin

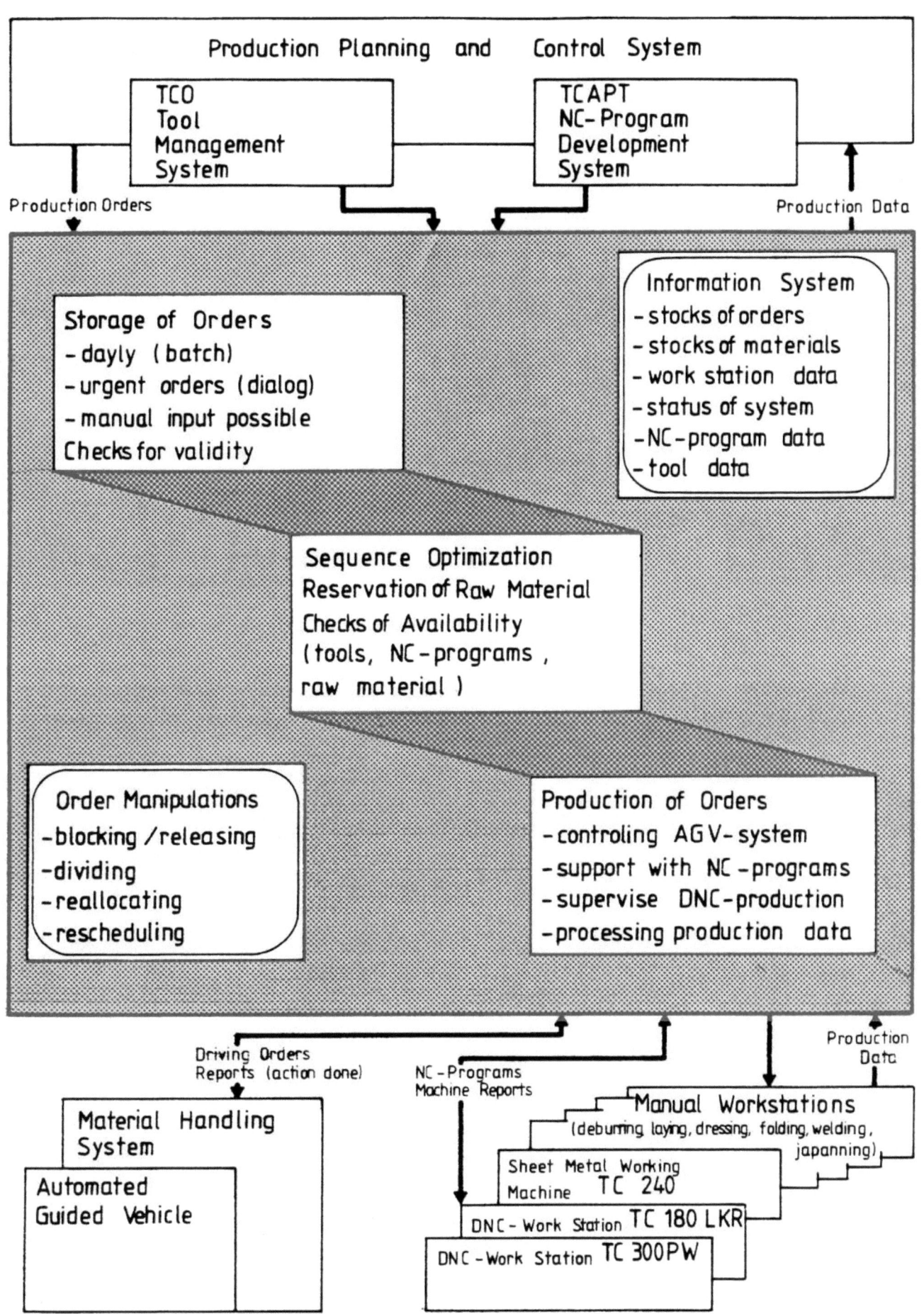

Figure 6

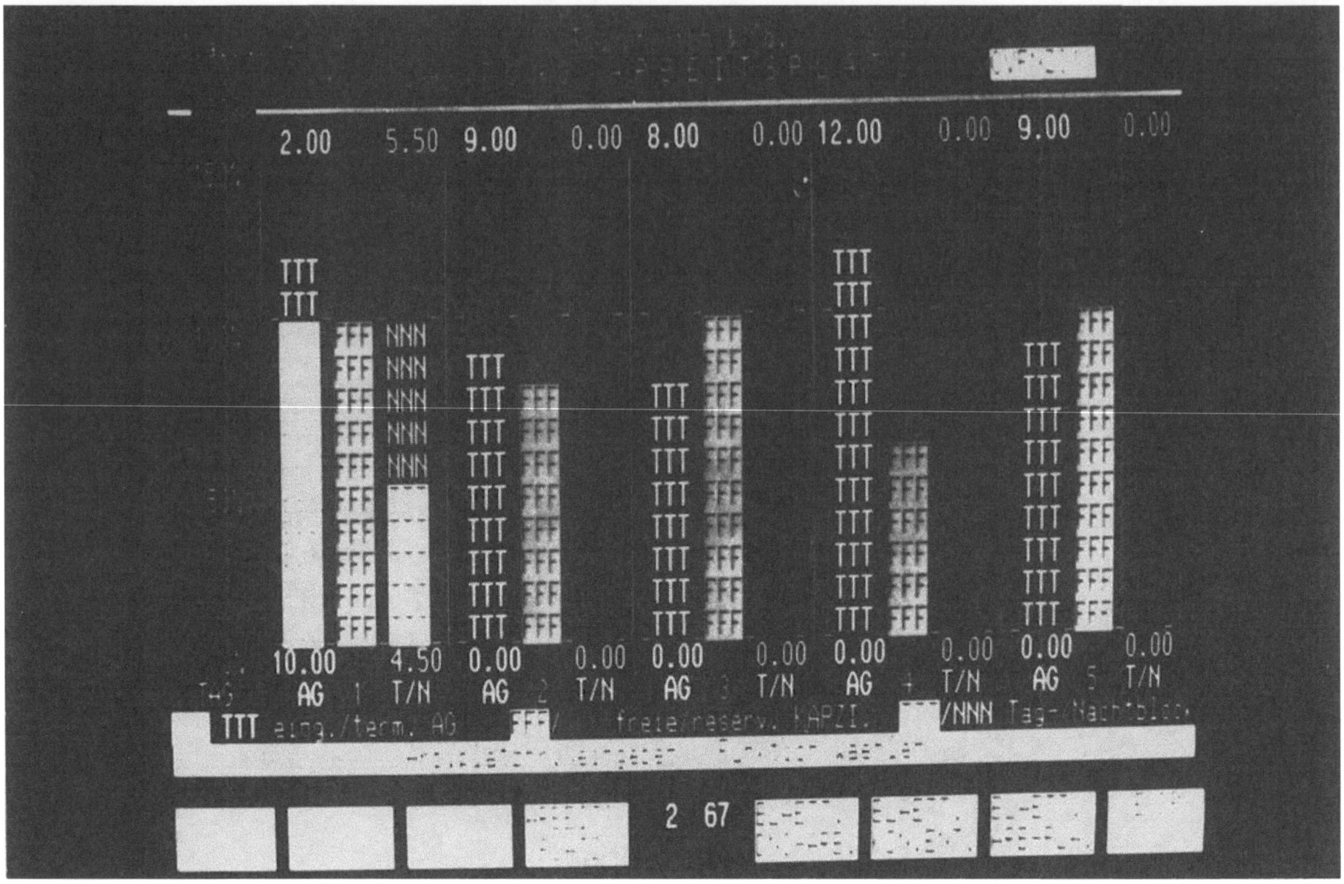

Figure 7   Load of a workingplace for a five days period.

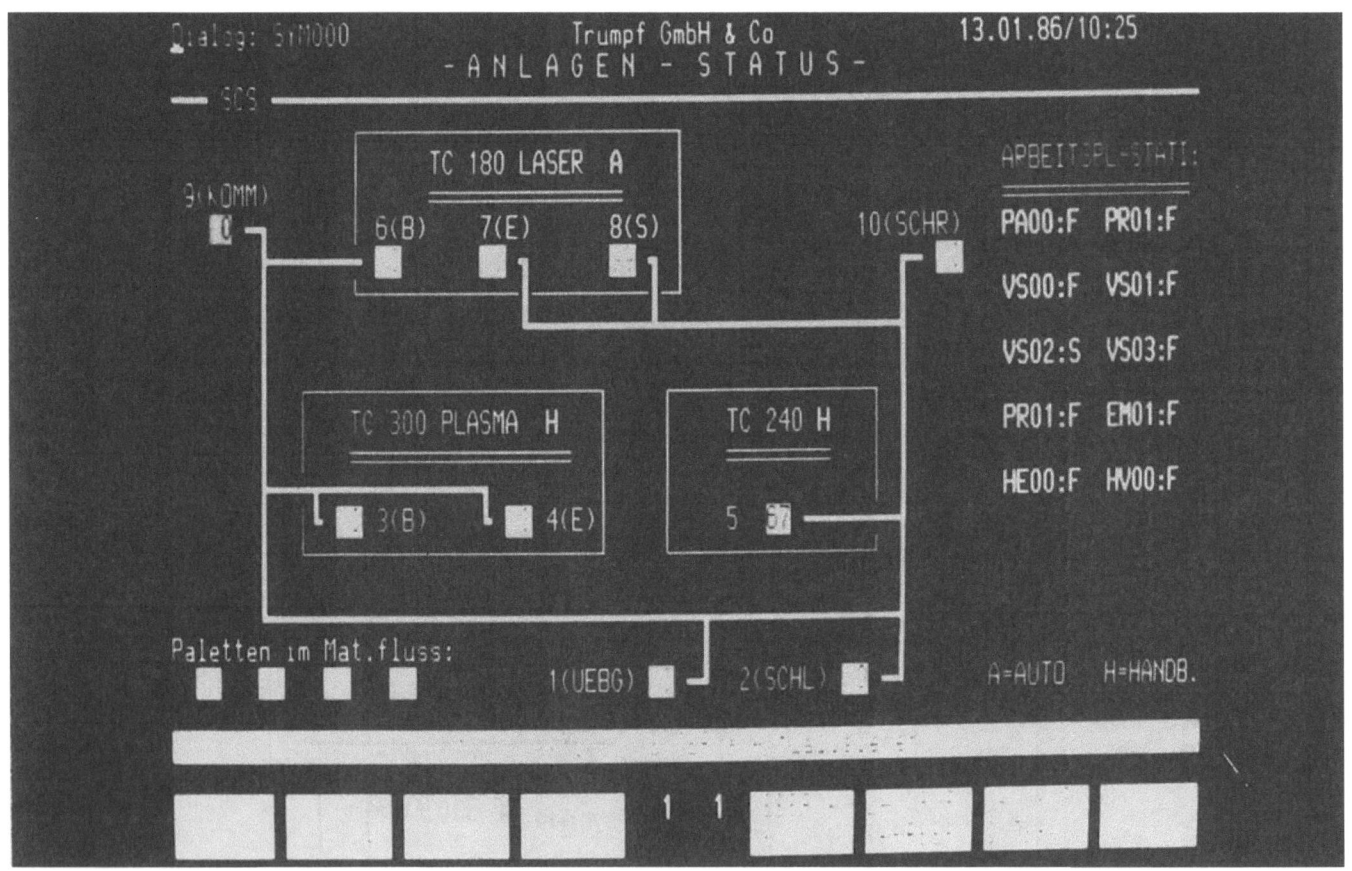

Figure 8   Status information on layout

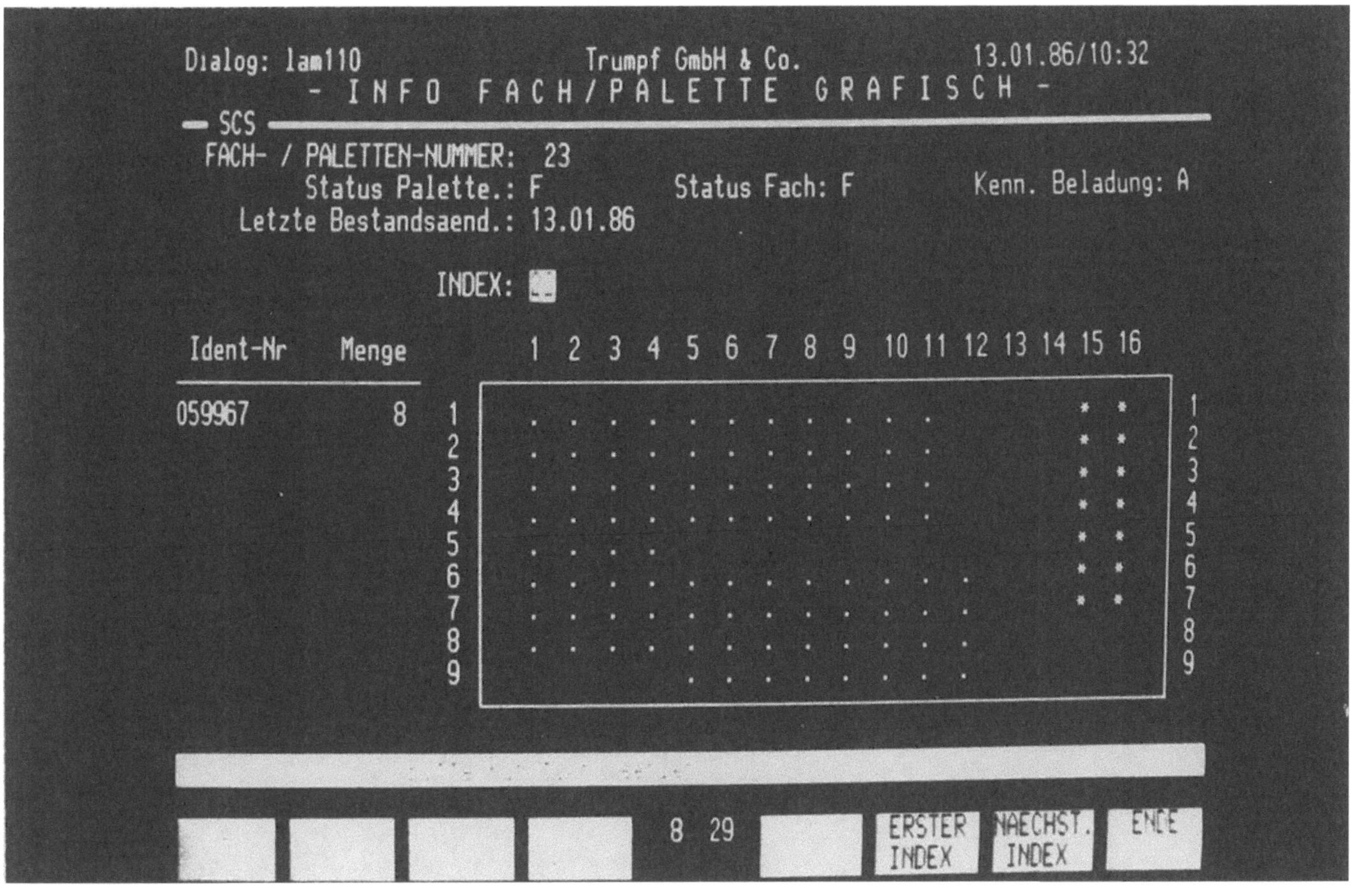

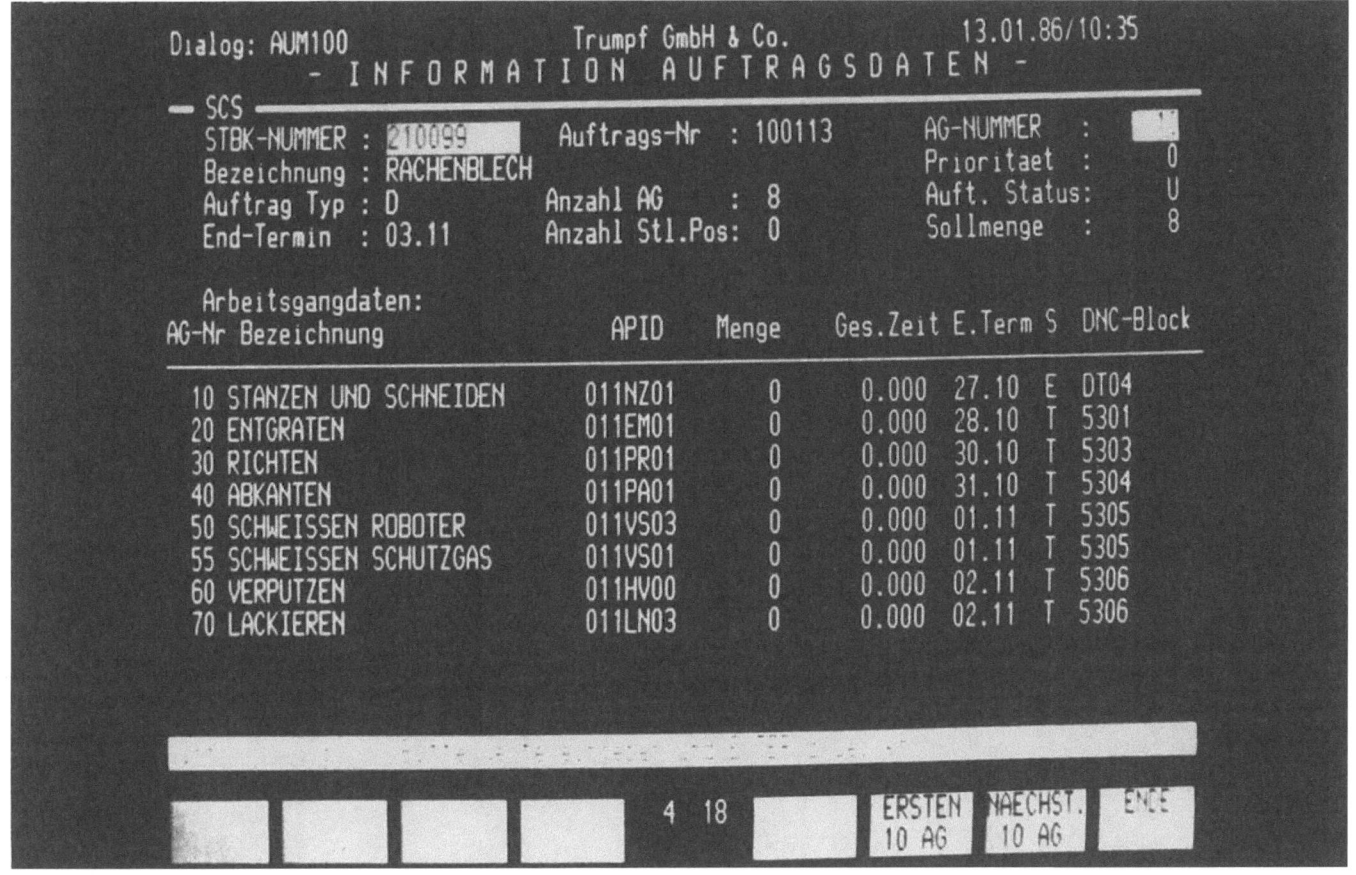

Figure 9　　　　　　　　　Graphical pallet load

Figure 10　　　　　　　　　Work order progress

Present status of sheet metal FMS in Japan

Y. Nakazawa
Mechanical Engineering Laboratory

M. Kiuchi
University of Tokyo

J. Endow
Tokyo Institute of Technology

M. Shinohara
Tokyo Metropolitan University

and

S. Matsubara
The Institute of Vocational Training, Japan

ABSTRACT

To promote the future development and application of FMS for plastic working, the survey and research activities have been done for two years. Of such activities, this paper covers the sheet metal working. The information about 42 systems (including FMC) were collected from the questionaire sheet sent to system suppliers and users, and/or visiting studies. Purposes and profits of installation of FMS by enterprises surveyed, pattern of work flow, object work, processing equipment, mean operating hours, accuracy of products, punch dimensions used and so on are discussed.

INTRODUCTION

As compared with Flexible Machining Systems, Flexible Plastic Working Systems have been slower in development, owing to various restraints. Still, efforts in various forms have been consistently made toward developing FMS and FA for plastic working in full recognition of the difficulties involved, and so far the employment of a significant number of such systems has been achieved, notably those that relate to sheet metal working.

A fundamental issue involved retarding the growth of such FMS is the absence, even today, of universally accepted recognition or interpretations relative to the definition, design theory, and evaluating process covering FMS for plastic working. To promote the future development and application of FMS for plastic working more realistically and efficiently, an in-depth study is required. Thus, survey and research activities have been conducted over two years to determine the present status of FMS for plastic working. The result of the first year's survey has been reported[1]. The definition of FMS for plastic working and the surveying procedure used were presented at that time. This report covers sheet metal FMS.

PRESENT STATUS

Purposes and Profits of Introduction

Tables 1 and 2 show the objectives and benefits of introduction of sheet metal FMS respectively. Two or more items were chosen by firms, consequently, the sum of the ratio exceeds one hundred percent in the tables. As shown in Table 1, introduction of FMS by the firms surveyed was primarily intended to achieve labor-savings in processing operations and to cope with product variety. Their secondary objective of slightly less importance was the desire to achieve production increases and faster delivery capabilities. That is, FMS have been employed by the firms for labor-savings and response to shortages of skilled labor and a lower volume of production of a greater variety of products with higher efficiency, and their faster delivery. In addition, unattendant operations for 24 hours and cost reduction constituted another major motivation for FMS introduction.

Presently, it is difficult to gain high profits through mass production as was possible in the past. Higher prices are not accepted despite lower volume production of a great variety of products, nor are lower quality levels tolerated. The demand for higher quality and lower prices still persists intensely.

Profits are generated by enhancing productivity that may be defined as the (output/input) ratio where output stands for the volume produced and input for the invested facilities, personnel, materials and other production means. To raise such a ratio, it is generally necessary either to increase the production volume if the invested production means are to stay at a fixed level, or to reduce the investment in production means if the production volume remains at a fixed level.

Table 1 Purposes of Introduction

Item	Ratio (%)
Labor-saving	93.3
Variety of Products	86.7
Production Increase	53.3
Faster Delivery	46.7
24 Hours' Operation	33.3
Cost Reduction	33.3
Inventory Reduction	13.3

Table 2 Benefits Of Introduction

Item	Ratio (%)
Labor-saving	85.7
Time Reduction	83.3
Cost Reduction	28.6
Rationalization of Production Control	23.8
Inventory Reduction	14.3
Other Benefits	28.6

Since FMS is a system that serves for manufacturing a great variety of products out of raw materials, both efficiently and economically, it will enhance productivity. By studying Table 2 from the standpoint of increasing the productivity ratio referred to above, the benefits achieved through the reduction of personnel (labor-savings) and shortened processing times are seen to be the maximum in a number of instances. Additional benefits which can be cited as found through the present survey have been more streamlined production management operations and reduced inventories. Figure 1 shows the relationship between the number of types of products and the mean lot size which is defined as the ratio of number of products and number of types of products. Each circle in the figure corresponds to a sheet metal FMS surveyed. Table 2 does not include the item of variety of products, but it is found from Figure 1 that the firms surveyed are successfully dealing with low volume and great variety production.

System Classification

Work flow in FMS is an important factor for grasping overall system features. With related work flows classified into four basic patterns as stipulated by the graph theory (as shown in Table 3), it is found that most work flows for sheet metal FMS fall under the line pattern category.
The situation in relation to Flexible Machining Systems, in contrast, reflects a majority of loop pattern systems with an increasing share of tree pattern systems employed more recently. Also, net pattern systems are employed where complex configurations are called for, but few that belong to the line pattern category. The reason for the above difference was described in a previous paper[1].
As shown in Table 4, a closer examination of line pattern processing reveals that it may be further broken down into a perfectly straight line type, and types that have branches in the middle. The latter may still further be broken down into three styles: A1 (parallels), A2 (branches), and A3, a combination of the first two.
Type A1 features the splitting of a flow into two or more (parallels), in each of which local processings are made. In type A2, on the other hand, the work branches off the mainstream flow, and after being processed locally at some point, again rejoins the mainstream. Figure 2 represents a typical example of line pattern sheet metal FMS.

Table 3 Classification of Systems

Pattern	Number of Systems
Line	39
Loop	1
Net	2
Tree	0

Table 4 Classification of Line Pattern

Type		Example	Number
Straight	A0		12
Parallel	A1		17
Branched	A2		6
Mixed	A3		4

Another feature of sheet metal FMS is the general use of feeders or the equivalent (Table 5). Since feeders or the equivalent are only rarely employed for Flexible Machining Systems, the above aspect may be regarded as a feature of a sheet metal FMS.
Systems employing robot manipulators have been found to be few in number, presumably because the use of a robot has been viewed as not justifiable to handle the sheet metal or products. However, where bending, welding, assembly, and other processes are incorporated in the system, more general use of robots appears likely in the future. In

the current survey, incidentally, only conveyor robots and handling robots have been taken up, and no welding robots were included in the count. Conveyorless systems have included those employing manual conveyance means and those incorporating coil strips .
Reasons for the general use of conveyors lacking self-propelling capabilities were shown in the previous paper[1].

Table 5 Work Transfer Equipments

Transfer Machine	Number
Conveyor etc.	26
Cart etc.	21
Robot etc.	4
Monolift etc.	5
None	1
Manual	1

Table 6 Application Field

Field	Number of Systems
Electric Appliances	24
Machines	3
Air Conditioning Equipments etc.	4
Elevators	2
Kitchen Utensils Steel Furniture	2
Vehicles	4
Vending Machines	3

Object Work

Typical object work processed by sheet metal FMS are frames, doors, side panels etc. Table 6 gives the field of their application along with the number of relevant systems, for contrast.
Features of the parts processed by sheet metal FMS include:
(1) Smaller lot sizes
(2) Mostly processing is possible with shearing, cutting and bending.
Figure 3 is an overview of the scale of production of sheet metal FMS surveyed. Eleven of the systems were found to handle 10 or less parts per lot per month, an extremely low production volume; even most productive systems were less than 3,000 parts. Thus FMS introduction has so far been focused on the processing of medium- to low-volume products.
Figure 4 presents an overview of the number of types of object parts currently being processed. Two systems were found to handle 10 or less part types, 27 systems 100 or more types, and some systems even as many as 5,000 types. FMS were originally intended for flexible processing, and this objective has been achieved by many of the existing systems.
Figure 5 shows a summary of our survey findings on the maximum dimension of materials, either transverse or longitudinal, that are currently processed by sheet metal FMS. Small material dimensions were found to suffice for various smaller parts processed out of sheet metal. Dimensions of materials processed by systems for one part at a time are naturally smaller, and those for multiple parts at a time are larger. Material dimensions of 1,600 mm or greater were found to be employed for the greatest number of systems surveyed, and even dimensions more than 2,600 mm are employed for some systems. Many systems were found processing pre-cut regular size sheets (sizes such as 20x110 mm, 1,200x2,438 mm, etc.) but some are used for processing coil materials. Other systems not only processed regular-sized sheets but were also equipped to handle coil materials as well.
An overview of the material thickness processible by sheet metal FMS is presented in Figure 6. Most of the systems surveyed were found capable of processing material 1.5 mm or greater thickness . The maximum processible sheet thickness is determined by the load bearing capabilities of the processor, the required accuracy level, material properties etc.

Basic Working Processes

The range of working processes incorporated in a system varies widely with the individual case. To incorporate sheet metal working processes in an FMS, computerized controls are required over the automation of plastic working equipment in linkage with material loading and unloading equipment, welding and assembling equipment and so forth.
The overview given in Figure 7 covers the working processes currently incorporated in sheet metal FMS. All such FMS incorporate cutting and shearing processes. Eighteen other systems additionally incorporated bending processes. Besides these, tapping, marking, and grinding processes were also performed as intermediate or after-processing. Welding processes were also incorporated in a singnificant number of systems, and two other systems have been found to incorporate surface coating processes.

Operating Status

In-Process Measurements. About 40% of the enterprises were found to fully employ in-process measurements for problem detection during system running. When including enterprises partially employing such in-process measurements as control pattern, work positioning, work shaping checks, tap breakage checks, etc., the above percentage increases to 80% for all related enterprises. Firms that employ no in-process measurements of any kind, therefore, constitute 20% of the total.
Firms employing in-process measurements for product shape, dimensions, etc. were found to approach 40%. Of these, the number employed for shape verification matches those used for dimensional measurements. Another 40% were found to be relying solely on visual checks, and about 20% were running no checks at all.
As remedial measures against system problems, all the responding firms depended on human intervention, and none of the equipment was found to be equipped with self-remedial functions. So even though problems are detected by in-process measurements, the equipment itself is not capable of remedial actions. These must be provided by humans instead, and no thoroughly unmanned system runs for a prolonged time period was found. Accordingly, the requirements will have to be analyzed in order to achieve systems with equipment having self-remedial capabilities in the presence of a problem. Consequently, the need for technical developments to atttain such objectives on an economical basis has been posed for the future.

Mean Oparating Hours. As shown in Table 7, the daily mean running time of systems was found mostly in the range of 12 to 16 hours. Fifty percent of the firms fall into this category. Some enterprises, however, were found to be running their systems 24 hours a day. In some 60% of the enterprises surveyed, systems were found to be running for 12 or more hours daily.
In contrast, 20% of the firms were running their systems for only 8 or less hours a day. The reasons for this were assumed to be either the low level of orders secured, or the lack of need to run systems any longer owing to the extremely high added value products manufactured.

Table 7 Mean Operating Hours

Operating Hours h/Day	Ratio (%)
h < 8	20.0
8 ≤ h < 12	20.0
12 ≤ h < 16	50.0
16 ≤ h < 24	5.0
h = 24	5.0

Table 8 Accuracy of Products

Accuracy (mm)	Ratio (%)
0.0 ~ 0.1	17.1
0.1 ~ 0.2	27.6
0.2 ~ 0.3	32.8
0.3 ~ 0.5	17.1
0.5 ~	5.3

Accuracy. Most of the ultimate products manufactured with individual systems fall in an accuracy range of from 0.1 to 0.3 mm, and the systems producing such products were found to amount to about 60% (Table 8). Slightly over 10% of the systems, however, were producing products of 0.1 mm or better in accuracy.

Processing Equipment. A review of the number of shearing/cutting machines incorporated in a system operated by various manufacturers revealed the incorporation of two to three machines that frequently include one or two turret punch presses and either one L blade shear or one guillotine shear. Table 9 shows an overview of the machine assortment. The number of machines of each individual type is shown there with the total number of such machines divided by 37, the total number of systems. This indicates that the average system is configured of 1.7 units of turret punch presses and approximately 1 shearing unit. About 40% of the firms have a function of bending process (see Figure 7). The equipment used were found to include press brakes in 10 instances, presses in 5, benders such as folding machines and tangent benders in 11 and roll forming in another instance.

Relative to the spring-back correction in bending processes, human settings (of a teach-in type) were found to be employed in many cases, and automatic settings in accordance to the material, sheet thickness, and bend length in a low number of instances. In a few firms is auto-correction exercised.

Nine responses were received related to the flexibility of bending processes, of which seven firms claimed the capability of bending any material of any sheet thickness at any bend length. The remaining two such feasibilities concerned only predesignated product types.

Handling systems for bending processes were found to be manual operations and handlers supplied with benders. The employment of a robot for material handling for a bending process has not been reported in the present survey. Even though a robot is technically capable of replacing a human, its employment for bending operations appears to be uneconomical and a resort to human labor appears still the current status.

Table 9 Shearing/Cutting Machines Used in a System

Machines	Number of Machines	Number of Systems	Mean number of Machines per System
Turret Punch Press	1 2 3 4 5 6	22 6 2 2 2 1	1.7
NC Cutting Machine	2	1	0.1
Shearing Machine	1 2 3 6	22 1 1 1	0.8
Laser Press	1 5	1 1	0.2
Laser Cutting Machine	1	3	0.1
Tapping Machine	1 2	5 2	0.2

Nesting. About 80% of the firms surveyed were found to be performing computer-aided (CAD/CAM) nesting operation, and 20% to be performing nesting operations manually.

Where the nesting CAD process was employed, 50% of the firms were found to have developed their own software, and about 20% had developed their software through cooperation of either Amada or Tokyo Electron. Approximately 30% of the firms purchase and use software developed outside.

Punch Dimensions Employed.   Tables 10 and 11 show responses to the inquiry relative to the punch dimensions installed in the punch stations. These tables cover circular and rectangular punches.

Regarding circular punches, maximum diameters fell in the 80 to 120 mm range in 50% of the cases, and fewer belonged to the 60 to 80 mm range. Some of the firms were found to employ maximum diameters in the 20 to 40 mm range.

In terms of rectangular punches, none of the survey objects fell into the maximum cross-sectional area range of 1,500 to 2,000 mm^2 or 2,500 to 6,000 mm^2. When classifying the maximum cross-sectional area in 500 mm^2 steps, about a quarter each thereof belonged in the 6,000 to 6,500 mm^2 and 500 to 1,000 mm^2 ranges. Firms owning punches larger than 6,000 mm^2 were found to be manufacturers of control panels and housings of vending machines, and all firms were using 80x80 square punches.

With regard to minimum diameter, 1/3 of the circular punches each belonged to the 1 to 2 mm and 2 to 3 mm ranges, and the remaining 1/3 to the 0 to 1 mm and 3 to 4 mm ranges combined. Punches of diameters 3 mm or smaller constituted 90% of the total.

With regard to the minimum cross-sectional area of rectangular punches, 60% belonged to the 25 mm^2 or less range, and about 30%, the 25 to 50 mm^2 range. Over 50 mm^2, slightly under 10% each of the enterprises owned punches with an area falling in every 25 mm^2 range thereabove.

Table 10   Maximum Dimension of Punch

Circular Punch	
Diameter d (mm)	Ratio (%)
20 < d ≤ 40	17.6
40 < d ≤ 60	17.6
60 < d ≤ 80	11.8
80 < d ≤ 100	23.5
100 < d ≤ 120	29.4
Rectangular Punch	
Cross-sectional Area s (mm^2)	Ratio (%)
0 < s ≤ 500	11.8
500 < s ≤ 1000	23.5
1000 < s ≤ 1500	11.8
1500 < s ≤ 2000	0
2000 < s ≤ 2500	29.4
....	....
6000 < s ≤ 6500	23.5

Table 11   Minimum Dimension of Punch

Circular Punch	
Diameter d (mm)	Ratio (%)
0 < d ≤ 1	17.6
1 < d ≤ 2	35.3
2 < d ≤ 3	35.3
3 < d ≤ 4	11.8
Rectangular Punch	
Cross-sectional Area s (mm^2)	Ratio (%)
0 < s ≤ 25	58.8
25 < s ≤ 50	29.4
50 < s ≤ 75	5.9
75 < s ≤ 100	5.9

Frequently Used Punch Dimensions.   Punch dimensions high in use frequency are shown in Table 12. There, the responses to the survey encompassing plural dimensional categories were equally divided therebetween and amounting to 1 when combined.

With regard to circular punches, 30 mm and smaller diameter punches were found high in use frequency, and in particular, 5 mm and smaller diameter punches were disclosed to be the most frequently used. Those next high in use frequency belonged in the 15 to 20 mm diameter range. Punches greater than 30 mm in diameter were also found, but appear to be used much less frequently.

With regard to rectangular punches as well, those with a smaller cross-

Table 12 Frequently Used Punch

Circular Punch	
Diameter d (mm)	Ratio (%)
0 < d ≤ 5	49.5
5 < d ≤ 10	8.3
10 < d ≤ 15	5.4
15 < d ≤ 20	23.0
20 < d ≤ 25	0.9
25 < d ≤ 30	12.7

Rectangular Punch	
Cross-sectional Area s (mm^2)	Ratio (%)
0 < s ≤ 100	23.5
100 < s ≤ 200	24.3
200 < s ≤ 300	6.6
300 < s ≤ 400	12.5
400 < s ≤ 500	0.7
500 < s ≤ 600	6.6
600 < s ≤ 700	12.5
700 < s ≤ 800	0.7
800 < s ≤ 900	6.6
....	....
1500 < s ≤ 1600	5.9

sectional area were found to be high in use frequency, such as those in the 0 to 100 mm^2 and 100 to 200 mm^2 ranges.
Beyond 200 mm^2, for every 100 mm^2 incremental cross-sectional range, some firms claimed their cross-sectional range to be high in use frequency, but for over 1,000 mm^2, no such claims were made. The one and only exception was one firm owning 80x80 punch who considered 1,500 to 1,600 mm^2 a range frequently used.

<u>Punch Replacement</u>. Although not tabulated, the responses to the survey also disclosed that only a 1/3 of all the firms surveyed were aware of their own individual intra-station punch hitting counts; the remaining 2/3 were not. Punch replacements are frequently made, based on visual or measured check for burrs on the product. Replacements made at a predesignated hitting count level were found in only 1/4 of all the firms surveyed. Even where hitting counts are known as mentioned earlier, not all such firms exercise punch replacement based on the hitting count.

<u>Punching and Nibbling</u>. To process work into a shape not achievable with punches at hand, either nibbling or laser processing operations were engaged in. The relative time of average punching operations in reference to that of nibbling (or laser/plasma cutting) operations is tabulated in Table 13. This shows an overwhelming number of cases where the ratio is 10 or less. That is, the high productivity of punching operations has not been taken full advantage of in most cases, and nibbling or similar other operations have had to be resorted to instead. Some enterprises, however, have achieved punching operations that are 50 times, or even over 90 times, as great as that of nibbling operations. This would appear to signify either the availability of a sufficient variety of punches, or conversely, fewer types of parts processed. The shearing of materials into a totally arbitrary shape is believed to mandate resorting to laser cutting or nibbling operations, but in view of productivity, if a relatively enriched variety of punches are available, the achievement of hige efficiency processing through punching operations is possible in certain cases.

Table 13 Ratio of Punching Time (P) to Nibbling Time (N)

P/N	Ratio (%)
0 < P/N ≤ 10	73.2
10 < P/N ≤ 20	6.7
20 < P/N ≤ 30	6.7
30 < P/N ≤ 40	0.0
40 < P/N ≤ 50	6.7
....	...
90 < P/N ≤ 100	6.7

Table 14 Number of Resetups per Day

Number of Resetups No	Ratio (%)
0 < No ≤ 10	82.3
10 < No ≤ 20	5.9
20 < No ≤ 30	5.9
....	....
No = 50	5.9

Resetups. Daily resetup frequencies of the punches are shown in Table14. About 80% of the firms were found to perform resetup operations 10 times or less per day (including four firms at zero), but one firm performs resetups as frequently as 50 times per day. There, resetups are done manually, but it takes only about a minute each time.

CONCLUDING REMARKS

It is estimated that there are more than one hundred sheet metal FMS working in Japan. As enterprizes want to make a secret of substance of FMS installed, the number of surveyed systems is not so large (42 systems). Interesting results, however, are obtained. The surveyed results about software will be presented at another chance.
It will be more difficult to obtain young manpower and there will be a growing tendency of unmanned factory. In addition to these facts, variety of needs will make FMS spread even in the field of plastic working.
A book showing examples of FMS for plastic working is published [2] and surveyed results are shown in detail there. It will be a good guide for engineers who are interested in FMS.

REFERENCES

[1] Endow,J., Kiuchi,M., Nakazawa,Y., Shinohara,M., and Matsubara,S.,"FMS for Plastic Working in Japan". Proc. 4th Int. Conf. on FMS, pp.69-79 (October 1985).
[2] "FMS for Plastic Working in the World (in Japanese)" edited by Association of Mechanical Technology, Machinist Publishing Co.,Ltd. Tokyo, 1986.

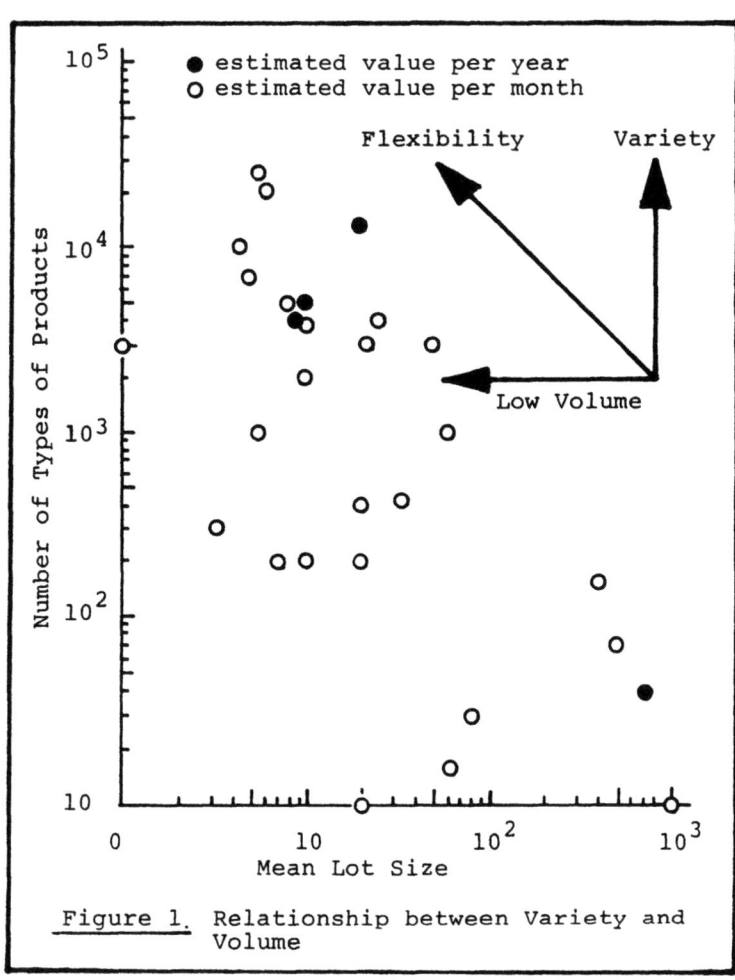

Figure 1. Relationship between Variety and Volume

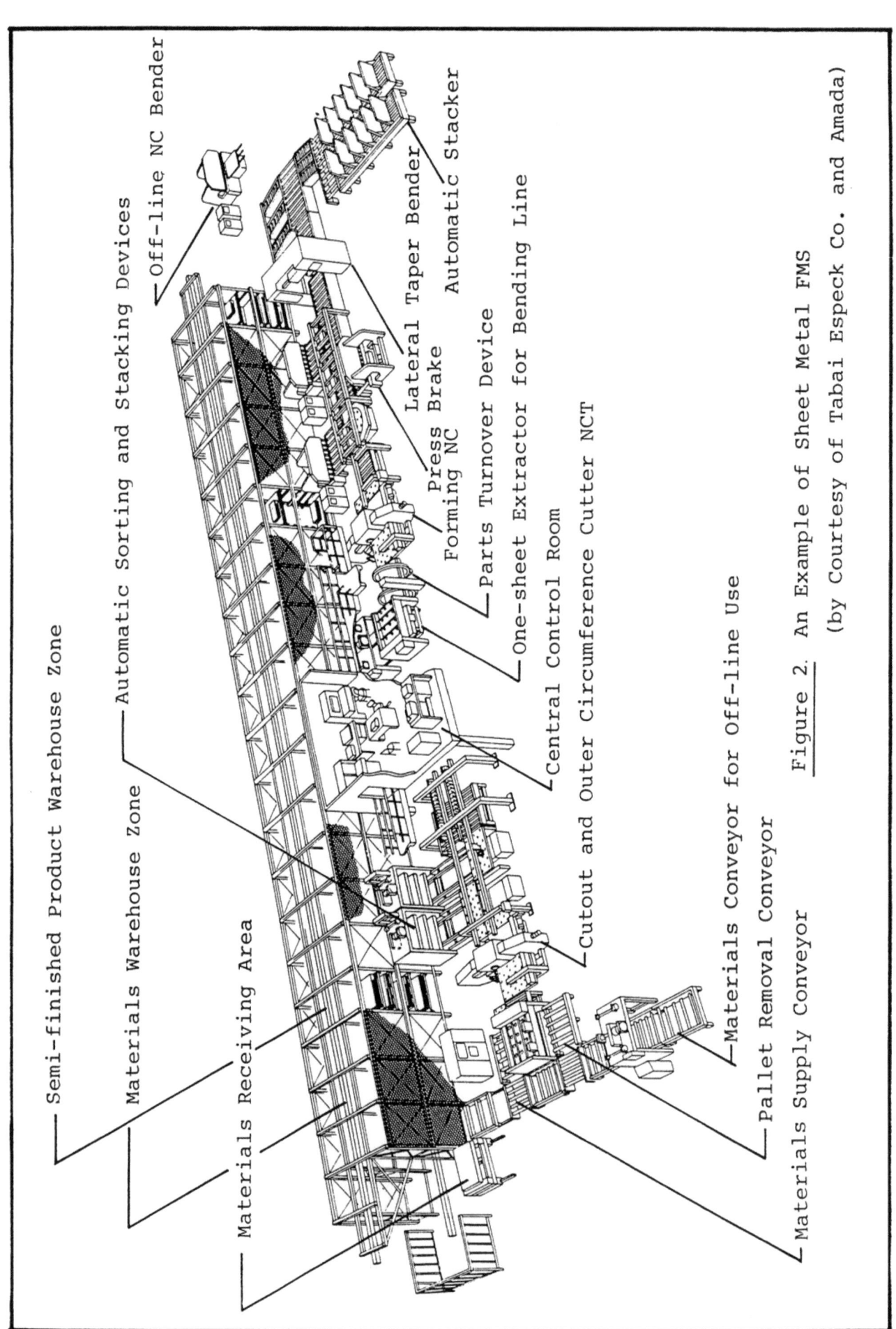

Figure 2. An Example of Sheet Metal FMS (by Courtesy of Tabai Especk Co. and Amada)

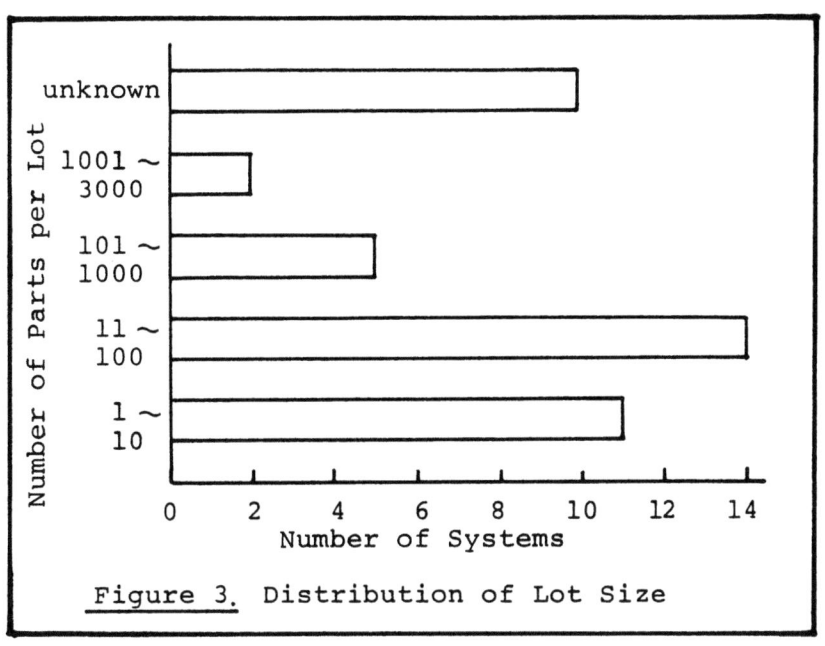

Figure 3. Distribution of Lot Size

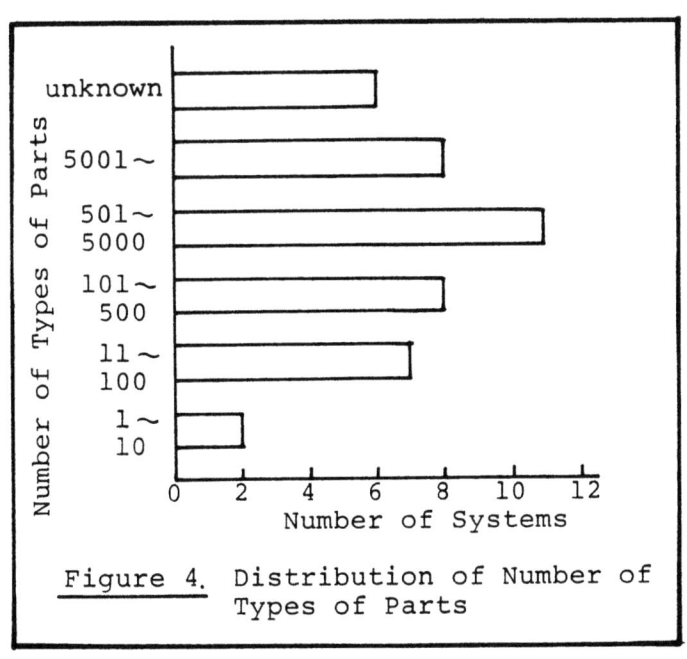

Figure 4. Distribution of Number of Types of Parts

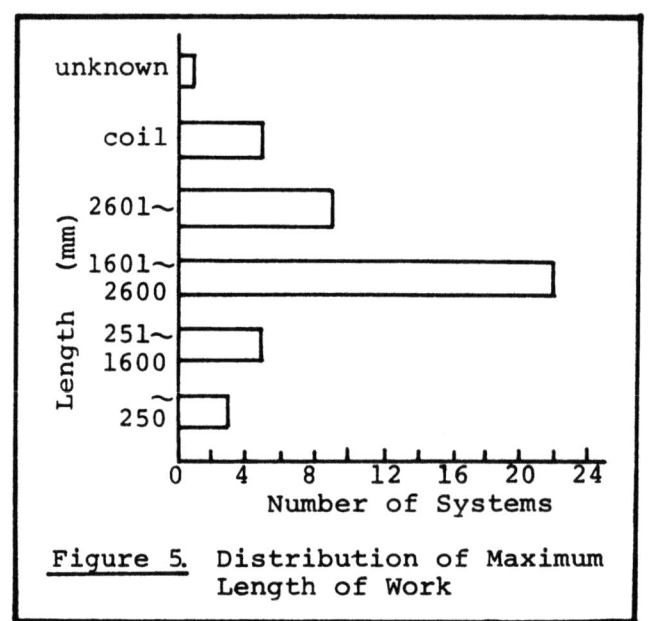

Figure 5. Distribution of Maximum Length of Work

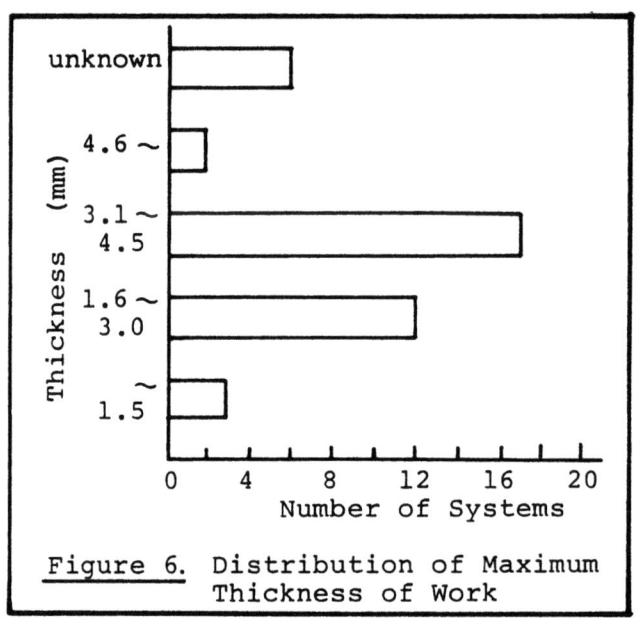

Figure 6. Distribution of Maximum Thickness of Work

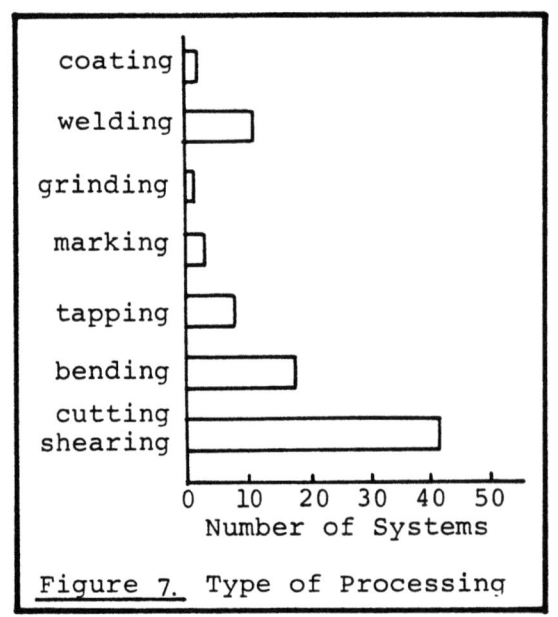

Figure 7. Type of Processing

Tooling concepts for FMS
P. Tomek
VUOSO, Czechoslovakia

ABSTRACT

The technical design in new FMS generation emphasizes the automatic tool transport.The reason is obvious - really unmanned operation with high level of flexibility is hardly possible without this feature.But what is the most efficient way how to control the tool flow and what are the concequences for FMS technology?These main problems are discussed and solution are suggested in the paper.New tool flow strategy based on random machining in FMS,where common tool pool is used and automatic tool transport matches transport of parts is described.Practical experiences from FMS operation together with basic information about new simulation model are given.

INTRODUCTION

Tooling creates a serious problem in FMS planning and application. It may often lead to severe production limitations and economic losses. No wonder then that tooling has become a topic discussed on industrial and scientific levels. Tooling is very expencive. According to our experiences it may reach 20% of a machining cell price, depending on the type of production. For FMS it may represent even higher proportion of capital investment as it covers cutting tools, technological pallets, fixturing systems, measuring devices, measuring machines and that all in numbers required for unmanned production.

For truly random production in FMS big capacity tool magazines are required. Normal magazine sizes used for machining centres are not sufficient for FMS. Mechanical design of the tooling side usually limits the application of machining centre in the FMS. Besides new technical approaches on FMS supplyers side there are necessary changes in production technology on FMS users side. Most production plans are to be changed to reduce number of parts setting, to enable measuring with touching probes on the machine and to machine the parts in FMS completely without external cooperation. The parts design is to be changed as well to reduce broad tool variety for the whole range of parts and to ensure automatic machining without manual interventions. On the tool supplyers side new cutting tools must be developped (boring bars with automatic diameter setting, special counterboring tools etc.).

The tooling problem in FMS (namely for prismatic parts) concentrates on cutting tools and their distribution. We will therefore focus our attention in next paragraphs to this aspect of tooling.

TOOLING STRATEGIES

We will now discuss three basic tooling strategies that might be used in FMS.

Batch of Parts - Group of Tools

This strategy copies the approach conventionally used in job shops: For a batch of parts a group of tools must be delivered to the machine tool matrix. Possible sharing of tools in consequent batches is ignored. For many current applications, where only several different parts are machined is this strategy convenient. The flow of parts is fixed in such FMS and the tools in machining centre matrix may stay unchanged for considerable time. Introduction of new part means usually new manual setting of machining centre. Thus truly random machining with low volume/high variety scheme is rather impossible.

Several Parts Batches - One Group of Tools

This strategy is more progressive in sharing identical tools among several batches of parts (or different machining operations represented by partprograms). Based on group technology a chosen mix of parts is delivered to the machining centre in a fixed period of time and this part mix is serviced by a common tool storage in machining centre matrix. Bigger tool magazines are needed (80 to 140 tools). The range of different machining operations which could be serviced by a storage of 140 tools may be especially with skilled group technology very large.

Common Tool Inventory Shared by Group of Machines

Contrary to the strategies described above, where individual machining centres are considered as basic element, this strategy applies to groups of machines constituting the FMS.The whole tool inventory creates a common tool storage,which is spred among machining centres.The composition of such pool is dictated by tool requirements for parts being machined in a fixed FMS production period.The fixed partflow is not required in such scheme any more.This is becouse the concept of grouping,where all machines have identical properties,makes it possible to address **group of machines** rather than single machines when controlling the part flow through the FMS.The automatic tool transport is necessity in such FMS and enables sharing of tools among machining centres.Before starting the machining operation on a chosen machine,the missing tools may be "borrowed" from other tool magazines,where they are not needed.Thus a small part of a common tool inventory of the FMS constantly moves among tool magazines.The worn tools are replaced by sharp tools from the tool room.The composition of tool storage for FMS is constant in a long production period.

In this approach the tool inventory does not reach the lowcount achieved by the previous strategy,but ensures the ability to respond to any unexpected situation which may arise in FMS production.The FMS is toolpreloaded for a production period to minimize the migration of tools between machining centres.The most often required tools reside in all matrixes and the chosen tools may be dubled.

A new logical problem arises – how to "tune" the FMS for a given production period with the tooling inventory shared by several machines.Compared with strategies mentioned above,shared tool inventory brings completely new requirements on FMS design:large capacity tool matrixes,automatic transport of **individual** tools between machines and the tool room. Such FMS design must be supported by the system software where tools flow must match parts flow.

We will now follow the last tooling strategy in more details and ilustrate it on several case studies and simulation model.

RANDOM FLOW PHILOSOPHY

As stated above a new way of tool management backed up by a new technical design of FMS enables the ralization of completely new strategies of manufacturing control.Groups of working station instead of individual machines are addressed.For every group of working stations a job queue is formed.This queue is servised according to actual FMS situation.The management is based on following principles:

a) The FMS production period is characterized by a fixed number of different parttypes which may enter the system.These parts may be machined in a given period of time,called **"steady period"**.During the steady period fixed tool inventory is used and no new fixtures are prepared.The tool life is monitored,worn tools are replaced by new tools but no new types are entering the system from the tool room.The steady period for a complex FMS with a large fixture storage may last for a long time (weeks or months).The technological pallets with fixtures are stored in the FMS and may serve for machining of large family of different parts.The batches of parts which are loaded on prepared fixtures may be arbitrary,that is they may start in one piece batch.In the steady period the batches may be repeated according to assembly line needs. For these complex FMS only a part of steady period is generally simulated.The FMS is **tuned** for the steady period,that is the tooling inventory (number of different tooltypes needed for machining and number of identical tools which are spred among tool magazines with number of pallet fixtures for all machining operations) is fixed.Transition from one steady period to onother requires refixturing and new tooltypes input,i.e. new tuning of FMS according to new parttypes production.The "retuning", new fixture setting, new tooltype setting and tool input into the FMS is realized during the previous steady period. Thus the FMS operation is continuous.

b) FMS production is based on random scheme. Workshop area, i.e. unmanned part of FMS is composed of groups of identical machines. The same philosophy is used for manned area (loading, reloading and unloading of parts on pallet fixtures). For every group of working stations type of operations which may be realized there is given.

c) Selection of most suitable job from queue waiting for machining (i.e. in fact selection of suitable technological pallet with a fixture with a clamped workpiece, which is waiting in a pallet storage) on a group of machining centres with given properties is based on following criteria:
- workpiece priority (special cases, where workpiece must go through the system as fast as possible)
- minimum of tool transport actions for machining operation on a given machining centre (best case is when all needed tools are present in the machining centre matrix)
- effort to complete the part (that is in a case when a part is machined in two or more setups)

d) It is not possible to identify a free workpiece. Any workpiece in FMS is identified by a technological pallet to which it is connected (by a fixture). Therefore storage of free workpieces is not allowed on reloading stations.

SIMULATION MODEL

Based on described principles the simulation model was created. It is used for FMS with automatic transport of individual tools, where work stations are devided in manned and unmanned areas. Unmanned area includes machining centres, washing machines and coordinate measuring machines. Manned area covers fixture setting, loading, reloading, unloading and the tool room. The desire is to reach the maximum production, i.e. to keep the utilization of machining centres as high as possible.

The simulation package is composed of following modules:
- Dialogue input of all necessary technological data into the database (production plans for all parts covered in a steady period, lists of tools needed for machining operations, tool library and so on).
- Tuning, creation of common tool storage for a steady period. It builds the starting position, when all tools are spread among tool magazines.
- Setting the FMS configuration (dividing the machining centres in groups for rough and precise machining, setting pace of work - opening and closing various workstations in a given time. For example manned area may be closed for afternoon and night shifts, washing machines may be closed for night shifts etc.)
- Control of tools and pallets flow in a given period of time - servicing queues for tools transport and pallets transport, servicing individual working places, i.e. selecting the best job from queues for group where a given station is a member, transporting the finished work from working station to pallet storage and placig the pallet to transport queue for next operation and so on.
- Analyses of system operation, monitoring all the changes in FMS and printing results in a chosen form.

The simulation may carry on for a given period of time, starting from previously reached state.

The simulation package is available in BASIC or SIMULA language and its basic software structure is described in more details in [4].

As implies from the above description the random tool and part flow is simulated.The following remarks clarify the simulation model functioning:

-By a random scheme an arbitrary routing of technological pallets and tools is meant.The operation may be realized on any workplace from the appropriate group - according to the type of operation.

-There are pallets with fixtures prepared for operations according to the job list.These pallets are waiting in the pallet storage.Once the pallet for the first setup is free,the next casting may be loaded.

-Selection of most suitable job from queue is based mainly on minimal tool transport.It means that all items in queue are tested and ideally the pallet with no tool transport requirement would be chosen.

-The tool flow matches the pallet flow.Tools are transported according to the pallets needs.When the particular pallet from queue is chosen for transport to the machining centre,the required tools borrowed from other matrixes are fixed into the tool transport queue.The machining will be started only on assumption that pallet with workpiece and all necessary tools for the operation are present on the machine.Using the tool in machining operation means that its life is reduced according to the cutting time.Only a tool with sufficient tool life may be used. When the rest of the tool life is 1/10 of the starting value,the tool is declared "worn" and is placed into the transport queue for the tool room for output from FMS.

-For subsequent simulations runs the conditions may vary in the number of pallets,in the time table of working places (namely in the manned area),in speeds for tool and pallet transport e.t.c.,so as to reach the optimum state of FMS.In simulation only one tool strategy is used.If very small number of different parts are being machined,then the FMS automatically turns to simple tool flow with no tool "borrowing" from other machining centres matrixes.Worn tools are replaced only.

-As the **tuning** procedure is concerned:it defines the starting conditions for FMS.It fixes overall number of tools and technological pallets.The starting number of pallets for various machining operations is limited by the pallets storage (whose capacity is chosen during the first technical proposal).The ratio of pallets number for subsequent machine operations for one part with several setups responds to time ratio of these operations.The starting number of tools is limited by the capacity of common tool pool.All types of tools needed for different machining operations must be present.(It means at least one tool for each type.)The number of identical tools from one tool type is fixed according to the "importance degree",or "weight of tool type",which says how much is a certain tool type needed for the forthcoming operation period.(Under tool type a certain type - as milling head,boring bar,taping head etc. with certain geometrical parametres as length and diameter is meant.For example two boring bars with the same diameter but different length represent two different tool types.)

-The **weight of tool type** (i) is evaluated as a function of number of pallets $P(i)$ used in the system for machining operations where the tool type (i) is used,total machining time $Tc(i)$ required for the tool type (i) for the number of machined parts according to the job list for the simulated period and starting tool life $T\emptyset(i)$ given by tool producer:

$$W(i)=f\{P(i),Tc(i),T\emptyset(i)\}$$

The **number of identical tools** of tool type (i) is a function of $W(i)$:

$$D(i)=f\{W(i),C\}$$

where C is the capacity of common tool pool used in FMS simulation.(The common pool capacity is not used totaly,approximately 15% is kept free for manipulation purposes.)

CASE STUDIES

Both following FMS case studies are using similiar structure:overhead transport of individual tools is used,two tool matrixes are added in tool room,pallet storage for unmanned operation is required.The lay-out of such system is on Fig.1.(the overhead tool transport may be seen on the left side,pallet storage on the right side).Basic machine,see Fig.8.

Flexible Manufacturing System 800x800

The lay-out of the system is on Fig.2.It is composed of five machining centres using technological pallet 800x800 mm.Box type parts from grey cast iron (mainly large gearboxes for mining industry)are machined.There are 12 different parts machined in 27 setups (this represents the job list for steady period).For machining operations 176 tooltypes are used. Everage machining operation takes 170 minutes (minimum 20 minutes,maximum 540 minutes).Everage number of tools used for machining operation is 15 (maximum 40,minimum 2).The workpieces are machined in two or three setups.After "tuning", the simulation uses 50 technological pallets and altogether 611 tools.

Fig.3 ilustrates tool type requirements and their usage in different machining operations.As may be seen tool requiremnts are very broad (176 different tool type used,but more the 100 types are used in two different operations only,less then 10 different tool types are used in more then 8 different machining operations).The result of tuning of FMS is ilustrated in Fig.4.The number of identical tools for different tool types is given.Most of tool types (123) are doubled,while 20 tool types have 10 identical tools each.

The main results of simulation of the first 20 shifts are on Fig.5.In the upper side is the time table of manned working stations (LD-loading,RL-reloading,UL-unloading).The aim is to close these stations for afternoon and night shifts without causing any harm to the machining system utilization,machining cells work continously.The middle part of the Fig.5 ilustrates pallets queue forming for machining centres.In the lower part of the Fig.5 utilisation of all machining centres as a function of time (shifts) is given.It may be seen,that the machines are gradually more and more employed,reaching a steady state after 13 shifts with 85 to 90% utilization.

Flexible Manufacturing System 1400x1250

Lay-out of the system is similiar to the previous case:only four machining centres are used together with washing machine,coordinate measuring machine,four reloading stations,input/output of parts and tools.Technological pallets 1400x1250 mm are used.As in previous case the pallet transport is realized by using rail guided trolley.Overhead tool transport connects six tool matrixes,each having 144 tools capacity.There are only three different workpieces machined in the system,but the total amount of different machining operations is 17.Complicated gear housings for mining industry are machined,all castings from grey cast iron.For all machining operations 135 different tool types are used.Machining time for operation is 20 to 400 minutes,everage number of tools for operation is 12 (2 to 53),the workpieces are machined in 5 to 6 setups.For the operation 50 pallets and 481 tools are used. Fig.7 ilustrates how the tooling storage was fixed after the tuning of the system.Most of tool types (53) have three identical items in the system.For 12 tool types there are 10 identical items for each type. The simulation results are on Fig.6 .For the pace of manned area working stations (in the upper part of figure)the pallets queue forming and utilization of machining centres as a function of time (20 shifts) is given.

CONCLUSION

Behaviour of FMS where automatic tool flow matches the random flow of technological pallets is influenced by a number of factors which may be named as internal and external.All these factors influence the FMS performance.It is practically impossible to express that influence analytically.

The external factors are independent on FMS structure and represent the technological data which determine the type of production (number of different part types,machining times,number of setups,number of different tool types,tool lives etc).The internal factors characterize the FMS structure and are dependent on production which is desired in FMS: number of working stations in various groups,number of identical fixtures for individual setups,number of identical tools,time table for manned area which should suit to FMS users and at the same time to FMS performace etc.

The external factors are independent on various control strategies.Internal factors represent in fact variables which through **tuning** of the system may gradually reach optimal values for the best FMS performace. The setting of these values may be found using the simulation model described in this paper.

REFERENCES

|1| Hankins,S."The inpact of tooling in FMS".Proceedings of 2nd Biennial International Conference,Chicago 1984
|2| Tomek,P."Tooling Strategies Related to FMS Management".The FMS Magazine ,Volume 4 No 2.April 1986
|3| Tomek,P,Zelený,J."Machining Technology in FMS for Prismatic Parts with **Automated Flow of Tools,** Proceedings of the 2nd International Conference FMS 2,London 1983
|4| Mojka,A.,Weinberger,J.,Tomek,P.,"Optimal Design of Flexible Manufacturing Systems Using SIMULA",Proceedings of the XIV SIMULA Users Conference,Stockholm,August 1986

Figure 1.Lay-out of FMS with automatic overhead tool transport

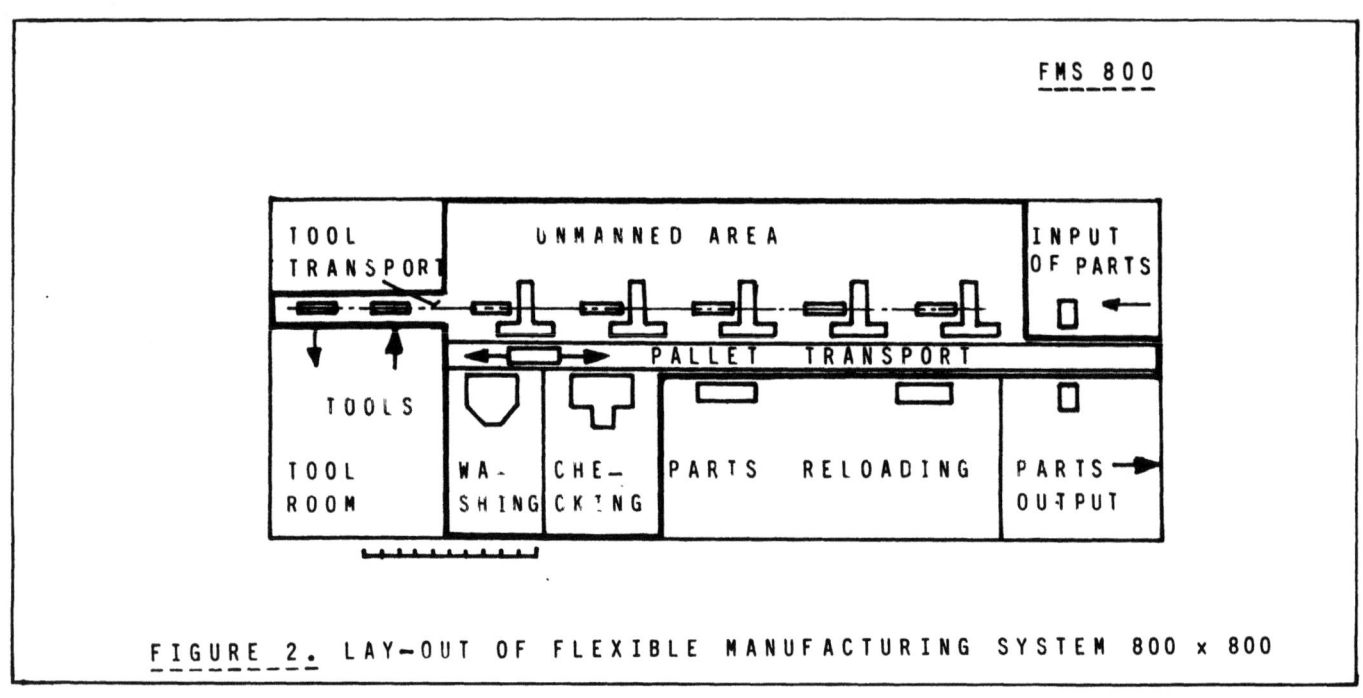

FIGURE 2. LAY-OUT OF FLEXIBLE MANUFACTURING SYSTEM 800 x 800

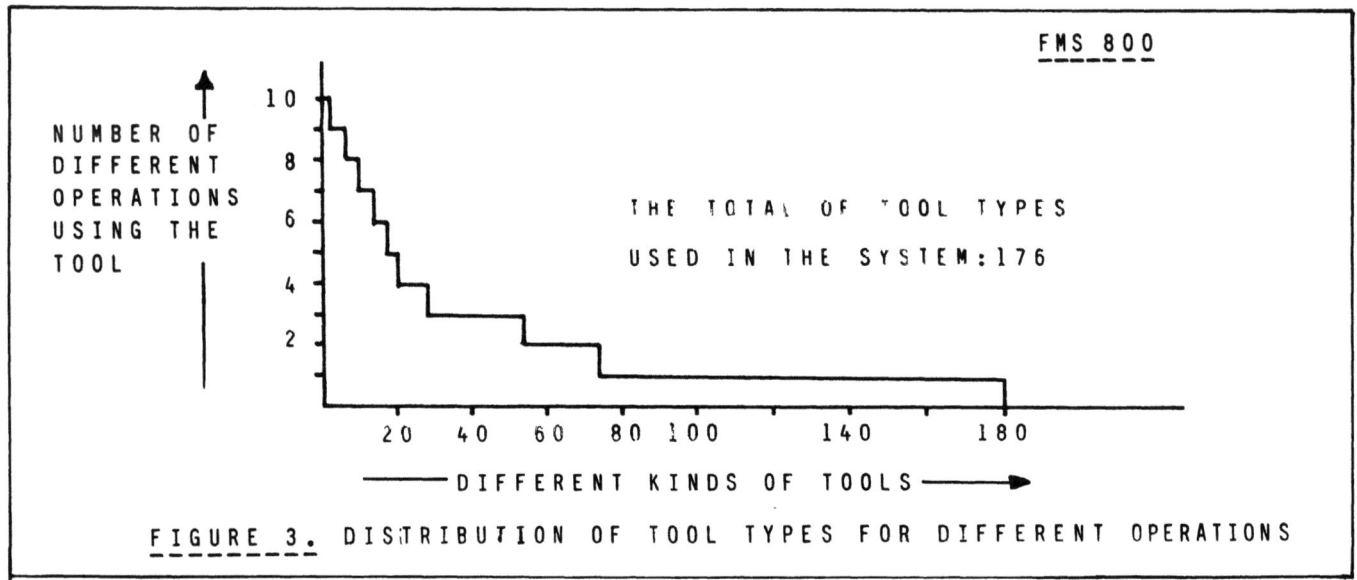

FIGURE 3. DISTRIBUTION OF TOOL TYPES FOR DIFFERENT OPERATIONS

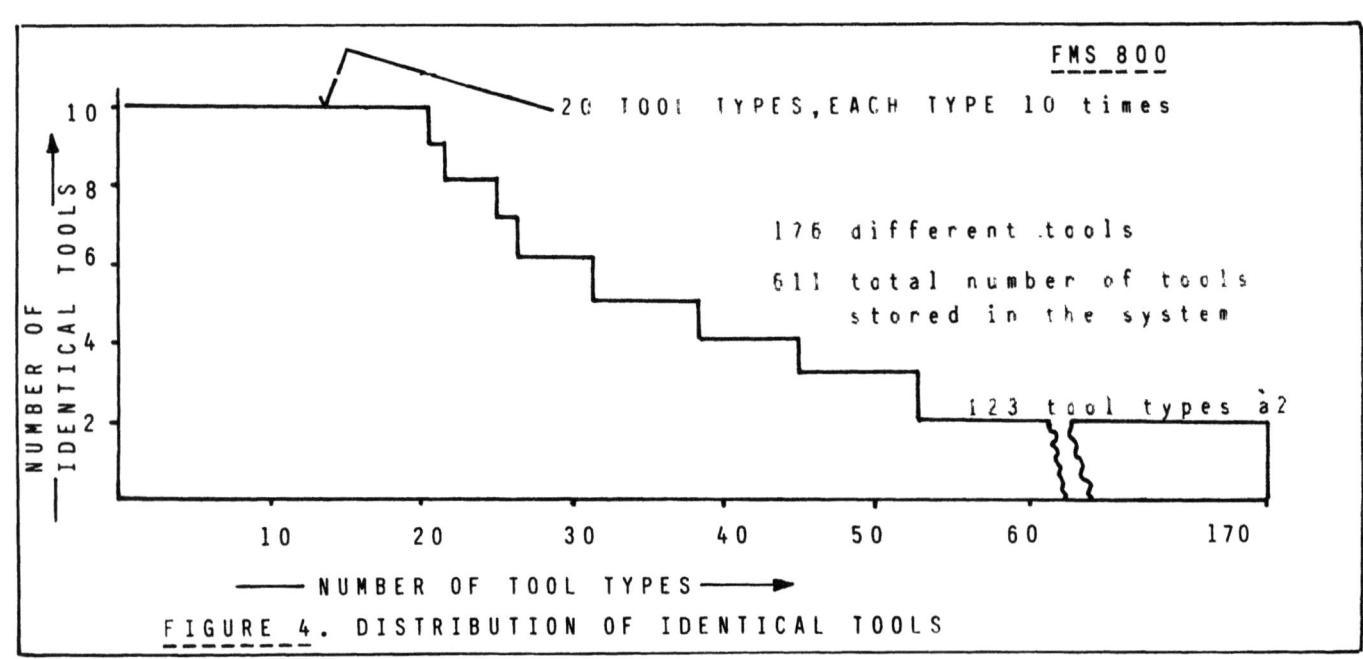

FIGURE 4. DISTRIBUTION OF IDENTICAL TOOLS

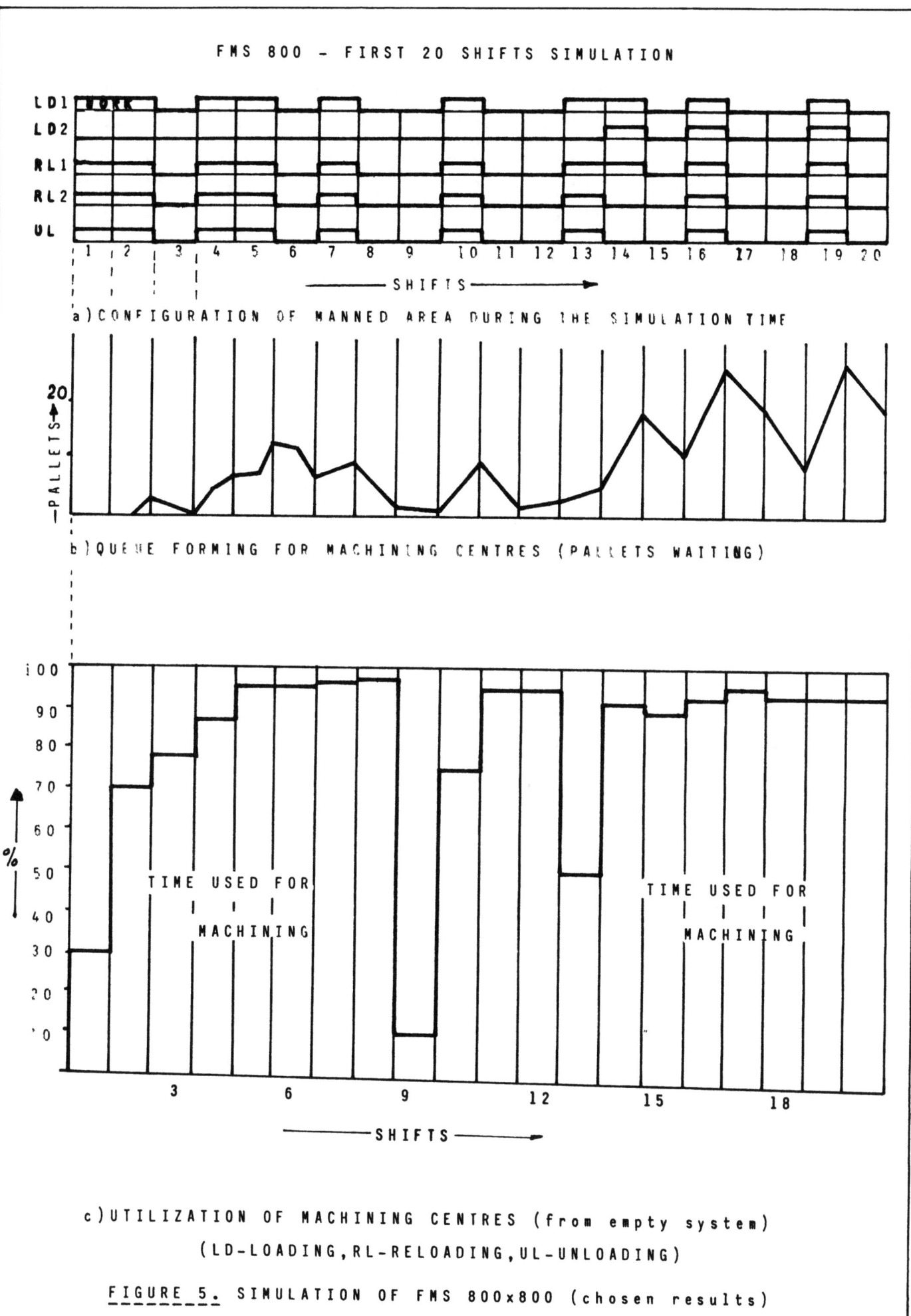

FIGURE 5. SIMULATION OF FMS 800x800 (chosen results)

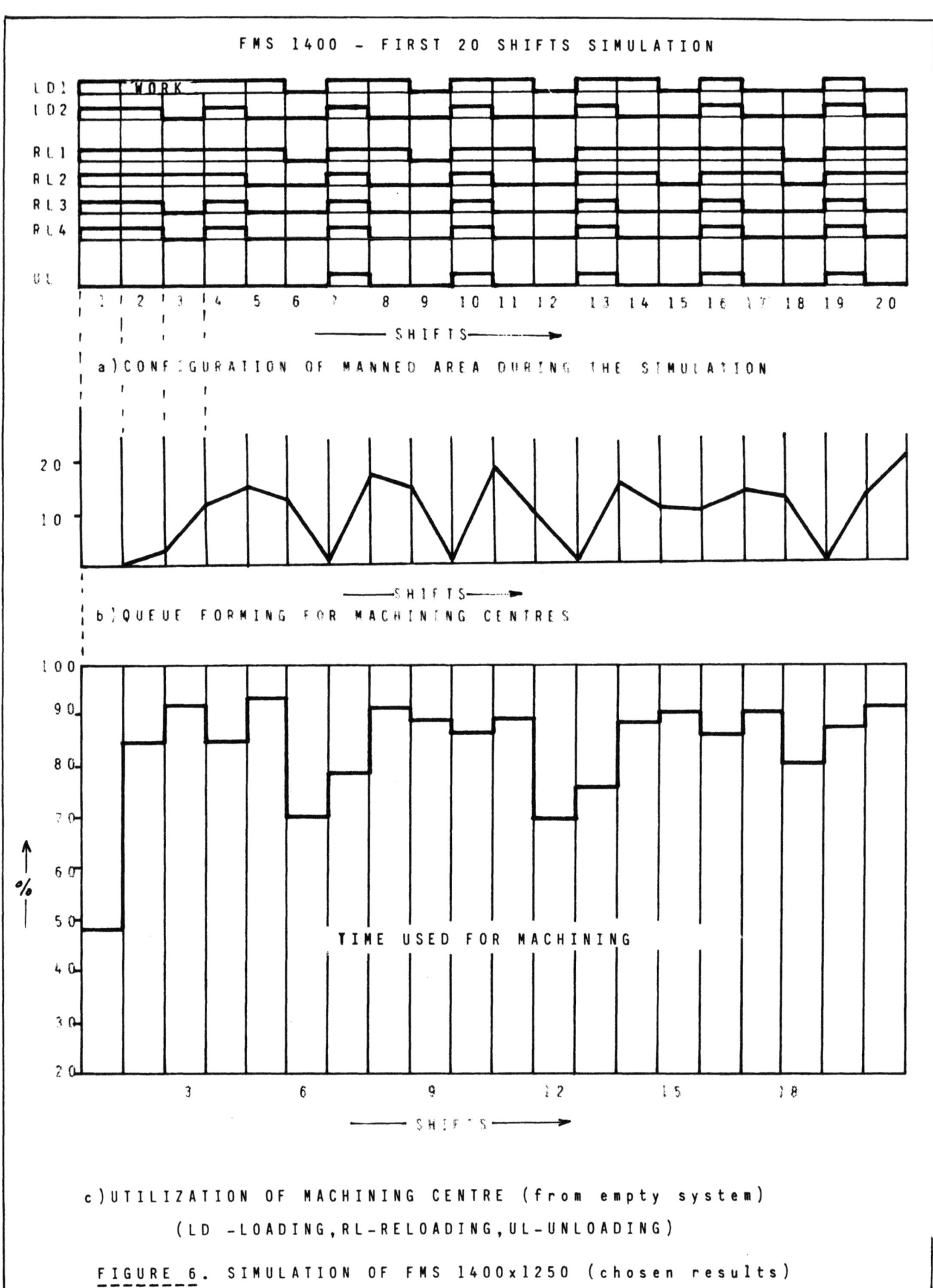

FIGURE 6. SIMULATION OF FMS 1400x1250 (chosen results)

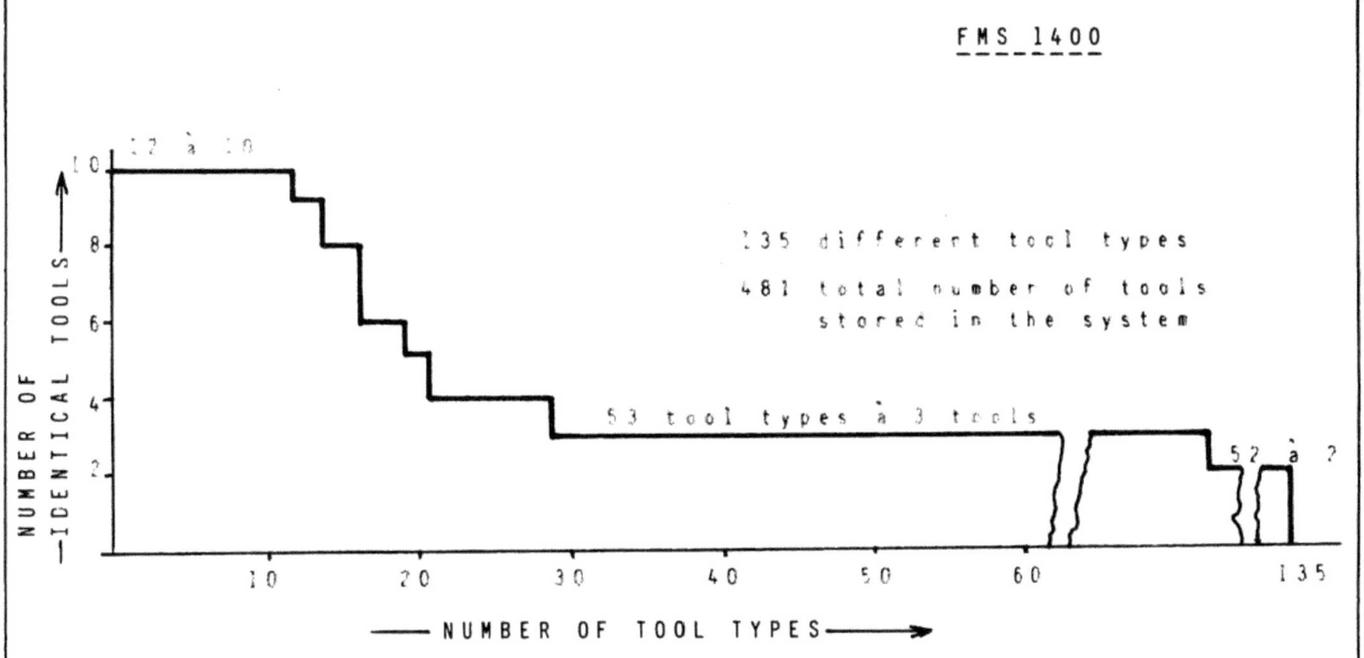

FIGURE 7. DISTRIBUTION OF IDENTICAL TOOLS

FIGURE 8. BASIC MACHINING MODULE WITH LARGE CAPACITY TOOL MAGAZINE FOR FMS

CELLS AND SYSTEMS — EVERYMAN'S FMS

boring machines in production with flexible manufacturing systems

Ch. Kesselburg
DIXI SA, Switzerland

ABSTRACT

JIG BORING MACHINES IN PRODUCTION WITH FLEXIBLE MANUFACTURING SYSTEMS

This presentation describes in details a practice-linked and fully operational FMS plant for the production of high precision housings, using for the first time, jig boring centres in practically unmanned operation. This opens a way to new production concepts for short term planned, small batch precision manufacturing of components where the accuracy of bores, surfaces and squareness to each other determines the quality of the final product.

JIG BORING MACHINES IN PRODUCTION WITH FLEXIBLE MANUFACTURING SYSTEMS

The pressures towards rationalisation and the short-term need to produce small quantities of components, linked as closely as possible with orders received, make it necessary to reconsider production methods and planning, without having to resort to highly scientific concepts. Waiting times for components are substantially reduced in the production process, leading to greatly increased use of the machins and consequently a considerable reduction in production costs.

Firstly the definition "FMS" should be clarified

By "Flexible Manufacturing System" we mean not simply a manufacturing unit or cell, but a self-contained system for the production of specific parts or part-groups, beginning with the take-over of raw castings, and continuing right up to the delivery of finished components, including the whole process of handling, transport, intermediate storage and loading, as well as the linking up and control of individual sequences and possibly of different machines.

A "manufacturing cell", on the other hand, describes simply a particular section within that manufacturing system.

What basic considerations should be taken into account when looking at a Flexible Manufacturing System ?

In contrast to the former preoccupation with improved production times, achieved by means of better performances by machine tools, as well as better cutting edges - which, however, did not always come up to expectations - today, to a far greater extent, the long component lead times, and also the average but generally rather poor machine usage times, need to be improved. This, as has been proved, is especially the case with boring and milling machines.

Starting from a possible annual working time of 365 days, i.e. 8760 hours, we are left, after deducting weekends, official days off, and holidays - which add up to 140 days or 3360 hours - with only 5400 hours available planned time. In two shifts this means only 3600 hours.

Closer examination of these available times reveals that approx. 30% is allotted to setting up and aligning the components, measuring and tool changing, and also 8 - 10% to rest times, personal breaks and other, unforeseen, interruptions. Consequently for machine functions - i.e. machining and positioning times - only about 60% is left over for continuous production. Tool changing and test runs on new series - whether large or small - require roughly another 10% of the available time, so that finally only about 50% of the two shift plan remains as effective machine utilisation time, or to put it another way, 20% of the total annual time reserves, or only 10%, using one-shift operation ! This should cause all production experts to stop and think !

These figures show quite clearly that, above all, the automatisation of tool changing, and especially component changing, must be considered of prime importance.

Since boring machines are now available with tool and pallet changers as well as with efficient CNC controls - no matter whether for rough or finish operations - a flexible manufacturing system can be set up. Profitability accounts have shown a 30% cost decrease per piece with two-shift production and even 55% on triple-shift production. Four conventional boring machines could be dispensed with, and instead of eight machine operators, only two people now need to be employed at the central setting-up area and for supervising the whole production area.

A further important aim achieved by this investment is the significant reduction of lead times per piece.

The most important decision when setting up an FMS installation is the choice of the correct machine tools, to meet the needs of the components to be produced, and not, as unfortunately often happens, simply to go for the choice put forward by the machine manufacturer who offers the greatest experience and the most attractive high-tech system. For what use is a long-term investment of this type, if the end product simply does not meet many of the required specifications ? (if, in other words, you have to remachine most of the parts again, on the quiet, on conventional jig boring machines ?)

Precision in Production has also become a reality in the flexible manufacturing system. Where the normal accuracy of traditionally designed machines no longer suffices, then DIXI jig boring machines or high-precision machining centres are the right answer. They can maintain a high degree of accuracy over long working cycles, they are of rigid design, in order also to cope with large and heavy precision components, and they offer long life performance under constant accuracy.

Such machines, even though a costly investment, can also some times be economically times on minor tasks, since in an FMS system it is not just the hourly rate of the machines that counts, but also that of the **entire** system. And, moreover, anything that has to be provided outside the actual machining centres will in any case cost about the same.

Be that as it may, it must in any case be considered whether the precision machines should take over the whole of the production, i.e. rough and finish-machining - which the DIXI jig boring machines are fully capable of doing - or whether they should be used only for precision operations. If the first option is chosen, the standardisation of the production means - such as tools, fixtures, pallets, programmes, etc. - as well as flexibility in the production planning, due to **complete** interchangeability of the machines for any kind of operation, represents an important cost factor in favour of the DIXI jig boring machine.

If rough machining and operations of minor importance clearly constitute the larger part of production times, then a split-up between the appropriate machining centres and the precision machines can readily be made. The stress-relief of the components after the rough machining operations could be a good phase at which to do this. A subsequent new setting-up is in any case necessary.

Once the final choice of machines has been made - which represents the most important element in the system, and which determines the accuracy of the end product - then the peripheral equipment can be chosen : the automatic buffer storage equipment, the robot vehicles (or AGVs), the central setting-up station, possibly a washing installation, a measuring machine and also a tool pre-setting centre. Finally, of course, we get to the system-technology, which has to control the entire plant by computer, including the automatic material flow between all the sections, even to the non-automatic machines, should there be any.

In the example of the FMS set up at the firm of Westfalia-Separator in Oelde, Germany (Fig. 1), the whole system technology was worked out by Schindler-Digitron of Biel in Switzerland in total collaboration with Westfalia-Separator, whereby DIXI had to co-ordinate the interfacing between all peripherals and the machines. DIGITRON also supplied the automatic high rack storage system, the robot vehicles and the control computer.

This fully-operational FMS, conceived, produced and installed in a period of just twenty months, with two DIXI 410 precision machining centres, operates independently of the in-house NIXDORF computer, which serves as the central production planner, as well as programming and storing all part programmes for the machine tools. Any kind of link between NIXDORF and the PDP 11/23 FMS computer would have caused endless problems, and, above all, would have seriously limited flexibility in production planning. Production planning in the FMS area is programmed on a daily basis on the system computer in dialog form by the works foreman himself, with the aid of the information contained in the general process planning sheets.

The FMS computer is linked to the Allen-Bradley 7320 CNC of the DIXI machines and requests, when necessary, the appropriate part programme (Fig. 2).

If this is not available in the CNC memory, then the NIXDORF computer will be asked by the machine control to send it "on-line" immediately.

In fact, only very simple information is exchanged between these three computers.

At Westfalia-Separator, the range of parts to be machined (Fig. 3) consists of over 40 different separator housings, i.e. high-speed centrifugal machines, whose bores and surfaces must be very accurately aligned and perfectly square to each other. Each housing is usually machined in three set-ups, taking 60 - 90 minutes each.

This time determines the number of machine pallets needed, which will have to be passed through during the night-shift. The central setting-up area (Fig. 4) consists in this case of 8 pallet stations altogether, of which 2 are used for day-shift working and 6 for the night-shift.

Besides a relatively large number of wooden pallets on which all the components are temporarily stored and subsequently conveyed to the setting-up area, there are 10 high precision pallets altogether for the two DIXI machining centres, of which two are always on the machines. Two more pallets are either at the day-shift setting-up station or on stand-by on the pallet exchanger of the machine. The remaining six pallets are restocked during the day with new components, which have then to be machined during the night-shift. The night-time production is carried on entirely without human intervention, and the pallets are transported automatically by the AGV's from the setting-up stations to the machines and back.

Since the system computer knows the position and the machined stage of each component - right from the start, and up to its exit from the FMS area - neither the pallets nor the components are identified externally. After each machining cycle the parts are immediately conveyed back to the buffer storage area by the AGV's, so that there are never any components lying around the workshop. The system computer would in fact just lose the track of them.

Precision in Production with a Flexible Manufacturing System is not only dependent on the machines (Fig. 5), but also on the design and the execution of the machine pallets and their holding and centering system.

Accurate machining - which means correct alignment, parallelism, squareness of axes and faces - making use of the table reversing method, can only be achieved if, in any angular position of the rotary table, its plane remains even within a few micrometers over the whole working area of the cross-table saddle. This is the remarkable attribute possessed by all DIXI jig boring machines. In the case of our machining centres with pallet systems, no compromise is admitted on these precision parameters. Hence, all pallets must incorporate an increased degree of accuracy, in order to be perfectly interchangeable with each other, which means they must show the same geometrical features as the DIXI machines without pallets. Unfortunately this is not the case with the most common pallets on the market, which are produced according to standard accuracy parameters.

Instead of the normal rotary table, the machine incorporates a pallet holder or pallet base, in relation to which the mechanical machine origins and the rotary measuring system are calibrated. For example, with the DIXI 4400 CNC control system, in connection with the Renishaw MP3 electronic touch probe (Fig. 6), the component reference, **including** any possible pallet-centering error, can be registered and is automatically compensated for in the programme - for any angular position, up to the thousandth degree, of the rotary table. This is a standard feature of the DIXI control system.

As regards the **tooling,** Westfalia-Separator's experience has emphasised some very interesting points :

The aim was to cope with the production of the separator housings with a maximum of 140 different tools, which had to comply with DIXI's automatic tool changer with 144 tool spots capacity (Fig. 7).

Thanks to exhaustive analysis into construction and work preparation, it was possible to reduce the number of tools coming into operation over the entire range of components to 120. This even enabled a dozen reserve-tools, to be provided in the tool-changer. The fact that there is no need to load and unload part or complete sets of tools for each new workpiece series, means a great time saving.

The CNC control system enables the monitoring of each tool's life, which is transmitted on a print-out either on demand or when the time is up. Thus each individual tool can simply be exchanged whenever occasion demands.

Nevertheless, during the night-shift, in order to guarantee that the unattended production process runs as smoothly as possible, new or re-sharpened tools are loaded into the tool changer magazine as soon as remaining tool life drops below 50%. After two years of experience with preparing the un-manned night-shift in this way, no problems have been encountered.

The problem of tool monitoring with jig boring machines is nowhere near solved yet - if only because of the diversity of the tools - since it can be determined neither mechanically nor electrically, whether a cutting edge is still producing the desired surface quality - there could be a tiny break-out of the cutting edge or a build-up of chips. As of today, available tool monitoring systems are still only half-solutions, which can only establish specific types of tool breakage. Very often this difficulty can be overcome by the choice of special cutting materials.

Frequently used tools are provided in two or three sets in the tool changer magazine. On expiry of the programmed tool life, an auxiliary function calls up the twin tool and brings it into operation.

An overall examiniation of the FMS set-up at Westfalia-Separator (Fig. 8), enables following conclusions to be drawn :

1. The user must **himself** choose the main components of the system in such a way that they best meet **his** needs.

2. Only a small team of "down to earth" experts should work out the system technology, in close co-operation with the system and software supplier, and basing it as well as possible way on the existing works organization. This means that for psychological reasons, critical key-people, such as future operating personnel and possibly the works council should be made aware of the scope of the project, of its continued progress, of the possible consequences involved, so that there is collective responsibility for the successful run-off of the system.

3. Detail problems difficult to solve in hard- or software should be classified according to increasing degrees of difficulty and postponed as long as possible, since experience shows that they usually resolve themselves during or after run-off of the system. These "problems" which are frequently of a purely theoretical nature, can, in certain circumstances, seriously hold up or even wreck a project.

4. Efficiency calculations should be worked out on a two-shift basis, at most. Thus the aimed-for goal of a third, short-attended or even unmanned shift becomes a significant bonus factor.

This FMS plant with high precision DIXI machining centres is not a mere experiment for the firm of Westfalia-Separator, but a definite step forward to automation and efficient, modern production.

<u>ACKNOWLEDGEMENTS</u>

I wish to thank Mr. Franz Vennewald, Technical Director of Westfalia Separator, Oelde, for his assistance in procurement of case-study material for the availability of the FMS film production.

Fig. 1 General view of FMS plant with two DIXI 410 precision machining centres at Westfalia Separator AG, Oelde / Germany

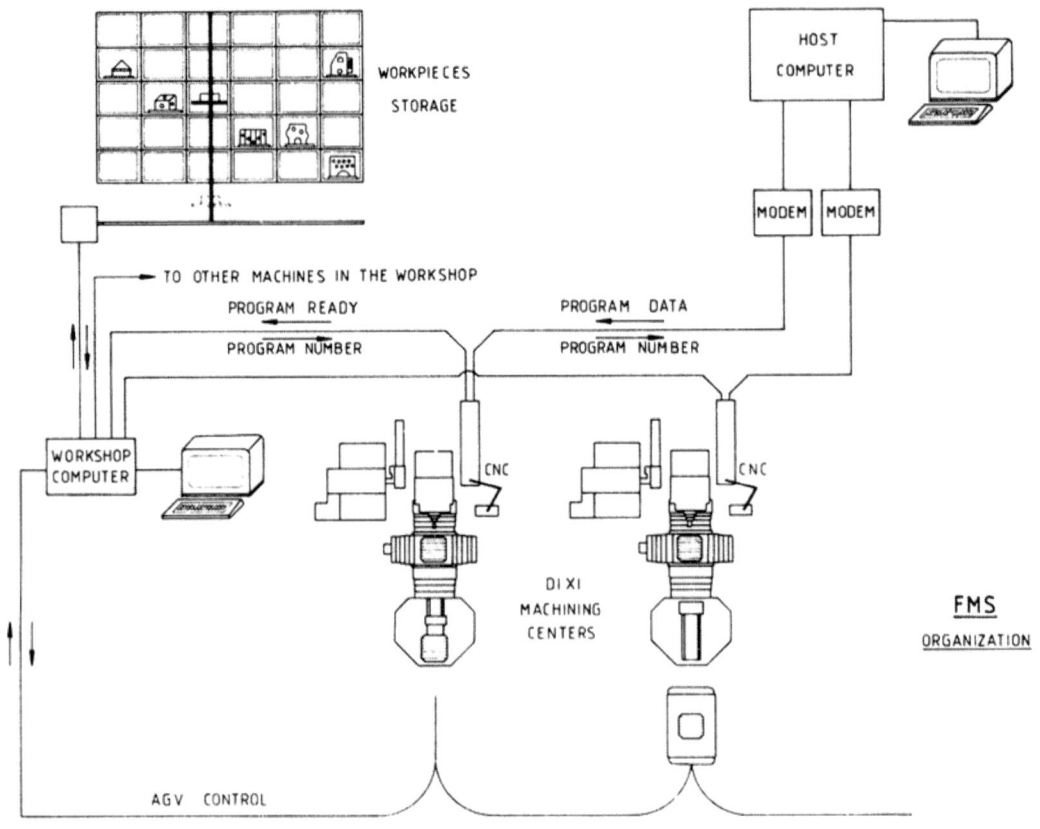

Fig. 2 Linking system between FMS computer, CNC machine control and in-house computer

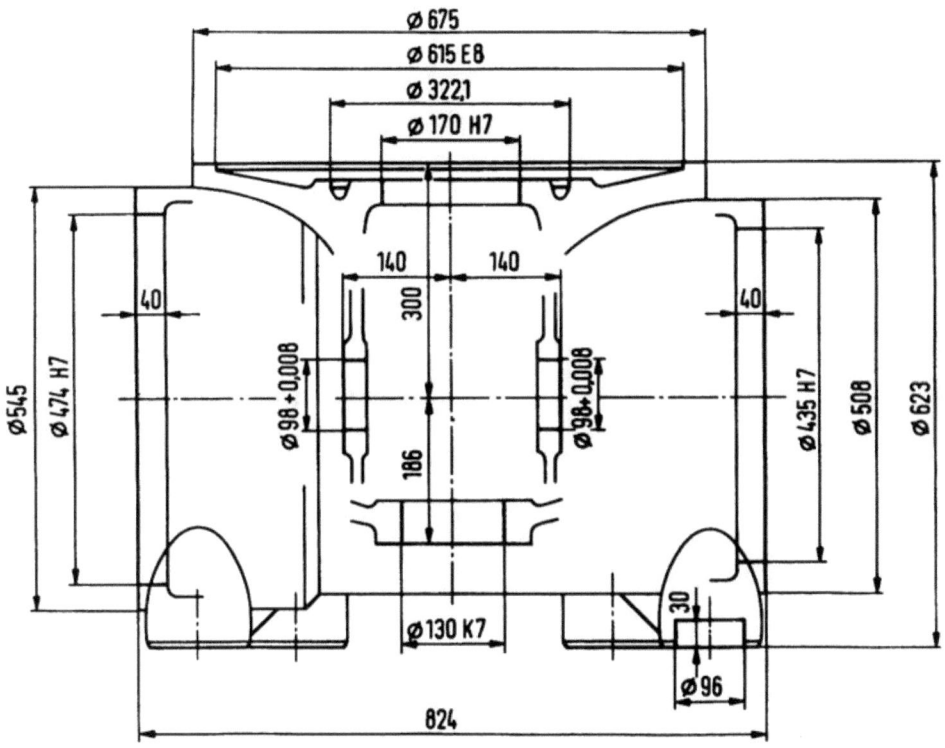

Fig. 3 Schematic drawing of separator housings

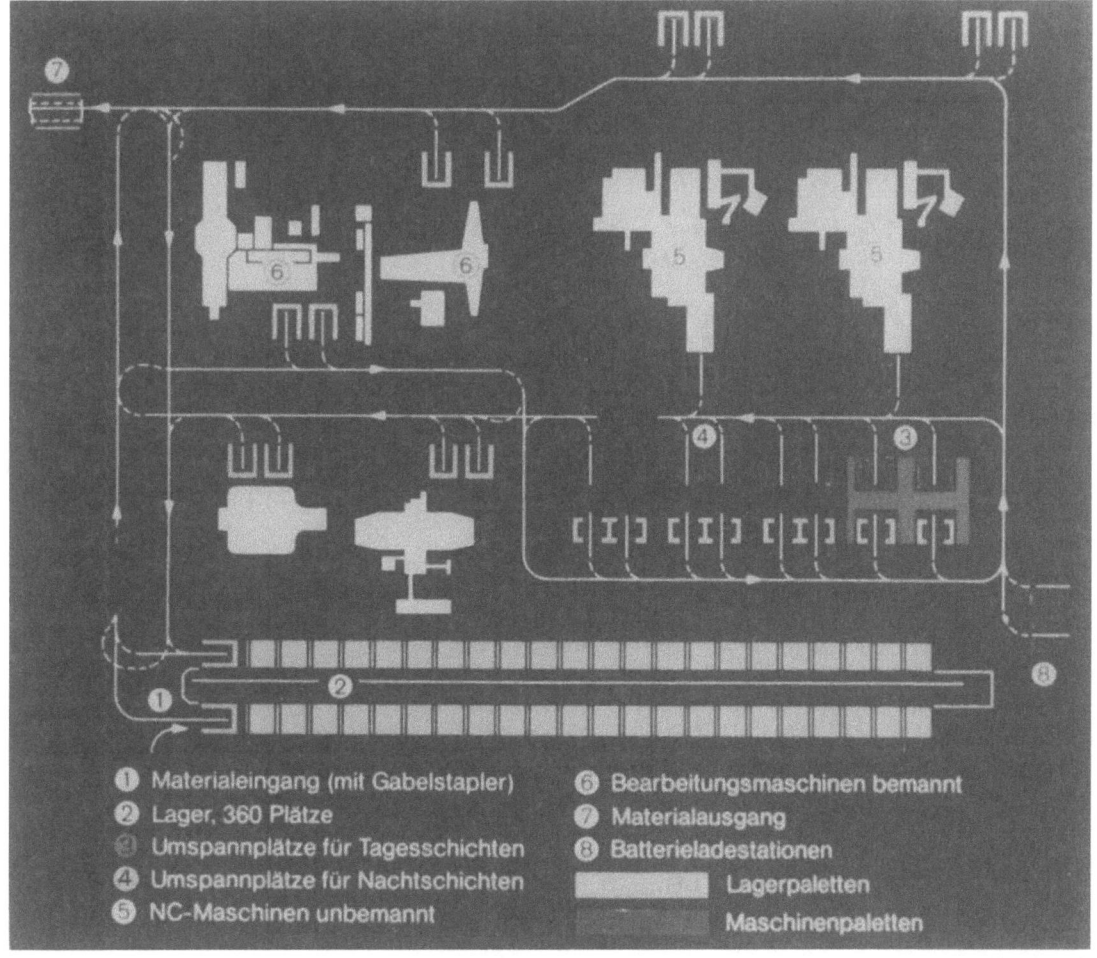

Fig. 4 Schematical view of complete FMS area

Fig. 5
DIXI Jig Boring Machine with pallet shuttle. The separation of pallet holder and pallet is clearly shown.

Fig. 6
Component's reference is picked-up by the Renishaw-MP3 touch probe.

Fig. 7
Part view of DIXI's automatic tool changer with magazine for 144 tools.

Fig. 8
DIXI 410 TPA/144 Precision Machining Centres with DIGITRON Robot vehicle.

Flexible manufacturing system for flywheels, bearing housing and gearwheel blanks

M. C. Aldridge
Lister-Petter Ltd, UK

ABSTRACT

An FMS application for component manufacture from raw material to finished part using robots as the material handling interface with the machine tools. The application of vision intelligence within the cell provides an extension to the robot capability eliminating the need for dedicated fixturing outside of the machine tools.

COMPANY BACKGROUND

Lister-Petter Limited, a Hawker Siddeley Company, is a new force in diesel technology, offering the best from two of the finest names in diesel engines.

The Company has a long history of mechanical engineering for it has brought together the world renowned names of R.A. Lister & Co. Ltd. and Petters Limited who between them have been manufacturing diesel engines for well over 100 years.

With an annual turnover in excess of £100 million and 60% of output sold in direct exports to more that 150 countries worldwide, the Company has established a reputation for the quality and reliability of its engine range, from compact single cylinder units right up to turbo charged six cylinder models producing outputs from 1.5 to 300 bhp.

Lister-Petter air and water cooled diesels engines can be found hard at work throughout the world pumping water, milling staple foods, generating electricity, powering fishing vessels or providing power for construction, oil exploration and countless other industries.

The Company has three production sites in the UK, at Dursley, Swindon and Staines and manufacturing facilities for local machining and assembly in twelve countries overseas.

Dursley, in Gloucestershire, features the headquarters of the Company where centralised Commercial, Sales, Design Engineering, Administration and Computer Services Departments are located.

In its endeavours to remain efficient and competitive the Company has progressively invested in automation and FMS during the last decade.

The FMS1 project for the manufacture of large turned components in the LT/LV engine range was initially conceived in 1982.

A complete reappraisal of the production philosophies identified this method of component manufacture to have the flexibility to respond to market fluctuations whilst maintaining maximum utilisation. The machines, robots, computers and associated equipment for the system were researched through to final selection.

The system was then endorsed by an external feasibility study prior to the application by Lister-Petter for a Government grant available through the Department of Trade & Industry FMS scheme.

Early in 1983 a team from Lister-Petter and IBM produced a joint study document which provided the foundation for the design and application of the FMS computer system. Every effort was made to create a standard modular approach which would benefit subsequent FMS projects.

D.T.I. grant approval and Lister-Petter Board authorisation for the £750,000 project was received and the briefing of capital equipment suppliers was undertaken not only to ensure their full understanding of the total FMS concept and timescale, but to secure the level of commitment required for success.

Commissioning and integration of the system commenced in January 1984 through until October 1984 with initial production test runs in August of the same year, leading into the multi-shift operation of the plant by the year end of 1984.

FMS INVESTMENT & EQUIPMENT

FMS1
Investment

	Total Cost	£739,600
	D.T.I. Grant	£246,533
	Payback period including Grant multishifted	2.8 Years

Equipment:
- Lathe — Swedturn 12, 4 axis CNC controlled.
- Broach — Rausch RS4/1000 2 speed, fully automated cycle.
- Drilling & Tapping — Beaver NC35 vertical machining centre CNC controlled
- Balancing — Schenck 200 WBAE vertical axis, fully automated cycle.
- Robots — Cincinnati T3-776 electric 68 Kg capacity (2 off)
- Vision System — British Robotics Systems
- Gauge — Tesa Metrology
- Computer System — IBM Series 1

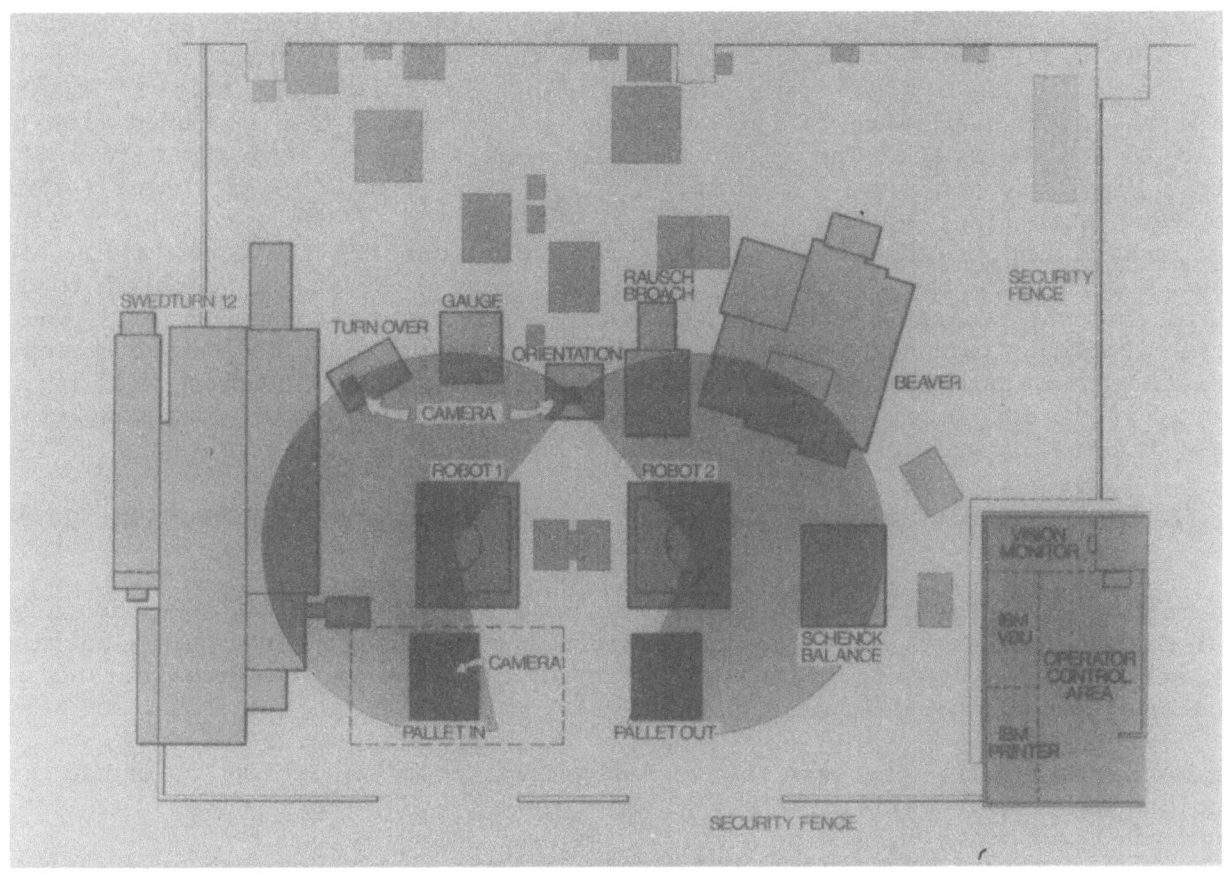

The sequence of operations for a flywheel commences at the input stillage station. The castings arrive directly from the Company Foundry in a standard stillage. Guide rails ensure that the stillages are repeatedly positioned to an accuracy of ±5mm from nominal to enable the vision system to capture the complete stillage profile.

As part of the set up for the cell a calibration routine is performed to establish communication between the vision system and the robot, translating vision co-ordinates into the robot 'world' co-ordinates. For each new stillage of components the operator enters the part number, and the number of layers via a keyboard, the system then produces a co-ordinate map of the component stacks identifying the number of stacks in the stillage. When the IBM host computer requests the next part the vision system processes the image of the stack from which the component is to be taken and sends XY co-ordinates to the robot. Components are sequentially removed from the stillage a layer at a time working from the centre outwards. The co-ordinates provided at the stillage picking operation are within 4mm of the centre of area of the component.

Equipped with the XY information the robot positions over the relevant stack and feeds into the stillage. The Z co-ordinate for the component stack is established by the use of fibre optic infra red proximity sensors. Having sensed the presence of the component the robot magnetic gripper is energised and the part removed to the regrip/centralisation station.

At the regrip station the vision system re-processes the component image and produces new XY co-ordinates, which have an accuracy to within 2mm of centre of component area. The re-gripped part is loaded to the lathe. Chucking of the component and release by the gripper is co-ordinated in a hand-shaking routine performed through the IBM host computer. The lathe then machines the first side of the component. The part programme allows for the component to be removed by the robot, turned over and re-loaded to complete the second side. Quality of component is maintained by the post process gauge which measures the critical dimensions and provides programme correction data. The first robot transfers the flywheel from the gauge to the orientation turntable.

The IBM signals the vision system that the robot is clear of the field of view and processing can commence. The component is then rotated through 360° in regular angular steps under the depth map camera. The turntable is controlled by the vision system computer. The system takes initial images over approximately 3° intervals. Data from the depth map image of the component so formed is then compared with taught data held in the database. The orientation is then known to approximately ± 3°. In order to achieve greater angular accuracy the system will scan a particular section of the component with finer angular resolution. The section of the component chosen will have to include one or more features that uniquely identify the orientation of that component. Data from the depth map image created will again be compared to taught data held in the database, and the orientation found to within ± ¼°. When the orientation of the part has been established the part will be turned through an appropriate angle, such that when the second robot picks up the part it will be in the correct attitude for loading into the next machine in the cycle. On completion of this task the vision system returns to Standby Mode.

Robot 2 collects the part and loads it into the vertical broaching machine for keyway cutting.

This operation is followed by drilling and tapping both faces of the components on the Vertical Machining Centre. The robot transfers the part to the Balancing Machine and on completion of the dynamic balancing operation performs a stacking routine of the finished components into the output pallet.

The whole operation is overseen by a management computer which co-ordinates all transactions within the cell.

MAIN TECHNICAL FEATURES OF SELECTED EQUIPMENT

Vision

The B.R.S.L. system for the FMS consists of 3 separate stations. At the pallet picking stage a solid state camera processes the image of components in the stillage and by use of a known calibrated reference provides X and Y co-ordinates of the central parts. A strong light source for both the first and second station is important to ensure definition. The centralisation station processes data of a single part from a close proximity, and is capable of providing co-ordinates within 2mm of its true centre.

Orientation is executed by projecting a light source across the component face. Rotation through 360° enables the line scan camera to trace the light source as it passes over the part. The angular relationship of features relative to a known datum is established and compensated for by correcting the angular position of the turntable. The processing of all vision data is by means of a DEC PDP 11/23+ computer.

Robots

Cincinnati Milacron model 776 electric robot having a 68 Kg load carrying capacity at the wrist face plate. Each robot has a reach of 2.5mm radius through 270° of travel and the roll wrist capable of rotation through 900°.

The robot control is designed to readily integrate with other machine tools and computers for FMS applications.

Grippers

Supplied by Taylor Hitec the electro magnetic grippers have been designed with a D.C. supply and incorporating a battery back up contingency.

The centre of the gripper houses a pneumatically operated cone which is used to push the component into the jaws of the lathe during the chucking operation. Lister-Petter engineers have developed the use of fibre optic infra red proximity sensors and retrofitted them to the gripper. Plugs, sockets and snap on connectors for proximity communications and pneumatics assist in reducing gripper to robot change over time. Each gripper has a common location fitting to the face plate of the robot.

Pneumatics

The pneumatic system for the robot grippers as supplied by Darenth Automation is fully programmable from the robots.

Lathe

SMT Swedturn 12, 4 axis, having 12 and 6 tool positions in the upper and lower turrets respectively. The machine is fitted with a contact measuring probe, force feed monitoring and a programmable tailstock.

A 400mm Forkardt chuck with quick change stepped jaws accommodates the Lister-Petter range of components. The CNC 300 control has additional software for robots and FMS computer applications. External data from a measuring system can be processed through the control providing dynamic tool offset adjustments.

Gauge

Tesa Metrology gauge consists of the machine and three sets of tooling, one for each family of components. Each set consists of a top and bottom half which open for loading, Electronic probes which are pneumatically retracted for protection are regularly calibrated to the respective hardened master. A Hewlett Packard computer processes the data from each measuring cycle and provides a wide range of statistical quality information. Measurement details communicated through the management computer to the lathe dynamically update tool offset parameters maintaining components within the design tolerances.

Broach

RS4/1000 Rausch two speed vertical pull broach with 1000mm stroke and equipped with automatic broach retraction. The component fixture is mounted on a shuttle type table to permit robot loading. This standard type machine had been electrically interfaced to operate in an FMS environment.

Vertical Machine Centre

The Beaver VC35 Fanuc 6MB control has been enhanced to provide the signals for FMS applications. The machine has an 18 tool position carousel, and a table mounted probe for tool length offset and broken tool detention. An extended column option permits the rotation of the large components on the machine. A full 4th axis enables the components to be accurately positioned for machining in more than one plane. The trunnion type fixture has automatic clamping and sensing incorporating quick release couplings for rapid change overs.

Balancing

Schenck 200 WBAE fully automated for flywheel balancing. The machine control computes the position and amount of unbalance, during the rotation of the part. The area of unbalance is corrected by removing metal using a special drilling head. After correction the component is run again to verify the balancing accuracy.

The fixture design allows the machine to be robot loaded and requires no alteration to accommodate the LT/LV family of flywheels.

Management Computer

The IBM Series 1 configuration consists of a 512k Byte Processor with a 30 MB disk, 3 VDUs and a printer.

The Series 1 has been programmed to monitor and control all activities in the FMS and is linked to all the Machine Tools and Robots in the cell using digital signals and RS232 serial links. The total application software comprising some 70,000 statements was developed by Lister-Petter and tested with an IBM PC (linked to the Series 1) emulating each of the Machine Tools and Robots. A data link connects the Series 1 to the Company main frame 3083 computer based at Dursley. On line management reporting is available anywhere throughout the Company computer network.

Interfacing

The Novonic interface cabinet provides safe electrical signal communications between each machine tool and the IBM Series/1 computer. There are two types of communication:-
1. Parallel digital signals.
    Digital - These are all 24 volt signals between the machines and interface cards. The cards provide the voltage conversion and opto isolation.
2. Serial communication.
    Serial - This is all to RS 232C, V.24 specification and is programmable from Series 1.

The installation is such that test equipment can be connected directly to the respective plugs and sockets for each machine.

of a flexible manufacturing system with high power laser
T. Ilar and C. Magnusson
Luleå University of Technology, Sweden

ABSTRACT

High cost investments and automation in small companies with many short series products can be justified if it is made with a high degree of flexibility. Material processing with high power laser is a typical high cost investment with demands for automation, on account of the very short machining time. Several flexible manufacturing systems or cells with high power laser have also been put in production, but merely for cutting applications.

Therefore it is a great opportunity for the Laser Research Group at the University of Luleå to participate in the development of a flexible manufacturing system including a laser welding cell, in co-operation with a small Swedish company. The laser research group have in the progress of the system build-up contributed with welding tests of the work-pieces, design of a flexible laser welding cell including fixture units and material handling equipment. We have also given recommendation at the purchase of a laser-system.

The purpose with this paper is to present the planning of this flexible laser processing cell, where the additional advantages with laser welding and flexibility are the main objectives with the installation.

1. INTRODUCTION

Material processing with high power laser is today an accepted method in several industries both in Sweden and internationally. This is true at least for big companies producing long series or small work shops having laser processing as its speciality. It have been more difficult for high power laser to be accepted as a machin-tool in small and middle size industries.

A reason for that is the way these investment decisions are made.

A common assumption is that it needs at least two shifts utilization to justify the laser investment, on account of the high investment cost. A rough calculation then shows that the laser during these two shifts can produce far more than the company needs, which often means that the investment won't be made. The laser process is thus punished for its high productivity.

This is most often a wrong way to estimate a laser investment, because laser processing offers a high number of other advantages, more important than high productivity. For instance, a laser in a cutting application can offer a later specification of the products as a side-effect. In a welding application one of the additional advantages is an insignificant temperature distortion of the workpiece, meaning less adjustments after the welding.

If these additional benefits are considered a laser investment can very often be justified even with a rather low utilization.

The very short machining time, which is represetive for laser processing, puts high demands on automation of the handling equipment etc. If this automation is made with a high degree of flexibility it is possible to produce a broad range of products in the cell. Such flexible manufacturing cells means that even small companies with many short series products can benefit from a laser installation.

2. THE STARTING SITUATION FOR THE COMPANY IN THE BEGINING OF 1985.

A small Swedish company, with a turnover of GBP 8 millions and about 100 employees, have faced problems with low productivity and low profitability during some years. In the begining of 1985 the company decided that something had to be done. They soon found out that the fundamental problem was the functional layout of the plant.

This functional layout meant long door-to-door time, vast storage areas and much material on the shop floor, on account of too many planning points and a difficult material flow. The avarage door-to-door time was about 3 months. The accounts for 1984 showed a loss of GBP 0.5 millions.

The main products of the company are valves for flow equipment in food industry. These are produced in several sizes and types and total some 30000 valves each year. These valves consists of four main parts; the cylinder with the air motor, middle section, upper valve and bottom valve (see figure 1).

In this paper we are going to concentrate on the production of the body for these valves, which include the following main operations:

* cutting-off of tubes.
* pressing of the bottom part of the valve house.
* punching of holes in the valve house.
* flanging of holes.
* face grinding of flanges.
* turning of seat-face and plates.
* TIG-welding of seat-face, connection tubes or/and plates on the four main parts.
* adjustment.

* face grinding.
* finishing cutting.
* shot blasting.
* assembly.

The punching, flanging and pressing is made by a subsupplier. The material is stainless steel with a maximum thickness of 2 mm.

3. THE FIRST STAGE, TO ACHIEVE A BETTER MATERIAL FLOW.

In the begining of 1985 the company started a project-group with the objective to find a better layout of the plant. It was decided that the best solution would be an automated warehouse with a high bay stracker crane and the workstations positioned on either side of the system (see figure 2). This configuration gave the company big storage space on a rather small area and much better material flow and control. The installation was made in october 1985, and took about two weeks including moving of machines, and could be carried out without any production stops.

The new layout of the machinery also involves a new grouping. Automated machines with high operation costs were grouped with small supplementary machines. Those supplementary machines makes it possible for the operator to add some extra operations to the workpiece during the main machines machining time.

The warehouse is handled by an operator, who loads the pallets and stores them in the racking under computer control. A signal from the workstations inform the operator that a new job is required, and he takes the next job on the planning list. In a near future the material management shall be integrated in a process planning system.

The total investment cost for the warehouse was GBP 150 000, and the pay-off time is estimated to 2 years. Main savings are obtained in terms of higher productivity and by the reduction of work in progress. The production quantity have been increased with 20% with the same labor force. They are also able to produce with shorter lead time, a reduction from 2 months to 2-3 weeks. The new layout also gave a much better working situation for the labor force.

4. THE SECOND STAGE, TO ACHIEVE A BETTER WELDING OPERATION.

The TIG-welding operation of the valve body, middle sections and cylinders leads to a temperature distortion of these workpieces. Therefore it must be followed by an adjustment operation and a finishing lathe operation, and those operations add high costs. The number of rejected workpieces and re-operations are also high, 3-4% and 20-30% respectively.

Facing those problems the manager decided to find out if it was possible to produce the welding of these bodies with laser. He took contact with the Laser research group of the University of Luleå, and we could after some tests confirm that the parts are suitable for laser-welding. The main reason for that is that the parts, which shall be welded together, are machined. Therefore is it easy to achieve the tight fitness, which butt-welding with laser requires. The welds of these bodies can be seen in figure 3. This application requires a laser power of about 1.5 kW.

The main advantages which laser-welding gives the company, compared to TIG-welding are:

* Much higher productivity. The typical production time for the welding of the valve-house is 20 seconds for the laser, as compared to 60-90 seconds for TIG-welding.
* Neither adjustment operation nor finishing cutting is required.
* Reduction of the number of rejects and re-operations.

> * The press-formed bottom valve-house is substituted by a welded bottom plate on a tube (see figure 4). The strength and ductility of the laser weld permits a deformation of the weld at the subsequent flanging operation (see figure 5). This enables the company to take home the punching and flanging operations, which today are made by a subsupplier. The personell that would have been laid of, on account the installation of the laser welding cell, can now be emploied in the flanging and punching operations.

The savings are obtained in terms of shorter tooling times and less rejections but perticularly by the fact that neither adjustment operation nor finishing cutting are required.

5. GRADUAL DEVELOPMENT OF A FLEXIBLE LASER WELDING CELL AND A FLEXIBLE MANUFACTURING SYSTEM.

The laser process will introduce a great change to the labor force. Therefore the introduction of a flexible laser system has to be realized gradually, to create the best conditions for a sucessful installation.

The first step is to introduce a new technology, laser welding, which is going to require a learning period. After this period it will be possible to start a gradual automation of the laser welding cell and to integrate other processes, so the final result is a flexible manufacturing system. The only prerequisite, that such a system will work satisfactory and lasting, is that the personal is given the opportunity to learn and operate the different components in the system.

This graudual development of the system can be described in the following stages:

> ### Introduction and verification of a cell for laser-welding of valve-house, middle-section and cylinder.
>
> A cell for laser-welding is purchased and taken into operation. The cell design shall make a gradual development of a flexible manufacturing cell possible. This first stage, including training of the staff and verification of the process, will lead to a suitable control and planning system. During the first stage also trimming of the material flow and components in the system will be carried out. The loading of the seat-faces, plates and connection tubes (see figure 3) will be automatic.
>
> ### Automation of the material handling.
>
> An industrial robot will be installed for the loading and handling of the valve-houses, middle-sections and cylinders. Those parts are fed into and out of the laser-welding cell on roller conveyors.
>
> ### Addition of other processes to the flexible manufacturing system.
>
> The turning, punching, flanging and blasting operations are integrated in the system, so that the production of the body of the valve-house, middle-sections and cylinders can be consided as one planning point.

6. THE DESIGN OF THE LASER-WELDING CELL.

Two TIG-welding stations shall be replaced by the laser-welding cell. The two remaining TIG-welding stations shall produce parts, not suited for laser welding and to maintain emergency operation. Position of the cell in the work shop can be seen in figure 6.

6.1 The principles.

The laser-welding cell has two work stations, which alternately can be reached by the laser beam. One station for the welding of the seat-faces and plates, and one station for the welding of the connection tubes. Each workstation consists of two fixture units with two supply-tubes loaded with faces, plates or connection tubes (see figure 7). These magazine tubes are loaded by the turning operator.

The valve-house, middle-section or cylinders are picked from a pallet on the roller conveyor and loaded in the fixture by an industrial robot. Thus must the workpieces be well defined and positioned on the pallets, to enable the automatic handling by the robot. A CNC controlled table (x-y and rotation) and laser processing head (z-axis) enables welding of the two seams. After this welding operation the valve-house is loaded into the next workstation, for the welding of the connection tubes. The cylinder and middle-section only requires this first welding operation, and are therefore placed on a pallet on a output conveyor. The connection tubes have to be face grinded, to obtain the tolerance, so after the second welding operation the valve-box is loaded into the grinding station. When this operation is finished they are placed on a pallet (see figure 8). Change of the magazine tubes will also be carried out by the robot.

In the first introduction and verification stage, loading, unloading and handling is executed by an operator.

6.2 The fixture unit.

A good solution of the fixturing of the workpieces is often one of the main problems to obtain a flexible manufacturing system, especially for welding applications. In our case the fixturing is not so difficult, as we only deal with machined and rotational symmetric workpieces.

The main function of the fixture unit, in the laser-welding cell, is to centre the seat-faces, plate or connection tubes and the valve-house, middle-section or cylinder on a rotation plate. A principle layout of the fixturing unit can be seen in figure 9.

The principles for the fixturing of the upper valve-house are as follows:

A bottom seat face is put into the fixture, from the magazine tube, by a shuttle. The expander is raised by a cylinder and a valve-house is put on the bottom seat face by the industrial robot (or operator). Those two pieces are centred by the expander.

Next steep in the fixturing is to place an upper face-seat on the top of the valve-house, which is executed by the upper shuttle. This face-seat is centred by a centre pin, which also puts some pressure on the workpiece. After this the shuttles and the expander travel back to their starting positions and the rotation and welding of the workpiece can begin.

The position of the upper shuttle, upper magazine and the dogs are adjustable, so that the valve-house, middle-section and cylinder and different sizes and types can be welded in the same fixture.

6.3 Control of the laser process.

As for all other automatic processing systems the laser cell must be equipped with sensors. The laser puts special demands on these control and safety system, on account of the invisible beam and high risk on long distance, to obtain a high safety for the operator.

Following conditions have to be met before the laser shutter can be opened:

- * The fixture is loaded and the work-piece is in right position.
- * The laser processing head is in right position.
- * The door in the shield, which surrounded the workstation, is closed.
- * The beam bender is in right position.

It can also be necessery to have an automatic control of the welding process, which can be obtain by an UV-detector monitoring the status of the plasma. If any changes are detected a signal from the detector will inform the operator or the cell computer that something is wrong.

All these statements are controlled by a programmable controller between the cell computer and the sensors. In the future the cell computer will communicate with a host computer controlling the entire flexible manufacturing system.

6.4 Invetsment costs and savings.

The total investment cost, for the first stage, is estimated to GBP 300 000, divided into following items:

Laser head, power and control equipment:	GBP 120 000
CNC and workstation, excluding the fixturing unit:	GBP 130 000
Development of fixtures, installation, training etc:	GBP 50 000

The total investment cost, including the industrial robot, is about GBP 35 000 higher.

The annualy savings are estimate to GBP 350 000 and divided into following items:

Neither adjustment operation nor finishing cutting are required:	GBP 250 000
Shorter tooling time, less work in process and less rejections:	GBP 100 000

7. OBJECTIVES FOR FUTURE DEVELOPMENT.

To obtain the objectives, that the production of the valve bodies can be considered as one planning point, the turning, punching, flanging and blasting operations have to be integrated in the system. All these workstations are not going to be totally unmanned, but shall permit production with limited manning.

This stage includes the introduction of a host computer, which controls the material management, pallet admistration, NC-programme admistration, workstation status and system status.

The final objectives is that the valve bodies shall be produced on daily orders from the assembly line.

8. ACKNOWNLEDGEMENTS.

First of all we wish to thank our colleague, Hans Engström, who has a great deal in this work. The authors also acknowledge the financial assistance provided by the National Swedish Board of Technical Development, and the willing assistance given by the other colleagues of the laser group as well as the industrial partner participating in this project.

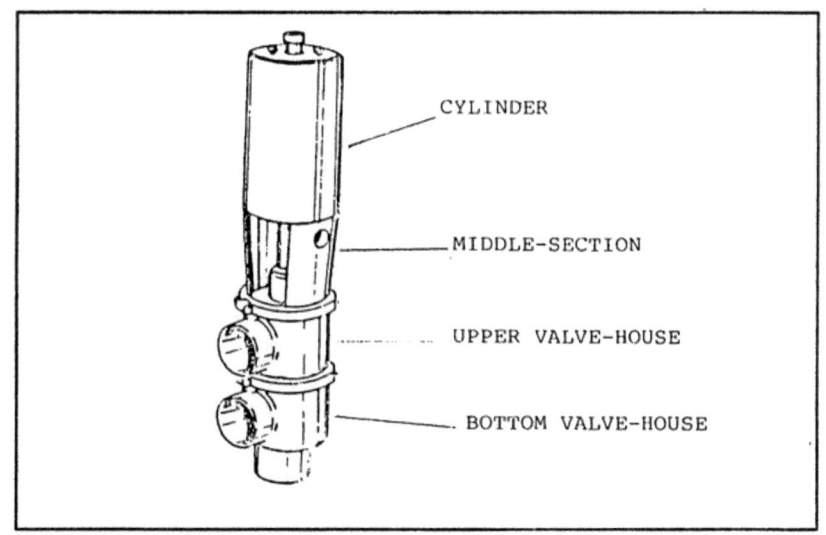

FIG. 1: The main parts of the valve.

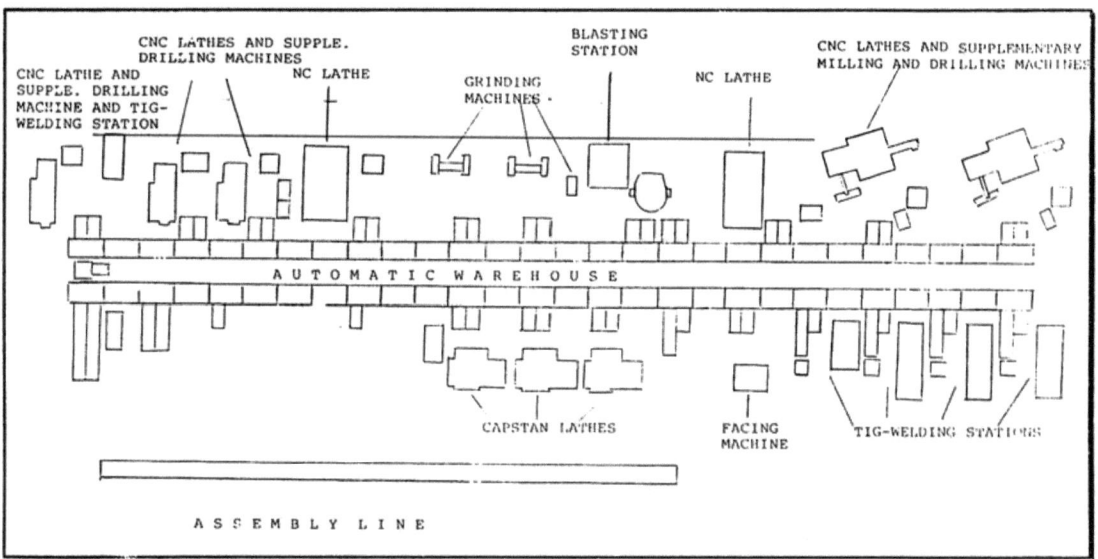

FIG. 2: The new layout of the plant.

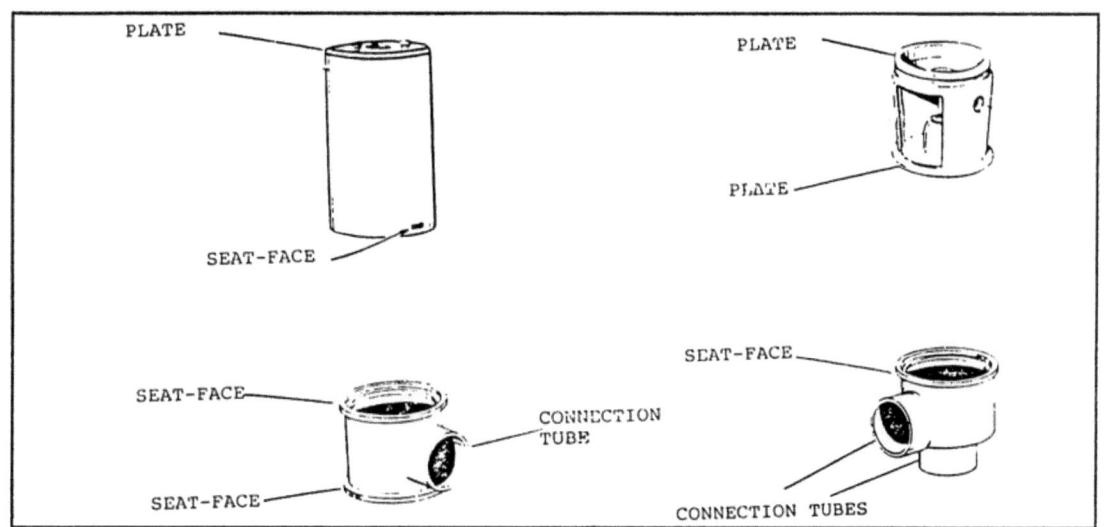

FIG. 3: Body-parts which requires welding.

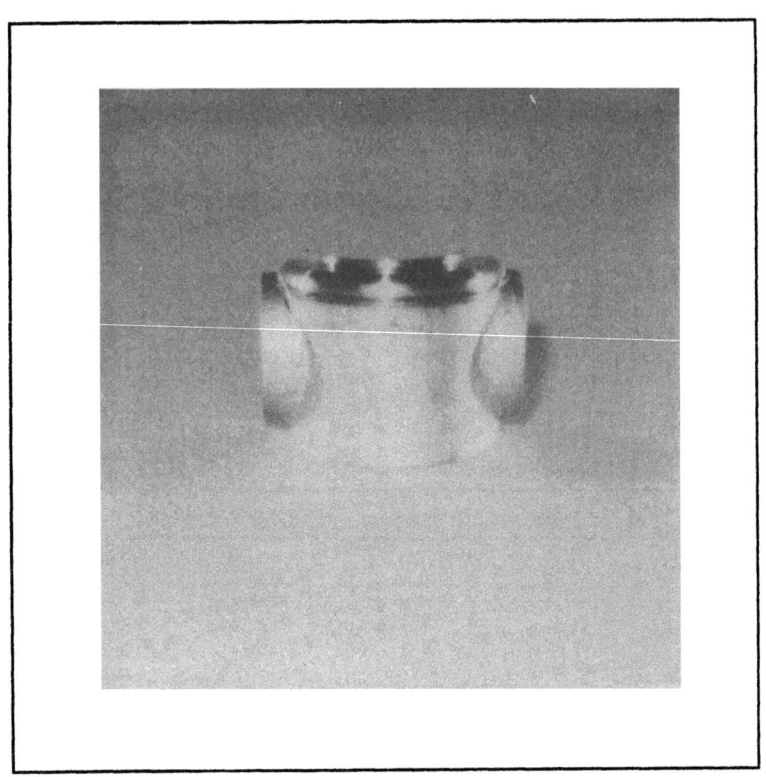

FIG. 4: The new design of the bottom valve-house.

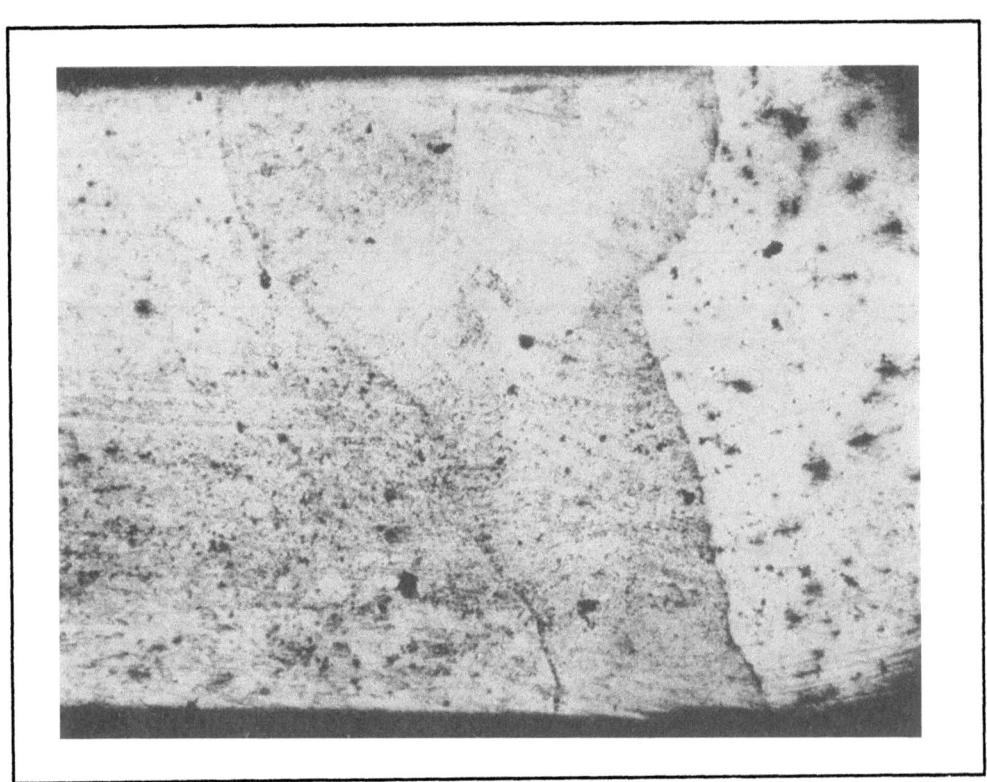

FIG 5: Deformed laser-weld, on account of the subsequent flanging operation.

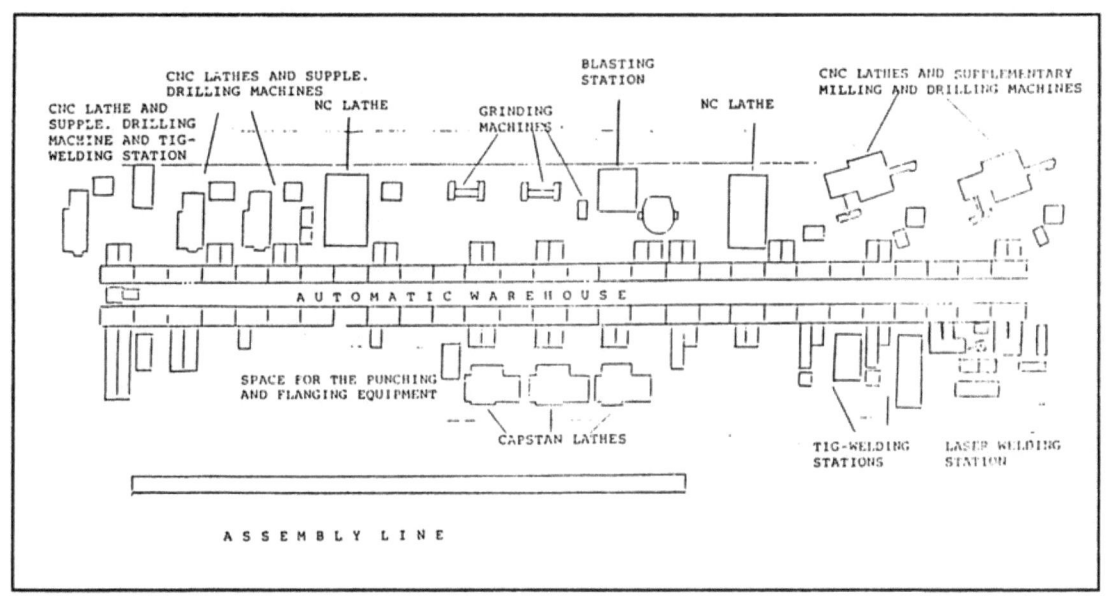

FIG. 6: Position of the laser-welding cell in the work shop.

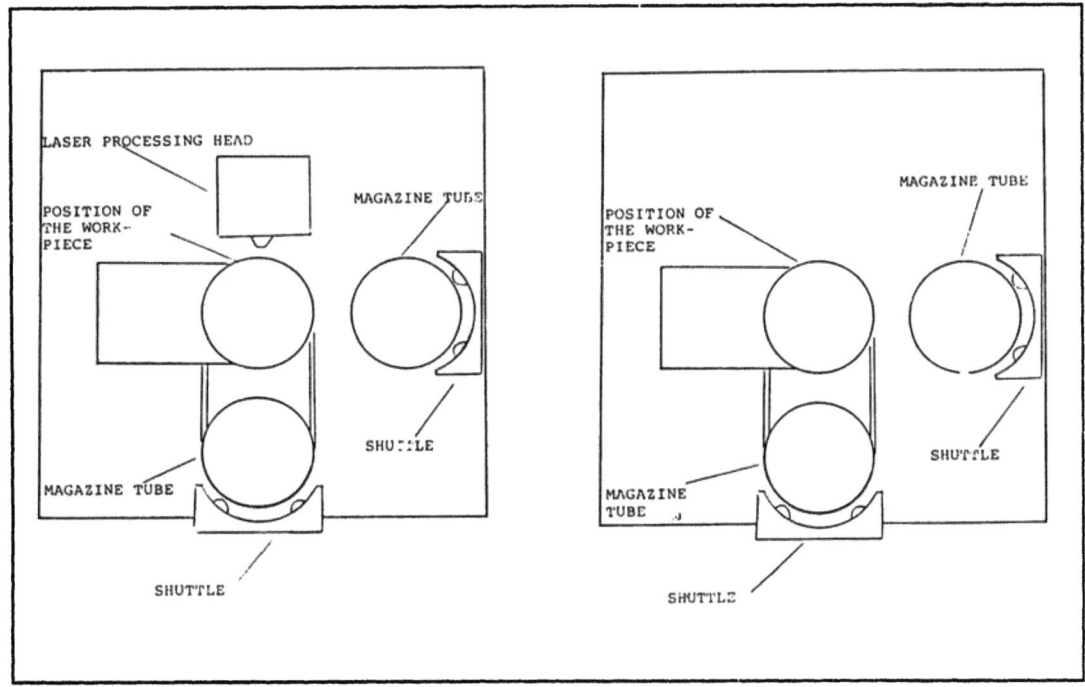

FIG. 7: The design of the work station.

351

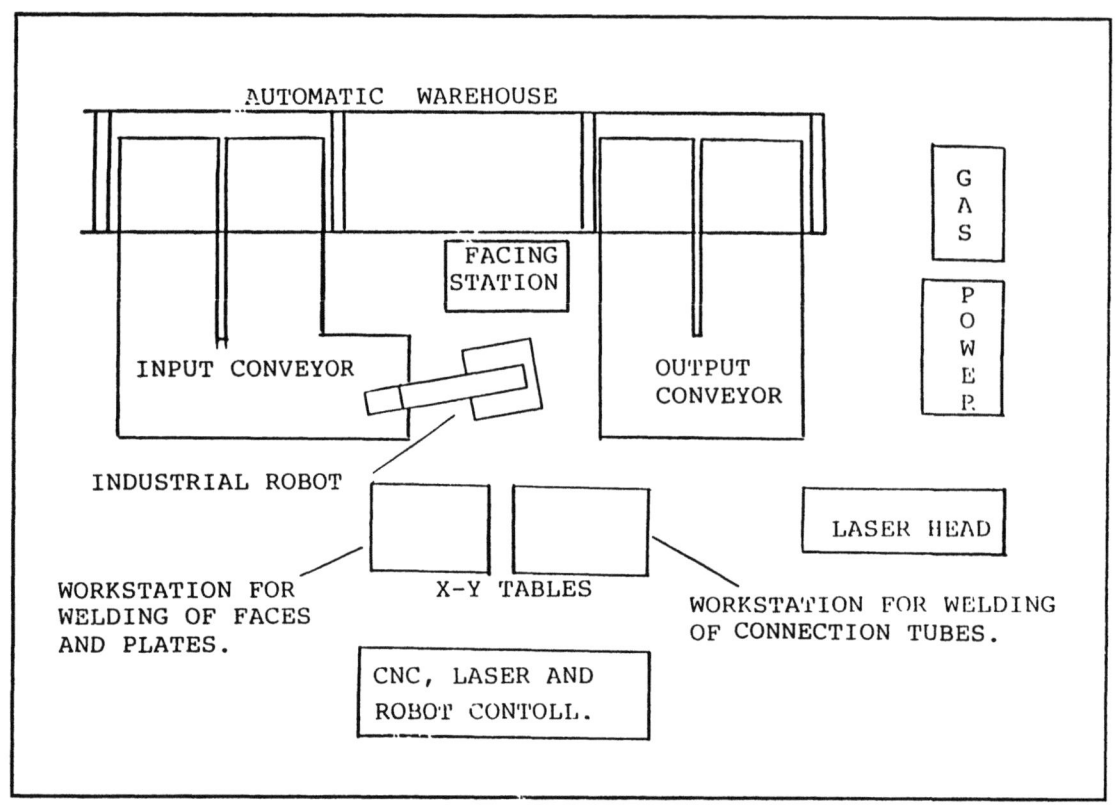

FIG. 8: The layout of the laser-welding cell.

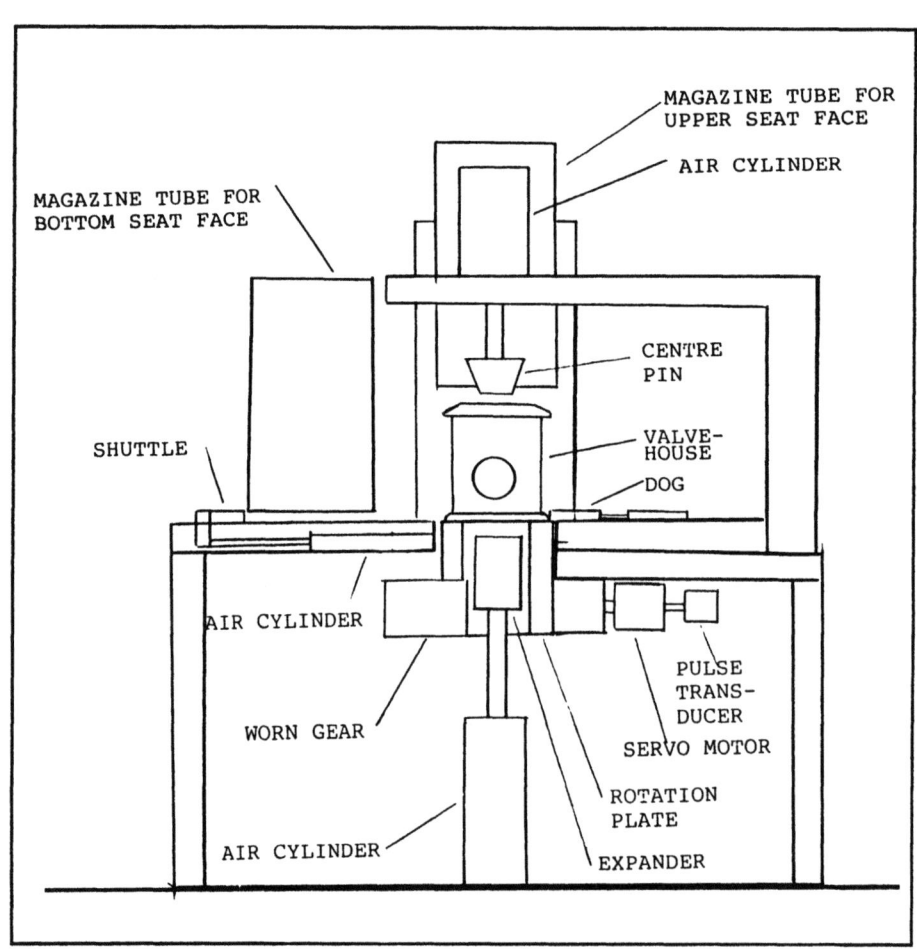

FIG. 9: Principal layout of the fixturing unit.

FM – the proven approach
F. Popplewell
Deckel Ltd, UK
and
P. Schmoll
Fr. Deckel AG, West Germany

ABSTRACT

It goes without saying that flexible Automation is - in our mind - providing the answer to growing market demands for short batch production, brought about by the need for more innovative products and their associated shorter life cycle.
Such machine tools and systems for flexible manufacturing are available today, even though a few problems still remain.
It is of paramount importants that the selection of such machine tools and systems is given full consideration, since total flexibility can not be bought without the full commitment and the involvement of the customer throughout the whole contract period.
Criteria associated with such involvement are indicated here. Certain case-studies are also detailed, showing how a measured approach - using a step-by-step approach - has resulted in success and "complete" flexibility for that customer.

INTRODUCTION

In todays economic environment with fierce competition in all areas of the world market, with a fast growing number of different products and with their shorter life cycle, each company needs to find its own economic strategy to tackle the problem of manufacturing new or modified products every day - in smaller batches - preferably down to one-offs! This needs to be done in the most economical way in order to achieve competetive price and delivery - and thus assure a full order book with a profitable margin.

Flexible machining is the answer - as everybody knows - or as he is told to believe. The profit will come almost automatically by smaller batches - less inventory - faster thro'put rate - just in time for assembly - optimum "up time" and so on.

The tools for flexible machining are available to day - as you will see - in great variety, in hard + soft + tech-ware and know-how, - but as more tools are available, the more crucial becomes their correct selection and the random manufacture may easily turn to manufacturing chaos. But let us first look at the available tools, their requirements, benefits, constraints - and also at some successful combinations for dedicated problem solutions.

THE TOOLS FOR FM

The basic Machining Centre: (FIG. 1)

The machining centre is now generally accepted and has become the universal basic tool of flexible automation - with many excellent products now available on the market.

The basic machine is expected to offer a multiplicity of features e.g.:
- rigidity and stiffness for high chip removal and stable cutting performance.
- accuracy for precision work.
- proven reliability for a high utilization rate.
- latest CNC- and PLC-technology for optimum use, both now and in the future.
- a design, sufficiently flexible, to ensure that retrofittable options can meet future demands for greater versatility and increased productivity ...

The automation periphery (FIG. 2)

The periphery for increased automation of the FM process can be devided into 4 basic categories, comprising the appropriate hard- and software. These are:
- optimizing the machining process
- safe-guarding the machining process
- work and tool-handling
- automation of the "data flow" or "computerisation"

Let us briefly look at these tools of automation, - at their benefits - and also at their constraints.

Options for the machining process (FIG. 3)

This group mainly includes equipment designed to optimize the cutting or machining process itself and the quality of the workpiece. Typical examples are:
- multi-spindle milling heads,
- angular milling and boring heads,
- high-speed spindle,
- coolant supply through the cutting tool
- spindle temperature control

Options for safeguarding the production process (FIG. 4)

The growing trend towards automating the production process thereby reducing the necessity for permanent human supervision, has speeded up the development of automatic process monitoring systems. This term denotes the different methods of checking the workpiece and cutting tool during the machining operation and includes such items as:
- tool life monitoring,
- power monitoring,
- tool breakage detection,
- probe technology,
- use of replacement tools ("sister tooling")

These options are all important - but difficult to use right from the day one. They require systematic collation of data, on the shop-floor, to collect all the relevant parameters. For example the "dullness" of a particular tool relative to an X % power increase needs to be listed, which will vary with different materials and tools of the same kind anyway!
The appropriate software features like "skip-routines" for selections of the programs or for aborting the entire pallet after tool breakage are of course available, but the programing then becomes more "process-oriented" and definetly not any easier. Thus a great amount of additional man-hours are needed.

Options for a automated work- and tool-handling (FIG. 5)

Flexible machining also means having enough work (large pallet-magazines), and the correct tool available at the right moment (large tool-magazines), to keep the machine working 24 hours a day. Many different parts, smaller batch sizes, more tool and fixture changes require even larger tool and work buffers to prevent the machine standing still, waiting for one or the other. Depending on work requirements, capacity needs, space availability and financial resources the potential user can select from various automation levels:

- carousel-type pallet-magazine (PS-8), used to continue production during breaks and, depending on pallet cycle time, for several hours during the unmanned shift. New set-ups do not interfere with machine "up-time".

- additional pallets for pre-setting away from the machine will reduce setup time, - an even more important feature on stand alone machines!

- modular linear pallet-magazine (MLS) with railguided trolley for multi machine link-up. This system, consisting of individual modules like rail and pallet-storage segments, loading stations etc. was designed to "grow with ones needs", it all allows the start of an FMS, even with a small budget and with one machine only - adding the second machine to the MLS and allways keeping the options to "add-on" pallet-storage and machining capacity open for future needs.

- AGV-systems allow full space flexibility as regards the arrangement of machines, their full accessibility, lowest possible restriction in the inter-plant traffic and highest economy in system expansion.

- Robot work loading systems by gantry or arm robot, for large batch production and the prerequisite of automatic work clamping equipment.

- Additional tool-magazine capacity up to 148 tools, where the corresponding PLC-software will find and present the necessary tool to the spindle.

- External tool supply systems which automatically provide additional tools by robot, from Euro-pallets, into or out of the machine magazine, and - of course - other less significant options.

If we now consider the vast volume of program-, tool- and workpiece-data necessary to machine a large family of parts at random, in a FMC, then one thing becomes apparent: There is no way around the "Automation of data handling" other than by means of a computer aided and computer-linked periphery.

Options for the automation of the data-flow (FIG. 6)

Depending on the complexity of the product family, the chosen FM-system, the level of acceptable manual interface and so on, the appropriate level of computer assistance can be decided.

- A "standard" personal computer
  will work as a programming station, it can assist in preparing tool-change-listings for new orders coming up, for training and of course simulation.

- Programming systems (e.g. Diaprog)
  with CAD features (e.g. MIN-CAD), geometry and technology processor, digitizing features and a software package for automatic order processing, could be linked up to the machine for DNC-operation, and even a small scale CIM system if required.

- Tool management and identification:
  Various systems for the electronic identification of tools exist. Tool-setting and tool-life data are recorded on the tool setting machine and transmitted to the NC or PLC on the machine by various means (e.g. the Bilz-system).

- The cell computer for local management:
  This is the "brain" of the FMS: It is linked to the machine and the machine periphery via terminals, fail-proof interface modules and procedures. The cell computer will automatically look ahead for the time-critical correlation of programs, parts, tools, transportes etc. in line with the production plan, its priorities and also its "emergencies". If the system will not perform according to plan. The cell computer will have to provide an interface to tie up to the host computer, which may have interfaces to large-scale CAD- or PPS-systems.

PROVEN FLEXIBLE MANUFACTURING SOLUTIONS

FLEXIBLE MANUFACTURING CELL (FMC) WITH CAROUSEL-MAGAZINE TYPE PS-8

The PS8-FMS seems to be favoured throughout the chipmaking industry. It is easy to manage, economical in terms of payback, retrofittable without problems and extremly flexible in application, as illustrated in the following case-studies.

Flexible manufacturing cell used for "one off" production   (FIG. 7)

Size of company	:	Approximately 60 employees.
Product	:	Tool manufacturing
Workpieces	:	Injection moulds and tools for plastic toys.
Material	:	Normally tool steel.
Parts families	:	Twenty five different parts families are involved.
Batch sizes	:	Every mold is different, therefore "batches" are "one".
Objectives for FMC:		A total of 25 different parts per day were required.
Previous methods	:	Conventional and NC milling and copying machines.
Deckel-solution	:	A Deckel DZ3-S machine with PS-8, tool life monitoring, measuring probe, 60 tools and management cell-computer.
Utilisation	:	A double day shift running approx. 20 hours essentially unmanned.
Pallet times	:	Varying with an average of 45 mins.
Operational mode	:	One setter/operator looking after whole operation.
Production methods:		Standard fixtures take mould blocks, containing different inserts, with tools in sets of 60. Parts finishing is by EDM.

Economic Considerations

Investment	:	DM 950.000,--
Pay back	:	approx. 2 years
Workforce	:	Reduced by 3 from 5 to 2.
Lead time	:	Down from 3 month to 3 weeks.
Comments	:	max. benefit lies in faster market response, consistency in quality and high utilization rate of machine.

System features the "one off" production in a mold shop on FMC.

A Flexible Manufacturing Cell for "just in time" production   (FIG. 8)

Size of Company	:	Approximately 150 employees.
Product	:	Agricultural diesel engines/agricultural machinery.
Workpieces	:	Engine parts - cylinder heads, blocks etc.
Material	:	Primarily cast iron, some aluminium.
Parts families	:	Two families of different sizes.
Batch sizes	:	In 100's approx. fifteen times per year.
Objectives for FMC:		To enable parts to be produced and assembled in sets, using the Deckel PS-8 cell enabling one set of parts to be provided per single complete PS-8-cycle.
Previous method	:	A combination of conventional and NC machines.
Deckel-solution	:	A Deckel DZ4 machine with PS-8 carousel, tool life monitoring, measuring probe, 60 tools and random pallet positioning.
Utilisation	:	Two eight hour day shifts with additional unmanned time (through brakes). 3rd shift is planned.
Pallet times	:	Averaging between 20 and 30 minutes.
Operational mode	:	One man per shift controls complete system.
Production methods:		Permanent fixtures for individual parts on each of eight pallets, catering for both sizes with replaceable units.

Economic Considerations

Investment	:	Approx. 700.000 DM
Pay back	:	After 2 1/2 years
Workforce	:	Reduced from 8 to 2
Lead time	:	Down from 8 weeks to 2 weeks.
Cost-benefit analysis	:	Main benefit resulted from additional use of 600 - 900 hrs p.a. by using PS-8, reducing piece part costs by approx. 20 %.
Comments	:	System features "just in time" production with standard FM-hardware.

A Flexible Manufacturing Cell for "mass" production (FIG. 9)

Size of Company	:	Approx. 25 employees
Product	:	Sub-contracting work, no end product.
Workpieces	:	Automatic rifle parts.
Material	:	Steel
Parts families	:	One
Batch sizes	:	30.000 parts per order
Objectives for FMC	:	To continuously and reliably produce with minimum manning.
Previous methods	:	New order
Deckel-solution	:	A Deckel DZ4 machining centre with PS-8 in standard execution.
Utilisation	:	24 hours per day throughout the week.
Pallet times	:	Approximately 40 - 60 minutes.
Operational mode	:	Unmanned, a part from loading and tool maintenance.
Production methods	:	Special fixtures furnished by Deckel allow multi-loading of parts ensuring quick change over.

Economic Consideration

Investment	:	Approx. 1.000.000 DM
Pay back	:	1,5 - 2 years
Comment	:	Repetetive orders have made the above investment most provitable. After completion of this production, system will be used for regular "odd batch" subcontracting work.

Flexible Machining Cell for the typical sub-contract application (FIG.10)

Size of Company	:	Approx. 85 employees.
Product	:	Solely a subcontract engineers.
Workpieces	:	Primarily Aerospace parts with other intricate parts included.
Material	:	Aluminium forgings/castings, steel and a little stainless.
Parts families	:	Seven regular types with complete variation on others.
Batch sizes	:	Regularly 25 - 30 with many others as low as 10.
Objectives for FMC	:	Primarily to achieve optimum utilisation, with 24 hour running and reduced setup times, with quick changeover.
Previous methods	:	Conventional (sometimes Deckel KF machines!), other NC and Deckel CNC.

Deckel-solution	:	A DC40/PS8/carousel type production cell with 10 pallets, 80 tools.
Utilisation	:	A double day shift with a completely unmanned third shift.
Pallet times	:	These vary from 15 - 360 mins. with regular components, averaging around 1 - 4 hours.
Operational mode	:	One operator for DZ4 and DC40 machines, with both machines running unmanned at night.
Production methods:		Generally within the cell, 4 fixtures remain permanently set, with others cycling, according to demand; new fixtures for new jobs are designed and made in house.

Economic Considerations

Investment	:	Approx. ₤ 205.000
Pay back	:	Anticipated 2 - 2 1/2 years.
Workforce	:	No reduction but output increased.
Lead time	:	Reduced by 20 - 25 % since similar CNC machines were already in use.

Additional Utilisation
Almost one complete unmanned shift.

Cost-benefit
analysis   :   The major benefit was in the full utilisation of the PS8 in the unmanned mode.

PROVEN FLEXIBLE MANUFACTURING SOLUTIONS WITH MULTI-MACHINE SYSTEMS

Whenever complete conventional production facilities are modernized, invariably 2 or more machining centres are necessary. When a regular, "long term" parts production exists (e.g. own product-line), when large numbers of fixtures remain permanently set on pallets for higher flexibility - and when floor space is limited:
Then this is when linked systems will probably be the solution.

However as investment for such a complete system is invariably too high, the organizational effort considerable and time consuming, then the "step by step" approach with "modular periphery" becomes inevitable.

FMS: The "step by step" approach (AGV-system at FD AG)   (FIG. 11)

Size of Company	:	2200 employees
Product	:	Precision machine tools
Workpieces	:	Columns, tables, spindle heads etc.
Material	:	Cast iron
Parts families	:	6 major families of 5 parts each.
Batch sizes	:	Approx. 30
Objectives for cell:		Reduction of lot size and inventory increase of throughput rate and faster market response.
Previous method	:	Conventional and NC-stand alone machines.
Deckel-solution	:	System with 4 machining centres and AGV-link.
Utilisation	:	2 extended manned day shifts, plus 2 hours unmanned: Total 20 hours per day.
Pallet times	:	30 - 90 min.

Operational mode	:	One operator for loading/unloading, one "part-time" systems engineer for set up and general control.

Economic Consideration

Investment	:	Approx. 5 Mio. DM
Pay back	:	Approx. 4 years
Workforce	:	50 % reduction
Lead time	:	Reduction from 100 to 60 days.
Cost-benefit analysis	:	Savings by reduced inventory 500.000 DM p.a.
Objectives for FMS	:	Have been full filled. System will be further developped in a step by step mode to incorporate additional four machines and production computer.

Flexible Machining System for the "un-controlled" third shift production
(FIG. 12)

Size of Company	:	1800 employees
Product	:	Precision machine tools
Workpieces	:	Housings, plates etc.
Material	:	Steel, CI, Alu
Parts families	:	11
Batch sizes	:	Average 50, 5 - 6 times per year
Previous methods	:	Conventional and NC-machines, subcontracting.
Deckel-solution	:	2 DZ4 machining centres, railguided trolley-link, 20 pallet-magazine stations, 2 washing stations.
Utilisation	:	3 shift operation, third shift unmanned
Pallet times	:	Average 20 mins.
Operational mode	:	1,5 operators per shift for loading and optimizing the system.

Economic Consideration

Investment	:	2.200.000 DM
Pay back	:	3 years
Workforce	:	Reduction by 3
Lead time	:	Reduction by 50 %
Comment	:	System features the full unmanned third shift with no tool monitoring systems available at date of installation, by running only technologically proven parts during night shift with "safe" feeds and speeds.

The Deckel "modular-link-system" MLS for automotive parts production
(FIG. 13)

Size of Company	:	180 employees
Product	:	Automotive brakes (heavy trucks)
Workpieces	:	Brake components e.g. back plates, cylinders, levers
Material	:	GGG49, hi-tensile cast iron
Parts families	:	4 (parts differ in size and customer specifications).
Batch sizes	:	1000 - 10000
Objective for Cell	:	More flexibility for machining an increasing number of varieties. New system to be step by step expandable and gradually replacing conventional machines.
Previous methods	:	Special carousel - and linear single purpose machines, which are still in use.

Deckel-solution	:	4x DC40, MLS-34 (linear rail-link system with 34 pallet stations) with tool and workpiece-monitoring systems.
Utilisation	:	A double day shift with completely unmanned third shift. 1 operator per shift.
Pallet times	:	70 to 90 minutes, achieved by clamping up to 12 components.
Operational mode	:	Fixtures remain on pallets permanently, - machines are dedicated to individual programs. Fast single spindle operation is more economical than multi-spindle heads.

Economic Considerations

Investment	:	Phase I+II (3 DC40 + MLS-20): approx. 2,6 Mio. DM Phase I-III(4 DC40 + MLS-34): approx. 3,5 Mio. DM Phase III still in planing state.
Pay back	:	4 years per phase anticipated.
Workforce	:	No reduction in manned shift.
Lead time	:	Reduction by 70 % for special orders, which was main objective.
Cost-benefit analysis	:	The MLS-system allowed the step by step installation of the FMS corresponding to budget available, gain in FM experience and changing market requirements.

THE USER-VENDOR COOPERATION

Proven tools for flexible machining are available, - as we can see from the above. However the selection of the right mix for an individual companies production needs, becomes quite often a major problem.
We are not only looking for an isolated manufacturing bottleneck which can be solved with an island solution but - as we are increasingly forced to look into more computer integration - the CIM-future - we have to look at the company as a whole.

We have to consider the companies present organizational status, financial and personel capabilities, present products and future developments and so on.
This information has to come from the user; He cannot buy his appropriate measure flexibility without giving his parameters to the vendor, who only then is able to come up with a dedicated solution. User and vendor have to come closely together, they have to apply sincere disciplines in selecting necessary and not available options.
The more you have got to select from, the more difficult it becomes.
So please, let's do it together!

GENAUIGKEIT ACCURACY	
ZUVERLÄSSIGKEIT RELIABILITY	MODERNE CNC- TECHNIK STATE OF THE ART CNC TECHNOLOGY
STEIFIGKEIT RIGIDITY	AUSBAUBARKEIT EXPANDABILITY

ANFORDERUNGEN AN DIE GRUNDMASCHINE
REQUIREMENTS OF THE BASIC MACHINE
(ABB./FIG. 1)

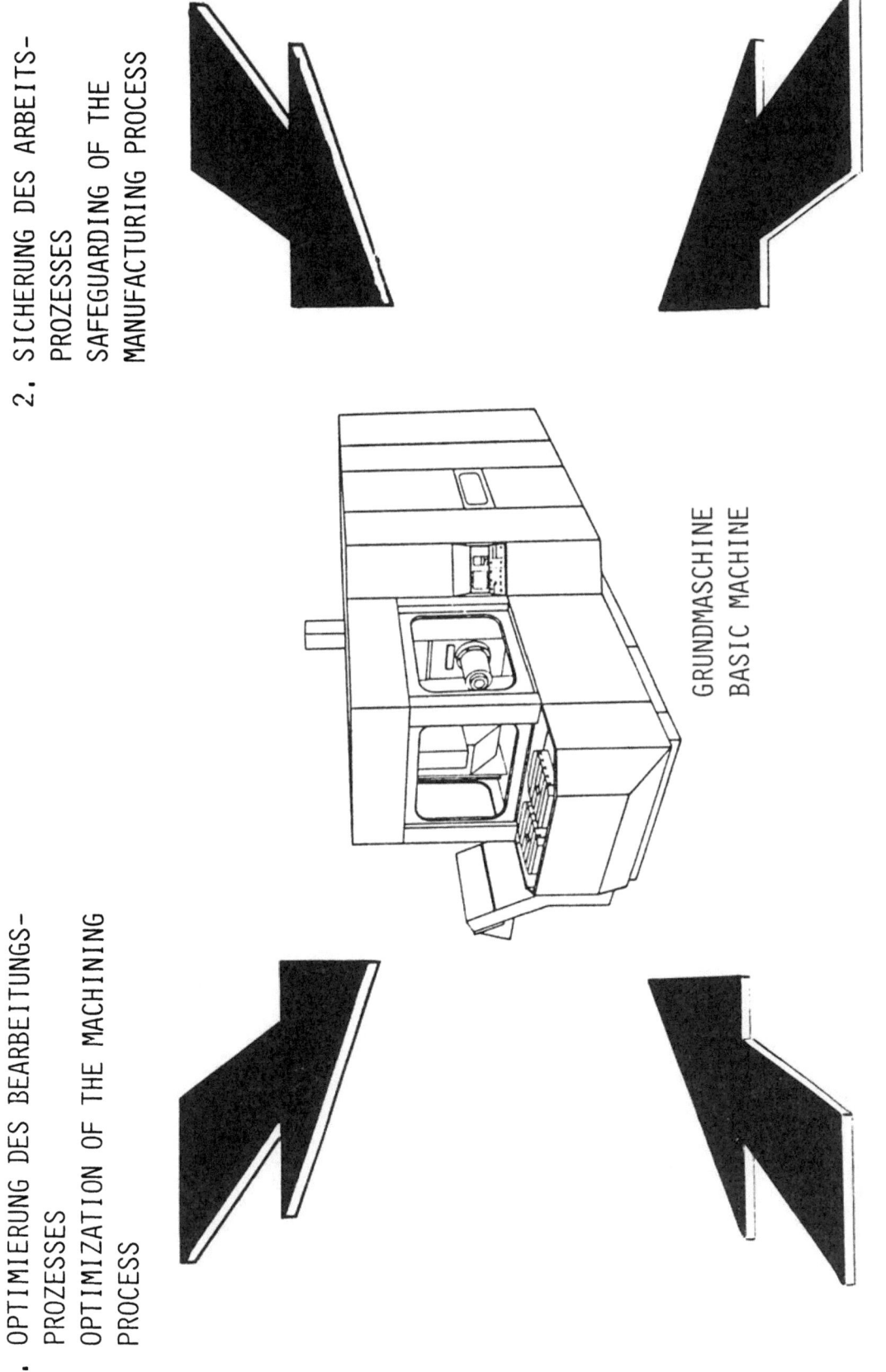

(ABB./FIG. 2)

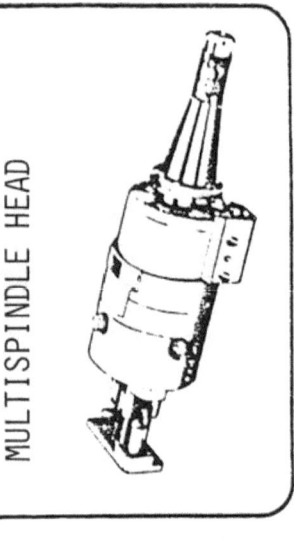
WINKELKOPF
ANGULAR HEAD

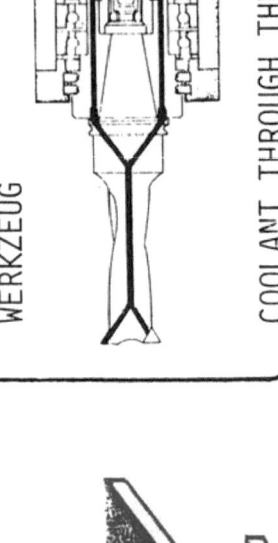
KÜHLMITTEL DURCH DAS WERKZEUG
COOLANT THROUGH THE TOOL

MEHRSPINDELKOPF
MULTISPINDLE HEAD

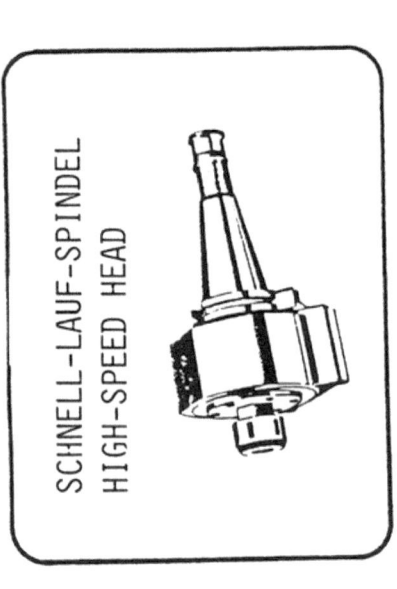
SCHNELL-LAUF-SPINDEL
HIGH-SPEED HEAD

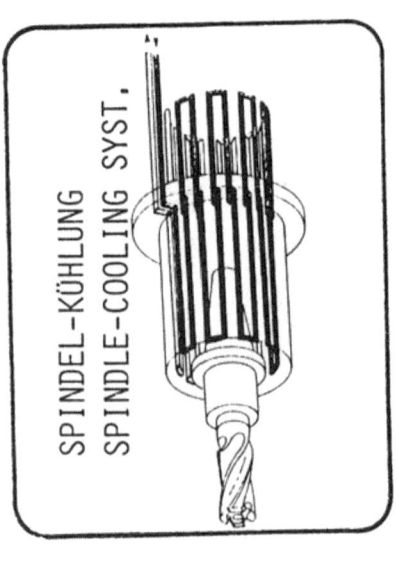
SPINDEL-KÜHLUNG
SPINDLE-COOLING SYST.

1. OPTIMIERUNG DES BEARBEITUNGSPROZESSES
   OPTIMIZATION OF THE MACHINING PROCESS

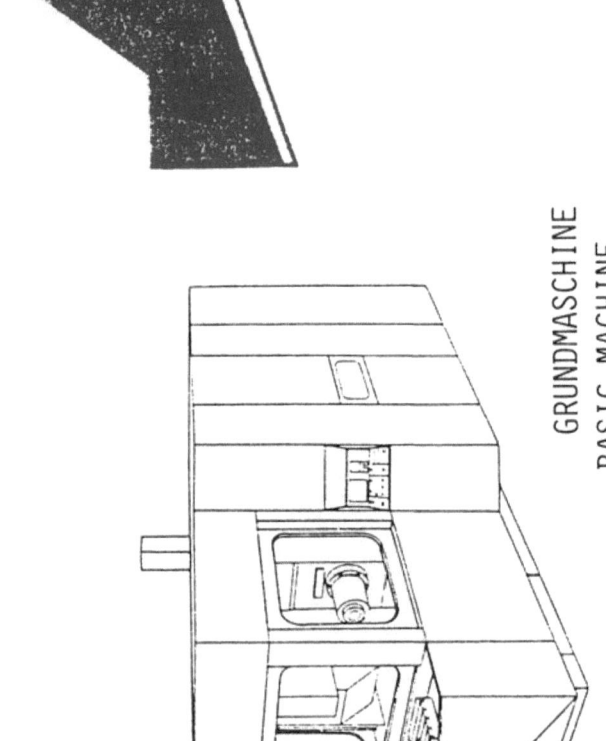
GRUNDMASCHINE
BASIC MACHINE

(ABB./FIG. 3)

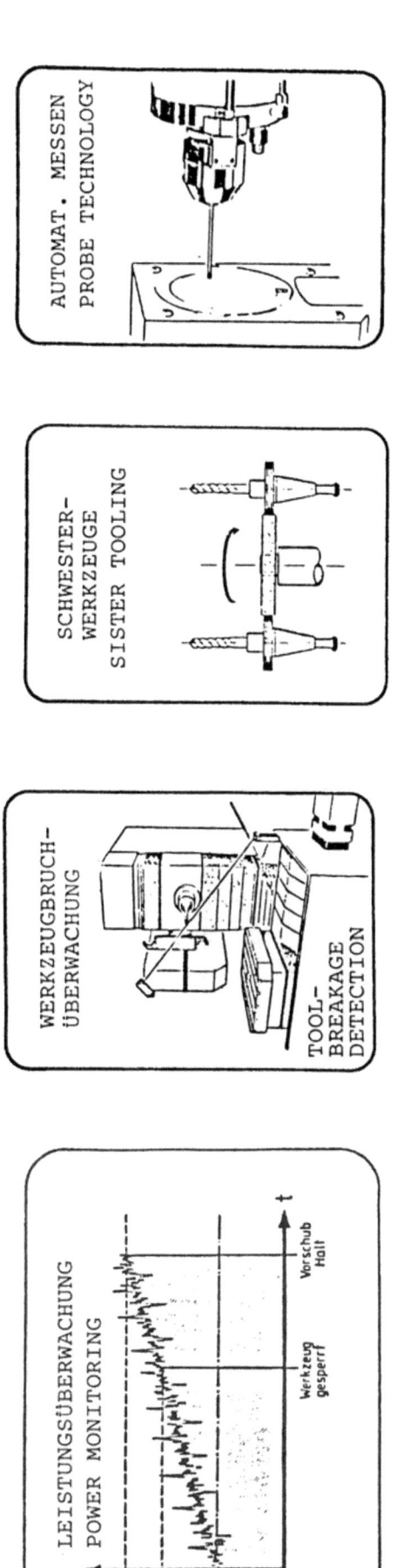

2. SICHERUNG DES ARBEITSPROZESSES
   SAFEGUARDING OF THE MANUFACTURING PROCESS

(ABB./FIG. 4)

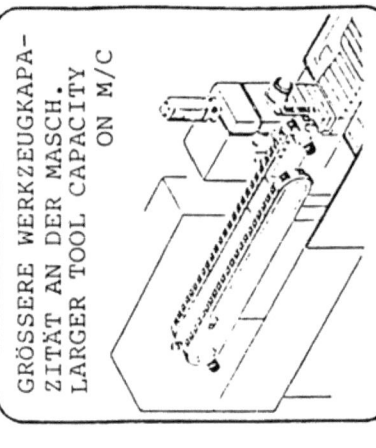
GRÖSSERE WERKZEUGKAPA-
ZITÄT AN DER MASCH.
LARGER TOOL CAPACITY
ON M/C

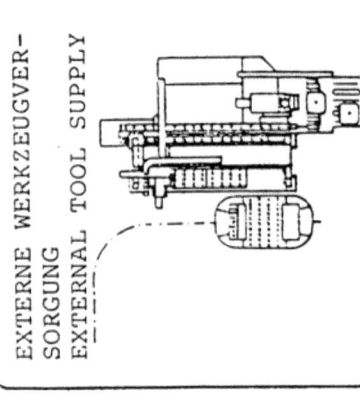
EXTERNE WERKZEUGVER-
SORGUNG
EXTERNAL TOOL SUPPLY

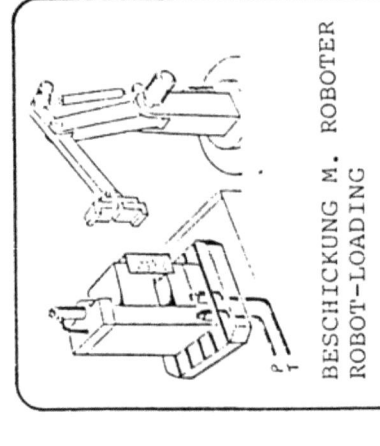
BESCHICKUNG M. ROBOTER
ROBOT-LOADING

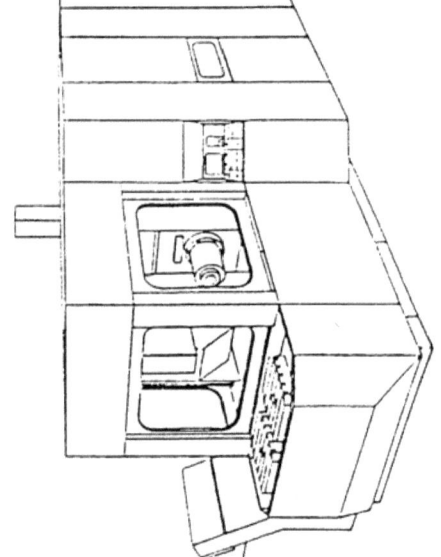

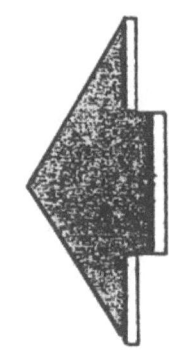

3. AUTOMATISIERUNG DER WERKSTÜCK-
UND WERKZEUG-HANDHABUNG
AUTOMATION OF WORK- AND
TOOLHANDLING

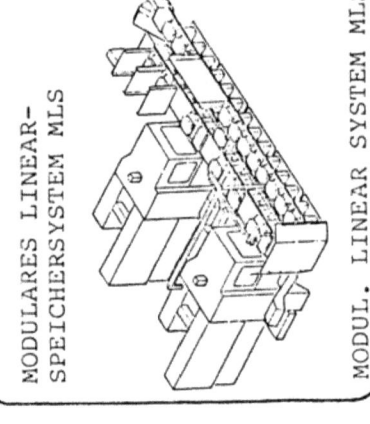
PALETTENSPEICHER IN
TELLERBAUWEISE (PS-8)
CAROUSEL-TYPE PALLET
MAGAZINE (PS-8)

MODULARES LINEAR-
SPEICHERSYSTEM MLS
MODUL. LINEAR SYSTEM MLS

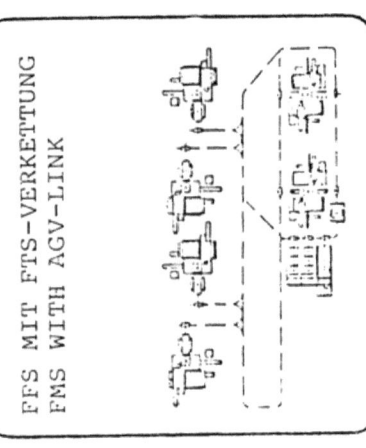
FFS MIT FTS-VERKETTUNG
FMS WITH AGV-LINK

(ABB./FIG. 5)

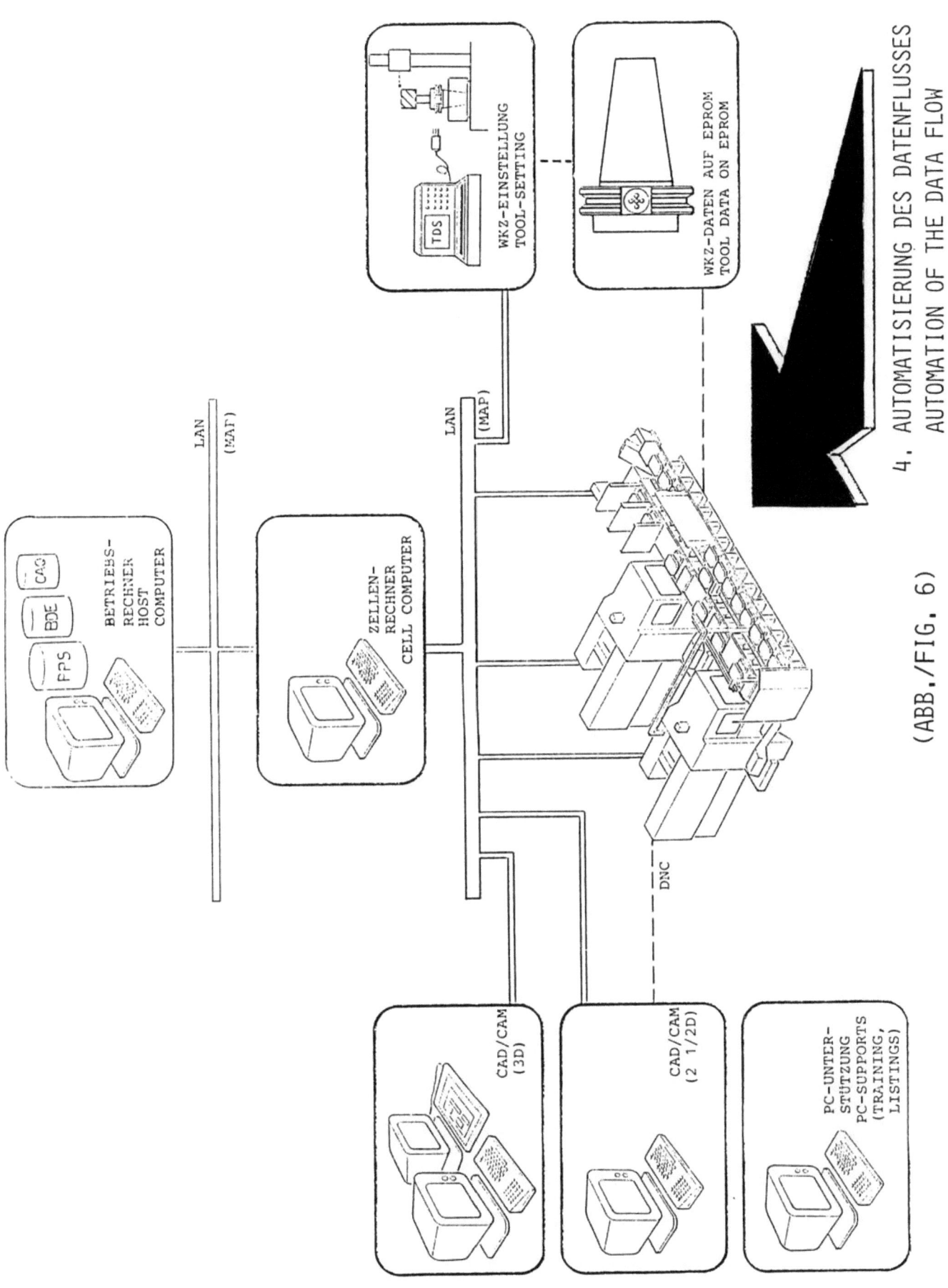

(ABB./FIG. 6)

FLEXIBLE FERTIGUNGSZELLE IM FORMENBAU
FMC FOR THE MOULD PRODUCTION

DECKEL DZ3 MIT PS8   DECKEL DZ3 WITH PS8

(ABB./FIG. 7)

FLEXIBLE FERTIGUNGSZELLE FÜR DIE "MONTAGEGERECHTE" FERTIGUNG

FLEXIBLE MACHINING-CELL FOR THE "JUST-ON-TIME" PRODUCTION

TEILESPEKTRUM
PART-SPECTRUM

DECKEL BEARBEITUNGSZENTRUM DZ4 MIT PALETTENSPEICHER PS8
DECKEL MACHINING-CENTER DZ4 WITH PALLET-CAROUSEL PS8
(ABB./FIG. 8)

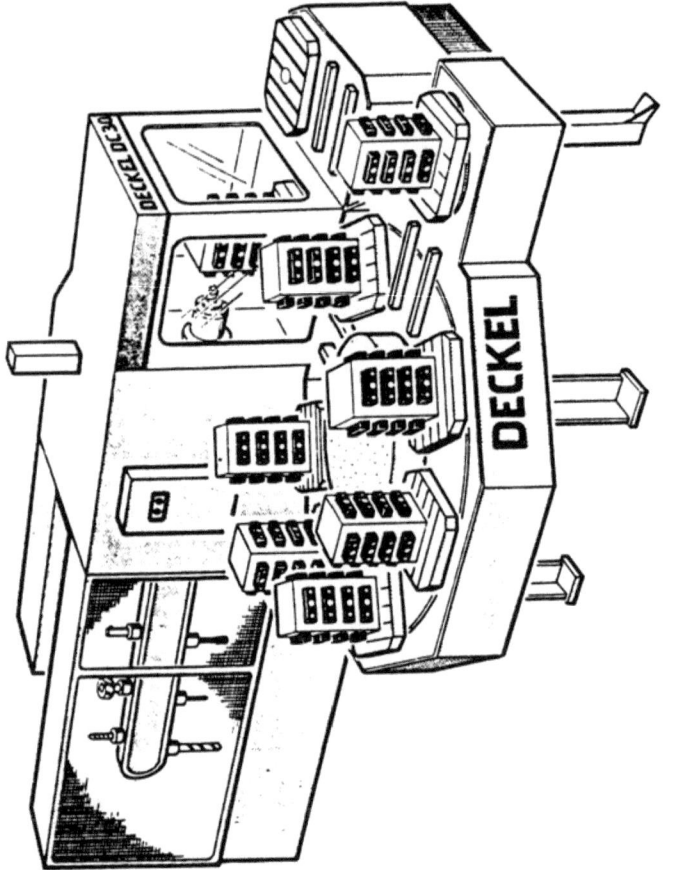

DECKEL BEARBEITUNGSZENTRUM DC30 MIT
PALETTENSPEICHER PS8

DECKEL MACHINING-CENTER DC30 WITH
PALLET CAROUSEL PS8

(ABB./FIG. 9)

FLEX. FERTIGUNGSZELLE FÜR DIE MASSENFERTIGUNG
FMC FOR THE MASS-PRODUCTION

VIELFACHSPANNUNG EINES MASSENTEILES
MULTIPLE-CLAMPING OF A MASS-PROD.

FFZ IM EINSATZ BEI TYPISCHEM LOHNFERTIGER
FMC APPLICATION IN TYPICAL SUB-CONTRACT CO.

(ABB./FIG. 10)

FLEXIBLES FERTIGUNGSSYSTEM MIT FTS-VERKETTUNG
FLEX. MACH.-SYSTEM WITH AGV-LINK

TEILESPEKTRUM
PART-SPECTRUM

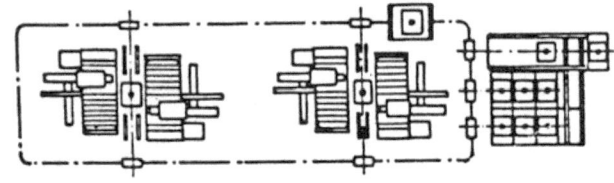

SYSTEM-LAY-OUT

FFS MIT 4 DECKEL S400
FMS WITH 4 DECKEL S400

(ABB./FIG. 11)

FLEXIBLES FERTIGUNGSSYSTEM MIT LINEAR-VERKETTUNG
DURCH SCHIENENGEFÜHRTEN WAGEN

FMS IN LINEAR LINK-UP WITH RAILGUIDED TROLLEY

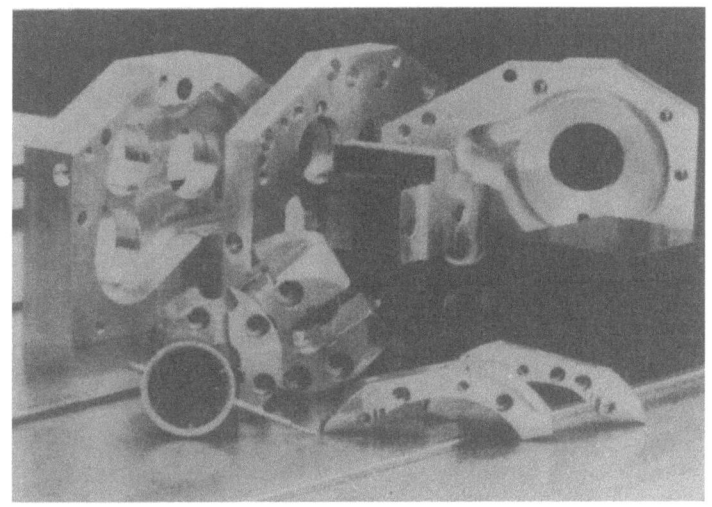

TEILESPEKTRUM
PART-SPECTRUM

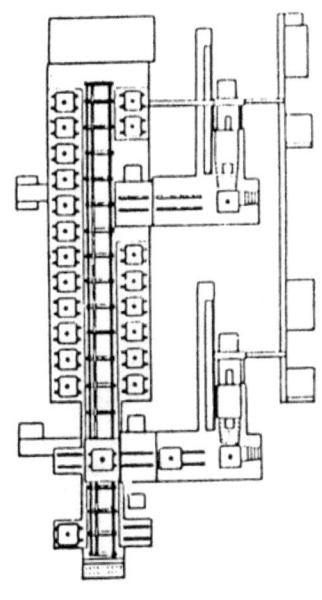

SYSTEM-LAY-OUT

FFS MIT 2 DECKEL DZ4
FMS WITH 2 DECKEL DZ4

(ABB./FIG. 12)

SCHRITTWEISER FFS-AUFBAU DURCH DAS MODULARE
LINEAR-SPEICHER SYSTEM MLS

STEP-BY-STEP FMS INSTALLATION THROUGH MODULAR
LINEAR MAGAZINE SYSTEM MLS·

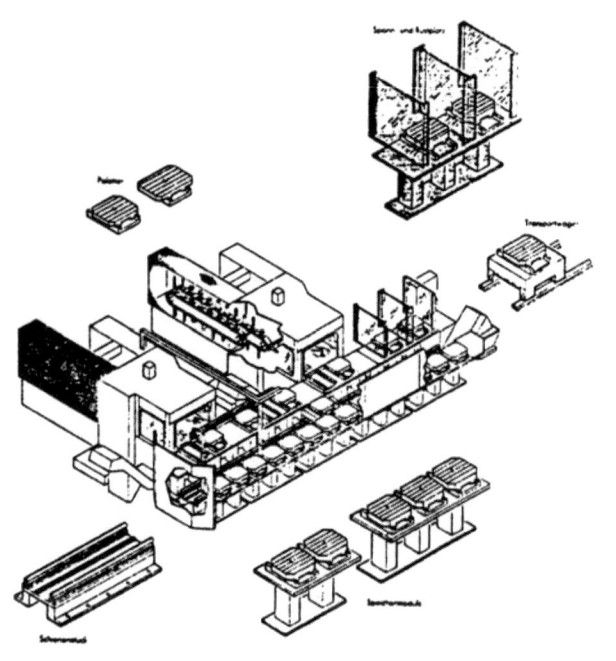

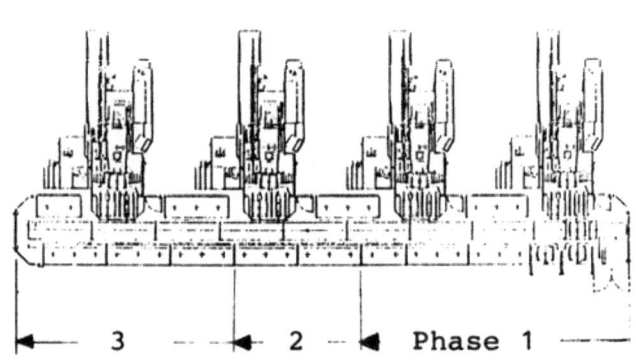

DAS MLS-KONZEPT
THE MLS-CONCEPT

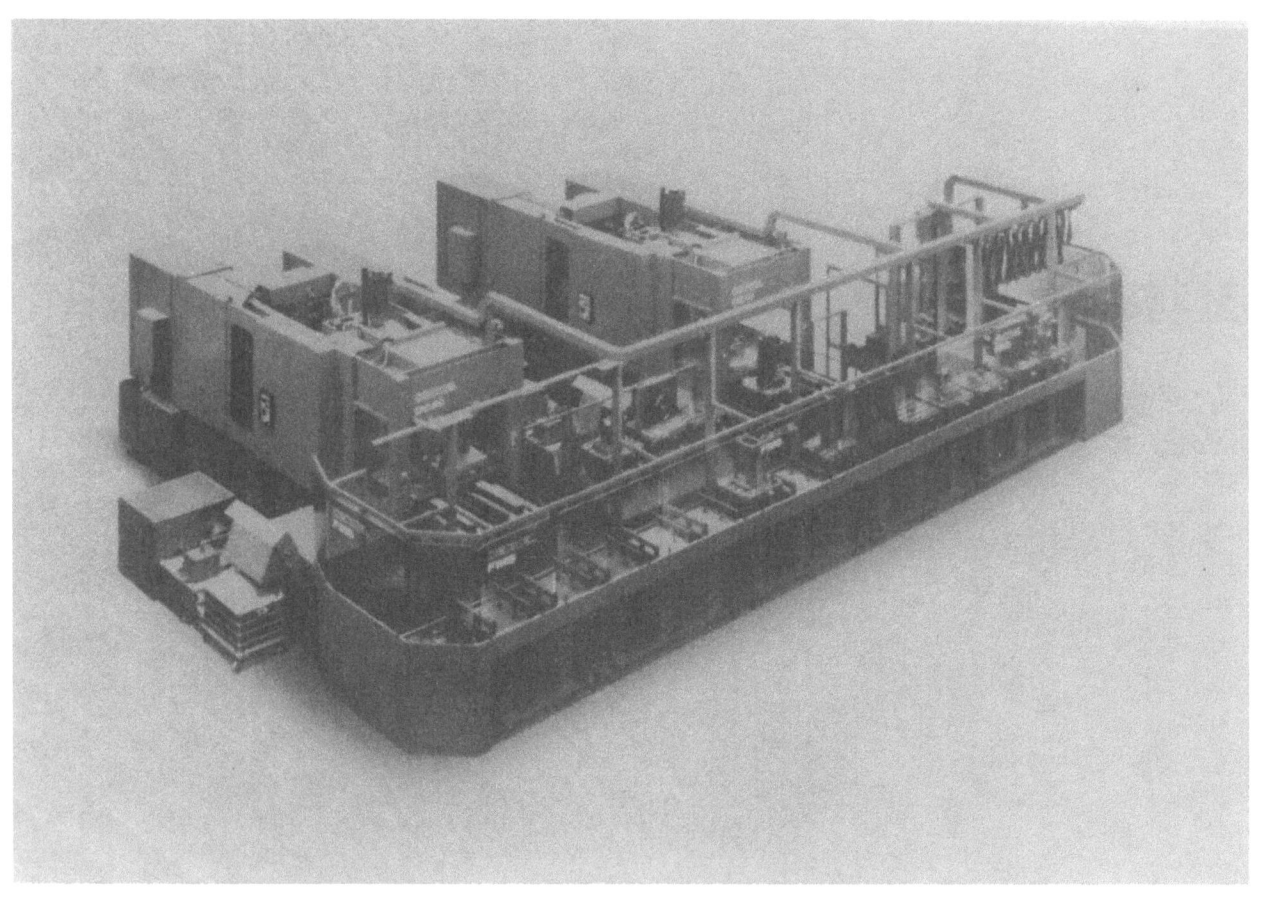

The use of CIM in window manufacture: a case study
T. E. Toth
PA Management Consultants, UK

The application of Advanced Manufacturing Technology and CIM is providing a crucial competitive edge for a diverse range of manufacturing companies. This paper describes the successful implementation of a computer integrated flexible manufacturing facility to produce made-to-measure wooden windows in unique batches of one.

1. INTRODUCTION

   A unique facility has recently been installed at the Lowestoft site of Boulton and Paul - a major UK joinery company - to produce made-to-measure wooden windows for sale to the trade or through their retail outlets.

   The company have long sought to be able to competitively enter a market, worth some £300 million per annum, which traditionally had been supplied by aluminium and PVC window producers or small joiners whose lengthy and labour-intensive methods of wood working would only result in a cost premium to the customer.

   By the strategic application of Advanced Manufacturing Technology to relatively unchanged and traditional methods of wood machining a unique computer controlled flexible manufacturing facility has been developed and successfully implemented with a capacity to produce several thousand window kits per week all in unique batches of one.

   This paper describes the facility and its operation, provides an overview of the computer control and its integration into the company's existing systems and finally discusses the benefits and the competitive edge that the application of AMT has offered.

2. PRODUCT RANGE AND DESIGN

   The first step in the development of the system involved a complete reassessment of the product itself in terms of profile variety and size. In many cases the successful implementation of automation in a process can depend on a structured review of the product adopting a design-for-manufacture approach to include:

   - a reduction in raw material variety and size

   - common physical characteristics

   - standardised tooling and jigs and fixtures

   - avoidance of unneccessary tight tolerances or specifications.

   A new design was produced dramatically reducing the variety of raw material sizes and finished profiles yet maintaining all the characteristics of the traditional British window in outward appearance.

   The final product range to be offered to the customer included 52 different styles of window of which any one could be ordered in any height and width between the limits of 0.5 and 3 metres in 1mm increments, see Figure 1. In addition a choice of hard or softwood was to be offered including options for fittings. These requirements were to form the basis of a specification for the machining facility.

3. DEVELOPMENT

   An extensive period of development followed the acceptance of a final design and product range involving Boulton and Paul senior management and technical staff, PA and the selected machine tool suppliers. The total commitment and involvement of senior executives and the close liaison with suppliers was instrumental not only in ensuring a successful implementation but developing an understanding of the technologies involved and the strategic implications for the company's position in the market place.

Advanced Manufacturing Technology was clearly understood to be not simply a piecemeal installation of automation but a major change in their approach to manufacturing involving the education and change in attitudes of people throughout the company towards existing working practices, quality and delivery performance.

4. OPERATION OVERVIEW

A key feature of the facility is the computer integration from order processing through to component manufacture and assembly. From the moment an order is accepted and processed at the company's head office to the despatch of fully machined window components no paper passes hands. This section provides an overview of the entire process and describes the principle concepts involved.

4.1 Order Processing to Manufacture

Figure 2 outlines the main elements of the computer systems and the nature of data transmission between them.

As orders are received they are input to the company IBM mainframe for invoicing and accounting. From here they are transferred to a MICRODATA computer running a proprietary Production and Inventory Control (PIC) system for manufacturing. Both machines are located at the company's head office.

As each window has its own unique dimensions bespoke software had to be written and integrated within the PIC system to produce and store a unique Bill of Materials (BOM) for each window from key dimensions input from the order forms. For each component in the BOM the software would:

- calculate its appropriate dimensions

- determine the size and type of raw material required

- attach all necessary Works Order numbers and position codes.

In addition other data relating to, for example, choice of fittings, finish, size of glass, etc is created and stored for use at the assembly stage.

On a daily basis all the BOMs are downloaded from the MICRODATA mini to a MICRODATA micro computer on the shop floor at the manufacturing site via a modem link. The micro acts as both an intelligent terminal link to the mini and interfaces directly with the machining line computer sending component data to drive the facility.

Other functions that are provided include documentation at the assembly site, work centre routings and a costing system.

4.2 Component Manufacture

As wood is a natural material a high level of rejects are suffered at all stages of manufacture due mainly to splits and knots. This can not be tolerated in one-off machining as any component rejected itself would have to be replaced by another one-off.

To eliminate this problem the raw material timber is partly machined and inspected to reveal any potential faults in the wood before it is fed into the machining facility bringing the reject rate down to tolerable single figures.

Each cross sectional shape is cut into a number of predetermined increments of length and stored. To produce a component the nearest length of blank is selected, cut to its unique length, joints cut into the ends and the profile finish machined.

Blank picking lists are produced from the daily production schedules by the computer with each list containing all the blanks required to make up groups of made-to-measure windows. The blanks are picked manually and transported to the machining facility.

The blanks, which have previously had a bar coded label identifying its cross-section and length attached, are fed semi-automatically into the machining line in a predetermined order. A laser scanner checks the bar code for verification and a mechanical stop releases the blank for machining if correct. Any faults or incorrect blanks are immediately displayed for operator action. Following final machining the components are inspected, sorted into window kits and transported for assembly.

Should a defective component be found at the inspection point an operator can use a light pen bar code reader to 'reorder' the particular part and reschedule it for machining.

4.3 Flexible Assembly

Following transport from the machining area the components are assembled, painted, fitted out and glazed. Some secondary machining operations are also carried out which, due to infrequent demand, would not be practical to have attempted to include in the main machining line.

A complete flexible assembly line provides the following facilities:

- fully automated sash and frame cramps. These machines press the components together into an assembled window repositioning themselves for different sized frames on a one-off basis

- extensive overhead conveyors to both carry the frames through the process and provide live storage

- automatic paint dipping tanks

- a bead cutting machine providing cut-to-length glazing beads from manual data entry.

After a final inspection the completed window frames are shrink wrapped, have a unique identification label attached and prepared for despatch.

5. MACHINING LINE DESCRIPTION

The main components of the line are described below with a layout shown in Figure 3.

1. Infeed - the blanks are fed onto a conveyor, scanned by laser and if verified transported lengthways down into the machine

2. End Tenoning - here, each blank has a joint profile cut into one end according to the type of component (head, cill or jamb, etc) and its position in the window. As the blanks are passed sequentially down the line they will stop and be cut at the appropriate tenoning station.

3. <u>Cut to Length</u> - each blank is gripped and under NC control passed through a saw to be cut to the required finished dimension.

4. <u>End Tenoning</u> - the required joint profile is cut into the other end of the blank in an identical process to (2).

5. <u>Mortice</u> - the blanks are indexed, in pairs, under two stations each with three morticing heads under NC control. If morticing is required the heads will position themselves (under direct computer control) and using chisel type tools cut down into the blank.

6. <u>Final Moulding</u> - the final machining process consists of being fed through a special purpose moulder to be finish profiled. The moulder has been designed for very fast changeover and will switch and position cutting tools in seconds.

7. <u>Inspection and Outfeed</u> - from the moulder the finished components are inspected and indexed along a wide belt conveyor for loading into boxes. It is at this point that rejects would be rescheduled.

5.1 Machine Computer Control

The interface from the MICRODATA micro computer to the machining line hardware is achieved through a special purpose SER computer which provides the following facilities:

- access of component data in the micro hard disc storage and translation into machine commands to PLC's and NC controllers

- operation and control of peripheral equipment, eg. laser scanner and bar code printer

- component tracking and database - this will trace each unique component through the system and store all recent component data for use if one has to be rescheduled following rejection

- machine status monitoring through sensors and PLC control

- full diagnostics displaying the nature and location of any faults

- management information - files are maintained to store throughput performance and rejected components data

- controlled start-up and shut-down procedures.

5.2 Manning

Manning at machine operator level is minimal and is confined to loading blanks, unloading finished components, day-to-day maintenance and housekeeping tasks.

One inspector at the outfeed position provides quality control of components prior to assembly. In addition a higher level technician acts as both a system supervisor and operates the computer controls to initiate start-up and shut-down procedures, pull down BOMs from the MICRODATA mini, and correct minor system faults.

To maintain a high level of utilisation a large degree of task sharing is practised with all operators expected to assist in dejamming or during any breakdown and to monitor any machine behaviour which could affect the quality of the finished components.

6. PROJECT MANAGEMENT

The FMS project involved many departments spread between various Boulton and Paul sites and numerous suppliers both in the UK and West Germany. Careful management and co-ordination of all activities was vital to ensure that; all tasks were being carried out according to an overall plan; clear, up-to-date, and unambiguous specifications were used by all parties; the project was on target both in time and expenditure.

Although all the activities during the 2 year span of the project could be said to have contributed to the success of the installation the key lessons could be summarised to include:

1. <u>Use of Simulation</u> - a computer simulation using OPTIK Visual Interactive Modelling was carried out to confirm critical details such as; process and conveying speeds, changeover times, conveyor lengths, effects of component variety mix and effects of breakdown.

2. <u>Attention to Detail</u> - every aspect of operation, design and contingencies was thought through, planned and documented in precise detail.

3. <u>Careful Progress Monitoring</u> using computer based project management tools.

4. <u>Strict Vendor and Acceptance Trials</u> - no item of hardware left the suppliers premises unless it had passed clearly specified acceptance trials.

5. <u>Staff Involvement</u> - in addition to extensive hands-on training under the supervision of the suppliers engineers numerous visits to the suppliers sites were made by Boulton and Paul operators; maintenance and technical staff; and line supervisors to attend meetings during both design and build phases.

6. <u>Maintenance Planning</u> - as part of their contractural obligations all suppliers had to provide; detailed programmes of scheduled maintenance activities; a complete set of spares to cover a 2 year period from commissioning; details of UK suppliers for all machine, computer and PLC spares.

7. <u>Installation Planning</u> - a detailed programme including provision of all services was prepared and confirmed by all the major suppliers. In addition several engineers were present, full time, to supervise installation and commission.

7. CONCLUSIONS

In all aspects the project represented a successful implementation of Advanced Manufacturing Technology. Its scope represents a major innovation in wood machining technology encompassing:

- FMS technology producing made-to-measure window components in batches of one

- direct computer integration of sales order processing, production control and shop floor manufacturing

- application of computer controls to wood working machines, full FMS diagnostics and real time bar code printing

- flexible assembly of the end product.

With such a low direct labour cost and the virtual elimination of all clerical and administrative effort the facility produces windows at a highly competitive cost and has provided the opportunity to enter an entirely new market.

This installation truly reflects the tremendous growth in manufacturing companies seeking to develop a competitive edge through the application of new technologies. In a fast-changing world of manufacturing, radical improvements in cost, quality, delivery lead times and due-date performance are all needed. AMT and flexible manufacturing are vital parts of the solution.

FIGURE 1 : WINDOW RANGE

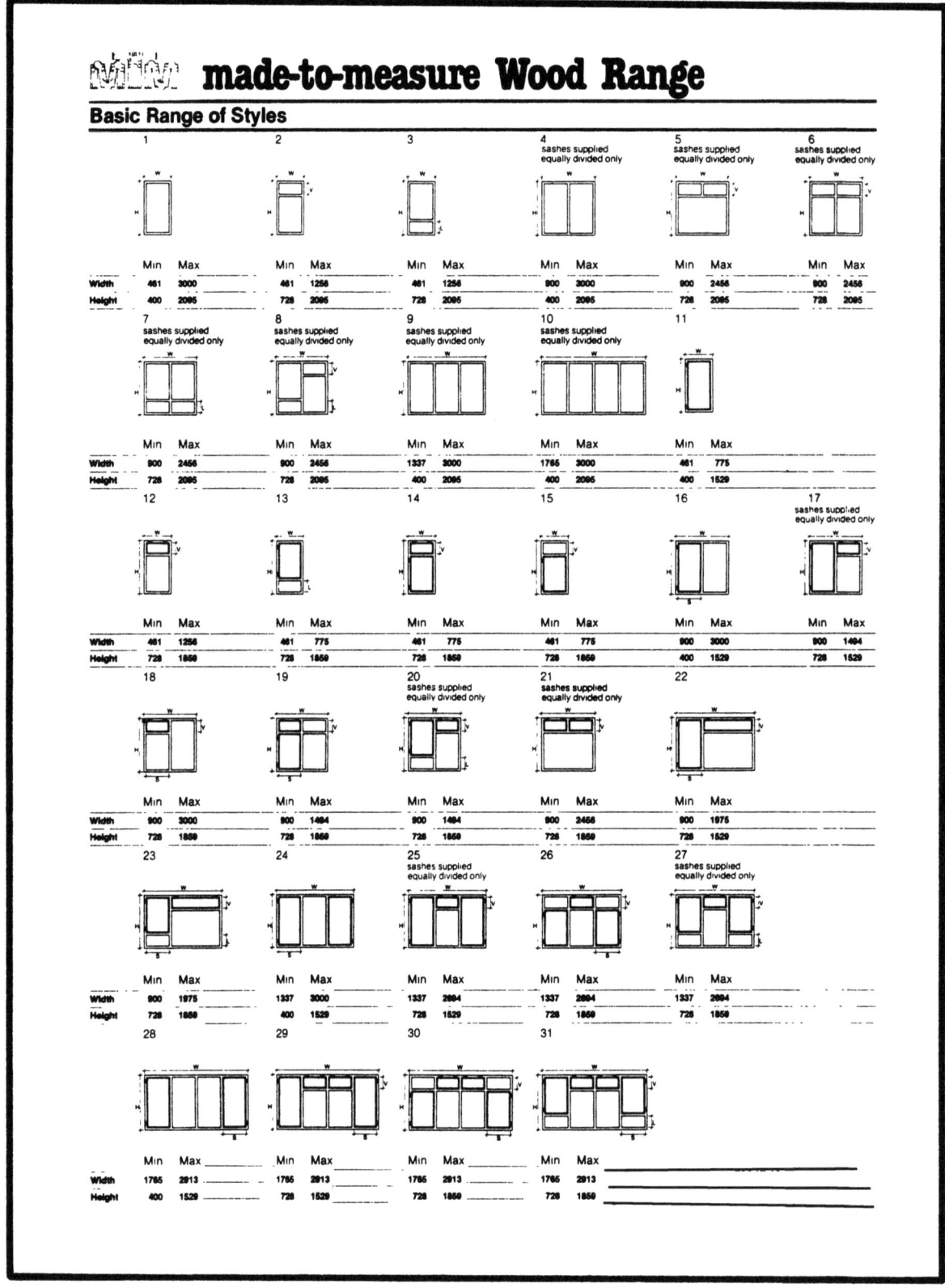

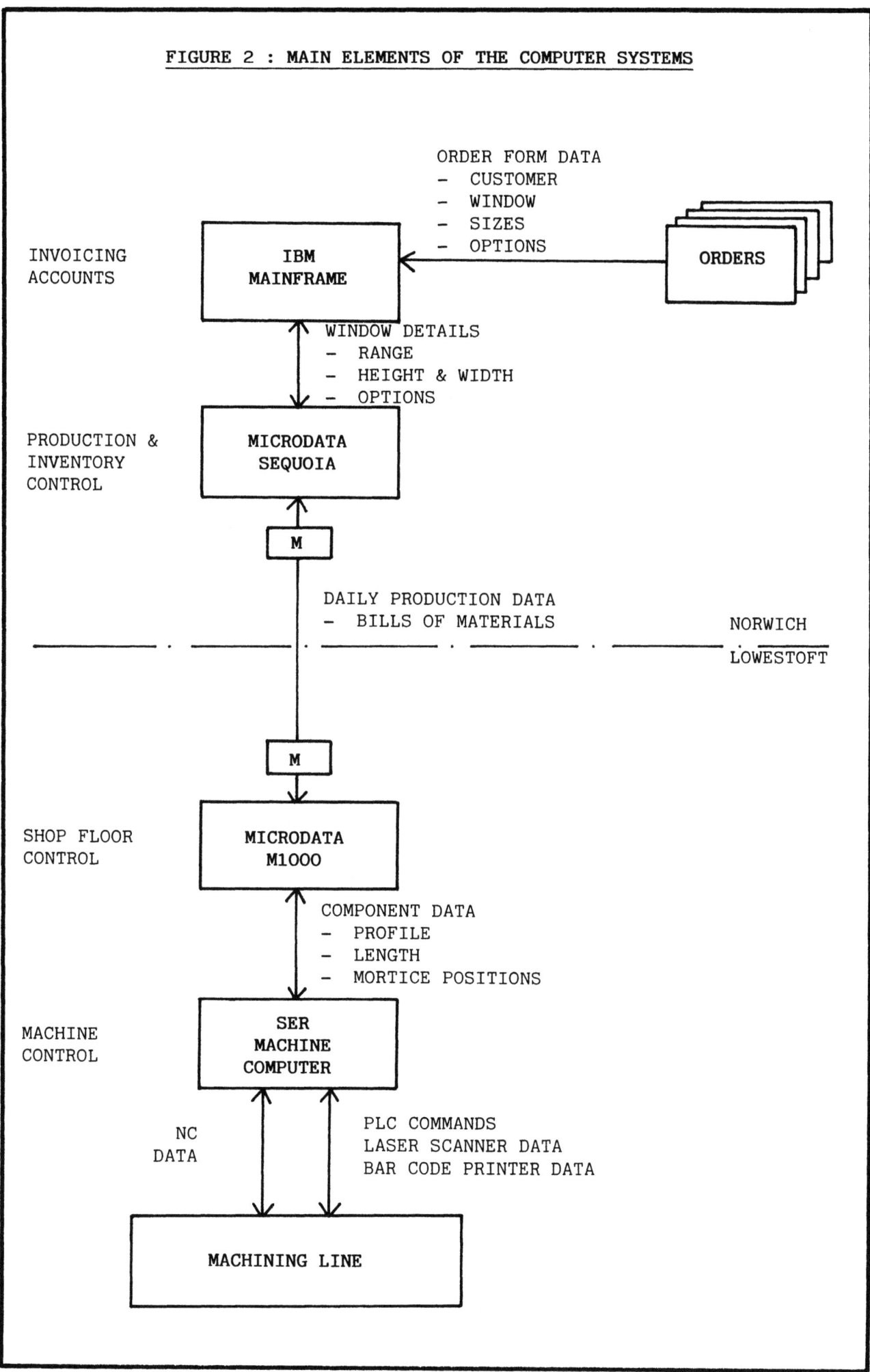

FIGURE 3 : MACHINING LINE

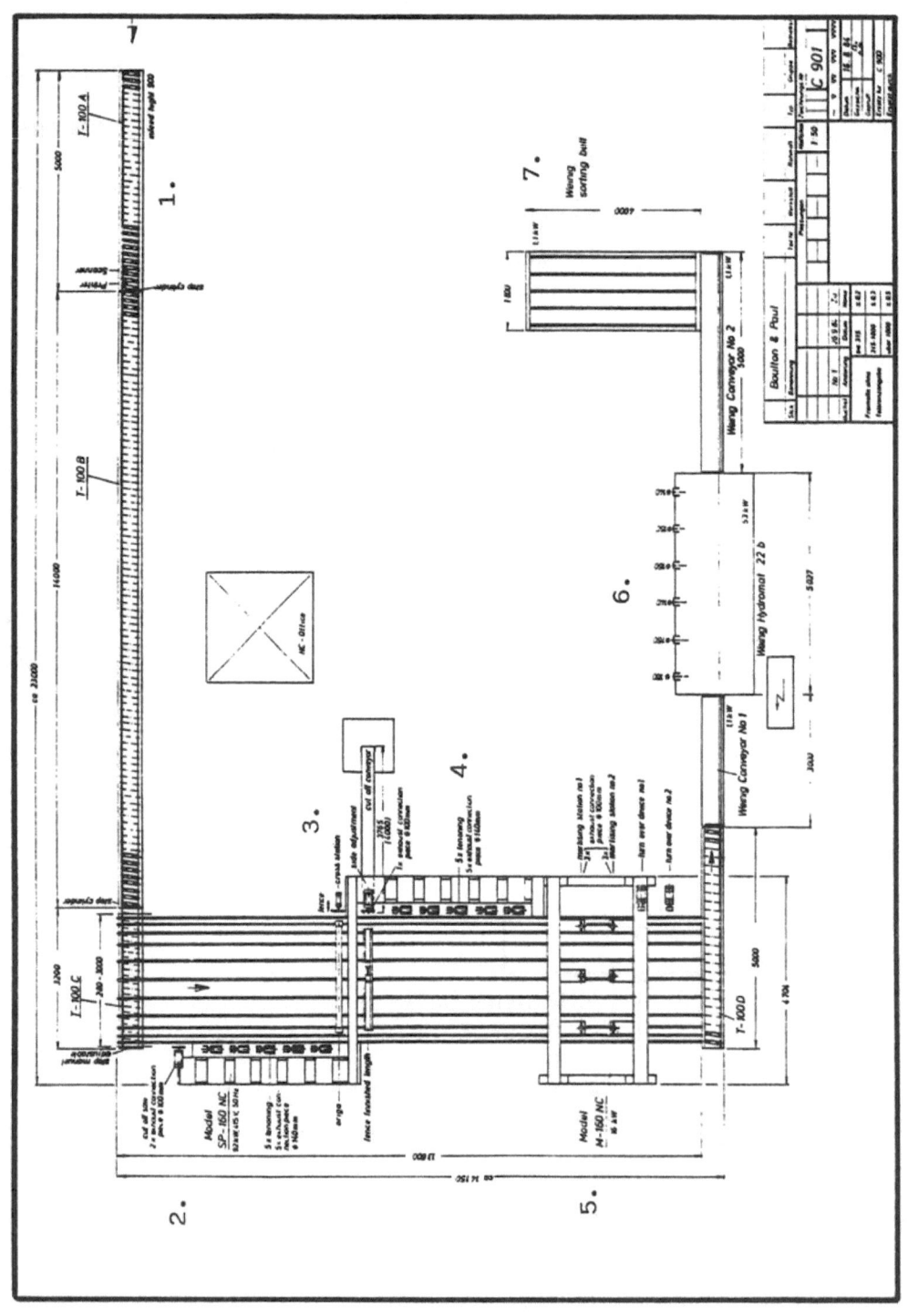

A modular FMS concept in practice for prismatic parts
R. Tuokko
Valmet Corporation, Finland

Abstract

Standardisation and modularity of system elements are key factors for the competitive application of FMS-technology. This minimises the need for one-off project-specific engineering and software development thus decreasing project lead time and ensuring higher reliability.

Valmet Factory Automation has developed a modular FMS-concept easily applicable for a wide range of production needs. Both circular and linear layouts are available for storing and handling pallets. The modular control system offers the possibility to choose control functions and modules according to actual needs and also allows for extensions to be made subsequently to meet changed conditions.

Valmet's FMS concept has successfully been used in a number of FMS-projects in Finland. This paper overviews the concept through four case studies, one is a circular type FMC and three are host computer controlled linear type FMSs. Preceding the concept overview some general aspects are presented which assert the importance of relieving the utilization of FMS technology discussing "The Way towards Everyman's FMS".

INTRODUCTION

Flexible automation adaptable for production technology is experiencing a high growth period within the manufacturing industry. Flexible automation and FMS technology offer a means to improve competitivity in market conditions where production requires greater adaptability and efficiency than previously.

FMS is rapidly becoming established also in Finland where the first FMS started operating in 1982 and where nowadays there are more than 10 FMS level systems operating. The lack of actual large volume production in Finnish factories reflects the lack of large systems. If we include robot cells and independent, flexible machining modules (FMM), capable of partially manned production we can estimate the amount of automation installations in Finland to be in the region of 50.

Valmet is one of Finlands largest industrial enterprises and in addition to being a potential FMS-user it is also a supplier of systems. Responsible for the delivery of FMS systems is the VALMET FACTORY AUTOMATION unit whose program includes various factory automation systems, from large volume production systems to small batch production automation modules and islands. Valmet is not just an integrator of systems but is also able to supply itself the majority of the elements from which the system is built up, manufacturing among other things machine tools, tooling, automatic storage and conveyor systems as well as computer controls with the necessary software. Valmet built the first FMS system in Finland, and has so far delivered in all 10 FMS systems of which the majority have been to Finland.

In the Valmet Factory Automation's main market i.e. in Scandinavia, the market for FMS-systems, when compared with central Europe, tends towards small concerns and small systems. This feature of the market has naturally also guided the development of Valmet FMS. In pilot systems for Valmet itself and for the domestic market it has been possible to develop and perfect the modular FMS concept for extending the marketing of it also for export. The aim of this concept is to lower the threshold for the application of FMS technology, and also on the other hand to offer the possibility of more extensive and functionally more versatile systems. One of the main objectives in the development has been to minimise the amount of customer-specific design work.

2 TOWARDS EVERYMAN'S FMS

2.1 Learning from problems

The ever increasing widespread usage of FMS technology is an indicator of the advancement of the level of technology, thanks to the cumulative experiences gained of FMS projects, and to the worldwide intensive and extensive research and development work within the area of FMS. The more widespread usage of flexible production automation is still however nowhere near its potential which in turn indicates that in spite of all the developments that have taken place there are still significant gaps and shortfalls in the technology. The majority of the systems that have been installed in the world could be compared with the first generation of cars (motorised carriages), and with the traffic of that period (dirt roads, no traffic lights). This gives perhaps a somewhat caricatured view of the development stage of flexible production automation.

The undeveloped technology is manifests itself in many ways. Despite the problems that have arisen FMS-technology is generally regarded as a "must". In various examinations there has indeed been found a clear connection between the success of a company and flexible automation. On the other hand information about the shortfalls that have appeared and the setbacks experienced has been - and will be - reflected in the research and development work in this field. There has been an awakening to examine the matter from an even broader perspective than previously, where, not only technology but also the role of people and the factors connected with this role are taken into consideration. Besides the suppliers viewpoint, there has been an awakening to examine also the important role played by the user and his readiness to adapt and adopt the technology.

2.2 **Simplification**

In the development of automation one should keep the main subject in mind, i.e. in this case "production". Automation therefore should serve production. In other words automate sensibly the work that is to be automated and leave unautomated what is difficult and expensive to automate as well as those work phases the workers regard as pleasant. To replace the last man by automation is extremely expensive and uneconomical.

The question in system delivery is not merely one of the skill and ability of the supplier, but according to experience, the very great problem of information transfer from the supplier's specialists to the user. It is quite natural that the more complicated the system in question, the greater the problem. The knowledge which the customer absolutely must receive includes, among others:

- a deep understanding of the operation of the system
- the operation of the system
- service and maintenance of the system
- modification of the system for changing conditions

The organisational readiness of the customer, and the confident cooperation between the supplier and the customer are key factors affecting the above-mentioned information transfer and system start-up.

The profitability of the FMS-system greatly depends on the degree of utilisation achieved, to achieve this the role of the operating personnel of the system has been found to have a decisive significance. Experience shows that the requirements set for personnel resources are in relationship to the level of new technology to be applied and the complexity of the system. The effect of complexity on utilisation manifests itself also through the technical reliability of the system. The fewer the number of components and elements that affect reliability, the greater the degree of dependability it is possible to achieve. Elements of equal value to the machines and equipment in this relationship are for example also the programs and software included in the control system.

On the basis of information about various sized FMS systems installed in the world, and on published reports it appears that in general the building elements and design and engineering tools for FMS systems have not yet reached the level required for complex systems. This can be seen from among other things by an overprolonged start-up time and by a poor degree of utilisation, because this kind of large entity cannot be tested on the supplier's premises and because, without exception, they include a large amount of one-off, previously untested and untried software and design work.

The technological shortfalls are apparent in modularity and in the mechanical and communication standards. For compact and small sized systems however it is easier to get full compatibility for the machines, equipment and controls, when the system supplier provides the major share of the machines and equipment involved.

In all it can be said that the current level of FMS technology and customer readiness support the building of relatively simple systems. Indeed, from such kinds of systems many have gained better experiences, also in terms of profitability, than from complex large systems. In small systems too it is possible to take into account future extension needs.

2.3 Software significance

The significance of the different software is emphasised in the integration of FMS-systems. The resources necessary for software work and the role of cost in the development and implementation work of the system has grown tremendously. The individual programs included in the software can, in terms of system reliability and utilisation, be regarded as a production component whose operational reliability and freedom from faults is just as significant as the reliability of the machines and equipment of the system.

Stringent requirements are set for the quality of the software and the software work. The software should be reliable, efficient and user-friendly. Software service and maintenance require the software to be easily testable, understandable, as well as being adaptable and extendable. In order that work once done can be utilised as efficiently as possible in different customer projects the software should also be portable. An essential part of the software quality is also the quality of the documentation.

According to a German research to examine software quality, the majority, over 60%, of programming faults were created at the definition phase. In the same context it came to light that over half the faults only became apparent after start up. Since faults are more expensive the later they are found it is not unusual that the majority of the programming costs (even as much as 60 - 70%) can be caused after start up. Examples can be found from the early days of FMS projects where the share of programming and engineering work has risen to as much as 60 - 75% of the total FMS project costs. In connection with these high costs are often decreases in the degree of system utilisation, operational malfunctions as well as difficulties in clarification of abnormal system status. These problems are highlighted in complex systems.

On the basis of what has been presented it is clear that the profitability of FMS projects can be substantially improved if a concept can be created where once designed and tested software can be adapted for customer specific projects with a minimum amount of work.

To decrease the amount of one-off, project specific software work, and to ensure the operational reliability of software requires experience through developed, highly structured software design. The software thus consists of various modules from which it is easy to build the control for different projects. The software modules should be independently operating and testable.

The adaptation and customisation of particular, project dependent software modules can be facilitated by efficient programming tools. Through making the work more systematic the quality of the programming and documentation will also improve.

Besides software, the reliability of the system is also affected by the equipment solutions adopted. Hierarchical control and equipment solutions and distributed operation lead to the situation where a fault in machines or equipment will not cause the whole system to stop . As a minimum requirement it can be stated that if any fault occurs in e.g. the host computer, the operation of the system should continue at least at a level of automation one degree lower.

3   THE VALMET MODULAR FMS CONCEPT

3.1 Development

The Valmet Factory Automation unit has a background of many decades' experience as a supplier of special machine tools for factories involved in mid and high volume production. This function is highlighted by the need for customer and project specific customisation. The emphasis in these deliveries has then been on machines and equipment . In order to minimise the design needs and the risks attendant with new constructions these were developed as modular, standard unit built machines.

When selling single standard machine tools one has been able to talk of "generic" machine tools but when dealing with the FMS area it is another matter. There is no such thing as a fully flexible generic FMS. Indeed in terms of customisation requirements for FMS system deliveries there are clear similarities with special machine tool deliveries. Thus right from the start of the designing of the first FMS system Valmet saw a clear need to modularise the FMS entity. The modularisation work has become more concrete stage by stage in connection with the construction of the pilot systems and the attendant development work. Of primary importance has been the possibility to build very different systems and to benefit from the experiences gained from these. The close interaction between the development work and practice has been an essential requirement for the achievement of a successful system concept.

Fig.1 describes Valmet's approach to the modularisation of FMS-systems. To clarify the presentation we will limit it to examining it in terms of the manufacturing of prismatic parts. Such systems are generally based on

machining centres, sometimes the use of vertical lathes might come into question if they are adaptable to the materials handling system.
Also flexible transfer lines are excluded from this examination.

Lowest level automation to increase utilisation of a single machining centre is represented by chain type 6 - 10 place pallet magazines. In this kind of flexible machining module unmanned operation is limited by the pallet magazine. The solution is not reasonably expandable to a multi-machine system. This concept is normally controlled by a programmable logic controller (PLC).

3.2 Alternatives in materials handling

In Valmet's approach FMS-system types are classified according to pallet handling and storage. A materials handling system serves not only the machine tools, but also such functions as for example loading/unloading, washing, inspection, tool pre-setting, deburring or assembly.

We can consider compact rack type pallet magazines as being closest to the concept of Everyman's FMS. The shelves can either be constructed in circular or linear layout.

Large scale system entities or special requirements set for material flow or layout can lead to a net-type or mixed system which normally requires a larger amount of project specific design or customisation work. For small material flow a net-type layout can also be easily achieved using one fork-lift AGV.

3.2.1 CILO Concept

One argument for the development of the Valmet CILO system was to find an economical solution for taking full advantage of FMS technology and philosophy in small cells and systems. The solution developed is based on a multilevel, circular form pallet magazine equipped with intelligent control and including a rotating and vertically moving shelf stacker for pallet handling (Fig.2).

The CILO pallet magazine can store a large amount of pallets (e.g. 20 - 64 pieces) economically using little floor space, and long periods of unmanned production are possible. One or two machining centres, a washing station and a loading station can, for example, be connected to the pallet magazine. The machines connected do not necessarily need a pallet changer as the CILO shelf stacker can replace it.

The CILO concept offers basic models for pallet sizes from 400 x 400 mm up to 1000 x 1000 mm weighing up to 3000 kg. One system can serve different machines with different pallet sizes. The operational reliability is secured by automatic, semi-automatic and manual operation modes.

3.2.2 Linear pallet magazine

Valmet's rack type pallet magazines are built from standardized profiles. The system is modular and enables step-by-step building and extending of the system   A railguided transfer cart (unilevel storage) or stacker crane (multilevel storage) equipped with telescopic lifting forks and an intelligent microprocessor based local controller takes care of pallet handling. Semi-automatic and manual operation modes in addition to normal automatic operation secure the operational reliability.

The vehicle has a maximum speed of 90 m/min. Quick pallet transfer allows shorter pallet processing times or the serving of a larger number of machines. As to pallet sizes and weights the same figures are valid as with the CILO concept. One feature is the capability of handling and storing also e.g. standard Euro-pallets for blanks, tools or fixtures alongside normal machine pallets. In case of smaller pallet sizes a gantry loader can be used for pallet and material handling.

3.3 Modular control system

According to Fig. 1 the FMS control system can be hierarchically divided into lower level cell control or higher level system control including more functions, with modular software in both.

3.3.1 Lower level cell control

Systems based on rack type pallet magazines or on one fork-lift AGV can most simply be equipped with just cell level control. The simplicity of microprocessor based control and its software is also evidenced in the price being of particular significance in small systems. Simple control is easy for customers to learn and adopt. The simplification of the system's operation facilitates and speeds up system start up.

In its basic form cell level control is made up of the following functional software modules:

- Pallet management. The control recognises the pallets within the system and controls their movement.

- Work queue management. The system is informed of the required production sequence in creating the work queue. Automatic manufacturing takes place according to this work queue. Changes to the work queue can be made manually or they can be done automatically under host level control.

- Status representation. From the status display the operating personnel and production management receive system status data.

- Fault diagnostics. One characteristic of the system is its ability to discover faults and to report these to the system user.

- Displays. Interactive displays facilitate system monitoring and operation and production management.

- Mini-DNC. This means a simple DNC interface e.g. for NC-program transfer.

- Set-up change. In connection with set-up change the system can be automatically given data on NC programs, tools etc. needed for the next batch.

3.3.2 Host level system control

When raising the automation level of the system, its flexibility or the amount of functions, control is carried out by higher level FMS control (Fig.1). The control is performed by a host computer equipped with a disc memory and continuously updated data base. The data base is split up into databanks according to the operation functions.

System level control is closely integrated to the lower level cell control and controls and monitors the operations of the cell level. A system previously equipped with cell level control can subsequently be complemented with host level control. Only those functions and modules which the customer considers necessary need be selected. When extending the system or integrating new functions the necessary new software modules can be easily added to the control. Project specific customisation work and its testing can be carried out on a modular basis without the need to go through the whole software.

Host level system control is constructed from the following functional software modules:

- Workpiece management. The control system has full data on the workpieces to be manufactured. With the help of this data prognoses etc. of production in the system can be made.

- Storage management. The system may include loading pallets for storage of workpiece blanks, fixture components etc. Storage status management is one of the functions of this module.

- Production control interface. The system has the possibility for integration to higher level data systems such as e.g. to production control.

- Reports. From the control one can get printed reports on production, malfunctions, reject costs etc.

- Automatic tool supply. The automatic replenishment of the machine's tool magazine from the central tool storage enables automatic set up change for the machine and thus in systems with several machines the manufacture of a wider family of parts with fewer tools.

- Tool management. Tool management includes full tool status monitoring and covers all tools used in the FMS. This requires a sophisticated DNC interface.

- NC-program management. NC-program management includes the maintenance of all the NC programs needed in the system in the host computer's memory as well as their up and down loading between the host and the CNC's.

- DNC. The DNC module enables real-time data transfer between the host computer and the machines' CNC controls. The data transferred includes NC programs, tool data and status data.

- Tool presetting. This module enables a tool pre-setting machine to be connected to the system and automatic transfer of pre-setting data to FMS-control and from there via a DNC-interface to the machine tools.

- NC programming. This module enables an NC-programming device to be integrated to the FMS.

4 CASE STUDIES

4.1 CILO-FMC for Bogies and Trailing Axles

The first case study is an example of the application of FMS technology and philosophy in a small workshop.

4.1.1 Demands for flexibility

Nummek Oy is a Finnish engineering company employing under 30 people. They manufacture bogie and trailing axle components for lorries. In 1985 the company had a turnover of approximately 20 million FMk. Their customers, Europe's leading heavy goods vehicle firms such as Iveco, Daimler Benz, Volvo, Man, Scania etc. not only place great demands on quality but also on the ability to deliver to tight schedules.

The axles manufactured by Nummek are used in heavy goods vehicles which are sold for use in tough conditions. The order for the axles is received only after the client has placed the order for the vehicle. Thus the axle unit, which still must be customised for the case in question must be manufactured in a really short time. The order can be for one axle unit or for a large number of axle units. The company's ability to stay competitive is based on its high quality and on its ability to deliver with short delivery times varying numbers of highly customised products.

4.1.2 System description

In spring 1986 the Valmet Factory Automation Group delivered to the factory a flexible manufacturing cell for the manufacture of the main components for the bogies and trailing axles (Fig.3). The cell is made up of an intelligent VALMET CILO pallet magazine, a loading station and, in the initial phase, a MITSUI SEIKI HR-5B horizontal machining centre. The system has been designed however to allow another machining centre and an automatic washing machine to be connected at a later stage. In this instance the VALMET CILO pallet magazine has 3 levels and 32 storage places.

The family of parts to be machined numbers over 20 parts, and the family of parts is continuously changing. The 92 tool tool-magazine in the machine contains sufficient tools and sister tools for th whole family of parts. For unmanned use the machines are equipped with tool life and spindle power monitoring as well as tool breakage monitoring. There is effective chip removal through powerful cutting fluid jets.

4.1.3 Automation level

In this case CILO is equipped with low level cell control which allows a freely selectable machining order. Workpiece manufacture is according to the given work queue. Operation is not based on a fixed program but is determined flexibly according to the status of the pallets, machines and loading station. The production program can be seen from the display screen of the control terminal and it is easy to make any necessary amendments.

The chosen level of automation has been easy for the customer to adopt. The solution offers however a considerably greater operational and capacity flexibility than a machine equipped with a chain type pallet magazine. Now they have an extendable system which allows longer unmanned operation periods, through which the production capacity can, according to requirements, be greatly adjusted upwards or downwards without adjustments in personnel.

4.2 FMS for vehicle axle components

4.2.1 General

The flexible machining system delivered to Oy Sisu-Auto Ab axle factory is an example of the modular linear type FMS supplied by Valmet (Fig.4a). In this case over 40 different vehicle axle components, e.g. various drive gear carriers, differential carriers, rocker arms etc. are machined in the system (Fig.4b).

Oy Sisu-Auto Ab is a Finnish factory within the metal industry, which manufactures heavy duty and special vehicles as well as vehicle transmissions and axles. The FMS was delivered at the beginning of 1986 for use in the new highly automated axle manufacturing factory. The factory manufactures approximately 3000 axles per year, and employs just under 80 people and has a turnover of 50 million FMk per year. The production comprises over 80 different axle versions and is designed to meet the needs for high flexibility. The production can be described by the following features:

- Short lead time, i.e. 10 days from order to delivery
- Production divided into sub-assembly based workshops
- Distributed assembly with the aid of an AGV
- Production control on the basis of assembly needs
- Unmanned production

4.2.2 System description

The FMS system for machining the key axle components consists of 2 horizontal spindle machining centres, an automatic washing machine, two loading stations as well as pallet storage and handling system using a rail guided stacker crane and modular shelving. The system elements are integrated by means of Valmet's modular host computer control.

The machining centres are Mitsui Seiki HR 5B type with 630 x 630 mm pallet size. Each machine is equipped with a tool magazine for 120 tools and options for unmanned operation. The machined parts are washed and dried ready for assembly in a washing machine designed for FMS operation. The modular rack type pallet magazine has storage places for 52 pallets on two levels. Extensions are possible both in the form of storage capacity and the number of machines served by the stacker crane.

The host computer of the FMS system is a PDP 11/73 with disc memory. The FMS system is so designed that production takes place if necessary also without the host computer.

4.2.3 Operational functions

Due to the DNC/FMS interfaces between the host computer and the machines, the system control has at any given moment precise information on tool status in the machines, machining programs and operational status etc. These can be followed from the display screen. The reports include data on produced workpieces, production lead times, degree of machine utilisation, tool status of the machines, tool need etc.

NC-program management is included thus avoiding expansions to the machining centre's program memories. The NC programming system is linked to the host computer via the NC programming interface. Also the tool pre-setting machine in the system will, at a later stage, be connected to the system control.

Group fixtures are used for the workpieces in the system to each of which can be clamped several components belonging to the same axle unit. In this way we can decrease the number of fixtures and handlings as well as reduce the lead time within the system. Production is controlled by axle assembly needs.

On feeding a pallet into the system from the loading station, the central control is informed by means of a push button whether the pallet is empty or whether it contains a part to be machined. The pallet code is read when the pallet is placed on the stacker crane. On the basis of the code data the control system assigns the pallet a storage location and the pallet is automatically placed in the right place on the work queue. If necessary the work queue order can be altered from the control terminal. The family of parts manufactured in the system can also be easily changed by entering the new data via the control terminal.

One of the features of the system's host computer control is that decision making for new jobs is automatic and is according to machine usage and tool status. Before a new pallet is brought to the machine the tool status is checked and also that the necessary machining program is available. In this way it is ascertained in advance whether the machine will be able to do the job. This feature increases the degree of machine utilization.

4.3 FMS for imaging equipment components

4.3.1 General

The Palomex group, part of the Instrumentarium concern, employs about 180 people, they manufacture and sell various imaging equipment for medical applications. In 1985 the firm had a turnover approaching 60 million FMk.

In the field of dental radiographic equipment Palomex is one of the world's leading manufacturers of panoramic X-ray equipment with a market share of over 25%. Over 95% of these Orthopantomograph products are exported.

In order to maintain and improve competitivity Palomex has invested in a Valmet FMS system (Fig. 5) for flexible automation delivered in 1986, for the machining mainly of components for the Orthopantomograph radiographic equipment. The family of parts consists of over 120 different aluminium and steel components With the renewal of the end product the family of parts to be machined is continuously changing. Through the increase of unmanned use and the planned extension of the system the company plans to extend their usage of FMS to include other products.

The company estimates the FMS system to be fully amortised in under 3 years. In addition to the achievement of flexible production the FMS brings about savings in direct and indirect wage costs, as well as savings through speeded up production lead time, decreasing the amount of work in progress and inventory levels. To operate their previous individual machines altogether 10 people were needed on 2 shifts, whereas with FMS 3 people are sufficient (2+1) on three shifts, with an additional decrease in supervisory personnel of 2. Thanks to FMS, for the products in question production lead time has fallen from 1.5 months to 2 weeks.

4.3.2 System description

In terms of equipment, the system supplied (Fig.5) is almost identical to that described in 4.2.2 for the Sisu-Auto factory system. One difference being in the load/unload station, where, instead of two turnover tables, two intermediate tables and special handling carts are used. Also the size of the pallet store is slightly larger, consisting in this case of 62 pallet storage locations. The group fixture technology used is however such that only 20 fixtures are needed to control the whole broad family of parts.

4.3.3 Operational functions

In terms of controls the system delivered contains the same functions as the system supplied to Sisu-Auto. Because of the unique nature of the production however, various new software modules have been included in the control, such as NC sub-program management and pallet version management.

Thanks to the NC sub-program management, sub-program technology can be utilised more efficiently than previously. In addition to the main program connected to the pallet to be machined only those sub-programs need be loaded to the machine's CNC which are not already there.

One feature of production in this case is the large number of different parts to be machined. Some of the parts are very small. Group fixtures are used in the system to which, depending on the parts, 2 - 100 parts can be clamped per pallet. The main principle being that the parts clamped to one fixture belong to the same sub-assembly. To reduce the amount of pallets the fixtures are designed to be suitable for several sub-assemblies. Thus there are many versions formed from each pallet, each having its own NC programs. These versions can be considered as different pallets in terms of control.

The different pallet versions can be controlled through the version control software included in the FMS control. On feeding the pallet into the system the control informs the possible versions for the pallet and requests a possible new version number. The version number is written automatically to the reprogrammable pallet code in the pallet. The extra features connected with version management are taken into account in the product bookkeeping and reporting.

4.4. FMS for CNC-turret punch components

4.4.1. General

Lillbackan Konepaja is a Finnish company employing about 120 people which manufactures CNC turret punches and presses for sheet metal work and swaging machines for hydraulic hose assemblies. The company is experiencing a strong growth in turnover which is currently at the 50 million FMk level over half of which comes from exports.

As background for the acquisition of the new FMS system the company has had long experience with NC machines of which they have 20 altogether. The positive experiences they had from their earlier acquisition of a machining centre with a 6 pallet pallet-magazine made the company interested in increasing the level of automation and flexibility with a new investment to increase production capacity.

The FMS system (Fig. 6) delivered by Valmet in 1986 was suitable for the existing production environment by connecting to it two previously acquired machining centres. In the FMS system delivered over 200 different parts are machined (Fig. 7a). Of these, the majority, i.e. over 90%, are parts for CNC-turret punches and presses. The parts include aluminium, cast iron and steel components.

Of the generally recognised benefits of FMS technology the company highlights in this instance the elimination of set-up work. The stored set-ups with tested NC programs guarantee an even quality without the need to manufacture large batches for stock. Thanks to FMS they can also quickly serve any possible unexpected part needs arising in assembly.

4.4.2 System description

The FMS system supplied consists of three horizontal machining centres, an automatic washing machine and a rail guided stacker crane based storage for machine pallets and special loading pallets (Fig. 7b). The system is controlled by Valmet's modular host level system control.

Two of the machining centres are old, being of the Mitsui Seiki HR-4A type. One has a 32, and the other a 46 tool tool-magazine. The new machining centre in the system is a Mitsui Seiki HR-5B type equipped with a 120 tool tool-magazine. For FMS use modifications had to be made to the old machines. The system has storage locations for 74 machine pallets (500 x 500 mm) and 90 special loading pallets which are both handled by the same stacker crane. The host computer for the system is a PDP 11/73.

4.4.3 Operational functions

In terms of control the system includes the same functions as the system described in 4.2 for Sisu-Auto. The differences are mainly connected with the use of the old machining centres as well as the pallet, workpiece and storage management.

Because of their small number of tools, the old machines are only used for machining certain volume components which require few tools. Their corresponding NC programs are stored in the CNC's expanded program memory.

The pallet priorisation possibility eases system operation in situations where pallets of varying urgency are being loaded simultaneously. The priority class, which is added to the pallet data, as well as the automatically forming work queues can however be changed if required from the control panel.

To facilitate fixturing for such a wide family of parts, the parts to be machined are prepared on the special loading pallets. Each fixture has its own corresponding special loading pallet for those blanks which are clamped to that fixture. 1 - 20 parts are clamped to one group fixture. The number of parts placed on one loading pallet is keyed in when the pallet is fed to the store. When, following the work queue order, the stacker crane brings a new pallet for fixturing it also brings automatically the corresponding special loading pallet with blanks to be clamped. The system control takes care of stock and storage location register.

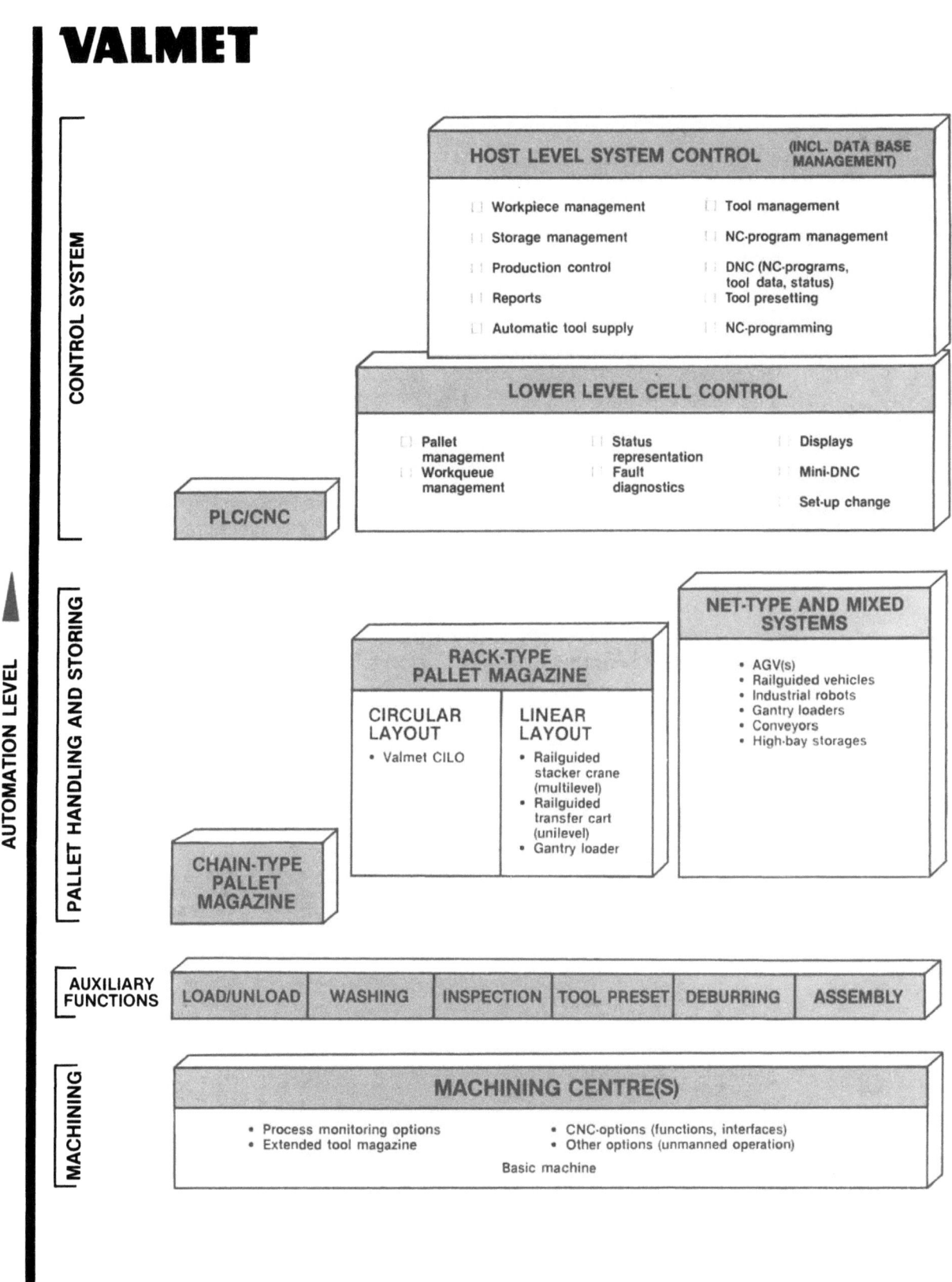

Fig. 1   A modular FMS concept for prismatic parts

Fig. 2   The Valmet-CILO-FMS concept, whose core is a compact, intelligent multi-level pallet magazine

Fig. 3   Flexible manufacturing cell built around a Valmet-CILO pallet magazine (Valmet/Nummek)

Fig. 4   a) General view of FMS delivered for machining vehicle axle components

b) Group fixtures used in the system with parts to be machined
(Valmet/Sisu-Auto)

Fig. 5 Close-up of FMS delivered for machining imaging equipment components (Valmet/Palomex)

Fig. 6 General view of FMS delivered for machining CNC-turret punch components (Valmet/Lillbacka)

(a)

(b)

Fig. 7 a) Example of parts to be machined in the system collected on a special loading pallet

b) Loading area and special loading pallet store of FMS for machining CNC-turret punch components (Valmet/Lillbacka)

DECISION-MAKING FOR FMS

Flexible manufacturing systems in Japan – an overview

H. Yamashina, K. Okamura
Kyoto University
and
K. Matsumoto
Nippon Denso Co Ltd, Japan

ABSTRACT

FMS was born in the latter half of 1960 as a means to improve productivity of small and medium volume production. Since then a number of FM systems have been installed in Japan. Recently comparatively more applications of FMS can be found in high volume production than in small and medium volume production, simply because FM systems become more effective when they are applied to large volume production. This paper seeks to give an overview of FMS in Japan from the historical point of view of rationalization, and presents current developments of FM systems for machining and integrated FM systems. To deal with the strategic aspect of FMS in Japan, required various types of flexibility are reviewed first. Then, some examples of different kinds of FM systems are given. And finally, this paper discusses the problems of under what circumstances what FMS should be developed and how the resultant FMS should be evaluated.

INTRODUCTION

Recently, we do not hear word FMS so often in Japan as we used to. Instead, we hear word FA more. This is partly because a number of FM systems have been installed in Japan so that FMS has become an old topic to technical journalists. This is also partly because FMS is now understood not only as a machining part of FA but to be almost similar to FA in Japan in a wider sense. In other words, FMS is not any more a mere idea but a necessary reality. FMS is strongly believed in Japan to be a major means to overcome various manufacturing problems caused by increasing difficulty of forecasting customers' demands, diversified customers' needs and shorter product life cycles.

The purpose of this paper is to give an overview of FM systems in Japan, to present current technological problems of FMS and finally to discuss Japan's FMS strategy.

HISTORY OF RATIONALIZATION IN JAPAN AND FMS

From Point Automation to Solid Automation

Fig. 1 shows the history of rationalization after the World War II. As is clearly seen from the figure, Japan mainly sought scale merits from 1950 to the first half of 1970 when there were enough demands in the market, rationalizing step by step from a single process (this can be called point automation), a production line (line automation) and a product line (plane automation) to an entire factory (solid automation). In the course of rationalization, automation, unmanned operation, consistency and integration were constantly pursued.

After the 1st oil crisis, the market started to show saturation as shown in Fig. 2. Japan went into the low economic growth age. Accordingly customers' needs have become diversified and forecasting customers' demands have become quite difficult consequently. On the other hand, the product life cycle has become shorter. This is partly because of accelerated technical innovation and partly because of built-in obsolescence policy by makers. For instance, a Japanese word processor, which cost 2,000,000 yen on average in 1980, cost only 280,000 yen on average last year. Nowadays the product life cycle of a word processor is said to be only a half year.

In order to cope with such problems, a new element i.e., flexibility, has become a crucial concern for making manufacturing systems.

Birth of FMS for Machining

FMS was born in the latter half of 1960 as a means to improve productivity of small and medium volume production which covered about 70% of the total production volume of the manufacturing industry. Since then a number of FM systems for machining were installed from the latter half of 1970 to the first half of 1980.

To date there is no precise definition of FMS. Through the development of FMS for the machining area, FMS is understood in a narrow sense as the automated manufacturing system consisting of NC machine tools, material handling equipment and in-process storage facility which are all under the control of a computer system. Such a system may better be called Flexible <u>Machining</u> System rather than the Flexible <u>Manufacturing</u> System.

The required flexibility was variant flexibility which meant the possibility of producing many different variants. The movement against FMS implementation in the machining area is said to be partly through for the time being.

Current Developments of FMS for Machining

(a) System:
- Unmanned operation in the evening, night and even day shifts.

- Tool management system.
- Networking (LAN, MAP, etc.).

(b) Software:
- Scheduling and sequencing.
- Highly sophisticated FMS simulation.
- Integration of CAD/CAM softwares.

(c) Elements:
- Automatic setup equipment.
- Chip disposal.
- High speed machining center.
- Better material transportation.
- Equipment to detect abnormality and to make a diagnosis of the system.
- Improvement of the functions and performances of various sensors that detect the occurrences of troubles.

Economic Justification of FMS

The main objectives for installing FM systems are to achieve effectiveness in small and medium volume production, to reduce operators, to improve labor productivity and to achieve unmanned operation in the evening and night shifts. Compared with the Western countries, Japan may particularly be characterized by the special demand for unmanned operation in the evening and night shifts.

By implementing an FMS, it is said that the number of operators can be reduced, roughly speaking, to about 1/5 and that the primary cost to about 1/2, compared with the conventional system. But in reality few cases have been economically justified.

Since in many cases it is difficult to justify FMS installation economically for small and medium volume production, there has been a growing concern for FMC i.e., a smaller scale of FMS.

For these FMCs and universal-purpose machine tools, the installment of an automatic transport system to the vicinity of them may be justified.

DEVELOPMENT TOWARD INTEGRATED FMS

FMS beyond Machining

Recently comparatively more applications of FMS can be found in high volume production than in small and medium volume production for which FMS was originally meant. Although flexibility is an essential feature of FMS, FM systems become more effective when they are applied to high volume production. For this reason, FMS is developing toward the integrated FMS including fabrication, assembly and inspection in Japan. From the meaning of FMS, an FMS may consist of various processes in addition to machining. FMS is apt to be understood in Japan in a broader sense: It may include fabrication, assembly and inspection.

In order to be economically justified, integrated FM systems need high volume. Reversely speaking, the rationalization process from point automation to solid automation mentioned above took in the concept of FMS to cope with new various manufacturing problems caused by increasing difficulty of forecasting demands, diversified customers' needs and shorter product life cycles. In the course of pursuing rationalization, Japan has accumulated, on top of special production techniques proper to each product, advanced technologies for repetitive small lot production control, materials handling and parts supply, quality assurance and maintenance. Thus, many good examples of FMS can be seen in this category in Japan.

Flexible Assembly Automation

Owing to the development of programmable inserting robots and teaching play back type robots, great progress has already been achieved in flexible assembly automation, especially in the areas of PCB assembly and assembly of tape recorders, video recorders and home appliances such as electric fans.

But there is still a basic problem left to be solved: that is, high breakdown rate. Typical figure of MTBF of an automated assembly lies between several minutes and several ten minutes, while that of FMS for machining is between 150 and 200 hours. Because of this problem, unmanned operation has not been realized in automated assembly.

In determining the number of operators to run an automated assembly line in Japan, the following items are considered:

(a) Time required for starting and stopping machines.
(b) Time required for feeding parts.
(c) Time required for exchanging tools.
(d) Time required for checking quality such as sizes.
(e) Time required for inspecting machines and time for supplying oil.
(f) Time required for setups.
(g) Time required for re-start after the stoppage of machines because of machine breakdowns and part conjestion.

Japan may particularly be characterized by a large number of machines each operator covers, compared with the Western countries.

In any case, the improvement of reliability of automated assembly is left to further research.

Other research topics in flexible assembly automation are:

- Economically justifiable robots with a higher speed.
- Easy off-time programming method for robots.
- Fast vision sensors, accurate and reliable force sensors and tactile sensors, etc.

Computer Aided Inspection

Inspection still relies mainly on human sense. Thus, one of the major problems toward the integrated FMS is computer aided inspection (CAI). There are three steps in achieving CAI as follows:

(a) The first step: automation of sense inspection itself.
(b) The second step: analysis of the information obtained by inspection by a computer. When a defective has been produced, its causes are to be analyzed based on the inspection information. In case of the processes which require adjustment, the amount of adjustment is to be calculated using the inspection information.
(c) The third step: corrective countermeasures against the signs of troubles based on the inspection information. Control items are constantly inspected and the previous processes to the inspection process are to be controlled in such a way that troubles do not take place.

There are some highly advanced factories which have reached up to the second step, but in general, CAI has just started in Japan and many problems are left to future research.

Issue of Networking

Japanese machine shops have already started implementing LAN in order to rationalize the communication networks, aiming to achieve central control and immediate processing of information and enable multiple applications of information. In this sense, MAP is receiving much attention in Japan. Some FMS manufacturers react positively to MAP although they see some problems concerning communication capacity and ability to handle the Japanese language. Other FMS manufacturers are dubious about rapid diffusion of MAP in Japan since they see fundamental differences between the American and Japanese ways of thinking about automation. In the U.S.A., they pursue automation from the top level to the bottom at a stroke by using computers. While in Japan they improve the level of automation from the bottom level to the top step by step. Moreover, there are quite a few companies in Japan who consider it sufficient to establish communication within each production system. Thus, they do not see any necessity of MAP. In this way, there is a gap in recognizing the necessity of MAP between the U.S.A. and Japan.

As is shown in Fig. 3, basic structure of the communication network of a company consists of the three levels: the upper level, the intermediate level and the lower level.

Light rings, which are the major means of LAN in Japan at present, are useful in high speed, long distance, limited and fixed communication. Thus, it is considered that they are still going to be used for the upper level network.

For the lower level network, it is considered to be sufficient to keep up communication within each production system. Thus, individual network systems will be used case by case in order to leave room to make improvement on the production floor.

When it comes to the intermediate level, its network system requires the following:

(a) A large amount of information can be handled (information capacity).
(b) Various devices can be connected (diversification).
(c) There can be connecting points at any place and the extention of network can be easily made (flexibility).
(d) No problems should take place even under bad circumstances.

Since coaxial broad bands satisfy these requirements, MAP is likely to be used for the intermediate level of network when MAP is going to be used in Japan.

In-house made FM Systems and Their Improvements Even at the Stage of FMS Construction

In Japan, the number of companies who develop in-house made FM systems has been increasing recently. This is because they strongly believe that it is quite important to be able to make in-house machines in order to develop the most appropriate machines for their products and to challenge positively new technology as is shown in Fig. 4 since those who have the know-how of manufacturing products are the manufacturer of the products itself. This is especially the case when it comes to integrated FM systems, since implementation of such FM systems requires a reform of the constitution of a company. High competitiveness cannot be achieved solely by purchasing an FMS from an outside FMS maker. It is nowadays believed to be vitally important to have in-house hardware and software techniques.

Investment of an FMS is apt to be costly, so the system must thoroughly be checked before it is going to be used. Fig. 5 shows a typical system used in Japan to assure the reliability of the system at the stage of FMS preparations. Japan may be featured by the fact that maintenance people and operators are already asked to participate in the construction of the FMS and that even at this stage improvement of the system already starts together with the people on the floor.

Applications to Other Areas

Receiving an impetus from success in the development of FM systems, various production areas started to develop their own FMS systems. Especially the following areas have already realized good results:

- Sheet metal forming, welding and extrusion areas.
- Manufacturing fields such as apparel, shoes, confectionaries and glasses.

FMS STRATEGY

Required Various Types of Flexibility

Although we require various types of flexibility, we do not intend to produce, say, from refrigerators to color T.V. sets by the same flexible manufacturing system. In other words, we look for various types of flexibility in the framework of a specific product, say, VTR to the extent that as far as VTR's are concerned, they should be produced by the same flexible manufacturing system.

If required various types of flexibility are viewed in terms of the number of variants, production volume and product life cycles, they can be summarized as follows:

(a) Variant flexibility, related to the mix of products (parts) which the FMS can process.
(b) Volume flexibility, related to the ability to change production volumes without causing difficulty to the FMS.
(c) Short product (part) life flexibility, concerned with design changes or introduction of new products (parts) to the FMS due to short product life.

Obviously, the FMS fully equipped with these types of flexibility would become quite expensive and could not be economically justified. This means that these types of flexibility must be ranked depending on the product. To which flexibility priority must be given heavily depends on how the product is recognized from the business viewpoints in terms of production volume, the number of variants and product life cycles. The design of the FMS will differ depending on this recognition of the product group. Then, under what circumstances what FMS is to be established, and how should the resultant FMS be evaluated?

Examples of Various Types of FM Systems

Fig. 6 is drawn to analyze various types of FM systems, taking production volume on X axis, the number of variants on Y axis and product life cycle on Z axis. In the past, the production volume was a main factor for determining a production system. Up to recently, the production volume and the number of variants were major concerns for making a production system. Now, we need to consider the production volume, the number of variants and the product life cycle when making a new production system as shown in Fig. 6.

For FM systems with high production volume, a large number of variants and long product life cycle (point 1 in the figure), substantial investment could be justified because of the production volume and long product life cycle. In this case, FMS strategy is to give priority to variant flexibility. A good example of this case is Nippon Denso's radiator FMS producing 60 kinds of radiators of different sizes. The cycle times of producing different radiators are different. To produce them by the FMS without causing much loss to the FMS, the different radiators are scheduled by a computer. This is one of a few cases in Japan where scheduling of jobs are actually made. The outline of the FMS is shown in Fig. 7. Incidentally, this FMS won a Japan Society of Mechanical Engineers' Award in 1985.

For FM systems with high production volume, a large number of variants but short product life cycle (point 2 in the figure), investment may be still possible to a certain degree due to the production volume. In this case, FMS strategy is to give priority to variant flexibility and short product life flexibility. FM systems could be equipped mainly with universal machines such as robots. Nippon Denso's cooler unit FMS belongs to this case. This has been meant to be used for cooler units for several generations. The outline of this FMS is shown in Fig. 8.

The above two examples are the cases where the position of the FMS can be located precisely in Fig. 6. However, there are many cases where the necessary production volume, the necessary number of variants and estimated product life cycle are not clear. That is, there are cases where we need to develop an FMS under the circumstances where the position of the FMS cannot be located in Fig. 6. In other words, on top of those three types of flexibility, we need to consider in such cases long-term volume flexibility, long-term variant flexibility and uncertain product (part) life cycle flexibility.

In case of epoch-making new products, there will be demands of them for a certain period, but it is difficult to predict the future growth of production volume. In such a case, FMS strategy needs to consider long-term volume flexibility.

Like personal computers, there are items whose production volume is difficult to predict and whose product life cycle is uncertain because of rapid technical innovation and tough competition in the market. In such a case, FMS strategy needs to consider long-term volume flexibility and uncertain product life cycle flexibility.

One of the methods to cope with such cases where we cannot locate the position of the FMS in Fig. 6, is the introduction of modularity in FMS.

The attitude toward modularity in FMS differs depending on the manufacturer in Japan. Companies manufacturing and selling equipment related to FMS tend to tackle positively with modularity from the business point of view. On the other hand, companies making in-house FM systems are not so keen about modularity since they try to design and construct an FMS in an optimum way to suit each concerned product group. They are afraid that modularity may prevent technical progress especially in the area facing intensive technical innovation. However, the use of modularity has started even in such companies since it is an effective way to increase flexibility and to reduce overall cost. Modularity will be applied step by step first to the FMS dealing with technically mature products.

In order to cope with short product life flexibility and uncertain product life flexibility, the FMS may be designed in such a way as to process various kinds of similar products. Labor intensive area may be left because it is difficult to justify its automation economically due to short product life, but materials transportation can thoroughly be automated since it can be used even if new products are put in. This idea was already applied to Fujitsu Tatebayashi FMS producing word processors.

There is no suitable method to cope with long-term volume flexibility so far except increasing or decreasing the number of FM systems (i.e., the use of separate FM systems) depending on the production volume. In the future, this type of flexibility will be achieved by combining all possible methods such as separate FM systems, modularity and rationalized materials transportation.

As is apparent from the above mentioned FMS examples, FMS strategy depends on how the product is recognized in terms of those six different types of flexibility.

Evaluation of FM Systems

The comprehensive evaluation of the resultant FMS is made in Japan concerning the following points:

- How was the product group recognized in terms of those six types of flexibility? What priority was given to those six types of flexibility consequently? To what degree were the required various types of flexibility actually achieved?

- To what degree was production control made easy and advanced? How much were important new technologies created and accumulated?

- What results were achieved for the following standard evaluation items?
  (a) Production capacity.
  (b) Number of variants.
  (c) The effective lifetime of the FMS.
  (d) Amount of investment.
  (e) Pay back years.
  (f) Efficiency.
  (g) Diffective rate.
  (h) Total manpower per unit.
  (i) Number of reduced people compared with the conventional method.
  (j) Value added/operator.
  (k) Value added/space.
  (l) Thruput times.
  (m) The number of buffer capacity in the processes.

Among those evaluation items, pay back periods would be one of the major concerns. Pay back periods accepted for rationalization is normally between two and three years in Japan, but in case of FMS investment, longer pay back periods between three and five years seem to be accepted.

What kind of FMS should be constructed also depends heavily on the accumulation of technology. As more and more various kinds of FMS are established, more know-how of designing and constructing a proper FMS will be obtained according to the characteristics of the product group and its conditions in terms of the six types of flexibility.

CONCLUSIONS

(a) The movement against FMS implementation in the machining area is partly through for the time being in Japan.

(b) Recently comparatively more applications of FMS can be found in high volume production than in small and medium volume production because FM systems become more effective when they are applied to high volume production. For this reason, FMS is deveoloping toward the integrated FMS including fabrication, assembly and inspection in Japan.

(c) Although great progress has already been achieved in flexible assembly automation, there is still a basic problem left to be solved: that is, high breakdown rate.

(d) CAI has just started in Japan although there are some highly advanced factories which have reached up to the level of analyzing the information obtained by inspection by a computer.

(e) Japanese integrated FM systems may particularly be characterized by the fact that maintenance people and operators are already asked to participate in the construction of the FMS and that even at this stage improvement of the system already starts together with the workforce on the floor, contributing to raising both the reliability of the system and the moral of the people.

(f) FMS strategy depends on how the product is recognized in terms of: variant flexibility, volume flexibility, short product (part) life flexibility, long-term volume flexibility, long-term variant flexibility and uncertain product (part) life cycle flexibility.

(g) The resultant FMS is evaluated especially concerning the following points: What priority was given to those six types of flexibility? To what degree were the required various types of flexibility actually achieved?

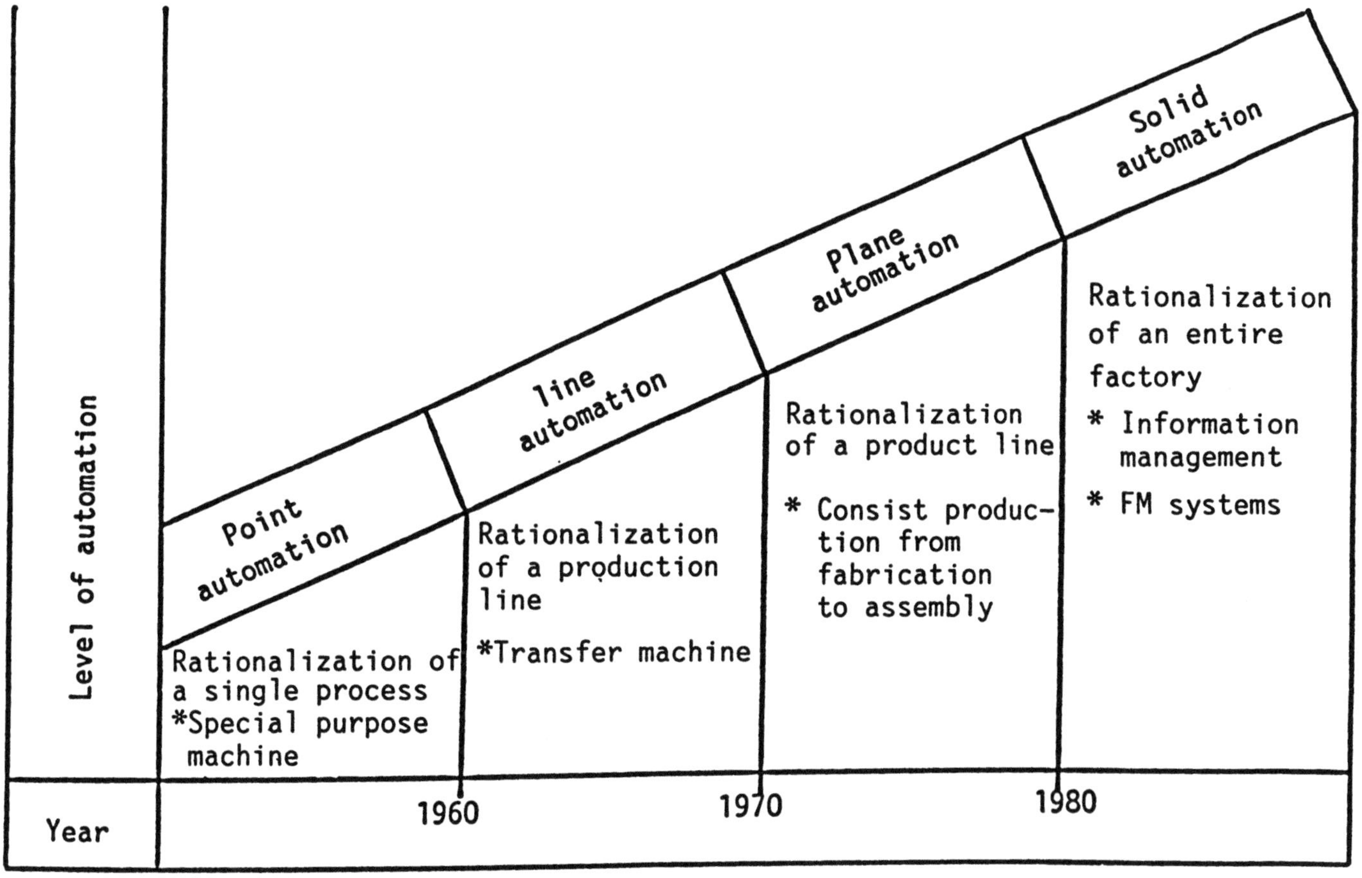

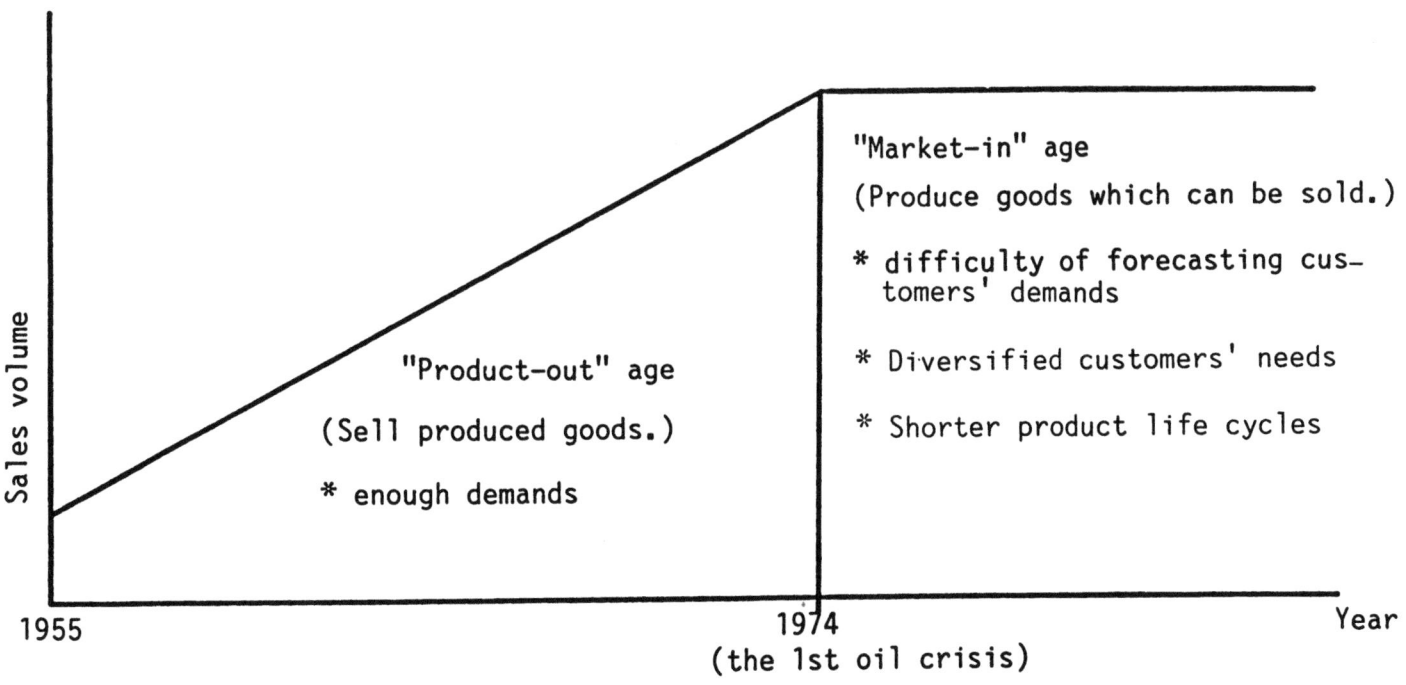

Figure 1. The history of rationalization

Figure 2. Market situations before and after the 1st oil crisis

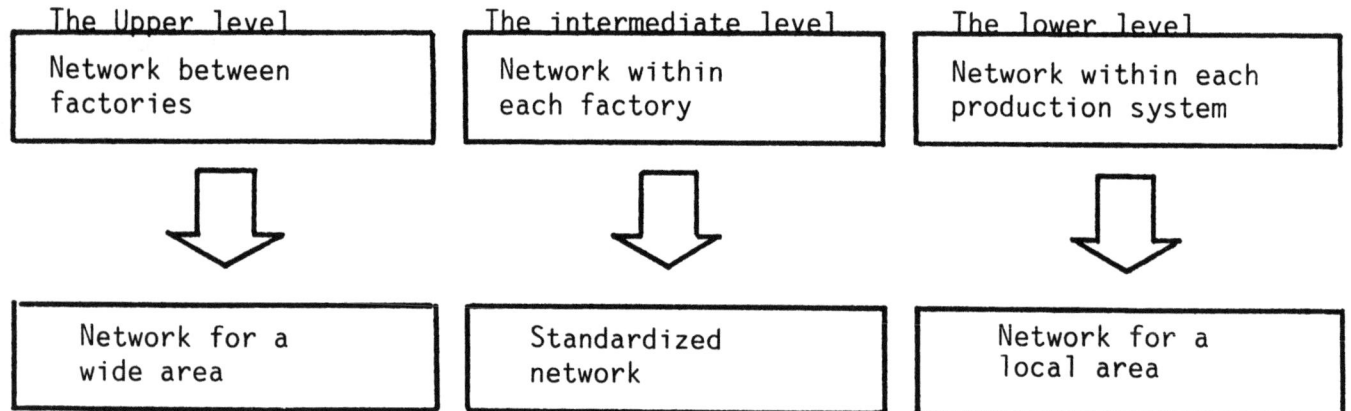

Figure 3. The basic structure of the communication network of a company

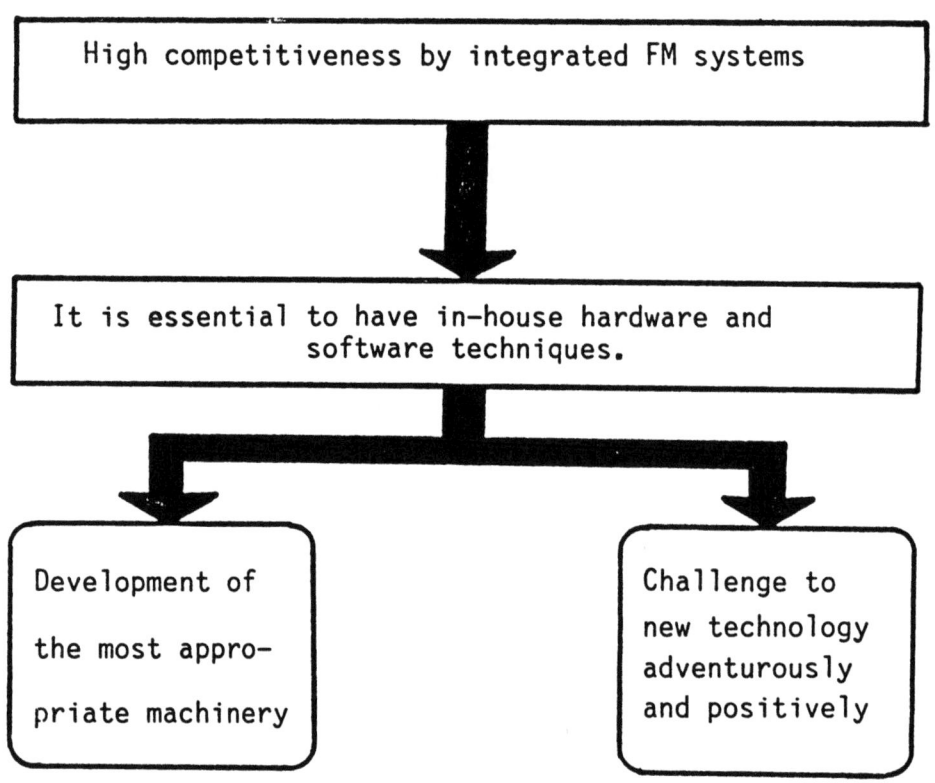

Figure 4. The necessity of making in-house FM systems

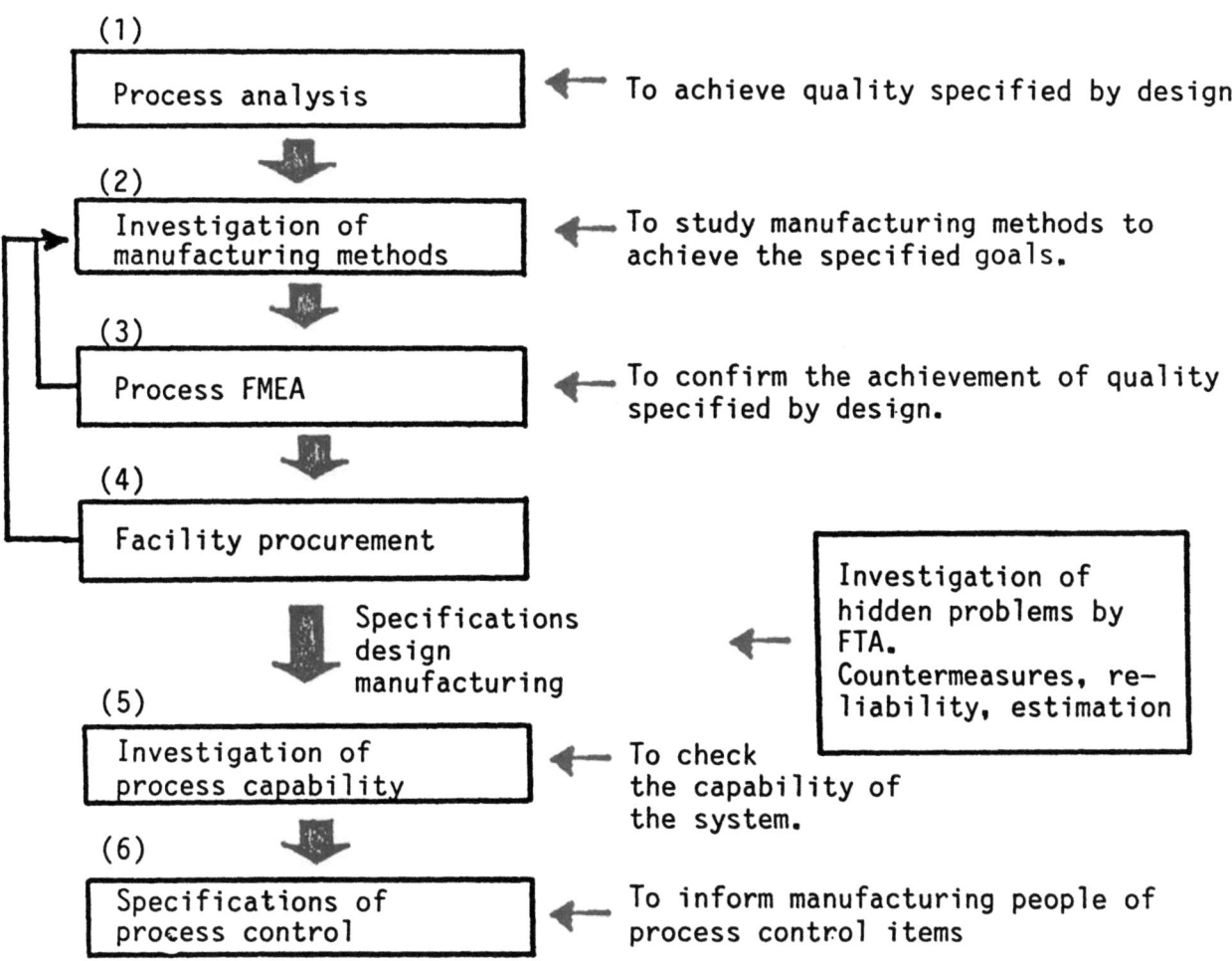

Figure 5. Assuring the reliability of the system at the stage of FMS preparations

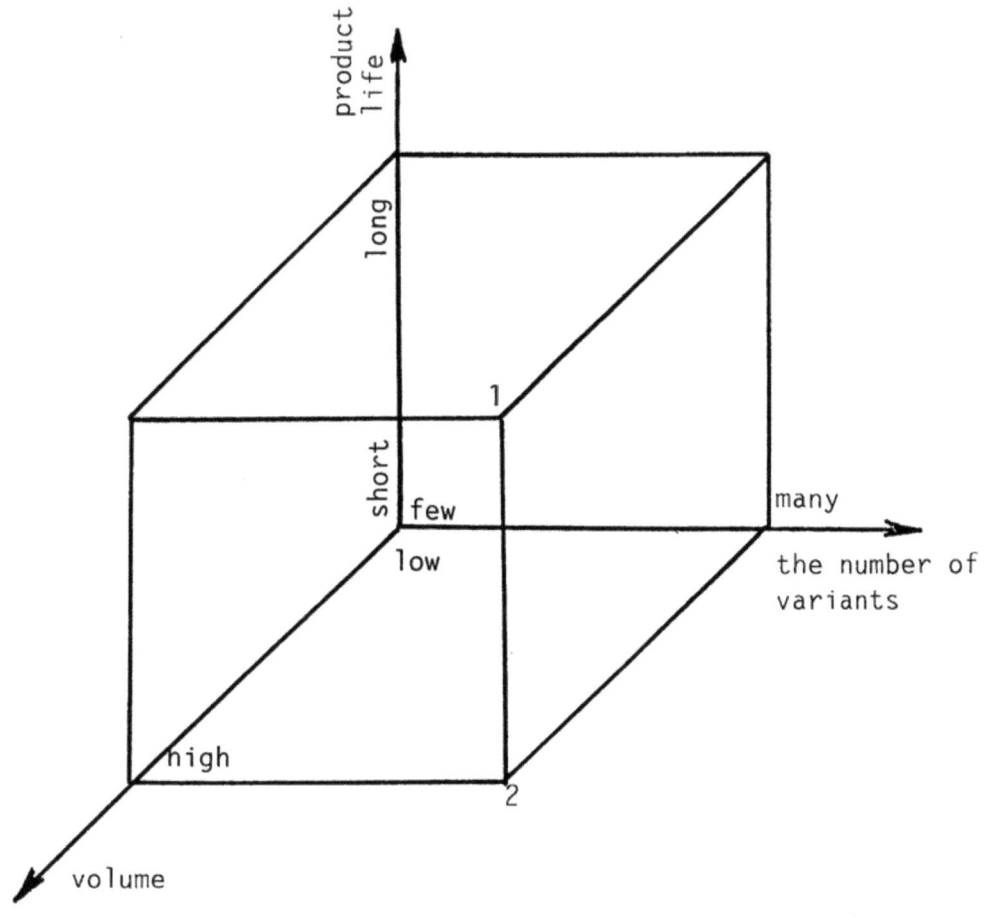

Figure 6. The analysis of various types of FM systems

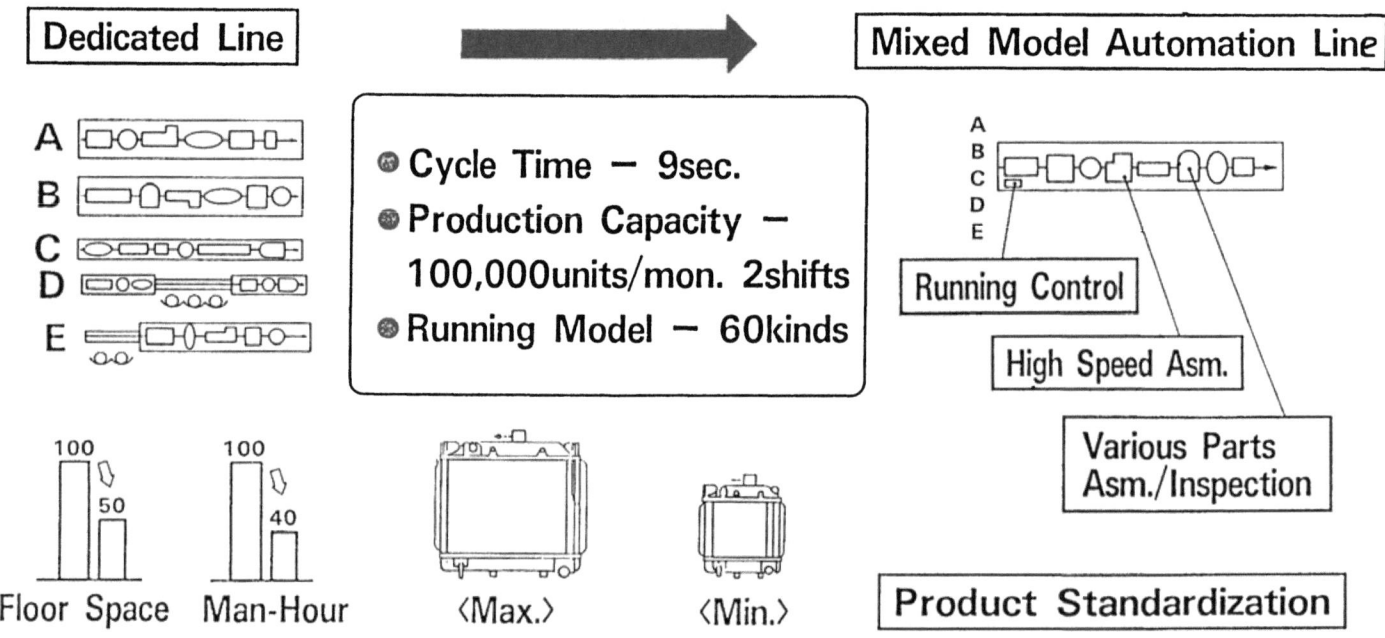

Figure 7. FMS for variant flexibility

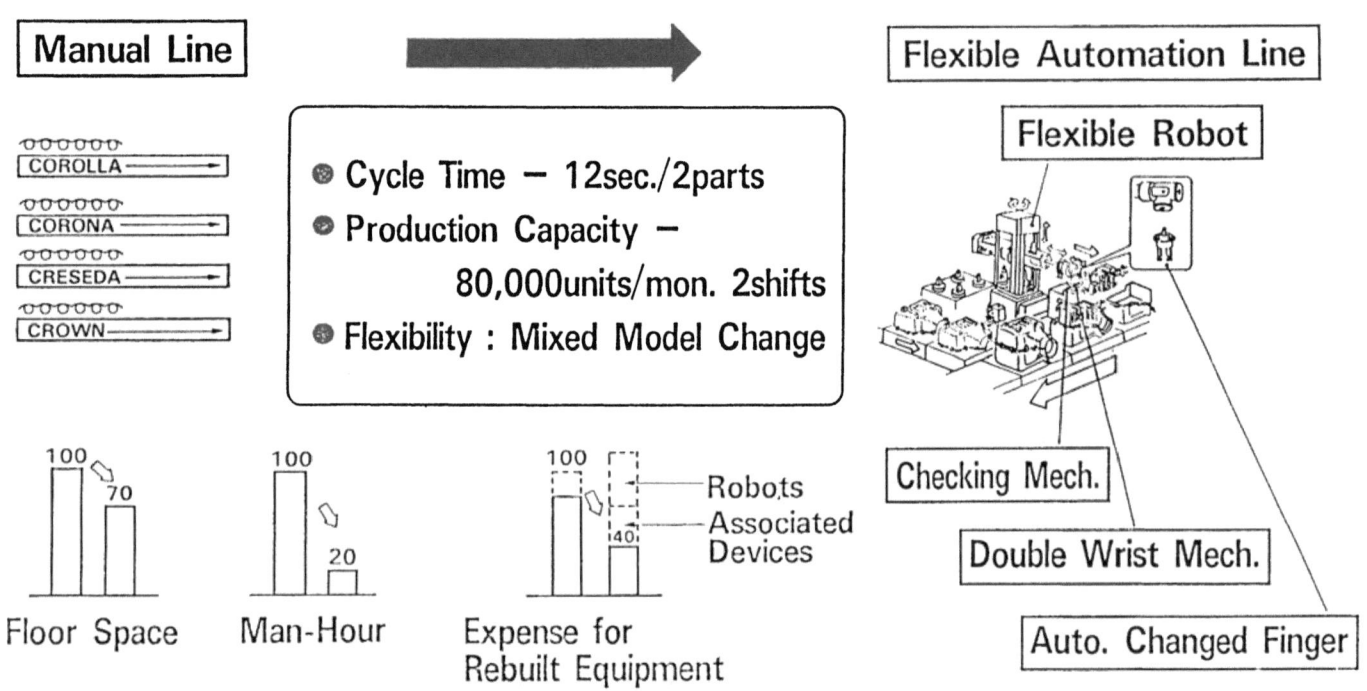

Figure 8. FMS for short product life flexibility

Justifying FMS to provide a competitive edge
A. A. Hunter
Cummins Engine Co Ltd, UK

To effectively compete with "Third World" countries and their very low Labour costs, we require to produce goods of a higher quality, lower cost and on shorter leadtimes. This can be achieved, in many instances, if we strategically apply A.M.T. to consciously provide a "Competitive Edge".

With the ever-increasing rate of change and radically reduced product life cycle, it is essential that the improved flexibility which A.M.T. can provide becomes an essential ingredient in our investment strategy, but, without a satisfactory "process" to effectively handle major investment, no amount of enthusiasm on its own will succeed.

The successful application of A.M.T. depends upon meeting the strategic requirements of the Company, coupled with sound financial evaluation and Management judgement of the intangible benefits which are projected. I need hardly say there is no place in a well-managed Company for "sexy" Management projects.

INTRODUCTION

When first asked to prepare this paper, my initial reaction was to look at a myriad of facts and figures which we have used to justify Advanced Manufacturing Technology, however, on further reflection, I reached the conclusion that success in this field does not only depend upon the facts and figures, but also the intangible benefits of A.M.T.

The justification for investment in Advanced Manufacturing Technology usually falls into one of the following categories:-

a) Strategic
b) Financial
c) Intangible
d) "Sexy" Management Project

Strategic justification of projects where financial returns are below the cost of the Capital involved, will quickly lead to insolvency.

On the other hand, highly attractive projects from a financial perspective, applied in an 'ad hoc' manner, are unlikely to meet the overall strategic thrust of the Company in pursuing its business objectives.

The successful application of A.M.T. depends upon meeting the strategic requirements of the company, coupled with sound financial evaluation and management judgement of the intangible benefits which are projected. I need hardly say there is no place in a well-managed company for "Sexy" Management projects.

This paper endeavours to examine the process which enables a company to apply A.M.T. as a "Competitive Edge" to achieve their business objectives.

The major issues in the application of A.M.T., I believe, start at Board level and permeate right through the organisation to the Shop Floor.

Without a satisfactory "process" to effectively handle major investment, no amount of enthusiasm on its own will succeed. Based on this assumption, I propose to structure my talk today around the following elements:-

a) Decision Process
b) Financial Ramification
c) Practical Examples
d) Essential Elements for Success

In order to orientate people to the background of my company which has heavily influenced the content of my paper, I propose to commence by giving you a brief overview of Cummins and its products.

Cummins Engine Company Limited is a wholly owned U.K. Subsidiary of an international American company, manufacturing high speed diesel engines in the 50-1800 hp range.

As you will see from the slide, we employ approx. 4500 people in the U.K. occupying some 1,999,000 sq. ft., with an annual turnover of £250M.

Dealing now with our Engine Manufacturing Plants, these are as follows:-

Darlington Component Plant

This Plant occupies some 180K sq. ft. and manufactures Fuel Systems, Gears, Air Compressors and miscellaneous components for our diesel engine business.

Darlington Engine Plant

This Plant occupies some 270K sq. ft. and manufactures our new 'B' Series and Small Vee Engines. These engines have displacements of 4-8 litres and produce 50-250 hp.

Shotts Plant

538K sq. ft. Produces 10-14 litre engines ranging from 275-420 hp.

Daventry Plant

245K sq. ft. Produces 38-50 litre engines ranging from 900-1800 hp.

So much for background. Let us now examine what I consider to be some of the fundamental problems behind the lack of investment in A.M.T. and, consequently, the decline of many good British companies.

While fully acknowledging the importance of Marketing, R & E, Finance, etc., I make no apologies for the fact that this paper presents a "MANUFACTURING" perspective.

FUNDAMENTAL PROBLEMS INHIBITING THE APPLICATION OF A.M.T.

a) Lack of Positive Business Direction from the Top

   Without positive direction on long-term Company goals, how can we expect our Manufacturing Engineering staff, many levels down in the organisation, to respond in a positive and imaginative way to the challenge of an ever-increasing competitive environment.

b) Inability to Formulate Long-term Plans

   Without clear Business direction, Senior Managers can only speculate and, consequently, lack the confidence to commit plans to paper. Plans should be considered as a "road map", not only to give direction but also as a base line from which to deviate.

c) Lack of Detailed Planning

   Again, without direction from above and clear policies, it is unreasonable to expect coherent detailed plans from Middle and Lower Management.

d) Inability of Accountants and Engineers to Communicate

   Traditionally, in British Industry, Accountants and Engineers seldom cross the perceived barrier into the other's areas of expertise, hence, the Engineer with a good idea is unable to communicate it in terms that the Accountant understands and as he is only interested in sound financial projects, many good ideas are "stillborn".

e) Gulf between Management and Shop Floor

   Although admittedly the U.K. is making tremendous progress to close the gulf which existed a decade ago, communication between the two Groups leaves much to be desired in many companies.

f) Lack of an Effective Two-way Communication

   Without a strong two-way communication system, Management are unable to effectively communicate downwards and good ideas from the bottom seldom find their way through to the top. We should reflect on the fact that the ratio of Senior Managers to other workers is low and rightly so, but unless we nurture and

FUNDAMENTAL PROBLEMS INHIBITING THE APPLICATION OF A.M.T.

f) <u>Lack of an Effective Two-way Communication</u> (Cont)

encourage the volume of people lower in the organisation to effectively contribute, I would suggest there is no way that the few Senior Managers available can make-up for the loss of this latent potential at the lower levels of the organisation. Furthermore, I would venture to suggest that it is the younger people who are better equipped to wrestle with the complexities of modern computer technology and, hence, better able to generate the specific ideas necessary for the application of Advanced Manufacturing Technology.

g) <u>Manufacturing "Poor Relation"</u>

Unfortunately, it is a sad fact of life that few of our brighter people can be attracted into Manufacturing due to its poor image. This situation, I believe, is exacerbated by the fact that, due to ineffective structure and poor communications, bright young graduates fail to gain visibility and soon become disenchanted and leave to join the "gin and tonic brigade" in our Marketing divisions. The complexities of A.M.T. today demands highly qualified Engineers able to wrestle, not only with metal cutting technology, but also with the systems networking. As we drive towards "Computer Integrated Manufacturing", this need to attract highly qualified people into Manufacturing will become even more critical.

STRUCTURED PROCESS FOR INVESTMENT IN A.M.T.

Let us now examine a structured process for directing, discussing, encouraging, approving and implementing new investment in Manufacturing Technology.

Looking at each of these functions in turn:-

<u>Main Board</u>

- Type of Business
- Short and Long-term Objectives
- Company - Organisation - Legal & Financial Structure
- Product - Differentiation/Mix/Life Cycle
- Co-ordinate - Marketing/R & E/Manufacturing
- Financial Performance Expected
- Measurements of Success

<u>Manufacturing - Senior Management</u>

Clearly articulate and publish Manufacturing Strategy:-

- Volume Assumptions
- Cost/Quality/Delivery Objectives
- Flexibility
- People Organisation/Working Practices
- Payment Schemes
- Cost Targets
- Introduction of Technology
- Training Policy
- Hurdle Rate for new Investment (D.C.F., Payback, N.P.V., etc.)
- Preparation of Business Plan
- Preparation of 3-5 Year Investment Plan

STRUCTURED PROCESS FOR INVESTMENT IN A.M.T.

Manufacturing - Middle Management

In joint consultation with lower Management and shop floor business units, prepare detailed Capital Plan aimed at achieving the Business Plan Objectives through the strategy defined by Senior Management.

Having followed the process described above and re-cycled the plan, as necessary, to gain approval, we ultimately reach the point of having approved Business and Capital Plans. This information is then disseminated down through the organisation and becomes the operational "road map" for the new financial year or whatever period may be designated.

The process I have described involves all levels of the organisation in the approval process, hence, when the ultimate plan is approved, every level of the organisation should, hopefully, have their views, ideas and suggestions incorporated. This will lead to a clear feeling of ownership and, consequently, commitment for the implementation stage.

Dealing now with technology investment, our Engineers at the "sharp end" should now be aware where their main thrust must be directed, the type of projects that will be acceptable and the financial returns or cost objectives which they are expected to achieve.

Armed with this information, the bright young Engineer is now able to submit projects with a high confidence level of success at the approval stage. Success in this activity leads to increased job satisfaction and personal motivation. This, in turn, generates an ever-increasing number of projects aimed at addressing the Business Objectives.

FINANCIAL JUSTIFICATION

Much has been written about the difficulties encountered in using traditional investment analysis to justify the application of A.M.T. On the other hand, perhaps, we should reflect on the fact that if a company, even for good strategic reasons, consistently invests in projects where financial returns are below its costs of Capital, it will undoubtedly be on the road to insolvency.

I would venture to suggest that the challenge facing today's Managers is to improve their ability to estimate the true costs and benefits eminating from the application of A.M.T.

Unfortunately, the benefits or cost savings fall broadly into two categories:-

Tangible Savings

- Labour Savings
- Overhead Savings
- Increased Capacity
- Space Savings
- Reduced Scrap
- Reduced Warranty
- W.I.P. Inventory Reduction
- Unit Cost Savings

Intangible Savings

- Reduced Response Time
- Greater Flexibility
- Work Force Training/Experience

FINANCIAL JUSTIFICATION (CONT)

Dealing with the above Intangible Savings:-

Reduced Response Time Time

In numerous applications of Flexible Manufacturing Systems, we have seen response times reduced by a factor of 80-90%, representing reductions from 30 to 3 days. This obviously has a significant effect on Inventory holding costs, which can be quantified, but how do we quantify the increase in revenue which this improvement may generate.

Greater Flexibility

With the ever-increasing rate of change and radically reduced product life cycle, it is essential that the improved flexibility which A.M.T. can provide, becomes an essential ingredient in our investment strategy. To accurately quantify this benefit is impossible, but even Accountants would agree that the figure is certainly not zero.

Work Force Training

In looking around the world at wage rates, it quickly becomes obvious that "Third World" countries and emerging nations can produce and sell goods at prices which are lower than our U.K. manufacturing costs. This situation will undoubtedly be exacerbated by the rapid industrialisation of China and would seem to dictate that, to remain a dominant player, the U.K. must move up the technological ladder in order to offset our underlying higher Labour cost structure.

If one accepts the above argument as being applicable to their business, then there is a definite benefit in investment in A.M.T. as a first step towards C.I.M. Unfortunately, this benefit in training and developing the work force is impossible to define accurately in financial terms.

Intangible benefits, although difficult to quantify, must not be ignored. An article in the April 1986 "Harvard Business Review" by R.S. Caplan stated "Conservative Accountants who assign zero values to many intangible benefits prefer being precisely wrong to being vaguely right".

We have, I believe, two solutions to this dilemma:-

a) By using "creative Accounting" practices, endeavour to quantify the Intangible benefits.

b) Run the product D.C.F. R.O.I. omitting these benefits, but then calculate what actual value these must be to produce your Company's R.O.I. Hurdle Rate, thereby quantifying the "risk" factor.

In both cases, it is essential that a "sensitivity" analysis is undertaken as this will greatly assist Senior Management in evaluating the "risks" involved.

As companies make critical decisions about whether to invest in A.M.T., they must avoid claims that such investments have to be "made on faith" because financial analysis is too limiting.

Successful investment must yield adequate returns, thus the challenge for Managers and Engineers is to improve their ability to provide an order of magnitude cost benefit analysis for proposed A.M.T. projects.

I should now like to share an actual example of the application of A.M.T. and how it was justified and applied to provide the company with a "Competitive Edge."

FINANCIAL JUSTIFICATION (CONT)

This example applies to diesel engine connecting rods.

Flexible Con Rod Line

- **Background**

    Traditionally, Cummins have manufactured Connecting Rods on dedicated flow lines, however, we are now seeing a rapid proliferation in model mix, which makes dedicated lines with their high front end investment totally inappropriate from a financial standpoint. Furthermore, product life cycles are reducing from 12-15 years to 5-7 years.

    Forecasting of total demand always proves difficult while accurately forecasting model mix has proved impossible.

    The above business environment motivated us to produce the following broad design specification against which our Manufacturing Engineers have developed their proposal for a Flexible Con Rod Line.

- **Design Specification**

    a) Must have inherent flexibility for multi-model machining.

    b) Should ideally be capable of handling Rods from $6\frac{1}{2}$" to 18" centres.

    c) Facility should be capable of incremental investment related to demand, thus avoiding high front end investment and consequently high depreciation costs.

    d) Initial investment should be capable of producing 800 Rods/day of any mix of six models.

- e) The Line should be designed to run through meal breaks and shift changeovers. Changeover time between engine models should not exceed 15 mins.

    f) Overall Line efficiency should be calculated at 75%.

    g) Line computer control system should ultimately be capable of interfacing with our I.B.M. 3033 Mainframe.

After considerable work, our Engineers presented a proposal and compared this with a dedicated line of similar capacity. This comparison is shown below:-

	%	Dedicated	Flexible
Headcount Reduction	73	88 Men	24 Men
Floorspace Reduction	25	10000 sq. ft.	7507 sq. ft.
Inventory Reduction	95	3000 Rods	150 Rods
Unit Cost Reduction (6 Rod Types)	31	100%	69%

These improvements will yield annual savings of £1,847,000.

FINANCIAL JUSTIFICATION (CONT)

Flexible Con Rod Line

Cost Avoidance

Additional financial savings, recognised by Management as fully tangible and which were included in the ROI calculation as one-off items were:-

Assume a dedicated line would undergo one major re-tool during its useful life.	£ 750,000
Ability to add an additional rod variant (new dedicated line)	£2,200,000
Investment in major product design change of current line	£ 585,000

Total Savings Identified: £1,847,000 (Annual) + £3,550,000 (One-off)

Total Project Cost: £8,674,000

Project ROI: 32.8%

Intangible Savings

a) Response time reduced by 94%, i.e. 5 days to 0.3 days.

b) Provide a facility to produce low volume rods currently externally sourced.

c) Line could provide the facility for producing rods during model phase-out hence enabling dedicated line to be re-tooled.

d) Majority of equipment by value capable of being used to produce other components within meter cube envelope.

e) Drastically reduced leadtime in responding to major design change.

f) High utilisation of equipment.

Physical Design

Having achieved project approval, we then undertook the physical design of the Line, the result of which is shown in the following slide.

ELEMENTS FOR SUCCESSFUL APPLICATION OF A.M.T.

From our admittedly limited, but relatively successful experience in the application of A.M.T., I would highlight the following factors as essential elements for success.

Senior Management must be Actively Involved

They must formally and explicitly set the Company's Business Objectives and Goals. They must actively participate with all levels of the organisation and, thereafter, clearly articulate the Company's Manufacturing Strategy.

Ensure the development of a Capital Investment Plan, which the Company can afford and which supports their Business Objectives.

They must be constantly alert in identifying political, environmental and economic changes which may signal the need for a change in their current Business Strategy and Investment Policy.

ELEMENTS FOR SUCCESSFUL APPLICATION OF A.M.T. (CONT)

Senior Management must be Actively Involved

Senior Management must create an environment where the acceptance of change is a routine matter, as opposed to a threatening event, which is treated with much scepticism and suspicion.

A.M.T. Benefits must be Financially Quantified

Do not accept "Strategic Justification" alone. Always ensure that, wherever possible, the cost benefits are defined.

As regards intangible benefits, Accountants should be persuaded to apply some creative accounting practices. Should this approach be inappropriate, then the risks/benefits should be defined in order of magnitude and presented to Senior Management for them to apply a business judgement factor.

On no account, should you proceed with A.M.T. on a "blind leap of faith" basis.

Detailed Front End Planning

Once the project financial approval has been given by the Board, the real work begins. Only detailed front end planning can reduce, to a minimum, the time gap between the Capital expenditures and the flow of cash benefits.

The following are examples of issues which require considerable investment in front end planning resource for a typical F.M.S.:-

- Component Grouping
- Fixturing
- Programming
- Planned Maintenance
- T.Q.S.
- Training
- Working Practices
- Systems Data Collection
- CAD/CAM Interface
- System Networking

Attract and Retain Highly Qualified Engineers

Manufacturing companies must attract and retain highly qualified, articulate Engineers, capable of wrestling with the rapid advances in metal-cutting technology, computer systems and project financial appraisal.

Publication of Formal Business Plan

Whilst strongly stressing the need for formal published plans, I do not wish to convey the impression that these are "cast in concrete". On the contrary, as top Management signals business environment changes, the lower levels of Management should be encouraged to modify and change the plans accordingly. An important factor, however, is that they have a base line from which to deviate. Plans are merely a "road map" serving to communicate the appropriate direction. Once the "road map" is established, any deviation of route can easily be defined and charted.

CONCLUSION

To effectively compete with "Third World" countries and their very low Labour costs, we require to produce goods of a higher quality, lower cost and on shorter leadtimes.

This can be achieved in many instances if we strategically apply A.M.T. to consciously provide a "Competitive Edge".

In most companies, financial justification tends to be based on current practice and historical cost, assuming the market-place, selling price and cost remain constant. I would suggest that this is far from realistic.

I believe we require to pursue a much more imaginative approach to financial justification where the future business outlook, projected cost, future competition and inflation trends are considered more important than our historical data which was achieved in an environment that is unlikely to be repeated.

REFERENCES

Primrose, P.L. and Leonard, R. "Evaluating the 'Intangible' Benefits of Flexible Manufacturing Systems by use of Discounted Cash Flow Algorithms Within a Comprehensive Computer Program", Proc Instn Mech Engrs Vol 199 No B1, 1985

Kaplan, R.S., "Must CIM be justified by Faith Alone?", Harvard Business Review, March-April 1986.

Flexible manufacturing systems – some fact, some fiction
D. B. Ewaldz
Ingersoll Engineers, USA

ABSTRACT

It's generally accepted as fact that most Flexible Manufacturing Systems have failed economically, and often both technically and economically. Still others have been immensely successful, and the concept of a manufacturing resource capable of cost-effectively producing a wide range of products is too attractive and important to discard. Based on actual case studies, this paper will point out what went wrong in the failures and what went into the successes that made them successful. It will use the results of the analysis to develop realistic guidelines for conceptualizing and designing machining systems.

Perhaps the most exciting concept to be offered the metalworking industry since the industrial revolution is that of the flexible manufacturing system--the "FMS." The FMS is a welcome alternative to the extremes traditionally available: hard automation, in the form of transfer lines, or soft automation, in the form of general purpose equipment enhanced with loaders and numerical control. Figure 1 shows the projected demand for FMS in the US over the next five years. At first glance it might seem optimistic. However, after a bit of reconsideration, one's opinion might be different. Consider the manufacturing manager trying to plan a production resource five years out:

- Unsure what the demand for the resource will be, either in annual amounts or in lot sizes.

- Unsure what changes will be made in the product over time or, in many cases, even what the product will be.

- Doing this while trying at the same time to run a factory, which is said by many to be a full-time job by itself.

Considering those influences, it probably is even more curious that demand for "flexible manufacturing systems"--resources that can efficiently react to changes in demand, in product design, and in total product--isn't even higher than projected!

Sadly enough, if history is any indication, most FMSs aren't going to be successful:

- They won't be "flexible," at least not with the kind of flexibility needed. They won't be, because their designers and owners won't recognize the kind of process and product flexibility needed or perhaps even understand what an FMS is.

- The cost of the FMS will overrun. They all do, because few firms today have sufficient experience in designing and building them to anticipate the cost of doing so. Each one is different. There's no software package to take off the shelf and stick in place, and virtually no one can realistically estimate the cost of developing and debugging the complex custom software needed to control an FMS.

- Most systems won't be supported. When they finally DO work, there will be stacks of material at the output end (production capacity will exceed demand) and nothing at the input end (supplying equipment won't be able to keep up with the efficiency of the FMS). The systems won't be adequately supplied with tools and won't be adequately maintained. The idle FMS is only slightly less common than the dead robot, and that's probably mostly because there are a lot more robots.

There are two key issues that must be addressed concerning FMS:

- Getting the right kind of flexibility

- Getting the right level of "system"

Both are key in successful FMS design, if success is to be measured in financial terms.

GETTING THE RIGHT KIND OF FLEXIBILITY

There are different kinds of flexibility, and getting the right kind is critical. Which kind best suits the individual firm and product varies widely. For example, one firm might produce several products over the course of a year, and so need "flexibility" in its manufacturing resource: the ability to change over its production capability to accommodate the different products. However, if this changeover takes place only once or twice each year, conversion time is probably not awfully critical, and most likely production rates are.

On the other hand, another firm which produces a wide variety of goods in small lots would be better served by a system that was quickly convertibly from one product to another, regardless (almost) of the production rate achievable. Hence, as Figure 2 shows, either variety and lot size or both can create the need for flexibility, but what serves one well is a poor substitute for the other. The best solution is the combination of ease of changeover and production rate that best matches the needs of the specific product and environment.

Figure 3 looks at this quandary in terms of machine tool alternatives. In the case of a firm with high production demands and infrequent changeover, "hard" automation is a good answer, delivering high production rates with more difficult changeover. The firm can accept the changeover difficulty because it is infrequent. On the other hand, where part variety is high, conventional machine tools, perhaps with numerical control, provide better investment returns.

Figure 4 shows why this is true. The curves represent fully amortizing equipment ownership cost against production output; if one part is produced, the total money cost of ownership (i.e., the interest paid on the debt or lost from not investing in an interest paying account) is absorbed against that part. If two parts are produced, each bears half the cost, and so on. The amortization is continued up to the full production capacity of the machine tool. (Please note that depreciation is NOT included. Depreciation, at least as we practice it in the U.S., is a cunning way of offsetting taxes and not a cost at all).

"Hard" automation is usually more costly than "soft" automation, so the curve is shifted to the right. "Soft" automation is usually less productive, so the curve ends earlier. If a firm has production demands at volume "A" and buys soft automation--perhaps an FMS--part cost is at "X," and the firm has made a wise decision. Had they selected hard automation--perhaps a transfer line or a rotary table machine tool--real part cost would have been about 30 percent higher, at "Y." On the other hand, if volume really WAS at "B," the soft automation alternative would not have had sufficient capacity to produce that level of output, and the firm would have had to forfeit the business or invest in another machine, resulting in a much higher part cost.

Opting for the wrong kind of flexibility for one's FMS is not just a whimsical exercise decided by hardware preference. It has real and often dramatic cost implications that determine the success or failure of the investment!

We can suggest three principles in matching flexibility with needs:

- Understand one's products coldly and objectively before making hardware and production system decisions. It would be a miracle if someone else's successful FMS would be anything short of dismal for another firm.

- Understand the process needs, considering the product, product volumes, and lot sizes. It would also be a miracle if the production methods in place at all matched present or future needs. Too many firms are comfortable with what have been traditional methods though they may be drastically uneconomic given the characteristics of the current product/volume/lot size profile.

- Assure the problem that flexibility is going to cure is worth curing. Nuisances are nuisances, but solving them is seldom an attractive investment.

GETTING THE RIGHT LEVEL OF SYSTEM

Virtually everyone--at least all managers and executives--has a preconceived notion of what an FMS is. If any manufacturing manager at random was asked to describe one, he probably would compile a list of specifications, starting with "machining centers," going on to "automated material handling," "automated tool control," and so on. And, of course, he might be right or wrong, because what an FMS really is depends entirely on the needs and idiosyncrasies and unique properties of each individual firm. Fortunately, there are some reasonably standard components that one could expect in an FMS. How much, how little, what form, or what kind of each best makes up the system is thoroughly dependent on the needs of the individual firm. Let's go through these components and see why.

Figure 5 compares the cost, on the horizontal axis, with the value, on the vertical axis, of the alternatives in distributing control information to the elements of the system. If the equipment is numerically controlled, as it probably will be, at the lowest level of automated distribution, a person carries NC tapes to the equipment, loads them into the reader, stores them when complete, retrieves them when they are needed again, and protects them against damage in the meantime. All these things cost, and eliminating them has a value. It's directly related to the number of tapes needed. Handling two or three per year creates minuscule cost. Working with two or three thousand per year is extremely expensive. Tape costs money, too. Some manufacturers dealing with high part variety report costs as high as $10,000 per year just to supply mylar tape. Lost time due to manual data input is costly, too; journeyman operators are hard pressed not to adjust, tweak, or otherwise "optimize" machining parameters.

Those are the sources of value. The size of each varies according to the specific situation. The cost of automatic distribution varies, too, according to the situation. The kind of distribution system is of major importance, and the kind is influenced by the situation. When NC programs are short and the number of controllers in the system is relatively few, control data distribution can accomplished by hardwiring each controller through an RS232 port to the mainframe computer where the data is stored and dumping it into a buffer, perhaps a bubble memory, for example, in the controller for use. If the program is long—greater than .5 Mbyte—then a true local area network will be needed. The cost can vary widely, from a few hundred dollars per controller to hundreds of thousands for a complex narrow band network. Getting the right one means matching economy with technical needs; buying an LAN when hardwire would have done means running up the system cost without gaining any benefit. Trying to do it with hardwire when an LAN is needed means the system, in essence, won't work.

Automated material and parts handling is often an important contributor to the success of an FMS. It's often the major contributor to the cost of an FMS, also, and worth careful analysis before spending a lot of money. Figure 6 compares the cost and value of a range of automated material handling systems.

The value stems from several sources:

- The manufacturing resource will be better utilized, because the material can be prepared for work in some remote location, ready to be worked on when the resource is available. Doubled resource utilization has been documented; that means the value of automated material handling can be as much as the cost of the resource being served.

- People who are paid to handle material are costly, too, and extremely difficult to monitor. Consequently, the real utility of material handlers is usually quite low, and their cost high, as is the cost of their equipment, such as forklift trucks and other devices.

- The people who are paid to handle material often lose it, place it in strange places, or run over it with their very expensive forklift trucks. Automated handling systems can capture material in the system and be aware of where it is and its actual state at all times.

These benefits often translate into large dollar amounts, and material handling automation is often a valuable part of a production system. On the other hand, it is not unusually the most costly part of the system. Wire-guided vehicles can be an economical answer where part sizes are not excessive and where handling frequency is not high. They also offer considerable flexibility in structuring and restructuring the system. However, where parts must be moved quickly and often, other alternatives offer more satisfactory—and often more costly—solutions.

The answer is not difficult; it is simply recognizing the characteristics of the particular situation and selecting the best solution from the host of alternatives.

Automated tool management offers similar kinds of values and costs. Studies have shown that machine tool operators spend about 2 percent of their work time placing cutting tools in the machine tool changer. They spend as much as <u>20 percent</u> of their time obtaining tools from remote locations. Clearly, the most important value of automated cutting tool management stems from getting the tool from the presetting department or the tool crib. The value of getting it into the tool changer is not so great, and the cost of doing so is often very high. Figure 7 shows the value/cost comparison under various conditions.

In many cases some form of automatic process control or automatic product testing offers important benefits where the manufacturing process is not completely reliable or where product quality requirements are higher than the reliability of the manufacturing process. The capability to automatically adjust a cutting tool or a formulating process as the work is being done to precisely meet end product specifications has some important implications in many cases. Values stem from reduced scrap and defect repair costs as well as from a reduced need for human intervention at the work station.

However, as Figure 8 shows, the current cost of such control systems is often very high, compared to their benefits in many situations. The economics of adaptive process control are often less than exciting. As further gains are made in sensor and control system technology, no doubt costs will come down and reliability will improve, and some form of automated process control will become a more practical element of a production system. Where the manufacturing process is fully adequate to meet the precision required, automatic process control offers limited value, however technically appealing it might be.

So the companies that make the FMS into a system can vary widely, and what satisfies one need will be a long way off-target for another situation. However, let's say we've done the analytical work very well and have identified and sized each component which will make up our FMS.

Let's assemble all the components we've covered into a flexible manufacturing system so that it just precisely meets the needs of our firm. As disquieting as it may be, there will very likely still be a need for manpower in the system. As Figure 9 shows, although some 80 percent or 90 percent of the work normally performed by people has been eliminated, studies have identified over 100 tasks that still will need human attention. These range from quite critical and demanding work, such as setting up the work pieces in their fixtures, to such mundane chores as directing the coolant nozzle so that the stream of coolant precisely floods the point where the cutting is being done. The system will be capable of LIMITED unmanned operation--a lights-out partial shift, for example--but won't be capable of operating completely free of human intervention.

TRANSLATING THEORY TO FACT

Each successful FMS is a unique resource, designed to fit the needs of one particular firm and set of circumstances and none other. On the other hand, we can easily generalize the steps needed to develop the unique FMS:

    o   One can't judge the "need" for an FMS from the feature stories in technical journals. The first step in developing a conceptual FMS is to develop a full understanding of one's products and processes:

        -- The process needs, including understanding the technology and arrangement alternatives available and possible.
        -- The flexibility needs, as we described earlier, considering whether the firm needs high production rates and limited changeover capability or relatively low production rates and high changeover capability.

- The system needs, also as we described, considering the value of each possible element of the system and the alternative forms it might take.
- The real cost issues, as distinguished from nuisances. There'll be lots of time and money to get to the nuisances when the real issues have been addressed!)

We wouldn't suggest taking the initial answers at face value; perhaps the best solution to flexibility is to change marketing strategy or the method of distribution or to prune "cats and dogs" product models, rather than adapting the resource to mistaken market strategies!

A conceptual FMS should take shape from this step, if in fact a "flexible manufacturing resource" is really the best solution to cost and responsiveness problems. If it isn't, THAT should fall out of the data, also.

Let's assume we find that an FMS will give us real benefits at a justifiable cost and that we can clearly define the concept and reliably size the benefits.

o The next step will be to sell the idea to top management. The technical journals have moderated their deluge of FMS articles, but the residual effect may help with the selling. The best bet, however, is to have a clearly defined, well conceived, well documented, realistic plan and to believe in it. It's important that the plan be broken into distinct steps, with the economics of each also clearly defined. The plan must be convincing; we don't know of a low cost FMS, and top management must see and be confident of the benefits. They will have to sign off on a considerable investment, and reports of failed FMS are becoming common.

o When approval has been gained, install the FMS one step at a time as defined in the plan. There are a number of good reasons for such phased implementation:

- It's easier to iron out wrinkles in one element at a time than to try to sort out the whole thing at once. Most FMS are quite complex, both mechanically and electronically, and are unlikely to work perfectly at startup. By the way, one should be certain to install the low risk items first, to build credibility in the system and its designers.
- It gives the installers a chance to learn something about installing before they tackle a major portion of the work or the whole very complex system. It also lets one go back and alter the plan as better ideas or errors surface.
- It lets the investment be made in a series of rather more palatable steps, instead of requiring a large commitment at one time. One can back off if the market changes, accelerate if need be, or make alterations in the FMS itself as the situation becomes clearer.

- o Monitor performance; there are often cultural barriers to good FMS performance, such as manning the system as if it was a conventional production resource or not supporting it adequately in maintenance or product scheduling. The values which could be gained from an FMS are often lost because the designers turned away from it too quickly.

In conclusion and summary, the potential benefits a flexible manufacturing resource can offer are too great to ignore. They include:

- o The ability to quickly adjust to changes in product design.

- o The ability to respond efficiently to customer orders.

- o The ability to accommodate (over a reasonable range) changes in lot sizes and production quantities.

If these are important to the firm, the right form of FMS could prove an invaluable resource. Unfortunately, the wrong one could prove a costly mistake; it's worth taking the time to be certain what appears to be a technological dream is not a financial nightmare!

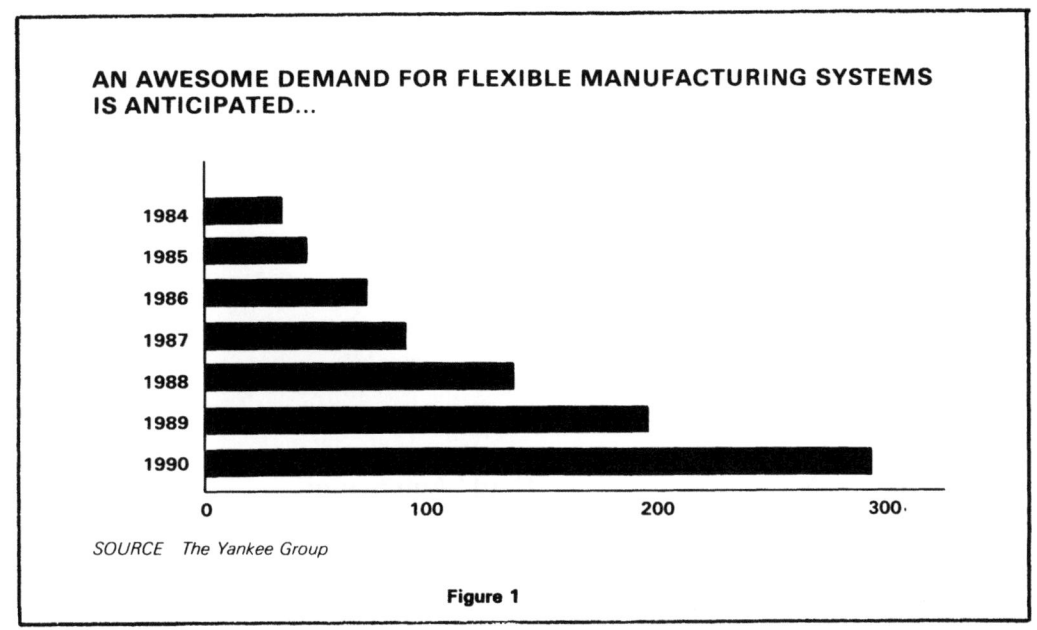

Figure 1

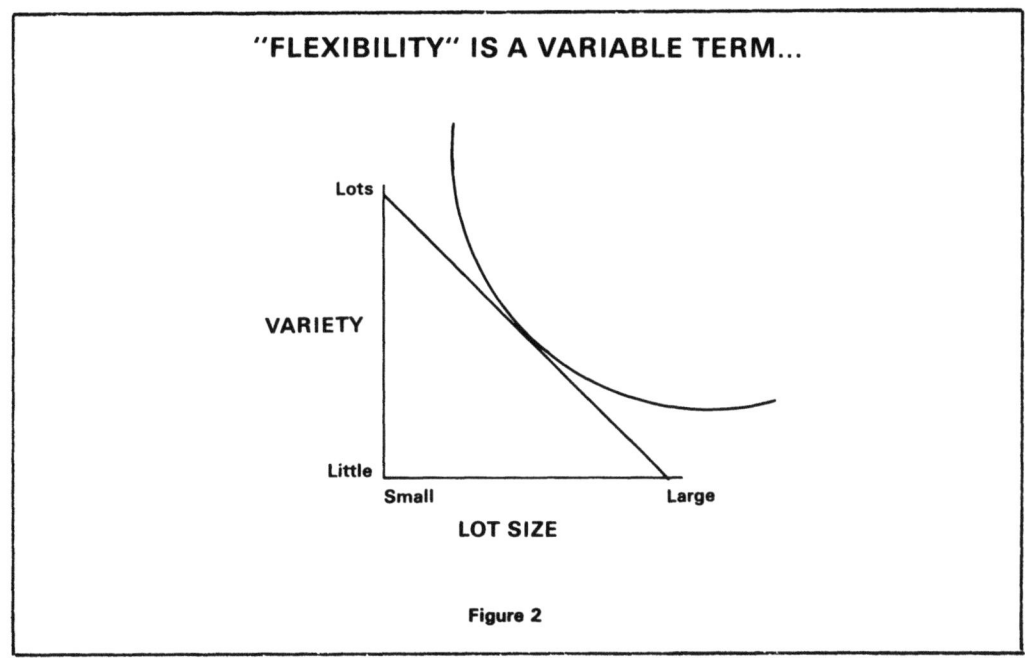

Figure 2

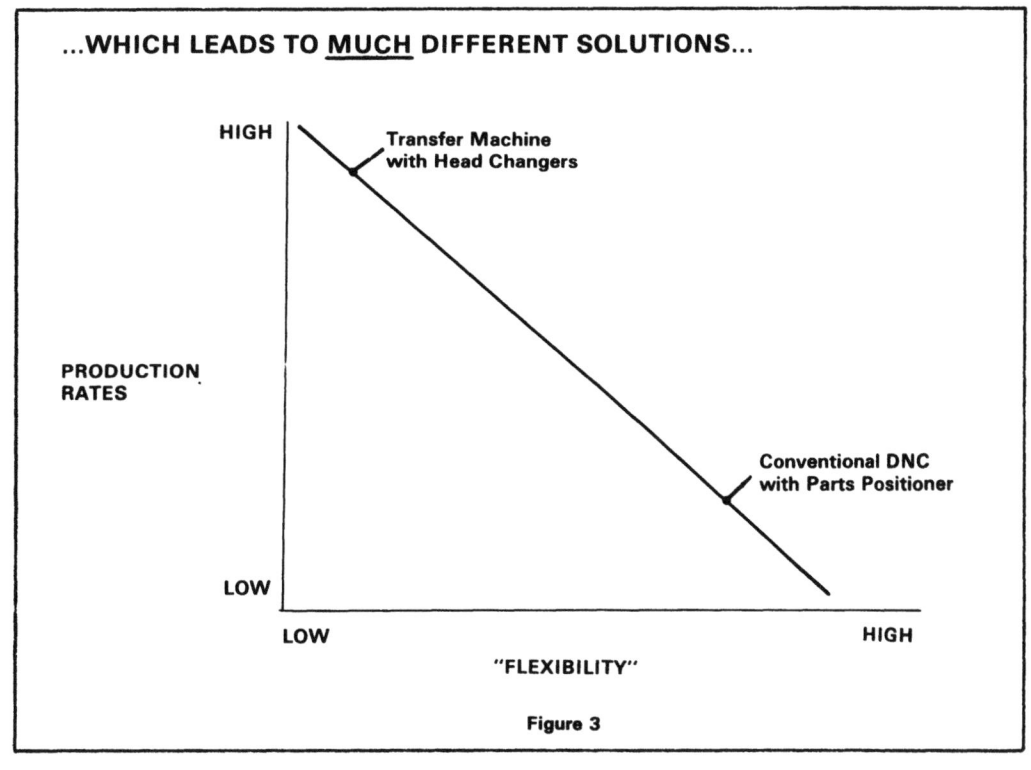

Figure 3

...AND DRAMATICALLY DIFFERENT PRODUCT COSTS

Figure 4

THE RIGHT LEVEL OF SYSTEM...

CONSIDER THE VALUE AND COST OF DIRECT NUMERICAL CONTROL...

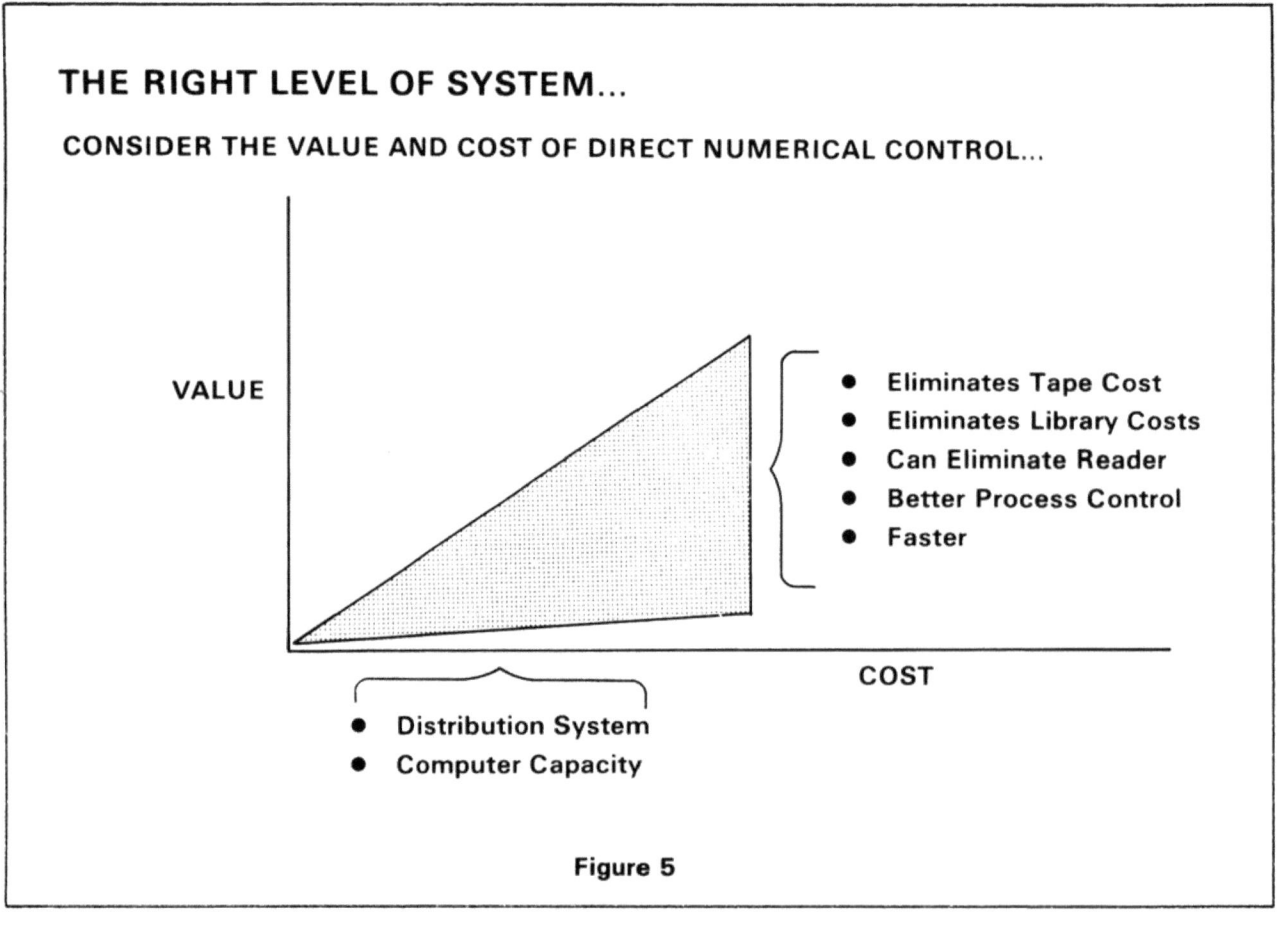

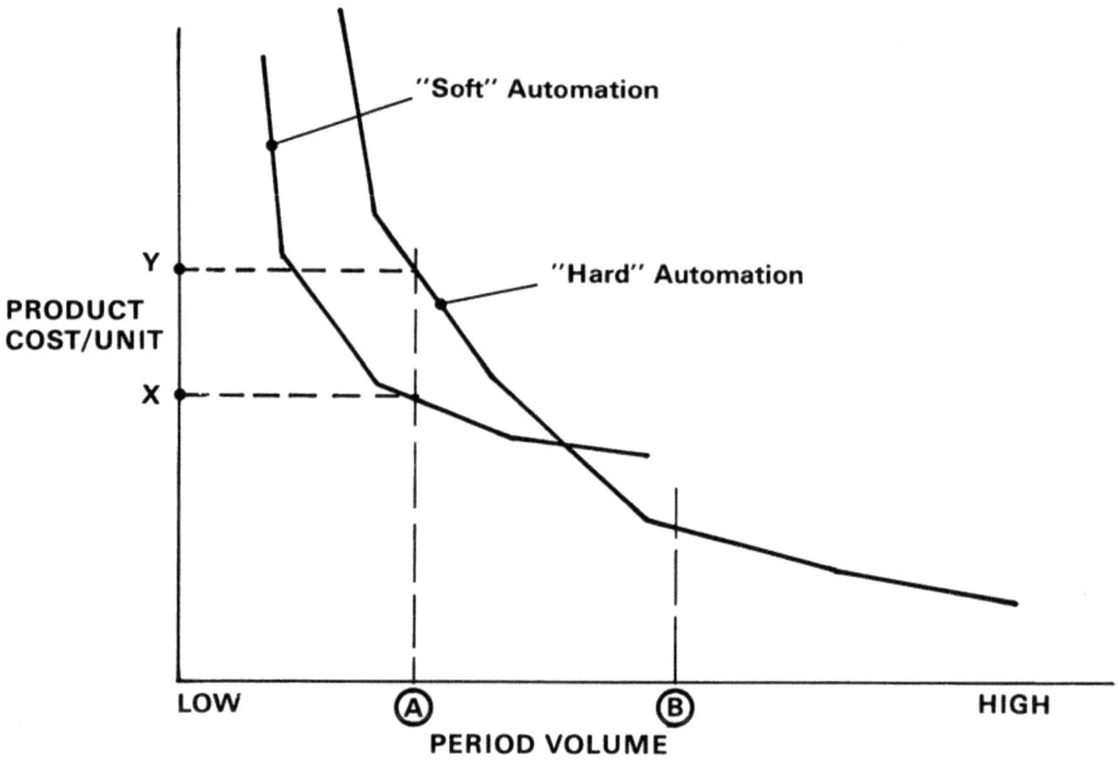

Figure 5

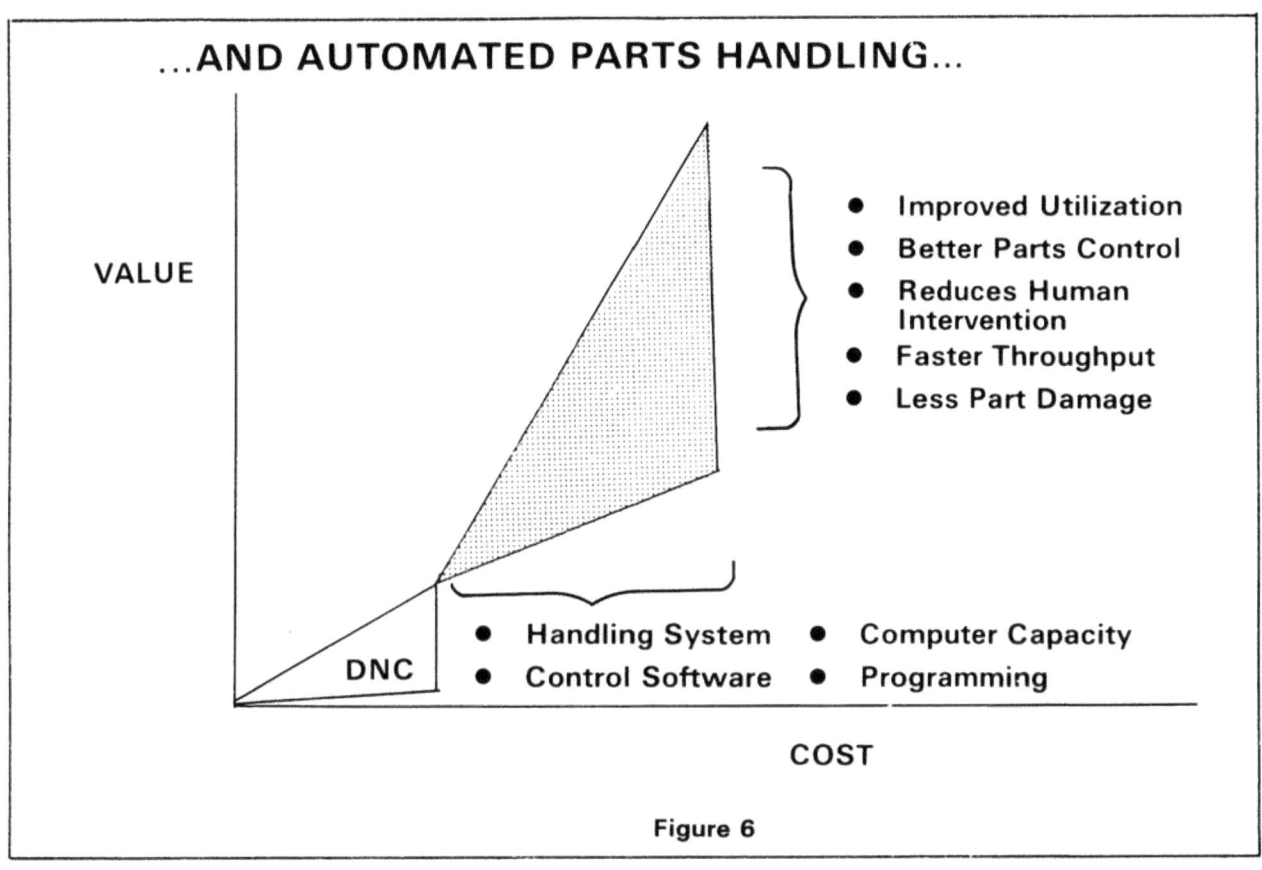

Figure 6

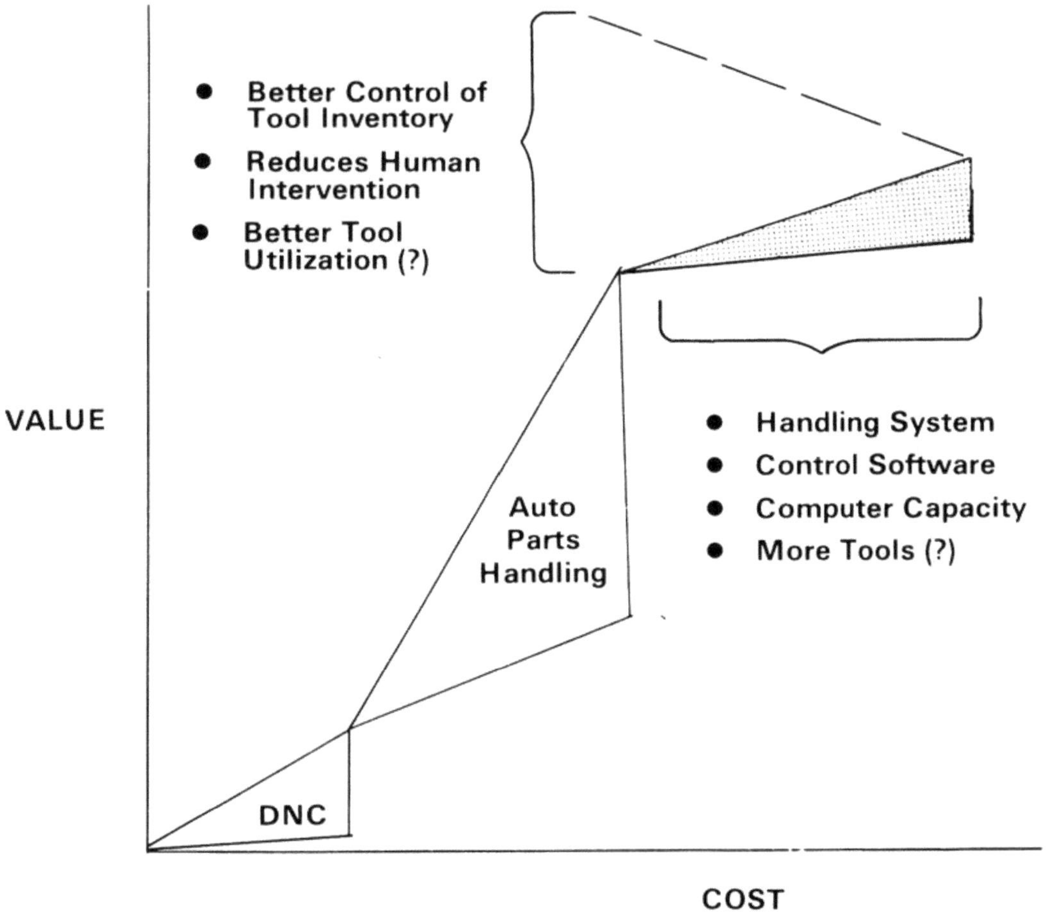

Figure 7

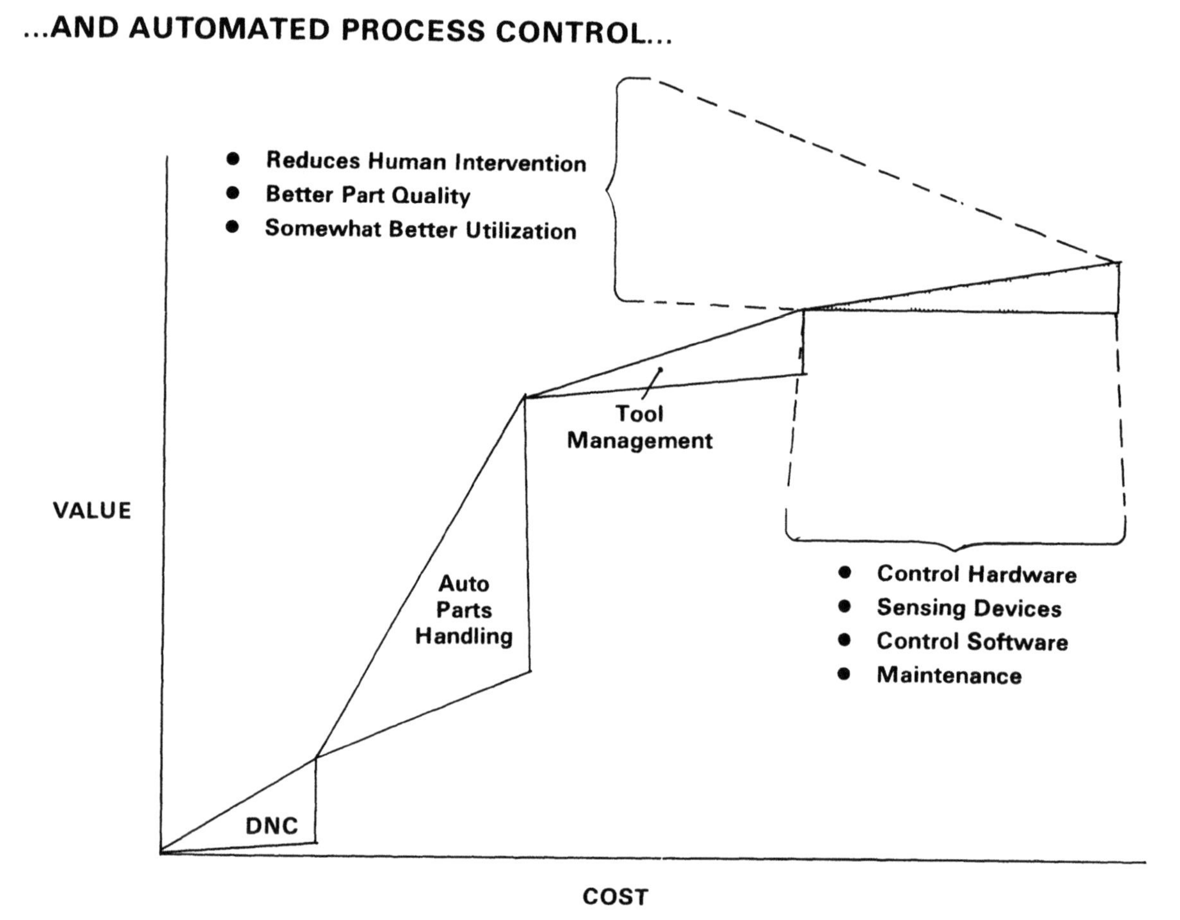

...AND AUTOMATED PROCESS CONTROL...

- Reduces Human Intervention
- Better Part Quality
- Somewhat Better Utilization

VALUE

Tool Management

Auto Parts Handling

DNC

- Control Hardware
- Sensing Devices
- Control Software
- Maintenance

COST

Figure 8

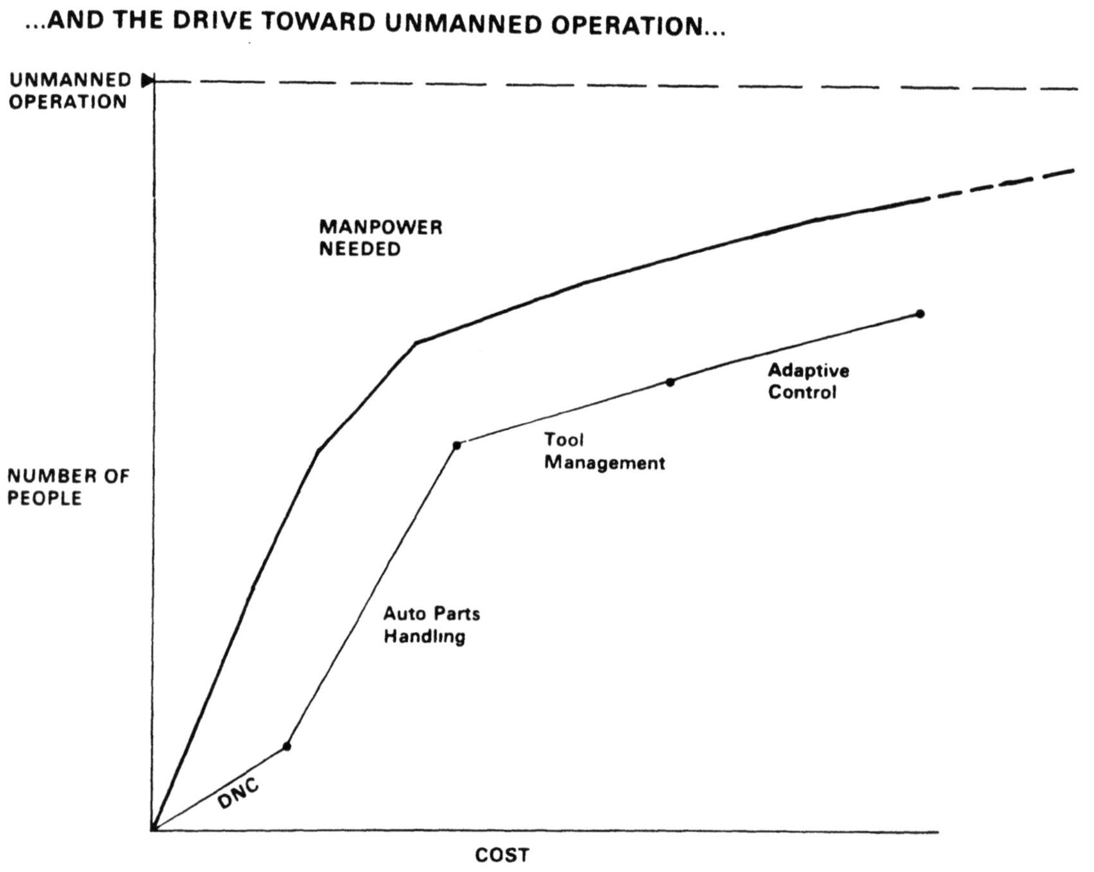

...AND THE DRIVE TOWARD UNMANNED OPERATION...

UNMANNED OPERATION

MANPOWER NEEDED

Adaptive Control

Tool Management

NUMBER OF PEOPLE

Auto Parts Handling

DNC

COST

Figure 9

The integration of FMS within existing factory systems
D. Little
University of Liverpool, UK

Whilst the number of cells and systems designed for flexible manufacture have increased dramatically in the last five years, the majority installed within the U.K. are essentially producers of components which have to be assembled to other items produced by conventional means within the factory. The implementation of FMS is unlikely to improve overall company performance to the level expected until a strategy is adopted for their integration within the existing factory systems.

BACKGROUND

The author is researching the development and use of methodologies for the design and outline specification of flexible manufacturing cell management systems as a member of a research group within the Faculty of Engineering, University of Liverpool.

Collaboration in this research is primarily provided by Texas Instruments, Case Tractors and Unilever Research Ltd.

THE GROWTH OF FLEXIBLE SYSTEMS

The importance of improving manufacturing performance and product quality within engineering manufacture is clear when one considers that this sector employs over 10 per cent of the employed labour force and contributes some 40% of Britain's visible experts. However, a recent analysis by NEDO [1] indicated that profitability remains inadequate to permit the long-term investment and expansion that is necessary to maintain competitivity.

The potential of flexible manufacture and its associated technologies to help bring about these improvements is highlighted by the report [2] of the Advisory Council for Applied Research and Development (A.C.A.R.D.).

> "New and advanced manufacturing technologies now being developed offer even greater scope for improved productivity and product quality. Those firms which do not make use of these technologies risk being overtaken by competitors achieving superior quality at lower cost".

A specific reference to FMS and some quantification of the first-year percentage of total potential benefits attributable to the implementation of FMS within medium-sized batch engineering companies is quoted by the N.E.D.O. Advanced Manufacturing Systems Group [3]. Following some forty case-studies into batch manufacture in the engineering sector. This first-year percentage is given as 20%.

An increased awareness of such benefits has lead to a rapid and sustained growth in the installation of flexible systems within the U.K. whilst this lags behind Japan and the United Stated, Frost and Sullivan [4] show some 33 systems installed within the U.K. up to the end of 1984 with a forecasted rise to 45-50 systems by 1990. This placed the U.K. second only to West Germany within Europe, see Figure 1, and indicates a clear reawakening of interest following the demise of Molin's SYSTEM 24 in the early seventies - arguably the world's first FMS.

Undoubtedly this vigorous interest in FMS may, in part, be attributable to the recognition of the main benefits of such systems, but the role of government grants must not be overlooked.

With an average capital spend of 2-4 million dollars [5] per system, grants have played an important part, I have knowledge of the critical nature of this support in the case of a large installation where the project could not be justified under existing company accounting procedures without the benefit of grant aid.

THE ACTUAL BENEFITS OF FMS?

The case made out for the implementation of FMS is a very strong one. Claimed benefits [6] for Advanced Manufacturing Technology as a whole, and these relate directly to flexible manufacture, include reduced lead-times, improved product quality and lower inventory levels - see Figure 2. Taken individually the benefits are significant and attainable. The claimed one million pound reduction in finished castings stock at Anderson-Strathclyde is a case in point.

THE NEED FOR A SYSTEMATIC APPROACH TO FMS DESIGN AND IMPLEMENTATION

Whilst there appears to be broad agreement in the literature on the need for an overall strategy for FMS implementation, remarkably little advice is forthcoming on how to approach this.

Flexible manufacture is commonly viewed as a logical development of numerical control via CNC and DNC with the addition of local workpiece scheduling and transportation. Such a view, whilst valid in hardware terms, misses the point and may lose many of the advantages of such systems.

A typical FMS, designed around part-processing and engineering requirements will improve local productivity by the provision of a comprehensive new stand-alone facility (at considerable increase in fixed capital) but may not optimise overall company performance due to the lack of systems integration.

Too often the approach is "bottom-up" when what is required is a total systems approach which must start from the top. Very few companies so far visited employ any formal design methodologies, such as S.A.D.T.

The size of the task in integrating an FMS should not be underestimated. Considerable effort is generally required to integrate and optimise the system components themselves without the additional problem of linking to existing factory systems.

It is understandable that this latter process is often left until time is available. That is likely to be too late. The system software will be largely developed and systems designers are unlikely to be willing to make the major amendments necessary at a later date. "Another island of automation" has been created.

THE NEED FOR INTERGRATION WITHIN EXISTING FACTORY SYSTEMS

The majority of U.K. FMS may be viewed as component producers. They do not produce end items but an output which must be assembled with components produced by conventional methods in the remainder of the plant.

The need for synchronisation of these sub-units becomes apparent if we consider the characteristics of an FMS compared with those of the more conventional plant. See Figure 3.

These characteristics produce the benefits indicated earlier and given a suitable supply of work the FMS will undoubtedly be able to reproduce them.

However, it is unusual for an FMS installation to be designed to be completely self-contained. On the contrary, the majority are serviced by the existing factory systems which may include:

- Requirements planning for materials
- Job scheduling
- Tool management
- Materials handling to and from the system
- Maintenance

The FMS is likely to place considerable demands upon these support activities, some of which will have serious operational implications because of the different nature of the requirements from those associated with conventional production. The result may be to reduce the level of support available to other areas or act as a brake on the FMS.

What is of concern is the lack of hard information coming forward to substantiate the overall benefits and the real cost of realising these. Indeed there is a rising tide of informed criticism which queries the ability of many installations to survive a rigorous cost-benefit analysis.

The technology may be new but there are sufficient established sites to provide the relevant data. It is unlikely that this lack of information is due solely to natural commercial reticence on the part of successful sites.

One is aware of projects which run well behind plan and others which meet unforeseen problems which have to be resolved on an ad hoc basis - tool contention is one case covered later. In another example of roboticised flexible cell, the available information suggests that the project would not be justified in financial terms although management saw it as a useful means of moving the company up the technolgoy learning curve.

The benefits, whilst demonstrably achievable individually, may be more elusive than is claimed. If the assumptions underlying those claimed by the Advanced Manufacturing Systems Working Group are closely examined, one discovers that such benefits may arise if:

- "... modern technology is implemented at the optimum rate"

- "... all the necessary people and financial resources are available at the right time"

- "... the business climate, management policy and employee relations remain conducive to continued investment".

One must question whether such assumptions will be valid over a prolonged period for a typical medium-sized engineering manufacture company with no previous experience of Advanced Manufacturing Technology.

Mr. R. Edwards of Ford of Europe added weight to this point [7] by stating that for Ford (not a company as described above) the shortage of skilled people was a major problem for the implementation of AMT. He suggested that the issue was "more of a management challenge than a technical one".

I would support the view of Herr Hotz [8], that most users grossly underestimate the time needed to plan, commission and debug an FMS. In consequence the initial priority is to make the system work. Most managerial and technical effort is put into this at the implementation phase and any integration issues become secondary. The money has been spent and the die is cast - often with major issues, for the plant as a whole, being treated as short term contingencies.

What is needed is firstly to move away from a simplistic justification of systems using traditional accounting methods as advocated by Primrose & Leonard (8), and secondly to develop an overall plan for the business which includes the technolgoy, its costs and implications. Effectively designed and implemented the technology can (and should) change the way a business functions. This should be the prime objective of its use.

To quote Edwards once more: "manufacturing is a strategic weapon" that "... brings cultural changes to the organisation". The technology can and should effect strategic business issues not directly related to shop-floor activities. Such benefits are difficult to quantify but may be enormous. An increasingly flexible response to customer requirements is an example.

Such considerations are too frequently given low priority at the planning stage or overlooked altogether. The results of this can be serious as one of the case studies shows, with the ability to achieve the real benefit of reduced work-in-progress in the FMS at the expense of a build up of completed components awaiting assembly due to lack of synchronisation and integration for the factory as a whole.

INTEGRATION AS A DESIGN ISSUE

While Allen-Bradley have established a five level hierarchy of computer control in building their computer-integrated manufacturing facility in the United States, for most batch engineering companies with comparatively low unit volumes, integration is a complex issue with much work to be done.

The Allen-Bradley system is a fascinating example of the "state of the art" in CIM. However, because of the self-contained nature of the unit which manufactures a specific product group in volume, it is not analogous to the batch systems discussed here. It is closer to a mixed model assembly line, as used by many car manufacturers. The problem raised by this paper is one of achieving the $4\frac{1}{2}$th level rather than the 5th.

D. S. Hoag of Allen-Bradley describe [9] this area, the integration of cell control and factory support systems as "the missing link". It is an area which must be viewed as specific to each facility and product group. Approaches may be general but solutions cannot be so.

A tool which provides a systematic approach to the design of flexible cell/system management systems needs to be developed. This would enable the major issues to be identified at the design stage of a flexible facility. A project engineer would use the tool to define the data elements and interfaces required both to control and integrate the system.

The need to integrate FMS installations within existing factory systems is a pressing requirement if the benefits available from such capitally intensive projects are to be fully realised.

CASE A

A medium-sized engineering company who produce precision durables in volume on a make to order basis for a wide range of customers. Orders are received and entered into an on-line master scheduling system. This drives an MRP system which utilises weekly time buckets.

The company is investing £4 million in an FMS which will handle one of five of the main manufactured parts which are assembled into the final produce - see Figure 4.

The lead-time for the conventionally produced parts is 3 weeks and the existing manufacturing control system copes well. The lead time for the FMS produced parts will be 2 or 3 days.

The problem

There is a need to synchronise FMS output with the assembly schedule. The considerably different lead-times between conventional and FMS production make this difficult. The existing method of loading the factory will not cope with the short-term demands of the FMS without batching-up on a weekly basis, which will create excessive inventory.

The company approached the FMS design from the "bottom-up', spending the majority of design time in identifying process requirements and selecting machining facilities to meet these. Discussion with production control personnel was minimal and no "top-down" analysis tools were used.

Serious efforts will have to be made to modify the manufacturing control systems to prevent a build up of FMS component inventory to utilise the flexibility available.

CASE B

A medium-large engineering company who manufacture a range of large capital items on a make to order basis. Again MRP is used but with monthly time-buckets - see Figure 5.

An FMS is employed to machine large castings which form integral parts of the products. The machine tools within the system have the capacity to handle a wide range of tools locally.

The problems

Again there is the need to synchronise conventional and FMS production and, in addition, one of tool management. Although the system was designed from the "top-down", tool management requirements were not fully recognised. The machining demands of the large castings using a conventional production engineering approach necessitate numbers of tools in excess of tool-changer capacity.

Simulation studies are now underway to find a solution to a problem which is causing down-time.

Either a means of exchanging tools in the tool-changers must be found or process plans rewritten to use fewer tools.

REFERENCES

[1] N.E.D.O. "British Industrial Performance". N.E.D.O., 1983.
[2] A.C.A.R.D. "New opportunities in manufacturing : the management of technology". A.C.A.R.D., October 1983.
[3] Advanced Manufacturing Systems Group. "Advanced Manufacturing Technology. The impact of new technology on engineering batch production". N.E.D.O., 1985.
[4] Frost and Sullivan, "Flexible Manufacturing machining systems markets in Europe." Frost and Sullivan, 104-1122 Marylebone Lane, London, 1984.
[5] Op. Cit.
[6] Op. Cit.
[7] R. Edwards, "Computer Integrated Manufacture". Lecture to Faculty of Engineering, University of Liverpool, July 8th, 1986.
[8] R. Leonard and P. L. Primrose, "Evaluating the intangible benefits of Flexible Manufacturing Systems by use of discounted cash flow algorithms within a comprehensive computer program", Proceedings of Institution of Mechanical Engineers, Vol. 199, 1985.
[9] D. S. Hoag, "The link between the plant control system and management information systems". Undated internal paper.

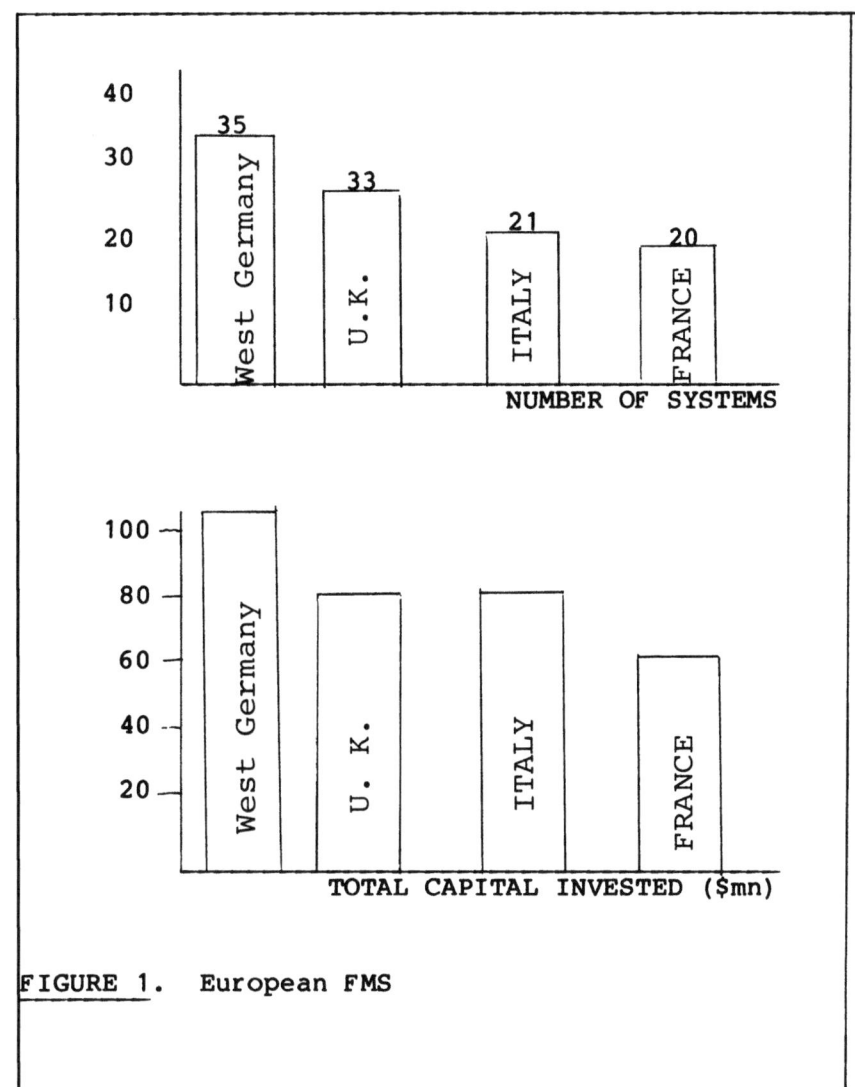

FIGURE 1. European FMS

- improved utilisation of capital equipment
- reduced levels of stock and W.I.P.
- shorter lead times
- better product quality
- reduced unit costs
- more flexible response to customers

FIGURE 2. Benefits of FMS

CHARACTERISTIC	FMS	CONVENTIONAL PRODUCTION
Lead-time	Short - hours	Long - weeks
Set-up time	Short - minutes	Long - hours
Batch Size	Small - customer orientated	Large - stock orientated
W.I.P.	Low	High
Throughput time	Short - can accommodate rush jobs	Long - rush jobs cause delays
Scrap	Low - zero defect policy	Too high

FIGURE 3. Comparison of Characteristics of FMS and Conventional System

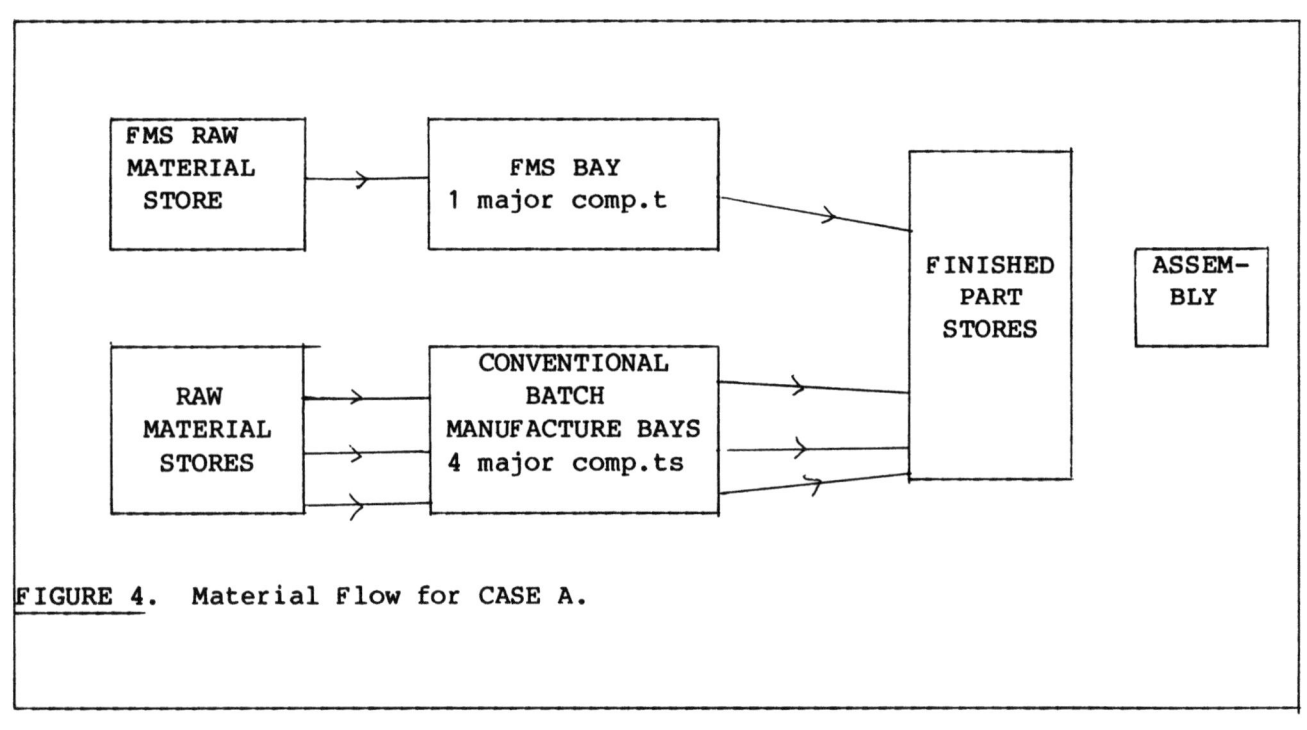

FIGURE 4. Material Flow for CASE A.

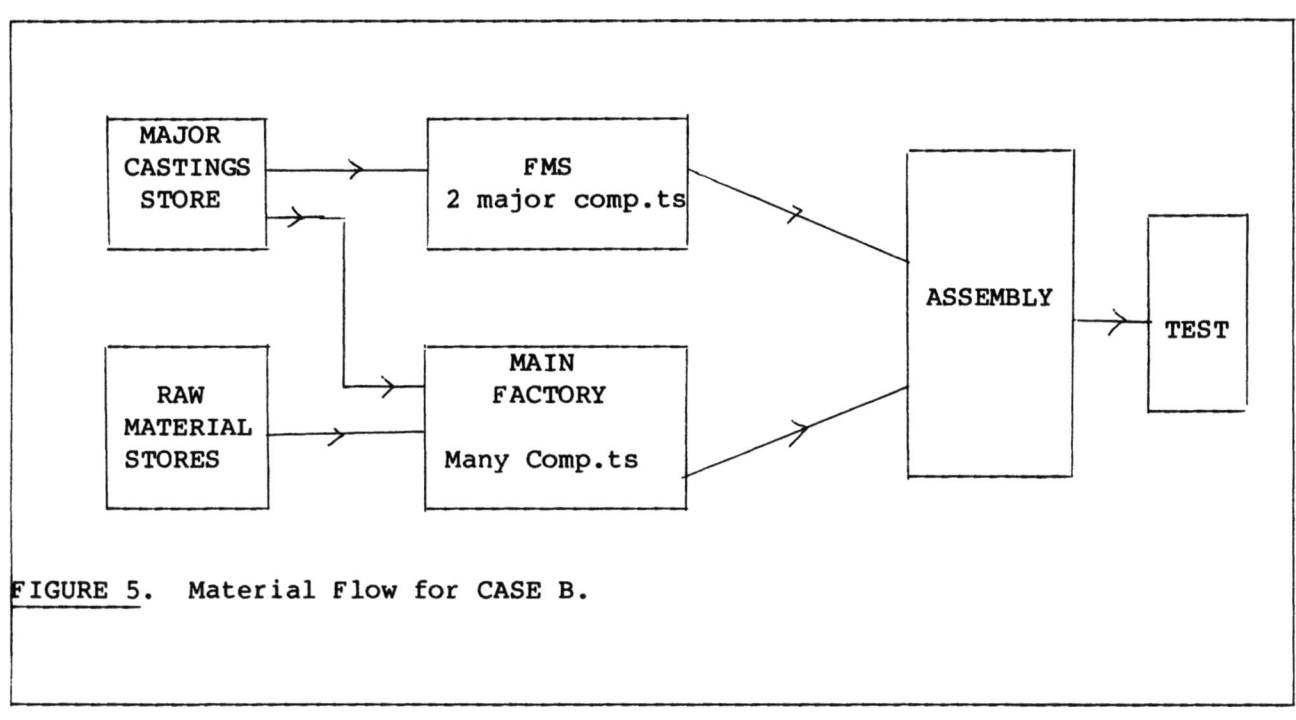

FIGURE 5. Material Flow for CASE B.

Planning for precision in flexible manufacturing systems

R. G. Hannam
University of Manchester Institute of Science and Technology (UMIST), UK

ABSTRACT

One of the elements that has to be considered in planning and using a flexible manufacturing system is how to obtain the manufacturing accuracies required by the parts. For many parts, this is not a critical element as the precision demanded is not high. For those parts with more demanding tolerances, however, how the required precision will be achieved must be considered in the planning of the system if the desired results are to be achieved. This paper reviews the factors which can affect the manufacturing precision of a flexible manufacturing system both at the machine level and at the system level. The factors discussed relate to the design of the machines in the system and to the design and operation of the system.

INTRODUCTION

Flexible Manufacturing systems (FMS) present users with many challenges, and although operating experiences from the early users of systems are now more readily available, the challenges still remain with every new system. Even experienced users accept that flexible manufacturing cannot yet be considered a mature technology in which most of the problems have been solved [1]. One aspect of FMS operation where there are still challenges is in the use of FMS to produce parts with low tolerances.

Unlike NC when that was introduced, FMSs have not been claimed to give substantial improvements in quality and discussions of precision and quality are often very brief (if present at all) in the literature on FMS. However, for many it is an important aspect of FMS operation. This paper reviews the factors that influence the precision and quality obtainable from an FMS so that those planning to install them can make more informed decisions on how to achieve the quality and precision they require. The paper is based upon the author's work investigating the accuracy and precision of both NC/CNC machine tools and flexible manufacturing systems.

Most modern FMS consist to a large part CNC machine tools and one might therefore assume the accuracy of FMS to be comparable to that of modern CNC machine tools. However, is this a reasonable assumption and if it is not, why is it not? In answering this question, it is worthwhile recalling users experiences when NC machines were first introduced. The early users of NC machine tools (and some not so early users) were puzzled and annoyed that their new and very expensive machines were not as accurate as their older manually operated machines. Their supposition that their older machines were accurate was, of course, false. It was the combination of the machine and the operator with his measuring instruments, his knowledge of the machine and his skill in setting it up and using it that achieved the required accuracy.

Early NC users assumed that the geometry of a part could be transformed directly into an appropriate set of instructions in an NC programme and that this would enable a part to be produced quickly, consistantly and accurately. The change of method from using manually set up dead-stops or a direct measurement by micrometer was far more significant than was realised. There were a number of reasons for this and they result from static and cutting errors which can occur in most machines.

Static accuracy

A machine's built-in accuracy is determined by its stiffness, its geometric alignments, the displacement or positioning accuracy of its axes and by the accuracy of rotation of its spindle. Specifications of values for these are covered by detailed tests such as those specified by Schlesinger [2] or by standards organisations [3]. A typical result of a test of positional accuracy of an axis is shown in Fig. 1 which shows lines representing mean cumulative errors. The lines are drawn dotted to indicate scatter and periodic errors have not been shown. The figure illustrates two aspects of potential error in the positioning of an axis. Firstly, uni-directional positioning differs from bi-directional. The differences in directional positioning (sometimes called the 'dead zone') can be due to slight slackness in a preload of a bearing or of a ball-nut or to an elastic deflection resulting from friction in these bearings and on the slideways. Secondly, the figure shows an almost linear cumulative positioning error. On many machines, this can be due to the combination of a heated ballscrew and the use of rotary resolvers driven by the ballscrew to measure axis position.

Axes are typically driven by precision ground ballscrews with a pair of double nuts, preloaded against each other, to ensure a suitable stiffness and the elimination of any backlash. Rapid traverse motions generate heat in the ballscrew nut, some of which is transferred to the ballscrew causing it to expand. Because resolvers measure ballscrew rotation rather than linear movement of the axis, positioning errors can result. While the errors involved are not large, in seeking part precision, small errors soon become significant. The use of linear inductosyns avoids some of these difficulties.

Cutting Accuracy

When a machine is cutting, its performance and accuracy are affected not only by its basic build alignments, but also by its dynamic behaviour while cutting. Factors of importance here are its static and dynamic stiffness and its thermal behaviour.

All machines have a certain elasticity (static stiffness) which means deflections will occur under load. Although machine stiffness is generally relatively high, tooling stiffness is often much less and up to 90% of a deflection due to cutting forces may occur in the tooling. Dynamic stiffness of machines (which if low can cause vibration and chatter) is generally adequate but affects surface finish and tool life more than accuracy.

NC machines were designed to exploit carbide cutting tools (in addition to offering programmable control) and installed kilowatts (hp), spindle speeds and cutting rates are considerably higher than those of older machines. All these elements are heat generating. As with the ballscrew bearings, preloaded rolling element spindle bearings as well as externally pressurised or self-pressurised spindle bearings are sources of heat, even with careful design and controlled lubrication. Electric motors are also a source of heat but their effect can be minimised if they are properly located. Hot chips from the use of high cutting speeds can heat up re-circulating coolant which in turn heats up parts of a machine. Additionally, hydraulic components within the machine can act as localised heat sources.

These elements are not fixed sources of heat but variable. Spindle speeds may vary from zero up to 5,000 rev/min. Cutting power varies from high values during roughing to lower values during fishing but often with higher spindle speeds generating more heat. These heat sources cause a machine to distort rather than just grow. The alignments which have been checked and proved to be within tight tolerances under no load are no longer within tolerance.

All these factors thus mean that a programmed position in 3-dimensional space, established in a component program with reference to the orthogonal axes of a machine under no-load will not, during cutting, be where it should be. Add to this the machine stiffness, tool deflections and tool wear and it can be seen why early users of NC machines had a problem on their hands when compared with machining a component which had been either marked out with reference to itself or measured by a turner with his micrometer. Many programs had to be modified to compensate for static and dynamic errors.

Fortunately, not all features on a component are critical, and tool offsets gradually became available on NC controllers which enabled the position of certain critical tools to be easily modified. Machine users realised that process capability studies had to be carried out and components' programmed dimensions were appropriately off-set within a program to make the alterations found to be necessary. Some machines incorporated on-machine tool setting [4] so that datums were established as close to a component as possible. Other means of maintaining accuracy will be discussed when FMS accuracy is discussed.

FMS OPERATION

As machine users and builders are aware that these aspects of NC/CNC machine operation need to be taken into account, one might wonder why producing precision parts on FMSs poses additional problems. It will be seen that FMS do pose more problems because:

> They involve machining components on a number of different machines.
>
> They are planned for minimal manning whereas many CNC machines still often have one machine, one operator;
>
> They have palletised components whereas few (though an increasing

number) standard CNC machines have this feature;

They operate continuously at utilisation levels of over 80%, whereas few CNC machines reach 50%;

They have complex functional specifications and in commissioning a system, there are so many other factors to review that process capability across system machines may be overlooked.

CNC machines are typically used in batch manufacturing environments and this enables an operator to get familiar with component programs on a given machine, with the quirks of the tooling, the off-sets necessary etc. In contrast, flexible manufacturing systems rarely have similar components follow each other at a particular machine and components are machined without recourse to an operator. The next component of a batch is as likely to be machined on a different machine with different tooling, as on the last machine used. The program used for one machine should, therefore, be usable by another and have common off-sets programmed into it, requiring only tooling off-sets to fit a program for a particular machine. If off-sets in FMS are tailored to a particular machine, the complexity of program management is considerably increased.

Thus, two of the fundamental concepts of flexible manufacturing, those of components being produced on any machine and without human intervention, lead directly to greater potential dimensional variability from part to part.

The mounting of components in fixtures and the fixtures on pallets and the pallets on machines provides many potential sources of difference from component to component, and yet the maintenance of a consistent component-fixture-pallet relationship is fundamental to the production of higher accuracy components. To help this, fixtures are invariably kept permanently attached to pallets. However, parts required in any volume are very likely to need at least two fixtures with again a potential for variability from fixture to fixture. Further, however well a machine is manufactured, the pallet to machine interface is critical and small differences are likely from mounting different pallets on different machines.

Another aspect of the use of pallets is their loading. In FMS, machine operators do not generally load the components into the fixtures on the pallets. This is carried out by a load/unloader. Aspects of the component location which are important which would soon become apparent to a conventional NC machine operator do not necessarily become apparent to the FMS component loaders. Thus although flexible manufacturing systems may bring many hardware and software elements together into a system, the machine-operator system which occurs with stand-alone machines is not integrated into the larger system. This can result in some of the benefits of having an operator by a machine being lost if care is not given to compensating for the change.

It is well-known that the greatest thermal variations in machines occur in the first 30 to 60 minutes after switching on. The continuous 3-shift operation of most FMSs can be a significant advantage as far as establishing thermal equilibrium is concerned. However, the continuous operation and the high utilisation puts pressures on machines in a system that stand-alone CNC machines do not have to withstand. These can be overcome as indicated in the following sections. An additional degree of robustness may be necessary that is not required for stand-alone machines.

ACHIEVING ACCURACY IN FMS

The previous sections have set the scene by establishing that machining accuracy cannot be taken for granted in modern sophisticated systems just as it should not have often been taken for granted when NC machines first appeared. The remainder of the paper seeks to show how accuracy can be planned for along with the many other aspects of planning that have to be carried out in the preparation for a flexible manufacturing system.

The points that are discuseed in the following section are all ones that contribute to improving accuracy. Some apply just to machines, some are more relevant to systems. The state of knowledge at present means it is impossible to quantify the benefit from any particular action discussed. However, the tolerances at which special care needs to be exercised were specified by speakers at a recent Chicago conference on FMS (May 1984). They were in general agreement that FMS could meet tolerances on position of 75 μm-125 μm (0.003"-0.005") and bores to 25 μm-50 μm (0.001"-0.002"). In making this statement, they were referring to systems with pallet sizes up to 1000mm square. Tolerances tighter than that need careful consideration.

a) Machine type

CNC machining centres are the most common machines in flexible systems and many systems only have machining centres. The reason for this is that the horizontal spindle machining centre (HMC) is the most flexible machine there is in terms of the range of operations it can carry out, particularly with large capacity automatic tool changers. The use of an indexing table means all four sides of a component can be machined at a single set-up and this helps to minimise the total number of operations required.

HMCs have accuracy limitations, however, for many of the reasons previously discussed. One approach to achieving better accuracy is to incorporate higher accuracy machines in the total system. The precision of bored diameters can be improved by the use of precision boring machines and the precision of bore locations can be improved by the use of head indexers or head changers because with fixed position tooling and the use of balanced cutting boring bars, higher positional accuracy is achievable.

b) Component size

FMS are generally designed to machine part families, but parts classified by machining operation may have considerable size variation. Users usually purchase machines with more capacity than they initially need to cater for future requirements but this may result in small components being located in the centre of relatively large worktable-pallets. This can result in spindle extensions and tooling overhangs which are undesirable when machining precision parts.

c) Machine design and location

One of the main problems in achieving high accuracy is thermal distortation. Thermal problems cannot be easily eliminated but they can be recognised and taken into account. A system with machines which operate continuously is going to have less thermal variation than those which start and stop. However machine shop temperatures can vary considerably. Most shops have a layer of cooler air at floor level and temperatures then gradually increase. To help component and machine thermal stability, requirements can range from a simple means of recirculating air to siting a system in a fully air-conditioned environment. Systems should always be sited away from doors which will bring different temperature air in when open and well away from blower heaters.

Thermal equilibrium on a machine can be enhanced by good design. Machine suppliers should be asked to highlight features of their machines which promote thermal stability. If the machine has a hydraulic system, this may require that the hydraulics are positively temperature controlled. Equally, the coolant supplied to a machine may need to be temperature controlled and it may be worthwhile spraying more coolant around than is necessary for cutting purposes to help maintain steady temperatures.

If a machine uses rotary displacement transducers driven by ballscrews, then means of compensating for thermal growth of the ballscrews should be incorporated. This can be done by taking an analogue signal from a linear cam arranged alongside the ballscrew and coupling this up with a probe which responds to ballscrew temperature. Software can be used to work out the appropriate correction to be applied to the resolver reading.

d) Software compensation

The large computational capacity of modern control systems can be exploited to extend the software approach of correcting a ballscrew error to correcting a whole machine. It is believed this approach has been taken by one system supplier and it is the best solution if it can be made to work. The approach is to build a thermal model of the machine tool in software and relate this model either to temperatures on the machine or to temperatures and known past machine thermal behaviour. The model should give a picture of the thermal errors in the machine and this error data is then used by the control system to postion the machine axes to compensate for the errors. While this is a good approach to tackling thermal problems because it accepts thermal problems occur; it is a difficult method to employ because of the problems of building the thermal model and it cannot be used to compensate for all errors.

e) Probing and Calibration

Renishaw and similar probes have been available for some years so that cutting machines can also be used as measuring machines. They are useful for reducing some of the errors which may occur in FMS machining but careful thought has to be given to what is actually being measured.

A probe is mounted in the spindle taper of a machine and by getting a null reading, uses the measuring system of the machine to determine the position of a feature of the component or the fixture. This reading is then used to 'correct' the machine by the differences between the programmed positions of the feature and the actual positions of the feature. The probe is thus correcting positional errors in the location of the fixture and these may have arisen in a number of ways, one of which may be due to the current geometrical alignment of the machine and another may be due to the differences between fixtures etc. While the probing of fixtures which have come straight from a load/unload station does not present too many problems, as soon as a fixture has swarf and coolant on it, probing measurements themselves cannot be guaranteed to be error free.

Probing is vital after a first operaiton if subsequent operations have to be related to earlier ones accurately. In such cases, the component can be used as its own datum with respect to surface features facing the cutter but washing and drying of the pallet assembly is necessary to ensure measuring accuracy. Some system manufacturers are supplying calibration fixtures which are pallet mounted and are transported to a machine every so often to calibrate it using a machine mounted probe. Such an arrangement avoids the problems of coolant, swarf and component interface. Probing procedures and software for such an approach need to reflect the thermal state of a machine when the measurments are made.

A significant disadvantage of all probing whether of components, fixtures or calibration cubes is that the use of probes takes time and thus the time available for machining is reduced. The more probing required, the more the time taken.

f) Manning and inspection

Most FMSs operate with some level of manning, despite the publicised few unmanned systems. A number of US companies using FMS typically have a ratio of one machine minder per 5 or 6 machines. This is in addition to those operatives involved in loading and unloading components.One way of achieving higher levels of accuracy is to use some degree of human involvement as and when it is required. This may involve the use of simple manual air gauges and the manual adjustment of tools or of tool off-sets.

Some users are having co-ordinate measuring machines (CMM) built into their systems to carry out post-operative inspection. Other users have decided that it is better to check the component is right while it is on the machine, rather than wait until it is transported to a CMM. There is no 'best' approach to this problem. The long term aim must be to use CMM effectively within FMS and this means developing means of using CMM to enhance the machining accuracy of a system.

DISCUSSION AND CONCLUSIONS

Achieving tight tolerances on flexible manufacturing systems is not easy.
How it is to be achieved and how it is to be checked it is being achieved needs careful planning and thought.

The four most common methods used at present to achieve higher accuracy are:

i) Accept the limitations on accuracy and plan to carry out the high accuracy operations as post-operations after the FMS. This approach has been adopted by some users seeking tight tolerances, but cannot really be classified as a means of achieving higher accuracy on an FMS.

ii) Incorporate higher precision machines within the FMS. This may seem an obvious method but it often means including machines which lack the flexibility that is desirable in machines in an FMS. If it involves specialised machines it is only likely to be a sensible approach for those FMS which produce high volumes of a relatively few part types.

iii) Plan how to assist the machines achieve the accuracy required by using manual assistance. This may mean including program stops prior to or after critical machining operations so that a machine minder can be summoned to check what has been or is about to be produced.

iv) Use a spindle mounted probe to check the machine against the fixture and component or against a pre-calibrated fixture. This method holds some problems as it uses the machine to measure itself. The data obtained ought reallly to be combined with temperature data.

The ideal approach to this problem is to find some satisfactory means of compensating for errors, rather than to try to design them out. Computer capacity is readily available to store programs using a thermal model approach. What will then be required is a series of thermocouples on a machine to input temperature data to the thermal model. This is not any easy approach and does rely on machines behaving consistently. This approach should be coupled with ensuring machines are designed with good thermal behaviour but there is a limit to what is posssible here. Sources of heat will always be present and the use of better cutting tools and high cutting speeds is likely to increase the heat input rather than reduce it.

It will always be beneficial to carry out process capability studies as if the variations in a process are known, so can they be the more easily controlled.
Co-ordinate measuring machines are best employed for this task even though they carry out post-process measurement. Using a machine tool for measurement is not as accurate (except perhaps for bores) and reduces the availability of a machine for cutting.

*   *   *

This paper has concentrated on discussing one aspect of FMS performance.
Most existing FMS are producing parts with adequate accuracy for their application requirements, although some users have had to accept less precision than they would have liked. As an example of what can be achieved with the right conditions and with care, it is worthwhile recording that one US system is holding positioning tolerances to 0.025mm (0.001") and bores to better than 0.01mm, (0.0004") TIR. However, it should be pointed out that the parts are small, the system is operated 24 hours a day, the shop is temperature controlled, the coolant is temperature controlled, process monitoring is carried out continuously and the parts are made from light alloy which is easy to machine.

Acknowledgements

The Whitwoth Foundation is thanked for providing funds which enabled the author to visit American FMS users to learn of their operating experiences. Engineers of the many companies visited are thanked for frankly discussing their experiences.

References

[1] Bergstrom, R.P, FMS Users and Vendors: Where we are today, Manufacturing Engineering pp48-53, (March1986)
[2] Schlesinger, G, 1978, Testing Machine Tools, (Editor) Burdekin, M, Pergamon Press, Oxford.
[3] Brooker, K (Editor), 1984 Manual of British Standards in Engineering Metrology, Hutchinson & Co (Publishers) Ltd, London pp.167-180.
[4] Macdonald, RL and Thorneycroft, M, 1971, "Developments in the Automation of NC Lathes", Proc. 12th International Machine Tool Design and Research Conference, UMIST, Manchester.

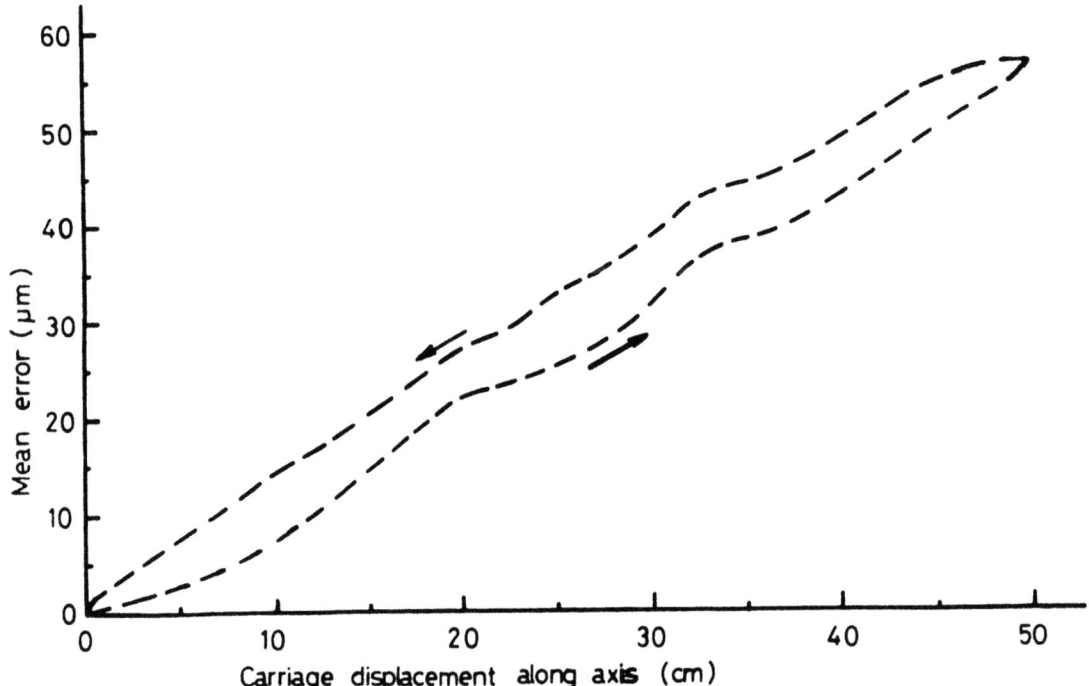

Fig.1 A typical positional accuracy measurement of an NC machine axis.

SUPPLEMENTARY PAPER

Flexible media of production automation
W. Apalkow
Instytut Mechaniki Precyzyjnej, Poland

Abstract

In the paper will be described a system and systematics of construction the flexible media of production automation, allowing to build the machines, sites, lines and flexible production system as well as to automate the many operation and interoperation actions, e.g.: transportation, changes the location of objects, control and measurements, completing and joining of parts, change of tools, palletization and transportation of containers, some cutting operations, painting, welding and pressure welding. The system is founded on unified and standarized moduls, sets and parts of machines, containing in the discussed flexible system of automation media of the industry.

1. INTRODUCTION

In machine industry the 80% of products is manufactured in small batches. At considering the piece times in production processes the especial attention should be given to handling operation and interoperation activities. At mechanical processes of production the time consumption for handling activity represents up to 70% of the total time, sometimes even more than that.

The existing achievements as well as development forecasts of production media and engineerings point to the fact that general automation and computeration are unavoidable. The automation, besides mass and large-lot production, has included also the small-lot production and even piece production. Flexible automatic methods of production are applied not only for single machines and one piece of equipment, but also for their groups connected by common system of transportation and feeding the elements, included in assumed technological process, having the common central system of the flexible control.

Flexible media of production automation are the machines, pieces of equipment or production objects built of unified, ordered in type-lines and standarized units, subunits and components. They have secured constructionally the possibility of the simple and rapid mounting and dismantling the structural new, integral different flexible media of production automation with kinematic structure and technical parameters accorded to technological process requirements - using the functional different units, subunits and components.

The building system the flexible media of production automation, supported on the elaborated systematics of units, subunits and components of machines, allows to build the machines, sites, lines and flexible production systems as well as to automatize the many operation and interoperation handlings, for example: transportation handling, changing the location of objects, control-measuring equipment, completing and joining of parts, exchange of tools, palletization and transportation of containers, some machining operations, painting, welding and pressure welding. The system is supported od unified and standarized moduls, units, subunits and machine components.

2. UNIT CONSTRUCTION OF THE FLEXIBLE PRODUCTION MEDIA

The future of industry automation is the automation supported on unified and standarized units, subunits and machine components, enabling to cover the requirements of mass, lot and piece production for quickly developing industry.

The construction of automatic working industrial objects, machines, essential technological instrumentation (e.g. turn-tables, linear-tables, cross-tables, settle-feeding equipment), supporting and linking structures (e.g. gate constructions, stand sets, beam sets, connecting assemblies, base sets, table-plate sets, reduction or adapting assemblies etc.), supported on unified and standarized units, subunits and machine components - enables the quick and economical construction (or dismantling) the unit flexible production media with structures defined by technological process.

2.1. Systematics of units, subunits and machine components

The base for construction of units, subunits and components, which are next used for building the unit industrial objects, is their ordered classification, called the systematics. For example, as a criterion for classification into groups (fig.1) the unified and standarized units, subunits and machine components, creating the flexible media of industry automation, was assumed their functional destinantion in structure of the flexible production media. The examples of systematized, unified and standarized units, subunits and components are presented below:

fig.2 - example of systematization the unified and standarized driving units. The driving units serve for putting into motion the other units, subunits or machine components.

fig.3 - there is presented the fragment of type-line (0100.017 ÷ 0100.093) of tilting motion units with joining and overall dimensions, defined max. moment and turn angel of the unit as well as approximately weight of this units.

fig.4 - example of systematization the unified and standarized carrying and integrating units. Carrying and integrating units are the body units, which are submitted to loading and connect mutually the units, subunits and machine components into a machine or another object of unit construction.

fig.5 - example of systematization the unified and standarized carrying and integrating components. Carrying and integrating components are the components, which are submitted to loading and connect mutually the units, subunits and other components into an unit machine or another object of unit construction.

Control units are the units by help of which the control systems of work cycle of the machine, equipment or another object of production automation are combined. For example the hydraulic distributor serves for control the work cycle of the rotary motion driving unit. The pipe-fittings set can serve for control the work cycle of the linear motion driving unit etc.

Feeding units are the units supplying an energy (hydraulic, pneumatic or electric ones) with required technical parameters to energy receivers, e.g. to driving units or control units.

Programming units institute the line of information bits, logically and essentionally connected one with another. These information connect the tasks which ought to be done by the unit machine or another technical object. At every unit, equipment or machine, entering into complete object of the production automation, the programme is finished with a question, which should be answered, or contains a gap, which should be filled up.

Various units, subunits and machine components are the produced standarized goods, achieved in shops, which are not put into the systematics of the presented direction. To the various units can be included, for example, standarized units of lubrication, cooling, chip removing etc.

2.2. Labelling of unified and standarized units, subunits and components

Digital labelling the units, subunits and components of the unit flexible production media enables a simple and synonymous record of essential data defining:

    2.2.1. - group labelling
    2.2.2. - size labelling

Assumed digital labelling permits for an extension of the record, e.g. the structure record of the unit manipulation machine or even the entire technological line which consists of the unified and standarized units, subunits and components.

2.2.1. Group labelling

For labelling of the group, to which the unit, subunit or component of the machine is assigned, a system of 4 digits from 0001 to 9999 was assumed. The unified and standarized units, subunits and machine components, which can be the parts of the flexible media of production automation, was divided into 7 groups (fig.1) :

    0001-0599 - driving units
    0600-1999 - carrying and integrating units,and subunits
    2000-4999 - carrying and integrating subunits and machine components
    5000-5999 - control units, subunits and components
    6000-6999 - feeding units, subunits and components
    7000-7999 - programming units, subunits and components
    8000-9999 - various units, subunits and machine components

For each group was given the close limited quantity of labellings, which is synonymously with with limiting of the max. number the various executions of the given unit, subunit or machine component with the same functional destinantion and with the same value of the basic technical parameters. A difference can be in the way of construction solution, driving sort, successive execution version etc.
The ssumed system of labelling permits for dividing the groups into subgroups. As an example can be dividing the driving units group (fig.2) as well as carrying and integrating units and subunits group (fig.4) :

Driving group :

    0001-0099 - rotary motion units
    0100-0199 - tilting motion units
    0200-0299 - linear motion units
    0300-0399 - multimotion units
    0400-0499 - gripping units
    0500-0599 - various driving units

Carrying and integrating units, subunits group :

    0600-0699 - stand sets and subsets
    0700-0799 - beam sets and subsets
    0800-0899 - connecting assemblies and subassemblies
    0900-0999 - base sets and subsets
    1000-1099 - table-plate sets and subsets
    1100-1199 - adapting assemblies and subassemblies
    1200-1999 - reserve labelling

Every set or assembly has close limited functional destination in the flexible media of production automation.
The construction of the flexible media of production automation should base in widest limits on unified and standarized units, subunits and machine components. Units and subunits should be also built of unified and standarized components. In fig.5 was shown examples of the unified and standarized components of the carrying and untegrating machines, which can be used at functional different units, subunits and flexible media of production automation.
Beginning list of unified and standarized components of the carrying and integrating machines is given below :

    2000 - base plate
    2001 - upper joining plate
    2002 - down joining plate
    2003 - table plate
    2004 - column
    2005 - beam
    2006 - adapting joining plate
    2007-4999 - reserve labelling

No		1	001	002	003	004	005	006	007	008	009
Characteristic size		2	0,010	0,011	0,012	0,014	0,016	0,018	0,020	0,022	0,025
1	010	011	012	013	014	015	016	017	018	019	020
2	0,028	0,031	0,035	0,040	0,045	0,050	0,056	0,063	0,071	0,080	0,090
1	021	022	023	024	025	026	027	028	029	030	031
2	0,100	0,112	0,125	0,140	0,160	0,180	0,200	0,224	0,250	0,280	0,315
1	032	033	034	035	036	037	038	039	040	041	042
2	0,355	0,400	0,450	0,500	0,560	0,630	0,710	0,800	0,900	1,00	1,12
1	043	044	045	046	047	048	049	050	051	052	053
2	1,25	1,40	1,60	1,80	2,00	2,24	2,50	2,80	3,15	3,55	4,00
1	054	055	056	057	058	059	060	061	062	063	064
2	4,50	5,00	5,60	6,30	7,10	8,00	9,00	10,0	11,0	12,5	14,0
1	065	066	067	068	069	070	071	072	073	074	075
2	16,0	18,0	20,0	22,4	25,0	28,0	31,5	35,5	40,0	45,0	50,0
1	076	077	078	079	080	081	082	083	084	085	086
2	56,0	63,0	71,0	80,0	90,0	100	112	125	140	160	180
1	087	088	089	090	091	092	093	094	095	096	097
2	200	224	250	280	315	355	400	450	500	560	630
1	098	099	100	101	102	103	104	105	106	107	108
2	710	800	900	1000	1120	1250	1400	1600	1800	2000	2240
1	109	110	111	112	113	114	115	116	117	118	119
2	2500	2800	3150	3550	4000	4500	5000	5600	6300	7100	8000
1	120	121	122	123	124	125	126	127	128	129	130
2	9000	10000	11200	12500	14000	16000	18000	20000	22400	25000	28000
1	131	132	133	134	135	136	137	138	139	140	141
2	31500	35500	40000	45000	50000	56000	63000	71000	80000	90000	100000

Table 1

Typical representatives of machine components with wide application are :
- upper joining plate (2001)
- down joining plate (2002)

The same plates can include into many functional different units or subunits, creating an unified part. It allows to connect the units and subunits by help of the unified and standarized connecting assembly (e.g. 0800), if all elements of the machine have the same second digital link (e.g. 2001.088, 2002.088, 0800.088). The number "088" means the coded size of the basic technical parameter of the given unit, subunit, component or the flexible production medium.

2.2.2. Labelling of sizes

Labelling the basic characteristic sizes of unified and standarized units, subunits and machine components was assumed basing on R-20 essential sequence of preferred numbers, acc. to Polish Standards PN-60(M-02100.

In table 1 were quoted the preferred numbers sequence and, adequate to them, values of maximal basic technical parameters, which characterizes the unit, subunit or component in the group. For example: the labelling 087 shows that basic technical parameter of the unit or the machine component is marked by number 200, it means:

- at the rotary motion unit (0001-0099) and the tilting motion units (0100-0199) the number 200 shows the maximal turning moment in Nm, obtained ny the unit;
- at connecting assemblies (0800-0899), adapting assemblies (1100-1199), upper joining plate (2001), down joining plate (2002) and adapting joining plate (2006) - the number 200 means a joining diameter, expressed in mm;
- at stand sets (0600-0699), beam sets (0700-0799), columns (2004) and beams (2005) - the number 200 means a length of the part, expressed in mm.

In fig. 2, 4 and 5 the horizontal line shows, which units, subunits or components can be directly joined one another by help of connecting assemblies (0800-0899).

Units, subunits or machine components with the same basic technical parameters are identical labelled without regard for multiplication of their appearing in the unit machine or the other technical object.

3. EXAMPLES OF COMPOSING THE FINEMATIC SYSTEMS OF THE UNIT FLEXIBLE MEDIA OF PRODUCTION AUTOMATION

Presented below examples of preliminary composing of the kinematic system of the flexible media of production automation by help of unified, ordered in type-lines and standarized units, subunits and components, will be based on "the producing method of industrial robots" (Patent No. P-224306).

The method allows for flexible unit construction of machines, technical installations or other media of production automation with almost any kinematic structure, because there are secured the possibilties to associate (mounting or dismantling) the functional different units, subunits and components into structural new, integral different the flexible media of production automation with kinematic structure and technical parameters according to requirements of the technical process.

Configuration the executive part of the flexible media of production automation, i.e. a form and dimensions of work room of the automizing pieces of equipment, is a function of the subject and purpose of automation, it meansa function of its real application.

Construction of the executive part of the unit flexible media of production automation relies on simple and easy mechanical (automatic) joining and diamantling the unified and standarized, ordered in type-lines units, subunits and components in functional new integrity.

The method of preliminary easy composing of the kinematic system is used, for example, by the producers of unit machine tools, especially during the offer elaborations. The method ought to be applied in the entire system including flexible media of productio automation.

Example No 1 :

In fig.6 is shown the unit industrial robot with kinematic structure $C_R B_{R1} B_{R2} B_{R3} A_L$ built of unified units, subunits and components. This structure have also the robots with compact construction, produced by main producers, e.g. CINCINNATI MILACRON, USA.

In result of assumed labelling system the record of the structure is simple and convertible, allows to mount the machine, equipment or another technical object without drawings as well as synonymously defines the elaborated technical documentation. Manipulating part of the unit industrial robot (fig.6) can be labelled as follows:

0900.094; 0800.094; 0001.121; 0800.093; 0600.092; 0800.093; 0100.117; 0800.092; 0100.101 (*); 0800.090; 0600.086; 0800.087; 0100.093; 0800.086; 0001.089; 0800.085; 0400.065.

(*) = with adapting insert in the unit

Example No 2 :

In fig.7 are shown examples of composing the unit machines and the record including characteristic sizes of the unit, subunit or component.

Example No 3 :

In fig. 8 are shown the following examples of composing the semi-automatic work stands with unit construction and with various technological destination, using in principle the same basic units, subunits and components :

fig. 8a - unit machine tool with turn-table, where can be done the operations of drilling, reaming or threading;

fig. 8b - welding of the tank bottom;

fig. 8c - smoothing of the weld, formed at welding of the tank bottom, by help of the grinder attached on a feed driving unit with straight-line motion,

fig. 8d - painting of the tank by help of the paint gun attached on the feed driving unit with straight-line motion, the same with in fig. 8a, 8b and 8c.

Example No 4 :

In fig. 9 are presented the examples of composing the automatic sites with unit construction, using in principle the same basic units, subunits and components. They are :
    fig. 9a - automatized working site
    fig. 9b - automatized mounting site

Example No 5 :

In fig. 10 is presented the construction of the flexible automatic mounting line of the unit machines with kinematic system programmed in computer memory and required by real technological process. Mentioned above line was also mounted of unified, ordered in type-lines and standarized units, subunits and machine components.

Index :

1 - overground conveyer track
2 - conveyer
3 - overground track of the travelling industrial robots
4 - industrial robot
5-12 - stores of tools, components, subunits etc.
13-20 - stores for unified and standarized subunits and units
21-24 - current auxiliary stores

4. FINISH

In machine industry the developing tendencies are going to achieving the production systems without human staff, to working out the new type of production plant, where the human will not be a slave of a machine. Architecture of machines and production rooms will be defined by function and economic calculation (e.g. vertical designing of production rooms and machine equipment(. Air conditioning and lightning will not be necessary and ergonomics science will not be decided and influenced on appearence and future of the industry. Full automatized production plants will be supported on machines and equipment of unit construction, built of unified and standarized units, subunits and components, ordered in type-lin securing automatic simple, easy and quick exchange of tools and devices, allowing for automatic construction and dismantling of the functional new object.

REFERENCES

1. Apalkow W.: Modulnyje promysiennyje roboty. Mezdynarodnyj Simpozium. Plovdiv 1981, Bulgaria

2. Apalkow W., Buc J.: Complex industrial automation systems. 11th International Sympozium On Robots. Tokyo, Japan, 1981.

3. Apalkow W.: Struktura kinematyczna z=ożonych robotiw przemys=owych. Przeglad Mechaniczny nr.23-24, 1982

4. Apalkow W.: Osnowy strojenia awtomaticzeskich manipułacjonnych maszy. Mezdynarodnyj Simpozium "MATAR", Praga, CSRS 1984

5. Apalkow W.: Systematyka i budowa zespołowych robotow przemyslowych. XXV Seminarium Robotow Przemysłowych. Warszawa, Polska 1985

6. Apalkow W.: Przyk=ady budowy maszyn zespolowych w oparciu o patent "Sposób wytwarzania robotow przemyslowych", XXV Seminarium Robotow Przemyslowych. Warszawa, Polska 1985.

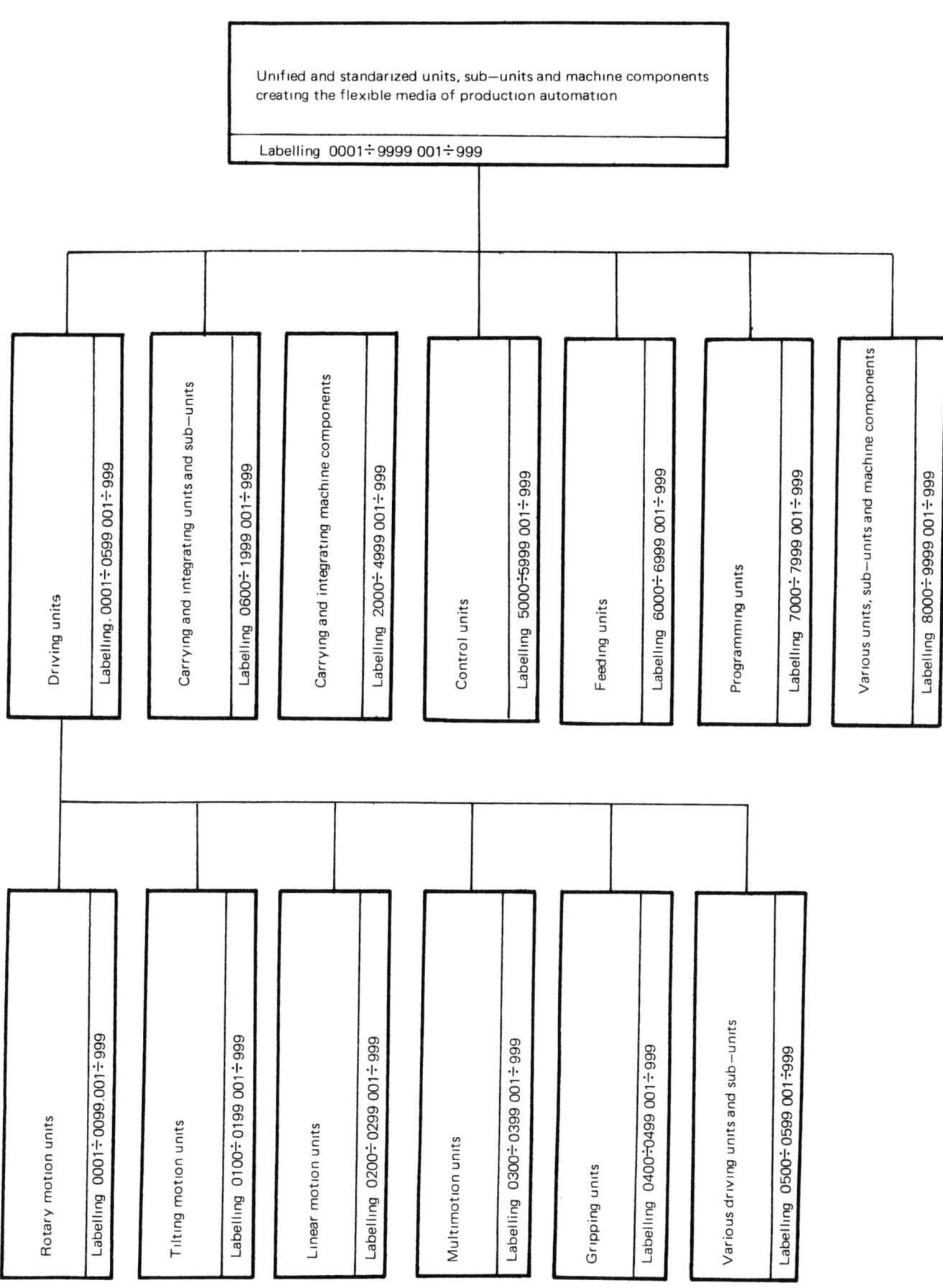

Figure 1

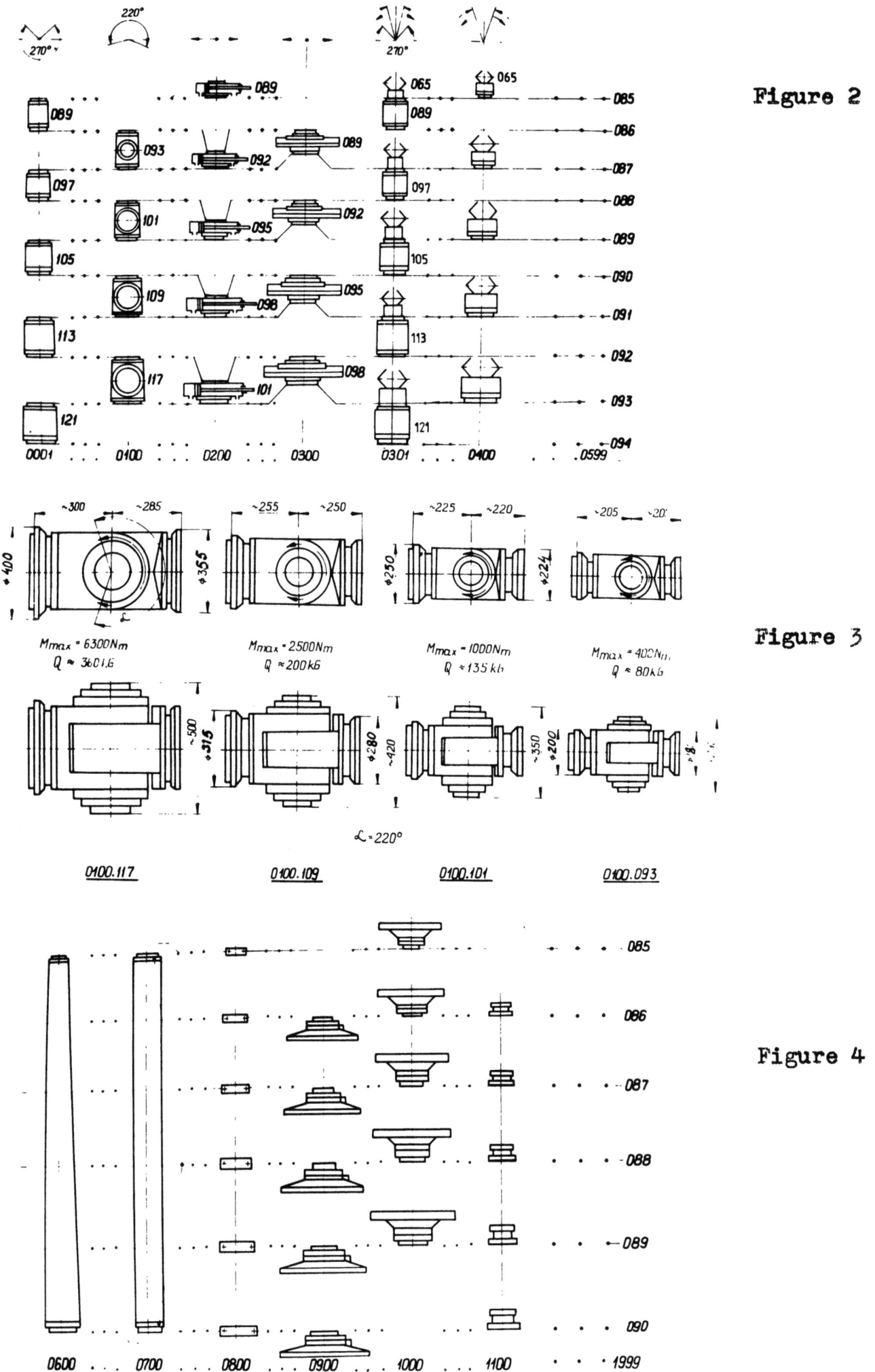

Figure 2

Figure 3

Figure 4

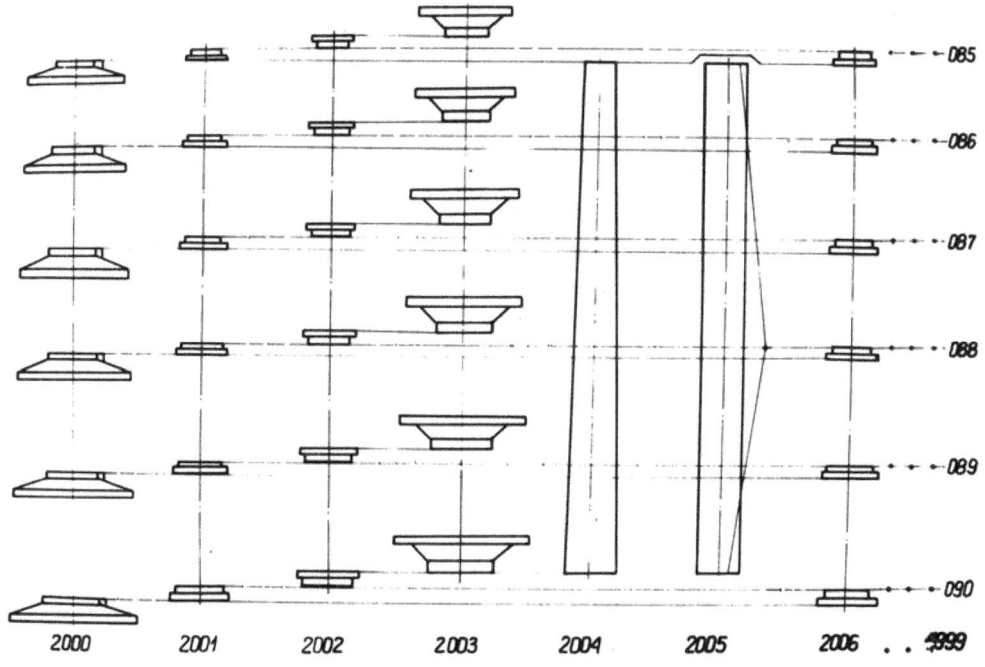

Figure 5

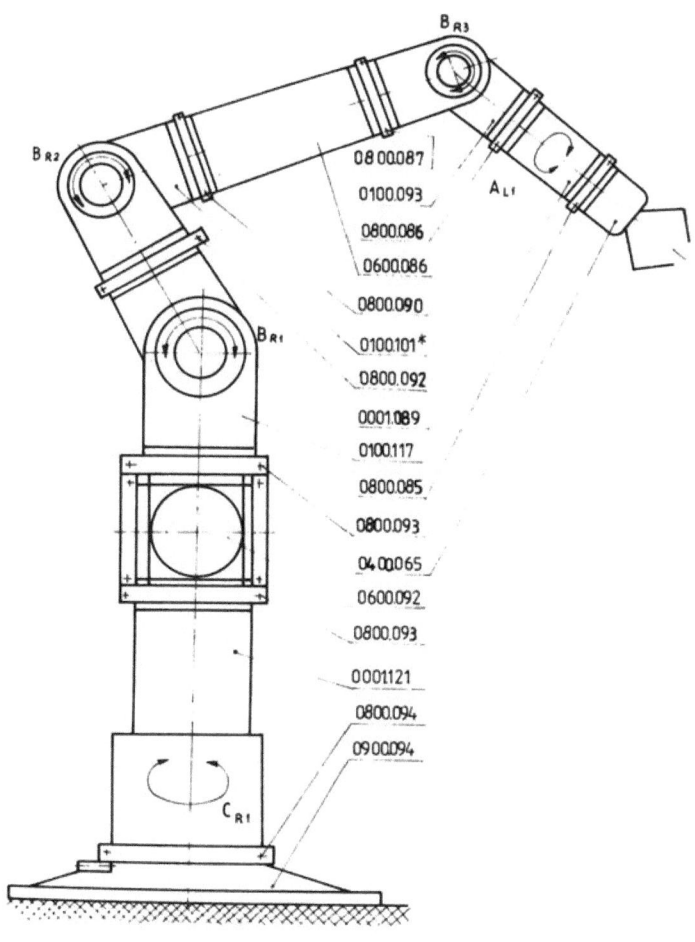

Figure 6

Figure 7

Figure 8

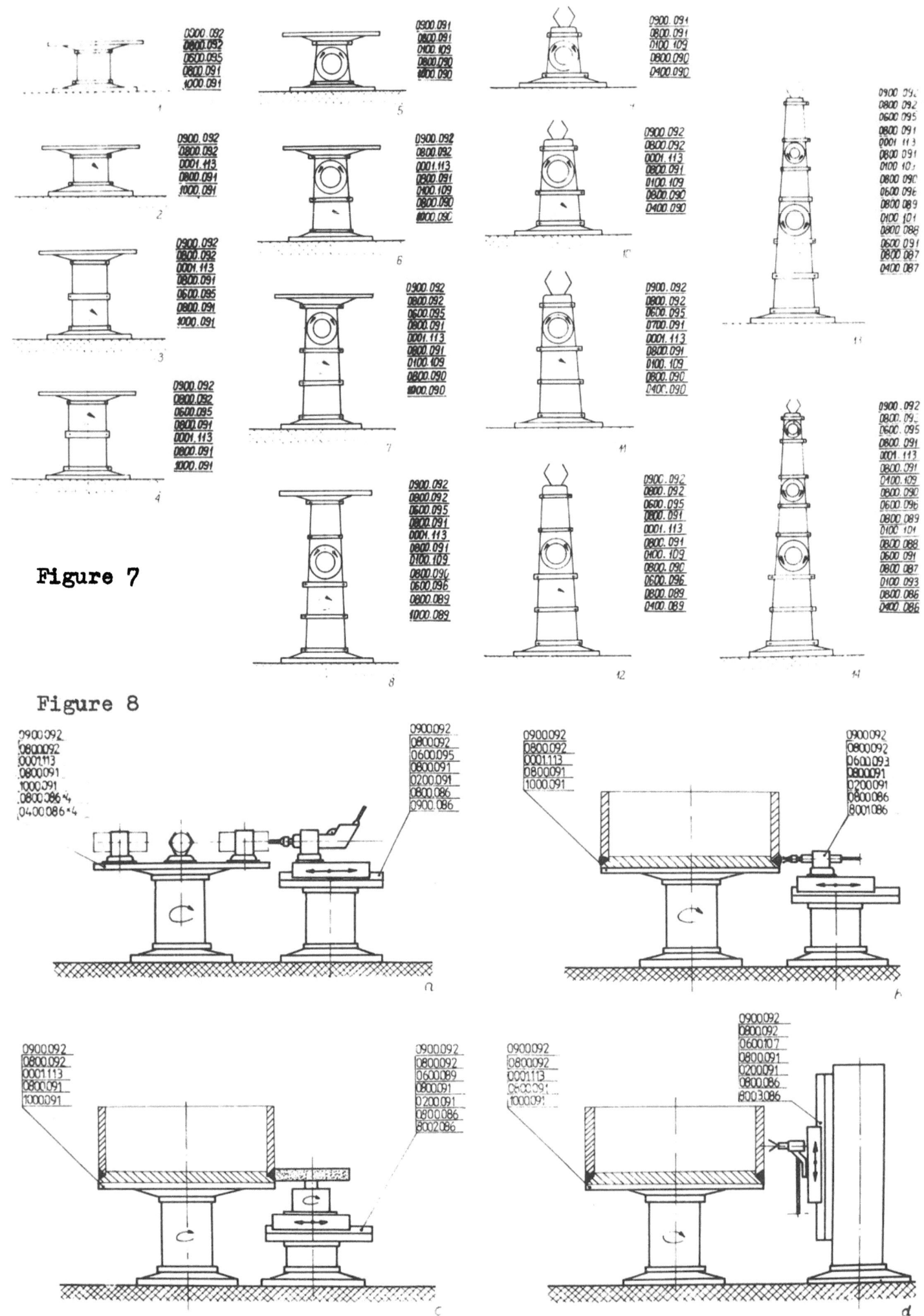

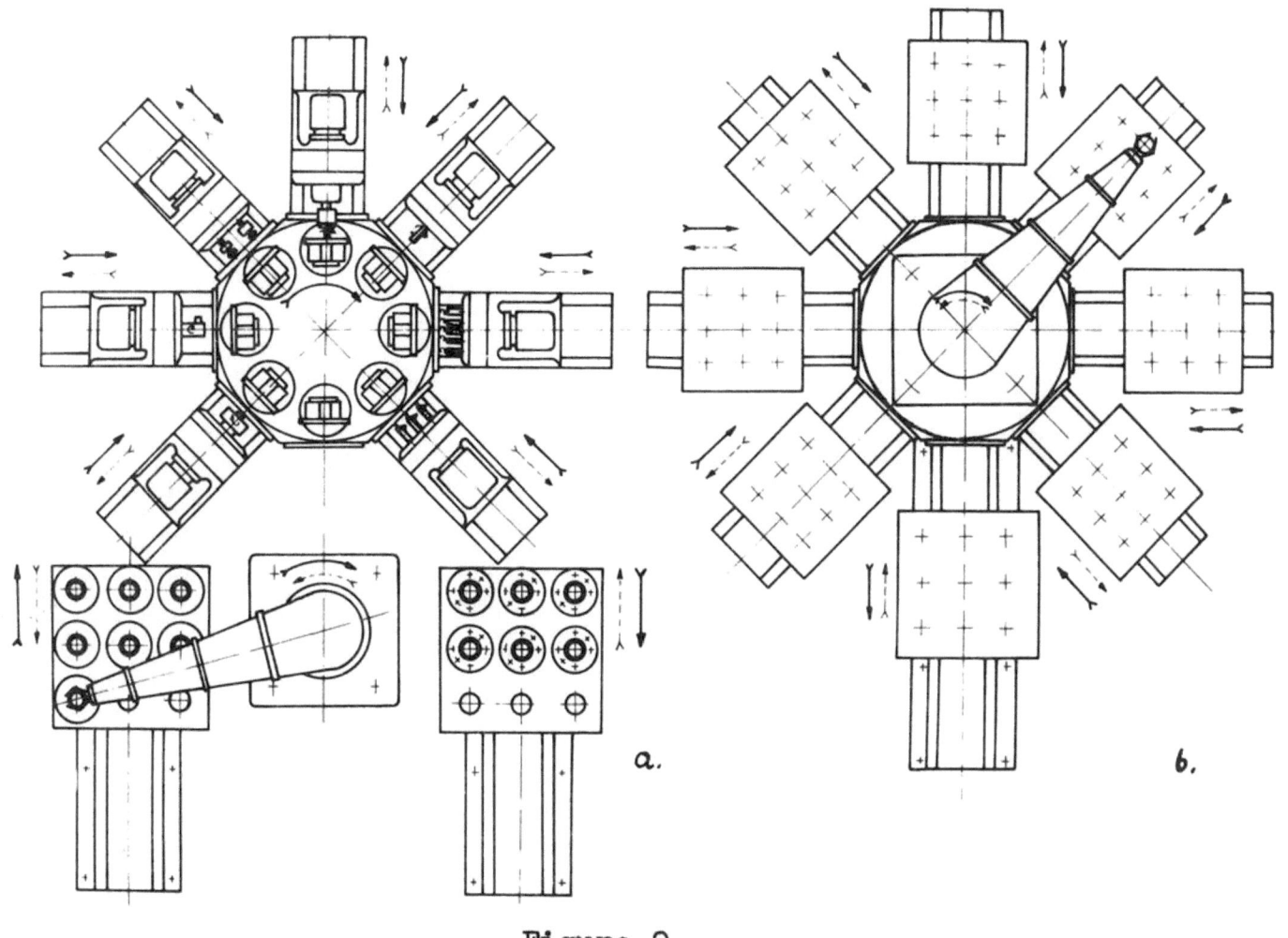

Figure 9

Figure 10

LATE PAPERS

European collaboration on FMS – developments in ESPRIT CIM

P. MacConaill
ESPRIT Directorate, Commission of the European Communities, Brussels, Belgium

ABSTRACT

The paper reviews the European Community ESPRIT programme with particular emphasis on its activities in Computer Integrated Manufacturing (CIM). Projects relevant to FMS and FAAS are briefly described, together with those project important to open system architectures and standards for communication. Plans for the second phase of ESPRIT are outlined.

INTRODUCTION

European manufacturing industry possesses many strengths - a substantial market, a long tradition, many instances of excellence in design - but it is now under intense pressure from North American and Far Eastern competitors. Whilst there are ample resources to meet this challenge, they are fragmented, and cooperation is inhibited by technical, linguistic and cultural barriers.

One key to maintaining efficiency and competitiveness in manufacturing industry is the harnessing of the opportunities presented by developments in information technology (IT). This is also an area where there are substantial European strengths but also strong external competitive pressures. Furthermore, incompatibilities between the equipment from different suppliers can inhibit the introduction of a computer integrated manufacturing (CIM) environment and favour the entrenched position of some established non-Community suppliers.

If European industry is to survive, it is vital to overcome all these barriers by pooling the available resources to build a common technological base and a harmonised open market in which indigenous European suppliers can compete.

ESPRIT

The European Strategic Programme of Research and development in Information Technology (ESPRIT) was initiated with a pilot programme in 1983. This was followed by the launching of a full-scale programme in February 1984. The objectives of the programme are to promote European industrial cooperation in precompetitive research and development in IT, to provide the European IT industry with the basic technologies it is going to need, within a five to ten years time frame, and to pave the way for the necessary international standards.

The programme addresses the following action areas:

* Microelectronics
* Software Technology
* Advanced Information Processing
* Office Systems
* Computer Integrated Manufacturing.

Although no one action area or project is dependent on another, there is a certain technological and chronological relationship between these areas. Advanced Microelectronics and Software Technology provide the basic techniques for the development of Advanced Information Technology which in turn can be applied to the two application areas - Office Systems and Computer Integated Manufacturing. In the second half of the ESPRIT programme we would expect to see many instances of technology transfer between the action areas.

Each project within the ESPRIT programme is undertaken by an international consortium which may typically include IT users and vendors, universities, and research institutions. Each project must have at least two industrial partners from at least two member states. Over the programme as a whole care is taken that Small and Medium sized Enterprises (SMEs) are adequately represented. To ensure the achievement of the strategic ESPRIT objectives, at least 75% of the programme is devoted to projects with defined objectives and measurable deliverables (Type A projects) as compared with those of a more exploratory nature (Type B projects).

The European Community contributes up to 50% of the cost of each project. 750 million ECU of Community funds have so far been authorised, so the total scale of the programme amounts to some 1.5 billion ECU. (The ECU - the European Currency Unit, in which all Community budgets are denominated - is defined in terms of a basket of member state currencies; its current value lies within a few percent of the US dollar.)

MANAGEMENT MECHANISMS

A programme of such magnitude, involving substantial public funds, clearly requires a number of formal management mechanisms.

Overall management responsibility lies with the European Commission. Member states are represented by the ESPRIT Management Committee (EMC) which decides both the overall workplan and individual major new projects. Advice on technical and strategic issues and on programme planning and operations is provided by an ESPRIT Advisory Board (EAB) consisting of representatives of IT vendors and users and research organisations.

Each year an updated annual workplan is published by the Commission, together with an open Call For Proposals. Proposals are confidentially reviewed by teams of external evaluators, according to published selection criteria. Project participants retain intellectual property rights over the results of ESPRIT projects, hence the possibilities of exploitation are an important factor in the selection process.

The volume of proposals received means that only one in five proposals can be accepted, although it is sometimes possible to combine the strengths of two or more proposals into a form that can be supported.

Once projects are under way, progress is monitored by peer review of contract deliverables.

PROGRESS

At the beginning of this year there were about 170 ESPRIT projects in progress, with 448 participants - 263 industrial organisations, 104 universities and 81 research institutes. Around 1300 researchers are working on the ESPRIT programme. Around 740 million ECU of Community funds has already been committed. (Further projects in this phase can therefore only be supported assuming that a certain number of projects already in progress will be prematurely terminated if it becomes clear that technical or strategic objectives are unlikely to be achieved.)

The general success of the programme is already apparent. A number of interesting intermediate technical results have already been achieved and have been reported at the three successful ESPRIT Technical Weeks that have so far been held. But probably the most important impact has been the forging of international partnerships, many of which will continue after the projects have been completed. An independent ESPRIT Review Board, in a mid-term review, reported that the programme was already beginning to have an effect beyond ESPRIT in the encouragement of European cooperation and an increase in confidence in the European IT industry.

RELATED PROGRAMMES

Many Community member states have national programmes complementary to ESPRIT. There are also some complementary international programmes. Two of these, RACE and BRITE, are also European Community initiatives, with responsibility vested in the European Commission and using the same management mechanisms.

RACE (Research and development in Advanced Communication Technologies for Europe) is a programme to develop the technologies needed to establish Community-wide integrated broadband communications by 1995. The programme is currently in the definition phase.

BRITE (Basic Research in Industrial Technologies for Europe) addresses R & D in new materials and new production technologies. The first phase of the programme was initiated at the beginning of 1985 with an agreed funding of 125 million ECU. There is some common ground between BRITE and the CIM area of ESPRIT.

EUREKA, an international cooperative R & D programme initiated in response to the US Strategic Defense Initiative, also involves many European organisations. It addresses many of the same topics as ESPRIT but is generally looking at developments a little closer to the market place. There is no common administration or central funding, each country chosing whether to support its own activities according to its own national policy. The programme is currently in the planning phase.

ESPRIT CIM

The specific objectives of the ESPRIT Computer Integrated Manufacturing programme are:

*   to strengthen the capability of European Community CIM vendors, and

*   to improve the competitiveness of European Community manufacturing industry.

The programme comprises the following areas:

*   Integrated Systems Architectures
*   CAD/CAE
*   Computer-aided Manufacturing
*   Flexible Manufacturing Systems
*   Subsystems and Components
*   CIM System Applications

There are currently 28 ESPRIT CIM projects, involving 1500 manyears of effort. The 150 contracting organisations include users from the automotive, aerospace, shipbuilding, wire and cable, and electronics industries, vendors of machine tools and robots, mini- and micro-computers, measuring equipment, sensors, and software products, and system builders.

To assist in the dissemination of results a specific infrastructure has been established for CIM to supplement the ESPRIT Technical Weeks. Operating under the label 'CIM Europe' it embraces publications, seminars, workshops and conferences. Activities are organised in a number of Special Interest Groups covering the principal topics of the ESPRIT CIM programme.

ESPRIT CIM PROJECTS

There are two main threads to the programme - the encouragement of an open market approach to integration by the development of open system architectures and standards for communication, and the advancement of selected CIM technologies.

Three important projects in the first category are projects 688, 955 and 322.

Project 688 is developing a CIM Open Systems Architecture, based on the ISO Open Systems Interconnection model. This reference architecture will consist of a Manufacturing Enterprise Reference Model, and a CIM Implementation Reference Model, each of which can be tailored to the needs of a specific enterprise. The use of the model will enable the vendor-independent evolution of CIM within an enterprise, reducing many of the inhibitions to the introduction of a CIM environment. The very wide participation in this project (there are 19 partners, led by CAP Sogeti) includes many leading CIM vendors and leading edge CIM users. This gives confidence that the results of the project will be implemented in available CIM hardware and software tools and in actual CIM implementations.

Project 955 is a closely related project developing a Communication Network for Manufacturing Applications (CNMA). This is an open system, multi-vendor approach to the development of standards for high reliability communication networks, at workcell, workshop, and plant levels, which can be easily reconfigured and which can be installed at acceptable cost. The project is complementary to the MAP and TOP projects and is associated with the European initiatives within CEN/CENELEC and SPAG. Pilot demonstrations of the network are planned for the 1987 Hannover Fair and for the flexible boring centre for A320 wing production at British Aerospace, Preston. British Aerospace lead this project, with GEC, Siemens, Olivetti, Bull, CGE, Nixdorf, BMW, Peugot, Aeritalia and the Fraunhofer Technical Institute.

The partners in project 322 are defining standard interfaces to enable different CAD and CAE systems to communicate with each other and with the downstream manufacturing processes. Neutral file specifications and pre- and post- processors are being developed, allowing: representation of 2D and 3D models of design objects, archival and retrieval of models, exchange of models over networks, storage of parametrised part libraries accessible from various CAD/CAM systems. This will permit the use of advanced modelling techniques for model generation, the standardised application of different finite element analysis programs, comparison of experimental and analytical dynamic analysis results, and dynamic model optimisation. The project is led by Kfk Karlsruhe, with the Universities of Karlsruhe and Leuven, the Rutherfield Appleton Laboratory and the Cranfield Institute of Technology of the UK, CISI of France, Leuven Measurement of Belgium, and BMW and GFS of Germany, as partners.

These three projects are major contributors to the development of national and international standards for CIM via the relevant standards committees.

FLEXIBLE MANUFACTURING SYSTEMS (FMS) AND FLEXIBLE AUTOMATED ASSEMBLY SYSTEMS (FAAS)

Particularly relevant to the subject of this conference are projects for Flexible Manufacturing Systems (FMS) and Flexible Automated Assembly Systems (FAAS). However defined, FMS and FAAS are essential components of CIM and feature prominantly in the ESPRIT CIM workplan.

Project 384, led by GEC, is demonstrating the principles of an integrated information processing system for the design, planning, scheduling and control of small batch assembly. The objective is to start with a high level description of the assembly process and to generate the data to automatically control the assembly cells. Artificial Intelligence techniques are being incorporated. There will be two integrated subsystems: computer-aided design and planning (using specialised data bases and feedback from simulation), and the production subsystem (assembly task scheduling, control and error recovery). The results will be demonstrated on a testbed system. The other partners in the project are AEG and Fraunhofer IPK of Germany, Telemechanique of France, and TNO of the Netherlands.

Project 504 is developing a systems technology for automating the management of discrete parts manufacture, optimising the tradeoffs between plant availability, product quality and safety. The project will demonstrate the integration of plant monitoring and diagnostics, within closed loop operational control, to achieve varying degrees of fault tolerant operation. A key feature is the use of a 'tactile' machine spindle giving three mutually orthogonal measurements of the forces on the cutter and on the drive train. This provides an early warning of cutter deterioration and of other fault conditions. The system incorporates a net-based tool permitting the application of expert knowledge to a real-time process surveillance system, giving fast response and use by non-specialist engineers. Initial results have already been successfully demonstrated on a single machine and work on a large scale demonstrator is under way. The project partners are Stewart Hughes (project leaders) and AMTRI of the UK, the Technische Hochschule of Darmstadt, the Battelle Institute and GRS of Germany, and Adersa/Gerbios of France.

Westland is leading project 534 with Dantec Electronik and the Riso National Laboratory of Denmark, the Free University of Brussels and the UK Medical Research Council, as partners. The project is developing a flexible automated assembly cell for the manufacture of mechanical assemblies of up to 0.5 m^3 size and 30 kg weight in low batch quantities (down to one). The project aims to integrate the key technologies of vision, manipulation and non-contact inspection. A study of human factors with respect to work design and organisation and hardware/software ergonomics is included.

Proposals were invited for a further FAAS project in the 1986 call. The aim of the proposed project is to advance the state of the art of robot based assembly in an FMS environment, including such tasks as the drilling and rivetting of randomly oriented objects in unstructured environments.

ESPRIT II

Planning is already well advanced to start the second phase of ESPRIT - ESPRIT II - in 1987, providing an overlap with ESPRIT I to build on its success and to maintain momentum. A substantially enlarged programme is proposed, involving some three times the resources of ESPRIT I. The programme will still concentrate on precompetitive R & D but with a greater emphasis on projects producing results which can be rapidly exploited in manufacturing industry and the market place. A number of Technology Integration Projects (TIPs) will be initiated, providing demonstrations of ESPRIT results.

The proposed workplan has been developed following consultations with expert Technical Panels and workshops. After consideration by the EMC and EAC it will be presented to the Council of Ministers for their agreement later this year.

ESPRIT II CIM

For CIM, a significant extension of the scope of the programme is planned to embrace the process industries. A specific process control workshop identified many commonalities and opportunities for synergy between the process industries and discrete parts manufacture, on which ESPRIT I CIM concentrated.

Activity will continue on open architectures and standards, with an emphasis on industry-specific implementations of the generalised concepts developed under ESPRIT I. In parallel, the work on selected CIM technologies will concentrate on the achievement of quantified targets, the integration of design and production cycles, and the exploitation of the enabling technologies - microelectronics, software engineering, and advanced information processing.

A CIM TIP will also be initiated, demonstrating state-of-the-art CIM techniques in several types of manufacturing environment.

Management of the tools resource in a FMS

L. Borghi, M. Briano and E. Parmeggiani

Comau S.P.A., Modena Division, Italy

INTRODUCTION

In this article COMAU does not intend to talk generically about FMS because with our many years of experience and over 20 FMS constructed we are convinced that is through the managerial optimization of the individual resources that the FMS reaches its highest utilization level.
Here we will concentrate on the concept of and the philosophy behind TOOL resource management. In more detail, we will illustrate the bar code encoding system for the individual tools which Comau has patented in the United States (N° 4.533.823 dated August 6, 1985).

THE COMAU PHILOSOPHY

The tool is the foudamental element for the functional character of an FMS. The following section will give a brief synthesis of the basic concepts underlying Comau's tool management philosophy. Our research goals in the difficult area of tool resourse optimization has led us the realization of a number of solidified to become the bases for Comau's tool resource management philosophy. These are:
- bad coding;
- management of each physical tool from the tool room to the spindle and vice-versa;
- management of tool transfer from one machining center to another while maintaining the data intact;
- tool change management in hidden time from primary to secondary tool magazine in a machining center.

A SUMMARY DESCRIPTION

The tool coding systems used to date have not been able to allow us to have all the information needed for complete tool resource management.
A typical example is mechanical coding using cams (figg. 1-2) which in addition to providing only a limited amount of information about the tool, requires a not-highly desirable physical modification to the standard ISO cone.
If we analyze a manufacturing system which uses an FMS, we are immediately struck by the fact that a large number of tools are used quite independent of the part-program they belong to. And thus the need immediately arises to be able to recognize them, to place them in their assigned slots for specific machining tasks and, finally, to know all their parameters.
This has become possible thanks to the evolution in the field of electronic optics which is not capable of automatically reading stamped codes and to transmit them in real time. Tie-in with a central computer allows all the data for tool management to be centrally stored.
Comau has patented a system of tool bar coding (fig. 3) precisely because we have seen a number of undeniable advantages over previous coding systems. In fact, these bar codes can be read (fig. 4) with relative ease and permit the insertion of such an abundance of checks that every possible reading error is eliminated. In addition, the system can be easily added to a center without having to use specialized personnel or costly machining.
Compared with the old systems of mechanical encoding, bar codes make a far larger number of codes available in a limited amount of space and without having to add mechanical plates to the tool. The savings in management costs and also the higher degree of reading reliability as compared with the old cam codes can hardly escape

our attention.

To conclude, the advantages offered by Comau with its addition of bar codes to the state-of-art production of flexible machining systems are the following:
1) the ability to offer total management of all tools (figg. 5-6) used in a workshop;
2) reduction in the costs of tools needed for an FMS;
3) the ability to create the tool bar code right from the tool layout stage by connecting the ticket printer to a computer;
4) reliability: if there is a read error the value read will not be interpreted with a number code.

Customer advantages from the use of bar codes are:
1) the use of coded tools without excessive tool cost increases;
2) the ability to use the entire tool resource without having to perform additional machinings on the tools;
3) ease of toolholder code change.

At this point the three foundamental goals in Comau's tool resource management in an FMS can be highlighted:
- management of each physical tool from the tool room to the spindle and vice-versa;
- management of tool transfer from one machining center to another while maintaining all data intact;
- management of tool change in hidden time from primary to secondary tool magazine in a machining center.

To be able to carry out this tool resource management the problem of handling heavy data flow concerning the various tools had to be faced (fig. 7).

To solve this problem a software/hardware package is distribuited locally along the different work stations and this supplies a number of support functions (fig. 8) for normal tool management by the NC.

This package operates with the data base where the stored information is classified into three types:
- fixed information: data which are input or up-dated in the host computer data base (and trasmitted to the package), e.g., tool and cutter codes, conformation and planned life, tool list and use in each part-program (tool mix). Alternatively (when the Host computer is down) these data can be input or modified by the operator;
- status information linked with the name of the different files (loaded, unloaded, etc.); the assigned characteristics keep tool placement in mind (primary or secondary magazine) and also tie-ins between part-programs and tool mixes;
- dynamic information: data on tool life consumption and offset which are updated by the s/h package by processing the signals generated by the NC.

The ability to use sister tools within the system is one of the particular features of FMS tool management and this brings about considerable tool movement between magazines (fig. 9).

Comau's tool management philosophy is to differentiate between the various tools in the primary and secondary magazines based on the following priority classification:
1) active part-program tools;
2) active part-program sister tools;
3) tools for the next part-program (holding for execution);
4) sister tools for the next part-program (holding for execution);
5) any other tools not covered by the first four classes;
6) dead tools.

If a tool belongs to more than one class it is assigned to the highest priority class. Class 6 is an exception and all dead tools are assigned here independent of the part-program they belong to.

A task in the s/h package distribuited at each work station is assigned tool movement management between the primary and secondary magazines.
Overall management of the FMS's tool resources is entrusted to the host computer which memorizes the information generated by the s/h package and, for new tools to be added, by the operator in the tool room.
It is at this level that we can begin to foresee the use of CAD to develop new tools and also the possibility of generating automatically the basic lists of the various components to be machined.
All the data on the individual tools are stored in a data base in the host computer and information exchange between the central control system and the individual work stations is via a number of files (fig. 10) which characterize the locally distributed s/h package data base.
These files enable the system to memorize any kind of situation concerning any tool within the system and this makes a very important goal in Comau's FMS tool resource management a reality:
- tool transfer management from one machining center to another while maintaining all data intact.

CONCLUSION

The tool is a foundamental element for the functional character of an FMS. It is in this context that we have stressed Comau's management philosophy for this resource. At the same time we should point out that behind this philosophy there is a whole series of hidden problems which make the full exploitation of FMS resources a matter of some difficulty. The overriding difficulty is the need to respond to customer demand for ever-increasing system flexibility at ever-decreasing prices.
In Comau's FMS management philosophy we have chosen the route of the highest level of standardization compatible not only with tool resource management but above all with such management integrated with all the various available resources where the tool retains its position as an extremely important element for overall FMS efficiency.

Fig. 1

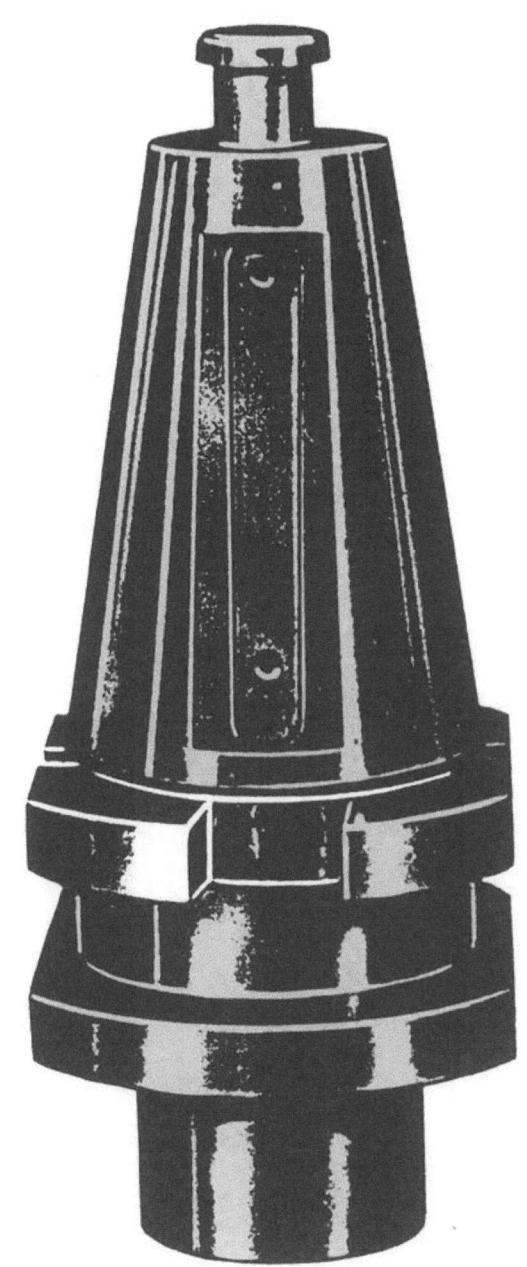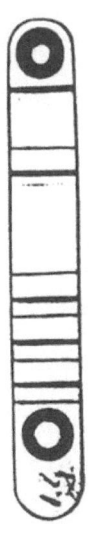

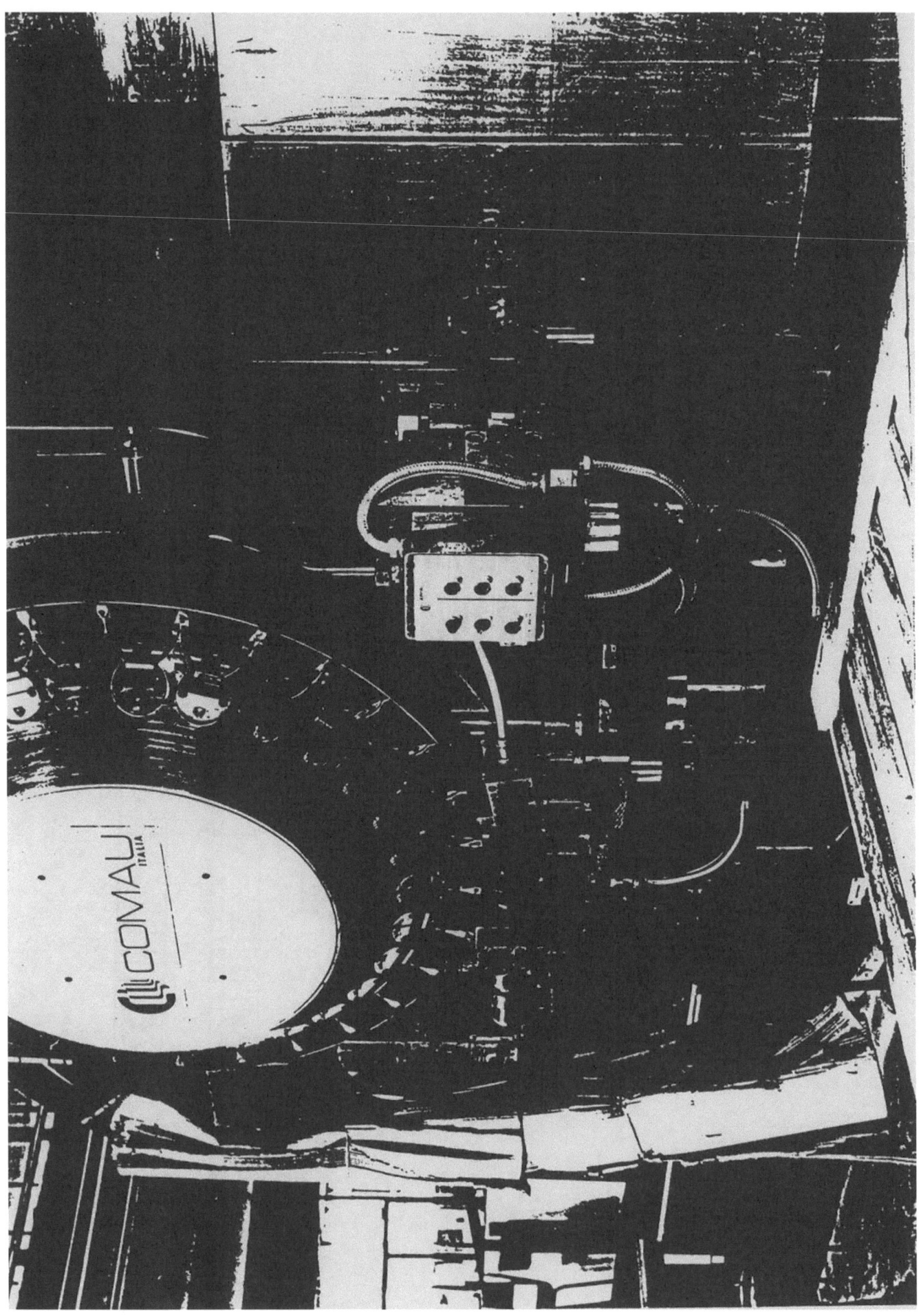

Fig. 2

Fig. 3

BAR CODE

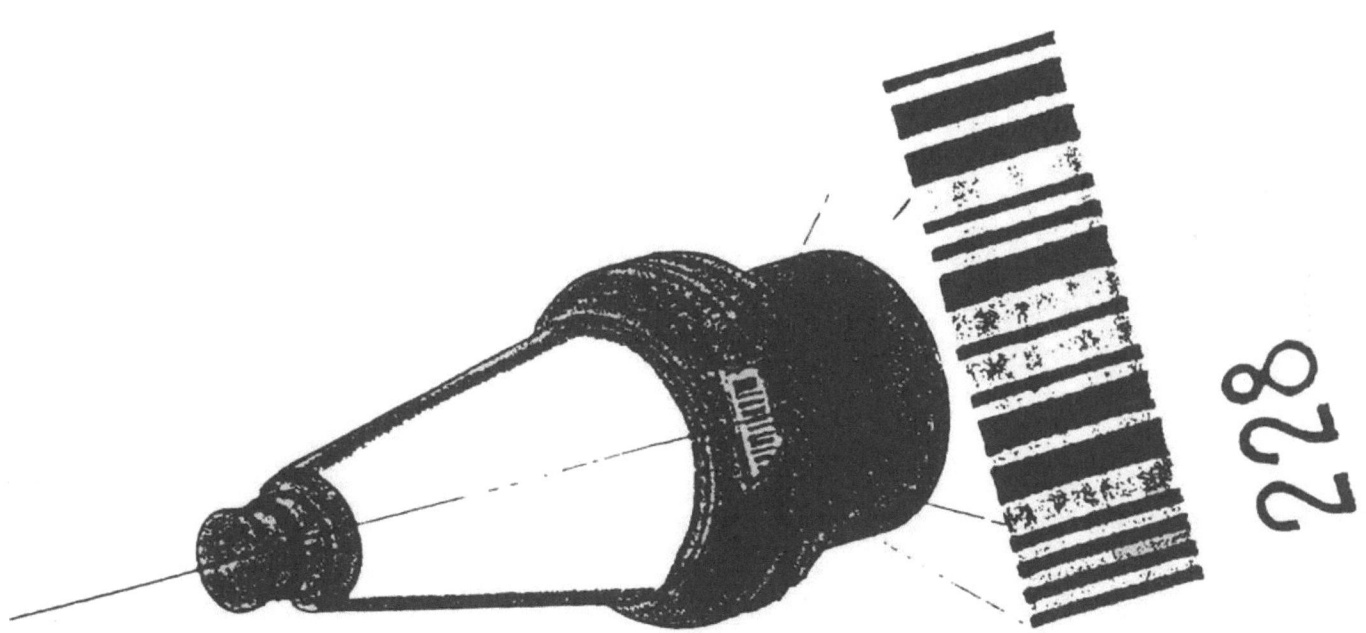

228

Fig. 4

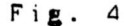

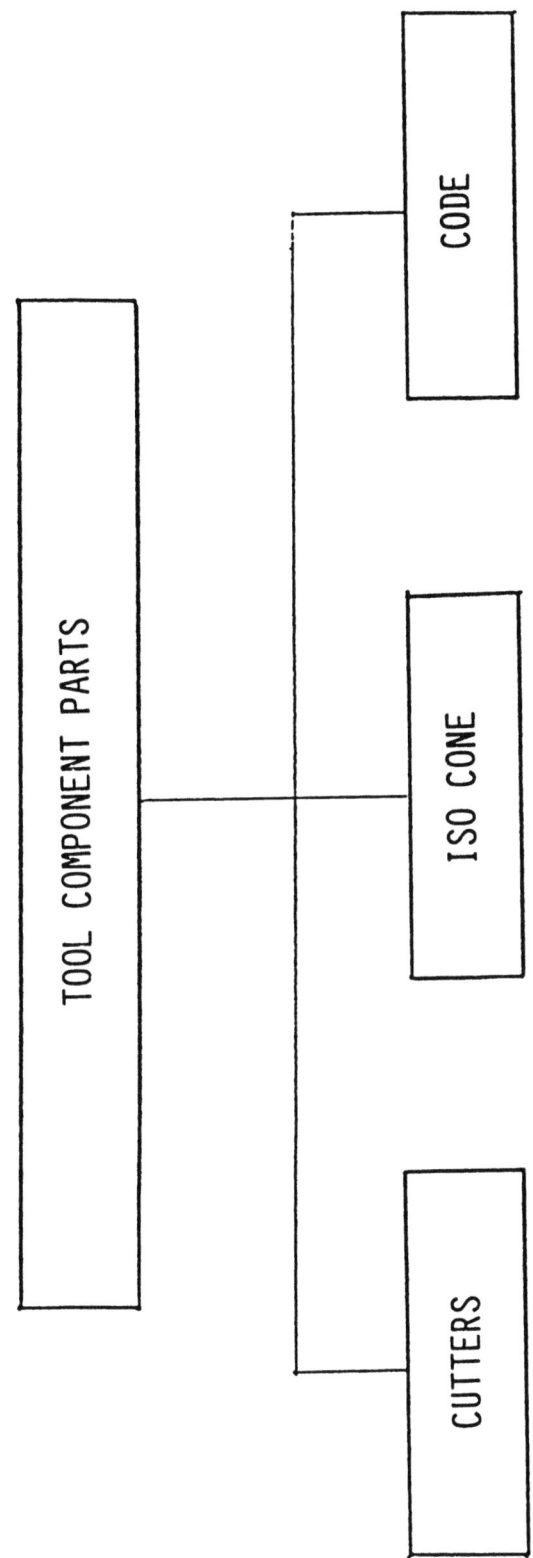

Fig. 5

Fig. 6

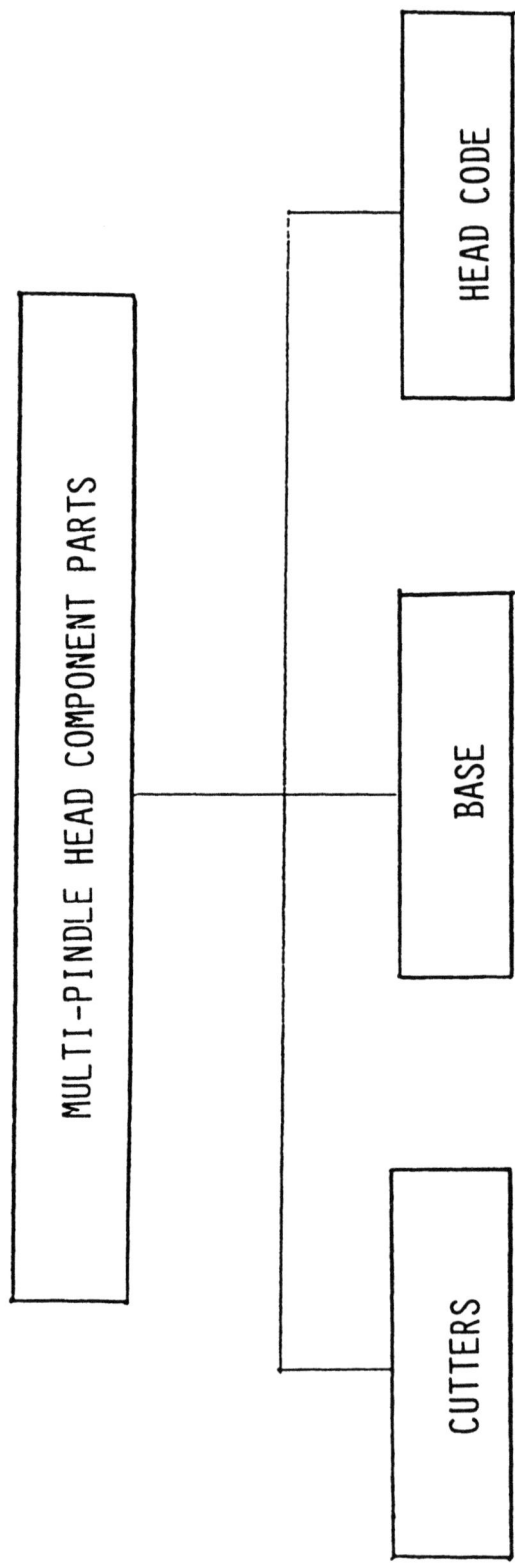

Fig. 7

```
┌─────────────────────────────────────┐
│      TOOL CHARACTERIZING DATA       │
└─────────────────────────────────────┘
```

- TOOL STRUCTURE CODE (TSC)
- TIPE CODE (TTC)
- IDENTIFICATION CODE (TPC)
- CUTTER NUMBER (NCUT)
- CUTTER IDENTIFIER (CUTID)
- ACCUMULATED TOOL LIFE (ATL)
- TOOL LIFE WARNING (WTL)
- MAXIMUM TOOL LIFE (MTL)
- LENGTH REAL (LREAL)
- LENGTH OFFSET (LOFF)
- MAXIMUM LOFF (MAXLOFF)
- MINIMUM LOFF (MINLOFF)
- MAXIMUM DELTA LOFF (MAXDLOFF)
- RADIUS REAL (RREAL)
- RADIUS OFFSET (ROFF)
- MAXIMUM ROFF (MAXROFF)
- MINIMUM ROFF (MINROFF)
- MAXIMUM DELTA ROFF (MAXDROFF)
- LOAD/UNLOAD DATE (LUDATE)
- TOOL STATUS (TSTS)

Fig. 8

S/H PACKAGE FUNCTIONS

- DATA EXCHANGE WITH THE HOST COMPUTER
- MANUAL TOOL LOADING
- AUTOMATIC TOOL LOAD
- MANUAL TOOL UNLOADING
- AUTOMATIC TOOL UNLOAD
- VERIFYING TOOL PRESENCE IN PRIMARY MAGAZINE
- VERIFYING TOOL PRESENCE IN SECONDARY MAGAZINE
- TOOL LIFE MANAGEMENT
- PHISICAL TOOL CODE REQUEST
- TOOL TRANSFER BETWEEN MAGAZINES
- TOOL OFFSET DOWNLOAD FROM PACKAGE TO CNC

Fig. 9

Fig. 10

```
┌─────────────────────────┐
│     S/H  PACKAGE FILES  │
└─────────────────────────┘

   ┌─ LOADED TOOLS
   ├─ UNLOADED TOOLS
   ├─ TOOLS LIST
   ├─ LOADED TOOLS QUEUE
   └─ LOG
```

Understanding the relationship between FMS and the total computer integrated manufacturing environment

E. A. Herring
Digital Equipment Corporation, USA

ABSTRACT

The major components of the total "CIM" environment have been identified, and drawn together into an universal model (strategy) for application to manufacturing industries. These manufacturing programs (FMS, J.I.T., TQC and MRP II), supported and enhanced by computer technology, are shown to be compatible and even essential to the success of the enterprise.

INTRODUCTION

The presence and necessity of the computer is almost a foregone conclusion in the minds of most managers and professionals in manufacturing today. In fact, we have advanced so far into the use of computers that it would be all but impossible to turn back. Manufacturers, looking beyond using computers for point solutions in manufacturing are looking for ways to integrate the critical functions of their enterprises through a framework of computer technology. Therefore, the term CIM suggests more than mere computer usage, but the planned integration of multiple functions of the enterprise through computer technology.

Information is the most critical element in the decision making process. Information access, quality and timeliness and commonalty are key factors in good information, however, the magnitude of information that is generated in any enterprise creates an almost impossible task of accessing and managing the "right" information at the "right" time. Computer integrated manufacturing is intended to assist in this information management function, providing linkages between information sources, then editing, compiling, reporting and or managing data to support the needs of the enterprise.

In todays world of graduate business schools, professional managers, professional associations, conferences and seminars, the direction to cost competitive manufacturing and growth is anything but clear. Each of the above organizations or schools of thought presents scores of approaches, ideas and information to assist manufacturers in achieving their goals. Some of these approaches are new, but others are mere restatements of methods that have been tried over and over.

We are beginning to realize that no one system, concept or approach (by itself) is the key to successful manufacturing. It is the cooperation of complementary concepts that will lead to success. The correct set of concepts and approaches may vary with each type of enterprise. This paper presents a view of FMS and other concepts (J.I.T., FMS, MRP II, TQC, etc.) along with the framework (CIM) which ties them together effectively.

WHY FLEXIBLE MANUFACTURING SYSTEMS

The nature of worldwide manufacturing has become focused on competition and growth. Cost, flexibility and quality are some of the major concerns (factors) of competition. The concepts of flexible manufacturing systems have evolved to address this factor of flexibility.

In previous years, much of manufacturing involved labor intensive processes and/or processes where labor was highly specialized and skilled. In these cases, labor became the most precious asset requiring management. Standards and incentives were successfully used as the method to control cost in this labor intensive environment. The need for flexibility due to changes in volumes, products and or processes was easily satisfied by the addition and/or reassignment of the labor force.

As the years passed, products and processes have become more sophisticated, relying more on equipment and technology than labor. Labor has become increasingly less critical in the manufacturers success equation. Managing the cost of these new manufacturing processes and equipment has become a real concern. No longer can manufacturers afford to maintain extra equipment for use when the volumes goes up/down or the primary equipment is broken. The cost of this equipment, for the first time, becomes a significant factor of success.

In addition to machinery cost, some other success factors have gained necessary attention. Raw materials inventory carrying cost, work in process inventory carrying cost and space utilization have also replaced labor as the key success factors in the success equation. Many programs have become popular as approaches to solving some of businesses obstacles to success. Flexible manufacturing systems have evolved as the major methodology for solving problems around wip management, capital asset and space utilization.

There is no "one" fms, nor is there an exact road map to reach fms. There are, however, concepts and ideas that contribute to the overall goal of providing flexibility to the manufacturing enterprise. Fms planners need to understand the basic charter of the enterprise. What are the products and technologies? What are the product mixes and volumes? What are the product life cycles? System planners must also define the flexibility opportunity. Where is flexibility needed? What does inflexibility cost? What type of flexibility is needed (materials movement, process, tooling, volumes)? Last but not inclusive, how much flexibility is effective and justified.

The actual design of the fms could vary from enterprise to enterprise depending upon the answers to the above questions. We can, however, expect some standardization or commonalty between systems to evolve within similar industrial applications.

FMS AND OTHER MANUFACTURING PROGRAMS

Before an enterprise undertakes a flexible manufacturing program, they need to fully understand the relationships, interdependencies and impact flexibility (FMS) has to other areas in the enterprise. Does FMS interfere with the functioning of J.I.T., MRP, etc.? Are FMS and CIM compatible? How does all this affect the success the enterprise?

As noted above, there are many factors which affect the success of any enterprise. Of these, changes in the product, process or volumes are key, but not the only critical factors. The flexible manufacturing system (FMS) has been identified as a major approach in managing these physical manifestations of change. Change also has a major effect on areas outside the shop floor, therefore, it is essential that the responses to change carried out on the shop floor also be supported by other functional systems.

Historically, changes on the shop floor have been supported by maintaining buffer levels of raw materials, process equipment and in process inventory. Also, materials were purchased and stored based upon the total production needs over long periods of time. The cost of maintaining this inventory, equipment and other resources were thought of as essential to maintaining and or producing a controlled output. The number (shipped) was everything. Early FMS helped us respond to changes and actually saved money in the cost of material/process changeovers as well as labor. Unfortunately, these early systems were designed for flexibility far in excess of legitimate business needs. They were designed to overcome operational problems that could affect output.

FMS and J.I.T.

Today, management has realized that the cost of flexibility must be weighed against the cost of maintaining under utilized machinery; the cost of carrying buffer inventory stocks; and the cost of ordering materials too far in advance. With these realizations, a new crop of programs and concepts became visible as tools to manage these realizations. J.I.T., or Just in time has become one of the major later day tools aimed at reducing waste in the process, specifically, raw materials and work in process inventories. Just in time causes us to look at the way in which we schedule and build in order to take advantage of the most direct route through the manufacturing processes (supplier to customer). Initially, It may appear that this direct approach to materials flow and inventory would be at cross purposes with FMS. This, however, is far from the case. We must remember that the true objective of both programs is the success of the enterprise and that overall cost supports that success. It is essential that the cost of carrying inventory and equipment be weighed against the cost of inflexibility in order to obtain the optimum set for the enterprise. This optimum set may require some degree of "buffer" inventory to support the flexibility required for legitimate changes to the shop floor. This optimum set may also limit the type or degree of flexibility in order to obtain the maximum benefit from both programs.

FMS and MRP II

As programs literally fell out of the woodwork during the past decade, all aimed at manufacturing excellence, the closed loop MRP system was the next major program to be promoted. As mentioned above, materials acquisition departments had previously purchased based upon gross requirements. They often augmented unreliable production schedules to avoid shortages. Both major MRP systems (net change and regenerative) required massive clerical efforts to support changes in the production schedule. Because of this, the MRP area was one of the first manufacturing functions to become computerized. This computerization allowed for the materials planners to "keep up" with the changes to the production schedules.

Even with the computer, MRP was still subject to massive jolts from the production floor. Inventory records showing stock present when none could be found, periodic adjustments to the record through physical inventories, limited visibility of the shop floor permitting miss use and or process inventories build ups while still ordering additional stock are some of the more problems of materials resource planning that have prompted the development of closed loop resource management (inventory, capacity, schedules, etc.). These programs, though not identical from enterprise to enterprise, have become fairly standardized.

The presence of an MRP II system is an asset to the effectiveness of an FMS. With the MRP II system looking at multiple variables throughout the entire operation, the number of interruptions and changes to the shop floor is sharply reduced. Once again, this reduction supports the aim of designing the FMS to support legitimate business changes. As the MRP II program eliminates some of the illegitimate changes to the shop floor, the FMS no longer needs to be flexible enough to handle them.

FMS and TQC

Total Quality Control is only concept, but has enormous implications in the design and management of an FMS. Quality is one of the major components of change. Poor quality creates demands through the failure of the product and/or materials. Failures create shortages and sometimes require additional routing within the process. Some element of flexibility is required to support the demands created when there is insufficient TQC.

Process Control. Manufacturing the product correctly is the basic tenet of Total Process Control. Supporting this effort will be a well defined and documented process with well trained personnel. In-order to "do it right" the first time, workers must know the differences between doing it right and doing it wrong. This training, however, can only come from well documented, mature processes consistent with the product to be manufactured. Historically, workmanship problems rarely come from well trained workers with well documented processes. They come most often when changes are made to the process, products and materials. During change, we usually fail to maintain the high level of training and documentation that we do for a mature process. Total Quality Control is dependent upon the good workmanship that comes from a well documented process and a well trained and motivated work force. Good workmanship, therefore, defect prevention, results in savings in material, labor and machine utilization dollars for the enterprise. It also reduces one more element of change that would normally require some degree of flexibility in the system.

Motivated Work-force. The answer to the first question is No, this subject is not in the wrong place. Regardless of the inspections, controls, documentation and training in place, manufacturing is still a people sensitive enterprise. A motivated employee is merely one who values the disciplines, training and skills that he/she has and respects the objectives of good workmanship. Today, technology and automation have generated some exciting work categories for many employees, but there remains a majority of monotonous, repetitive and "less than brainy" assignments. It can be difficult to maintain a motivated work force in this environment. It is, however, clear that motivation of the work force is in direct proportion to workmanship, and that good workmanship is essential to high performance.

Manufacturers have been looking for ways to motivate workers since the beginning of the Industrial Revolution. Studies in animal behavior have indicated that with most lower level animals in captivity, the promise of a reward (usually food or the absence of something unpleasant) motivates the animal to perform repeatedly. The animal seemingly never tires of this stimulation and can be counted on indefinitely to continue this behavior. In studies of human behavior, we have discovered that we soon become less responsive to the repeated application of the the stimuli. For many years, management has attempted to motivate human beings through reinforcements, praise, threats and money (to name a few). None of these has appeared to be successful for any extended time.

Behaviorist have concluded that, for the human being, motivation comes from within. Today, increasing numbers of managers are trying to provide an environment for encouraging motivation rather than trying to motivate directly. The creation of quality circles, hi-performance work teams, job rotation and creative job assignments are a few of the programs resulting from the increased focus on motivation.

Raw Materials. The old expression "no silk purse out of a sows ear" is appropriate to begin a discussion of raw materials quality. Having either the wrong materials or bad materials in the process causes major to the efficiency of the manufacturing operation. It would seem that the simple answer (which is historic) would be to inspect all materials coming in the door, reject the bad/wrong and maintain large enough stock levels to insure continued operation while these "bad" materials were replaced. This would certainly satisfy the needs of the manufacturing operation, however, the cost of following this practice has become prohibitive in many cases. In many manufacturing operations, raw materials components are extremely expensive and costly to carry. In other operations, changes are frequent and the potential for being stuck with large buffer stocks which have become obsolete is also increased.

As J.I.T. and MRP II programs develop within an enterprise, they provide some element of predictability and consistency in response to changes. An increased focus on reducing the level of raw materials inventory will be one ofthe prime targets of J.I.T.. Extreme care should be taken by management to closely link the planned reductions in inventory with the introduction of proven "new methods" for assuring quality parts are available for the shop floor. To do otherwise would cause immediate problems on the shop floor and actually increase total operating cost due to these inefficiencies. "Zero inventory" is utopian, meant only to encourage management to achieve the optimum balance for their operation. One day, when there is real "ship to shop floor" quality programs, Zero Inventories may become a reality. There are, however, tremendous dependencies on layers of vendor processes and transportation factors that will inhibit total "ship to shop floor" for some years to come.

In the absence of this "ship to shop floor" environment, an appreciable reduction in the levels of buffer raw inventory can still be obtained by 1) using the MRP system to more effectively plan purchases across product lines; 2) using statistical feedback (inspection and test data) to chart vendor/part performance and chose vendors accordingly; 3) working with design to establish effective tolerances and specifications on purchased parts, 4) Identifying in transit defects and work to eliminate them, and 5) implementing source inspection and requiring vendor process certification programs. These are a few of the efforts that can lead to reduced inventory and as well as reduced "bad" components on the manufacturing shop floor.

Total Preventative Maintenance. Second in importance only to a mature product design, preventive maintenance is perhaps the least implemented program in manufacturing today. Maintenance management based upon exception reporting has been the rule for many years, even in major machine intensive environments. Machines operating outside defined tolerances continue to be a major contributor to in process workmanship problems. The cost of maintaining large inventories of replacement parts (versus planned replacement) is prohibitive. The cost of excess machinery and floor space necessary to maintain production levels with high machinery downtime is increasing as machines themselves cost more. These are some of the major results of inadequate preventative maintenance. These cost are clear and easily understood. There is an additional major cost that is rarely considered. Down machinery has historically created the single largest demand for flexibility in materials movement on the shop floor. This "artificial" demand increases the purchase and operating cost of flexible manufacturing systems.

FMS Tomorrow

There are some who think that FMS is no longer appropriate as a manufacturing system concept. An embarrassing number of early flexible manufacturing systems failed to show a return on their investments or provide the benefits that were expected. Faced with massive investments and miss directed flexibility, many manufacturers have written of the concept (and equipment) and are looking for yet another "magic carpet" to success.

Most systems or concepts, weighted down with some of the artificial demands for flexibility identified above would fail. In fact, elimination of these illegitimate demands and careful selection of the direction and magnitude of flexibility is critical to the success and investment return of the system. The viability of the other manufacturing programs (MRP II, J.I.T. and TQC) is critical to the successful design and return from the FMS. It is also critical to the success of these same manufacturing programs that manufacturing system be able to respond to the enterprises legitimate changes efficiently and effectively.

FMS AND CIM

Management of change in the manufacturing enterprise has been identified as one of the major functions of the FMS, supported by the associated programs (J.I.T., MRP II, etc.). The FMS provides for a shop floor that is readily responsive to change and accomplishes this change effectively and efficiently. The directive to change may have numerous origins. Some of these are market changes affecting volumes, process changes, product changes and other sources of change which must be processed and acted upon within the enterprise. The directives to change are not always clear, they may come from feedback within the operation itself.

In addition to change, one of the other prime factors to success is control. Control is essential to maintaining processes, skills, reliability and other critical operating elements. Both maintaining control and managing change depend heavily upon the movement and availability of information within and between functions.

The third major factor in the success equation is information. The efficiency and effectiveness in which we obtain, process and share information is critical to maintaining control and managing change in todays competitive environment. The success and effectiveness of FMS and other programs (J.I.T., MRP II, and TQC) are directly limited by the system which manages this information. The wealth of data generated by and required for the enterprise identifies the need for a framework of information management.

Computer Integrated Manufacturing. So far there has been little discussion of computer integrated manufacturing. Each of the major programs discussed in this paper require (to some extent) the use of computers to for management and/or control. The computer is essential to sophisticated process controls, report generation, product testing, materials ordering and many other applications. By some definitions, this is computer integrated manufacturing. Clearly, this describes computers "in" manufacturing.

CIM is not the acquisition and use of computing technology, but rather, a much more deliberate effort to plan for implementing the technology in conjunction with other planning efforts, both short term and long range. Just as "zero inventories and J.I.T." are abstract concepts that do not establish a standard goals for each manufacturers, CIM is also a moving target. The rate of growth in CIM needs to be carefully coordinated with the needs and plans of the total enterprise. One optimum vision of CIM would be an enterprise where all the systems, processes and functions were linked electronically, sharing data, providing immediate feedback and control. In some factories, this is practical and certainly obtainable. In other factories, the stand alone use of computers is their CIM.

FMS, TQC, J.I.T., MRP II, Automation, Robotics and other manufacturing programs are supported and enhanced by computer technology. CIM is present if this computer technology is planned in conjunction with and consistent with the selection, implementation and growth of the enterprises selected programs. CIM is deliberate.

FMS integration projects

M. Dub

Research Institute of Engineering Technology and Economy (VUSTE), Czechoslovakia

ABSTRACT

Many industrial application of both the vertical and horizontal integration have recently been implemented in Czechoslovakia. These integration projects were mainly concerned with the discrete sub-systems identification, functional specification, computer network architecture design, information interface design and communication aspects. Conclusions related to particular results as to the nature, scope and form are open to discussion.

INTRODUCTION

The basic factor influencing an effective automation of engineering production is integration of manufacturing activities with activities of other departments - mainly design, process planning and material and resource planning. To a certain extent this was the basic idea behind an integrated system CAD/CAM.

The high degree of automation of product development processes in Czechoslovakia has been achieved by means of progressive HW and highly sophisticated SW implemented in design, process planning, manufacturing and other activities involved. This approach alone, however, cannot ensure the required degree of integration. Nor has it yielded the potentially obtainable effects.

This has been due - apart from other things - to the implementation of heterogenous HW and application of different programming languages within one enterprise, and incosistent data objects. The transfer of information between individual subsystems (activities) has not been paid adequate attention to. Consequently, isolated islands of automation have developed. They are characterized by inconsistent goals (optimization is limited to isolated activities), poor mutual communication and a number of unsystematic links.

The current study of design and implementation of integrated manufacturing systems in our country can be characterized as an effort aimed at integrating FMS into engineering enterprise environment.

This process has been carried out in the form of so called integration projects.

THE WAY WE UNDERSTAND INTEGRATION

There are many definitions of integration. Of these the following has been accepted in our institute:
Integration is a set of methods and means, the purpose of which is
- to represent all decision making and process activities in the form of data
- to have these data in such a form that permits their generation, transformation, application, transfer and storage within the computer environment
- to ensure a free transfer of all the necessary data between individual subsystems (activities), and free access for all users, who can make use of them, throughout the whole product development and manufacturing process.

The objective of integration is to create such mechanism of decision making, monitoring and evaluating functions that allows for multicriterial optimization of the final product and is based on digitalized information flows.
In this context integration of manufacturing processes in our conditions has been approached as integration projects dealing with two basic groups of interfering information flows.
One includes adequate methods and means supporting the flow of technologically oriented data, i.e. integration of activities connected with the product design and development, process engineering and the product physical materialization.
This group is referred to as the vertical integration of the manufacturing process. These vertically integrated activities can be considered in our conditions identical with the activities of a CAD/CAM system.

The other includes adequate methods and means supporting the flow of material oriented data, i.e. namely integration of activities connected with the manufacturing process planning, scheduling of manufacturing tasks, transport of materials, tools, coolants, chips, lubricants etc.. This group is referred to as the horizontal integration of the manufacturing process.

Any engineering enterprise with both vertical and horizontal integration can be considered a system that makes use of CIM philosophy, i.e. philosophy behind the implementation of open computer integrated systems in engineering industry.

THE INTEGRATED SYSTEM DECOMPOSITION

There have been at least two factors influencing the current approach to integration
- traditionally conceived content, structure and organization of engineering activities within the engineering enterprise
- design of integration links that does not take into account the conditions of their physical implementation.

It has become clear that integration of manufacturing processes in an engineering enterprise must be based on decomposition of engineering activities:
- with respect to information flow categories
- with respect to the physical implementation of integration links.

An integrated manufacturing system operates with basically three types of data
- product oriented data
- process oriented data
- customer oriented data.

The product oriented data are essential for product design and development. The application of CAD systems has made it possible to represent design solutions in the form of computer internal models that serve the subsequent procedures and processes. Manufacturing process planning and verification of physical materialization of the product are based upon these computer internal models. At this stage the process oriented data and process simulation methods and means are utilized. With their help the following information can be generated:
- operation sequence at technological and other workplaces
- identification of manufacturing aids
- data for NC and their verification
- data for manufacturing process control.

Flexible computer control of automated production makes use of the above information and ensures namely
- distribution of data to individual technological and other workplaces and systems
- acquisition of technological and operational data
- evaluation of the manufacturing process as a whole.

Customer oriented data penetrate the whole manufacturing system.

Furthermore the decomposition of the integrated system should allow a step-by-step implementation of the integrated system, i.e. a gradual application of discrete autonomous modules. This requirement has been met by a consistent application of principles of distributed intelligence. The discrete autonomous modules are defined by their functional characteristics and external information interface, where data content and formats correspond to the rules of communication with other subsystems.

INTEGRATION PROJECTS STRUCTURE

The first stage of the integrated systems application includes the following subsystems:

Product research and development

Is represented by a set of functions the aim of which is to create a concept of the new or innovated product, to develop design variants, to verify new technical or technological solutions, to choose the optimal product design variant and to prepare all resources needed for the dicisions about the physical materialization of the product.

Engineering design

Is represented by a set of functions the aim of which is, based on the product research and development and the technical and economic requirements, to elaborate and verify the comprehensive design documentation.

Process engineering

Is represented by a set of technical and organizational functions and measures assuring the technological readiness of the manufacturing process. In our projects the following basic pre-conditions were identified:
- comprehensive design and technological documentation
- lay-out documentation
- necessary standard equipment specification
- specification of standard manufacturing tools and aids
- design and fabrication of special equipment, means and aids

These resources have to correspond to the required production output.

Technological process control

Covers the control of the technological and assembly workstations (workplaces) including the local tools flow and technical inspection, the control of the mechanical interfaces (e.g. robots) between those workstations and the transport and storing systems, the generation of all data assuring the technological process, the control of supporting functions and maintenance.

Transport and storing

Covers all functions, methods and means connected with the material flow, i.e. namely storing, transport, handling and identification of the elements and the control of the movement of them in the space and time.

Process planning

Covers the planning of the manufacturing capacity (using the manufacturing process simulation), the functions of the hierarchical dynamic scheduling of manufacturing tasks, the co-ordination of material, resources and subcontractors deliveries for the manufacturing process, the choise of convenient manufacturing tasks according to the actual state of the manufacturing system and their sequencing, keeping records, etc..

INTEGRATION STRATEGIES

There are basicly three integration strategies that form a significant part in our integration projects:

Computer systems architecture strategy

This strategy concerns with the distribution of integrated system functions and deals with the architecture of computer and automation means.

Data strategy

Contains the rules for the integrated manufacturing database design and distribution and principles for data sharing by specific processors and procedures. Moreover the data actualization, security and maintenance methods and means are included in this strategy.

Communication strategy

Deals with the communication principles in the integrated system architecture, chooses the communication protocols and creates all necessary conditions for the integrated system internal communication security.

The overall structure of the integration projects is shown in Fig. 1.

The integrated manufacturing system is to be considered as a product-oriented system the goal of which is to produce products with the required parameters. The process is executed within the technological subsystem. All other activities in the integrated system can be qualified as supporting, preparatory, supplementary or conditioning functions. The technological subsystem is then the place of contact between vertical and horizontal information flows (Fig. 2). In view of this we have been devoting maximum attention mainly to the integration of the technological process into the vertical and horizontal structure of information flows within the engineering enterprise. In order to provide successful operation of an integrated system the functional autonomy of the system modules has to be secured. Throughout the whole integration projects solutions this principle must be maintained down to the lowest operation levels.

INTEGRATION PROJECTS CHARACTERISTICS

The engineering enterprises environment in our country is characterized by:
- a set of new and older machine tools and other equipment supplied by a wide range of vendors
- many types of computer hardware
- differring levels of process planning and engineering and manufacturing functions automation
- various organizational and information structures
- many specific attitudes and approaches, following from the manufacturing traditions and habits.

Two basic obstacles hindering integration have been identified:
- the absence of comprehensive data for the computer internal product and its processing representation
- the absence of the individual subsystems ability to communicate with each other.

When solving the integration projects the highest priority has been given to communication. The absence of the subsystems ability to communicate with each other currently leads to the application of extremely expensive conversion programs, lowers the efficiency of individual subsystems, does not enable to create and utilize the distributed data bases and consequently does not result in higher productivity of manufacturing systems. First of all we aimed at introducing suitable standards and know-how for the data communication between various computer and automation hw throughout the system down to the shop-floor level.

The basic computer systems architecture strategy of our integration projects is the LAN strategy (Figure 3). Solving the communication we have applied, the philosophy of the hierarchical multilevel OSI/ISO model.

The hardware which is commonly accessible to our users, does not suit the functional requirements specified by levels 1 to 4 of the above model. Up to now the majority of communication lines in our installations has not exceeded the distance of about 80-100 m and the number of connection has never been move than 16. These fact made us use communication lines based on 20 mA current loops and the LSV 2 procedure.

When implementing the communication functions according the OSI/ISO model we proceed in the top-down direction, i.e. starting from the seventh (application) level. We are aiming at introducing some of existing international standards (MMFS, CCITT X.409, ISO File Transfer etc.) for the semantics and syntax of messages.

From this point of view the MMFS (Manufacturing Messages Format Standard) specification, which was submitted as a standard of the MAP (Manufacturing Automation Protocol) application level and which is also working document EIA 1393, seems to us to be the most perspective way.

We assume that a MAP type standard will be internationally accepted in this category of applications. In view of current needs the data transfer between the office and shop-floor levels in our integration projects has been solved by standardized DNC systems that are in conformity with the MAP. These DNC systems have implemented the specifications of the OSI/ISO model lower levels and accomplish the following basic functions:

- the transfer of NC data to and from the technological workstations
- the acquisition of data about the actual state of the manufacturing system modules
- the transfer of arbitrary textual information.

The connection of all systems is executed by the use of special microprocessor oriented adapters.

In the field of the data complexity for the computer internal product and its processing representation we have been primarily solving the problems of validity and consistency of data in a manufacturing data base. In any case reliability and comprehensiveness of all data entered as a source of manufacturing information must be assured.

Validity in our projects means that all data entered into the data base obey any direct or calculated constraints imposed on them. Consistency means that a change in one data item is accompanied by changes in related data items.

Traditional data processing has had separate data files for each application program.

As a result data translation between files has been needed, and data have often been inconsistent. We have separated data files from particular application programs. It was essential to désign such a structure of data that would meet the requirements of the integrated system in terms of
- minimination of conflicting situations that are the result of multiple access of programs to shored data
- minimization of inter-program communications for updating data managed by a particular program or program packages
- minimization of access time
- simple data maintenance.

A compromise solution has been developed aimed at combining advantages and minimizing the short-comings of databank and distributed data base type of data organization. The solution is based on purpose oriented data files for which the following regulations have been stipulated:
- which programs can operate on a shored basis, securing quasiparallel run of these programs which preventing undesirable influencing of shared data
- which of these programs cannot operate on a shared basis, or, in other words, for which programs new data files must be created, the updating of which is done be means of interprogram communications.

We have been developing a common set of data management routines that preserve accuracy and consistency of data.

Whether data are dependent or independent depends on the point of view of the user. Our companies treat this as an organizational issue as inconsistency arises very often when changes by different people at different times are not coordinated. On the other hand techniques that provide for the multidirectional dependency are not yet well developed.

CONCLUSION

When implementing the integration projects in engineering enterprises two basic groups of problems have been dealt with:
- the unified structure of the integration projects
- the integration strategies.

The first stage of the integrated systems application includes the following subsystems:
- Product research and development
- Engineering design
- Process engineering
- Technological process control

- Transport and storing
- Process planning

The integration strategies form the basic approach to the integration systems step-by-step implementation. They cover namely the rules of:
- the distribution of integrated system functions in the computer hw architecture
- the manufacturing data base design and distribution and the principles for data sharing
- data communication in the integrated system architecture.

The integration projects applications can be characterized by the following features:
- computer and automation hw architecture of a LAN type
- distributed manufacturing database based on purpose oriented data files
- usage of communication standards in conformity with the MAP.

From the efectiveness point of view the projects which have been carried out in our engineering enterprises have made it possible
- to reach higher functional and quality level of new products
- to shorten the lead-times of the new products
- to increase the productivity of manufacturing systems.

These effects as to the nature and scope are considerably greater than those, that could be obtained by conventional means.

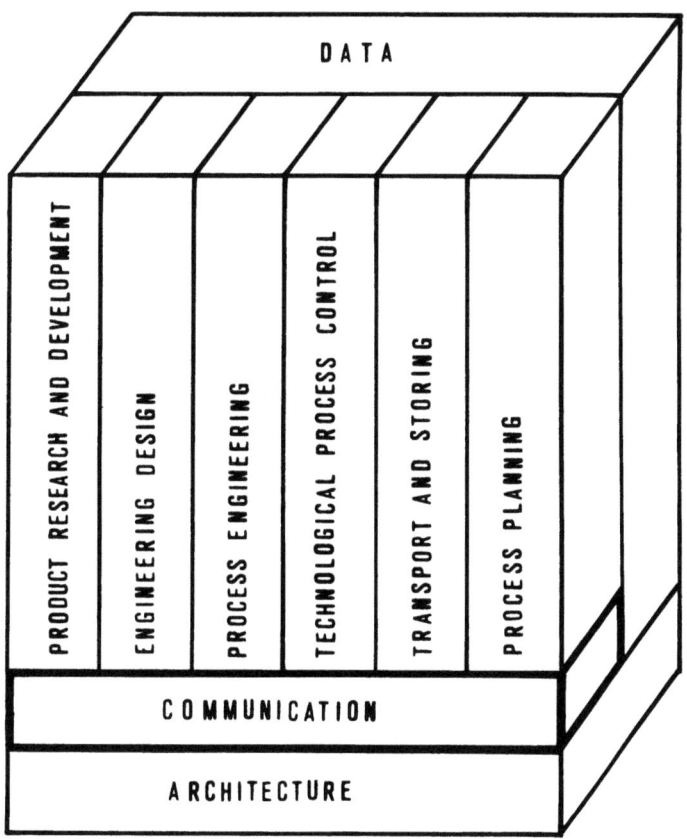

Fig.1. Integration projects structure

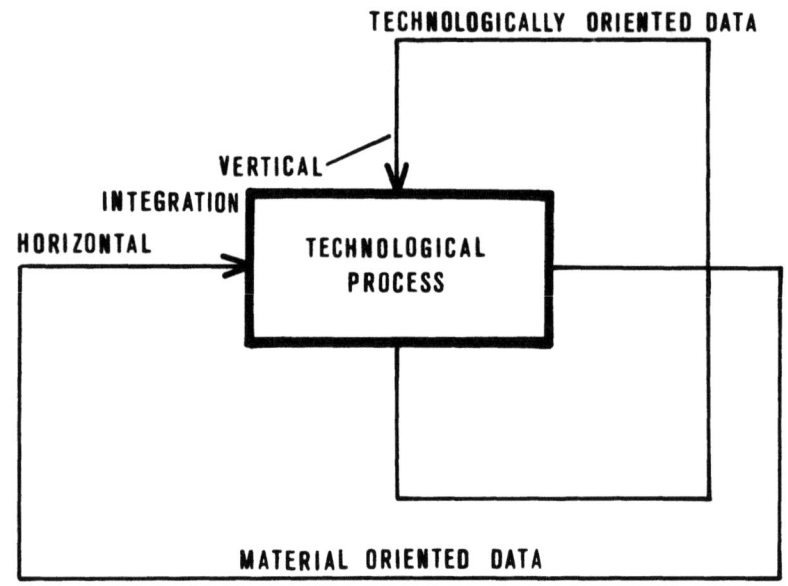

Fig.2. Vertical and horizontal integration

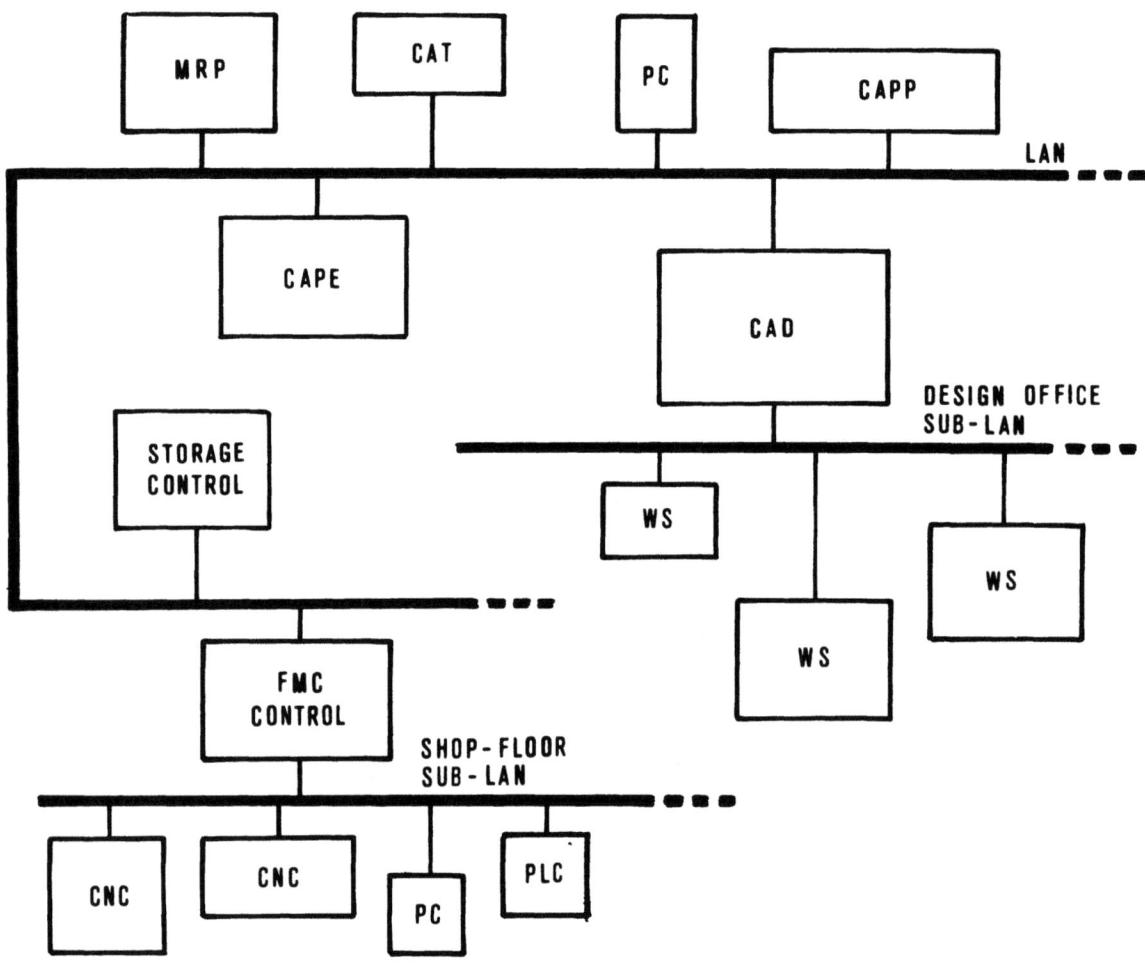

Fig.3. Computer systems architecture strategy

Robot cells and system for small part assembly
S. Ljung
ASEA Robotics AB, Sweden

The use of robots in small part assembly cells is adding flexibility to low and medium-size batch production. The use of intelligent six axes robots combined with flexible grippers and feeding systems can be optimized in a cell concept.

Multi finger grippers and a combination of active and passive feeders can accomodate several variants in a part family without prior change over of the set-up.

A supervisory Cell Control System can control the peripherals as well as provide production statistics for the production cell.

The short life-span of some products promotes the use of re-usable equipment like robots and standard accessories.

INTRODUCTION

The Robotic Cell Concept

The increased number of variants of a given base product, combined with an effort to cut costs related to work in progress and stock, is pushing technology forward in small part assembly. A number of variants combined with a short life cycle of the product call for flexibility in production now and in the future.

Robots can provide versatility and flexibility when used together with proven concepts and new ideas in tooling, sensors and communication.

Robots also provide the tool for quick change over between old and new product generations. In Robotics the self-supporting cell is the most flexible. There are several aspects of flexibility when it comes to robotics. In this paper I will discuss and relate examples from several installations in small part assembly. The examples used have certain aspects in common like the general lay-out, the robot model and the type of industry. (Fig. 1).

The Robot

A number of highly advanced assembly installations have been completed in the last two years featuring the pendulum type robot, like the ASEA IRB 1000. This type of robot accomodates very high speed over the whole work envelope with full access to all sides of the square shaped work area. Feeding of parts is easy to this system. (Fig. 2).

The Industry

The manufacturing industry most active in the use and development of flexible cell concepts is the modern electro-mechanical and home electronics industry. This industry has often a high degree of in-house manufacturing of small part feeding and orientation devices. The more complex but highly intelligent and flexible cell-concept is easy to integrate into this environment.

FLEXIBILITY IN HARDWARE

Flexible grippers

The modern type of robot can be equipped with a multi finger gripper system. (Fig. 3). The ASEA IRB 1000 is a six axes all servo robot where the Multigrip is adapted directly to the mounting flange of the sixth axis. This gives the robot the ability to work the grippers in all directions possible by a six axes machine.

The advantage of not being restricted only to the horizontal and vertical planes has been used in our installations to cut cost and optimize the design of part pick up positions from feeders and magazines. The advantage of being able to present a part in the most practical way for the material or the shape of the detail saves a great deal of money in engineering and hardware.

The reliability of the feeders will also increase working with natural methods. The possibility to fit several models from a family of parts in the same feeder with minimum rearrangement of the part pick up position will increase.

The ability to group feeders close together in different angles has been used in the installations to cut time picking up the parts. An average of 4-5 parts are normally picked up at the same time located close to each others. The robot moves to the fixture and assembles the parts one by one with very little motion in between sequences. This method saves a lot of time and wear on the mechanical arm compared to single part picking.

In a number of installations the robots are carrying 6-8 gripper fingers while only 4-5 are being used at the time for a specific variant. This means that in some cases no work has to be done to the robot hand when the product is changed.

In order to detect parts in the fingers of a swiveling gripper new techniques have to be used. One method is to communicate through Fiber optics. The swivel of the Multigrip (Fig. 3) is fitted with a sender/receiver while each pair of fingers is equipped with Fiber optic sensors.

The robots sixth axis is more than adequate to allow transmission of the signal through the free air between the individual fingers and the main fiber.

Flexibility in part supply

The combination of a well defined, square shaped work envelope and the ability to carry multiple fingers has been utilized for flexibility in a number of applications. Through program selection, one or more, simple feeders can be passive during parts of the day depending on the product mix. Some cells are equipped with up to eight (8) feeders while only 4-5 are used at a time. This means that neither the robot nor the feeders need to be changed when a new product in the mix enters the cell. Only the program selection has to be initiated by the specific variant.

Another way that can be utilized for flexible part supply is the use of bar coded trays in a standard pallet magazine. The pallets can be called up randomly by the robot in cooperation with a Cell Controller - PLC-type. (Fig. 4). The multigrip system with a maximum of eight fingers can be used by the robot for quick change over.

Sensor technology has naturally been used for identification and selection of parts for assembly. Smart sensors like Vision Systems and more traditional types like proximity and Fiber optics must of course interface to the robot in an uncomplicated way to simplify the installation.

Flexibility in the fixturing

By using an intelligent robot with six axes it has been possible to arrange automatic retooling in cells and systems for mainly the electronics industry. In an installation where transistors are fastened to a "heat sink" the IRB 1000 can do the physical change over of some of the part dedicated fixturing. The equipment is stored locally in this case.

The modern type of standard conveyor systems available on the market are very suitable for flexible assembly with robot. An intelligent robot in communication with a Cell Control System can very easily change either the whole pallet or reorganize the pallets to accomodate a new product with limited change over time. In complicated small batch runs this system can keep track of the different combinations of tooling in an accurate way in cooperation with the main production computer.

The robot as a tool holder

A variety of tools have been developed for robot use. A six axes robot of the pendulum type can copy almost any motion that the human arm and wrist can perform. A typical motion is the wrist-bend motion used when fastening parts to each other by means of metallic clips.

Other examples currently used, and very often developed by the user, are tools for the assembly of pre-activated springs or for routing and preparation of wires for terminal assembly. One of the most common tools in robotics is the automatic screw driver either pneumatic or electric, supplied with an automatic part feeding system.

Tools like these can either be carried constantly by the robot or picked up when needed by the robot through a tool changing system. These systems are common today supporting several pneumatic functions and occasionally signal connections to the gripper.

Flexibility through "Routing"

The concept of robot controlled cells can be used as a single cell or as several cells in a line combined with a standard conveyor like in the case of a gas valve assembly at Schlumberger in the Netherlands. (Fig. 5). In this case each cell adds on to the product a predetermined number of parts in a controlled preselected cycle. Several variants are assembled in highly flexible cells supervised by a Cell Controller. Another way is to route the product through a number of independent cells, each with a specific function, by guided vehicles or conveyors with alternate routing. The functions of the cells can be: Assembly stations, testing, rework, product idenfitication or packaging stations.

When connected to a production computer system this type of set-up can initiate the production of "product replacement" at point of sale.

Integrated quality control

Quality control is a very important part of the robotics cell. Quality can be achieved in the simplest way by acknowledging part present in grippers and magazines to prevent failures.

The quality of insertions can be controlled by using very simple methods. When dealing with small forces a preadjustable gripper plate can be used featuring two spring loaded plates connected through a parallel motion and checked by a very accurate mechanical switch.

A very accurate, non-contact, measuring check requires a higher level of sensors. The excellent repeatability of the robot makes it feasible to use a laser measuring system.

In an application at the Volvo Engine Plant a "Laser Optocator System" from Selcom is used to inspect the assembly of cylinder heads, and in particular, the insertion of the tappets. The laser optocator is fully integrated to the robot which in this case provides an extremely accurate and fast examination of the tappets. Sixteen points are measured in ten (10) seconds over the full length of the cylinder head.

Leakage of hazardous fumes and beams can also be checked with robots using probes of different kinds.

FLEXIBILITY IN SOFTWARE

A single cell as described earlier, with, for example, active and passive peripherals, as well as several cells in a system can not be flexible without some kind of supervisory control, internal or external. A robotics cell for the assembly of odd components for instance, can sometimes require batch sizes of no more than thirty boards in a highly flexible environment.

The Cad computer, working through the supervisory production computer, can generate component identification, position on the board and polarity of the component to the robot cell as well as other equipment in the production shop. All data can be received to the robot cell controller from the production computer while the robot is working. A library receives all information about the next batch while the robot is finishing up the batch already in production. Production statistics are also a part of this system. Small batch flexibility and consistent quality are the main advantages of robotics used in the production process. Programs can also be generated off-line to accomodate quick change over from olf to new products. The ability to do off-line programming is increasingly being used in installations featuring robots. This procedure quarantees better programs and shorter stops for changes in the production. Be aware though that the mechanical arm is still doing the job, not the computers.

THE RE-USE OF EQUIPMENT

The typical robot cell featured in this paper has a very high content of standards like robots, magazines, grippers and other peripherals. Unlike a dedicated machine, mechanical arms and standard peripherals can be re-used. This is becoming increasingly important.

The product life-span in the home-electronics and appliance industries is getting shorter by the day. At the same time the number of variants of a base product is rising. Standard robots, magazines, gripper bodies and tools like automatic screw drivers can be re-used as well as most of the modern type conveyors with plastic chains or pallets. One of the big advantages of the robot cell is the re-use of standard material.

Utilizing the old equipment in a new cell is effective. The personnel is familiar with the equipment as well as the basics of the assembled product, even if it has a slightly new design.

These facts make installation and start-up easier with a guaranteed high up-time as a result. Service and maintenance are also more effective from the start since familiar equipment is the base of the "new" system.

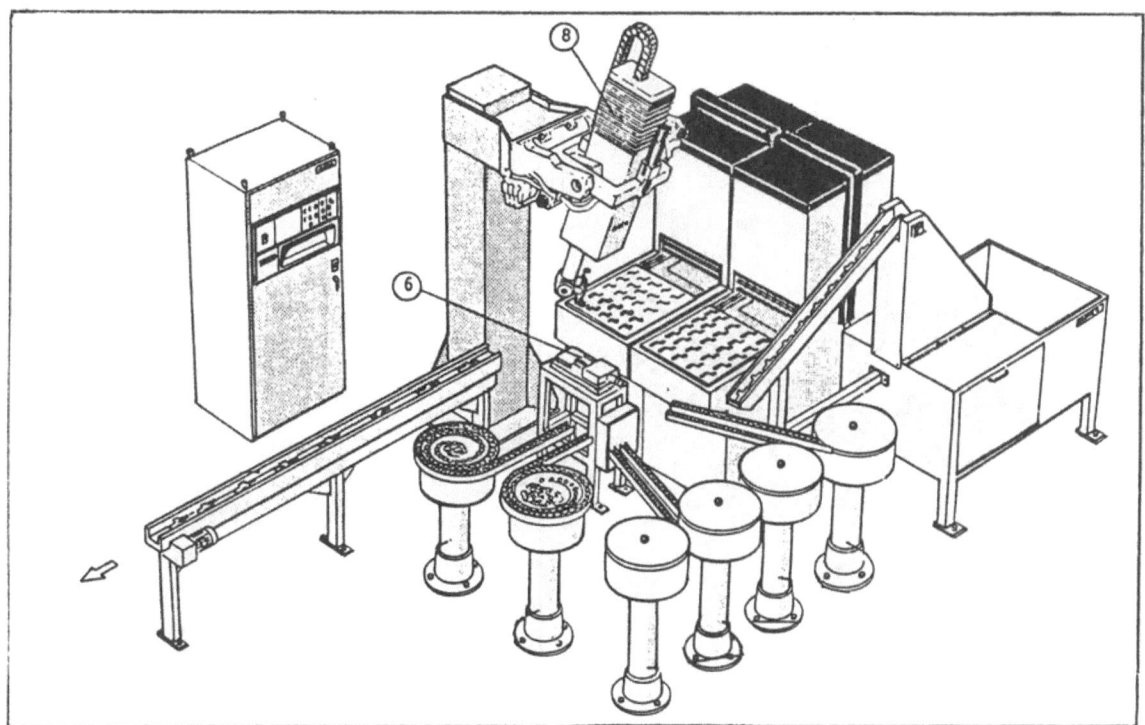

Fig. 1

A typical lay-out of a small part assembly cell, featuring the ASEA IRB 1000 robot and standard peripherals. (Wiper assembly at Electrolux).

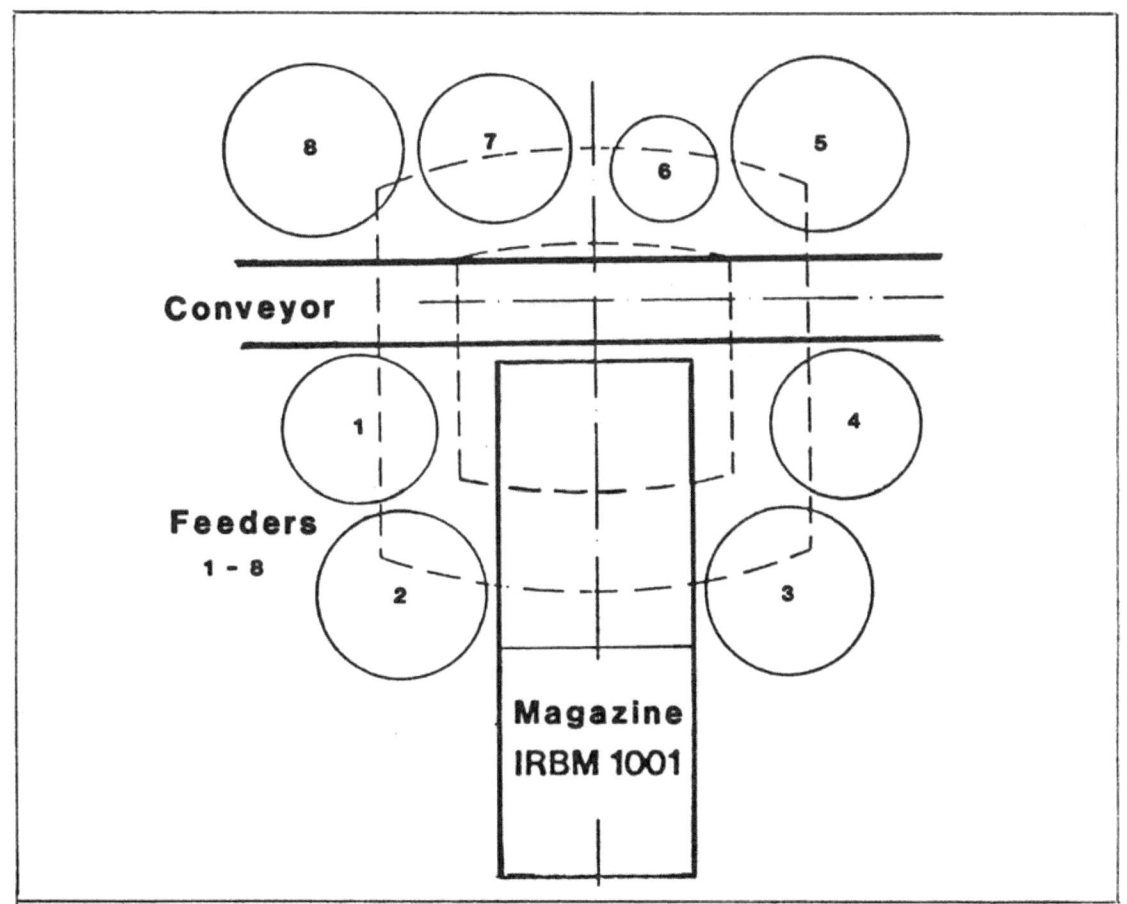

Fig. 2

A cell lay-out from the Schlumberger Group, with a ASEA IRB 1000 robot and the standard pallet magazine. The outer broken line is showing the work envelope of the IRB 1000.

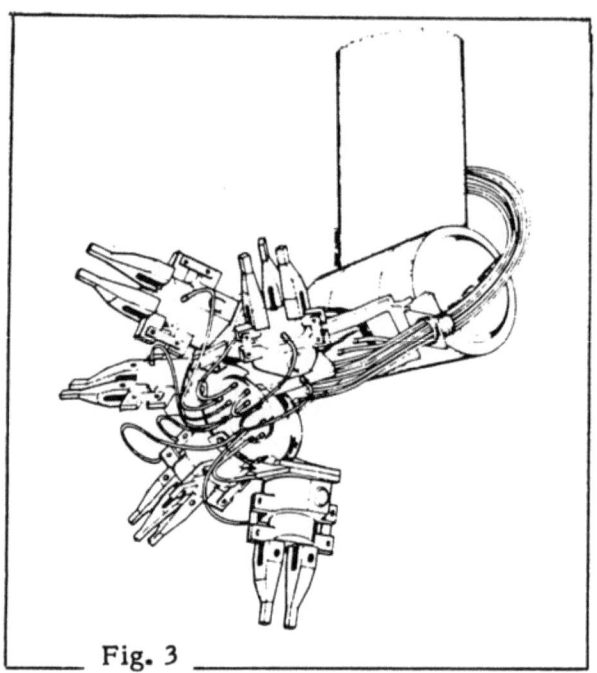

Fig. 3

The ASEA Multigrip System with 5 fingers. Maximum is 8 fingers

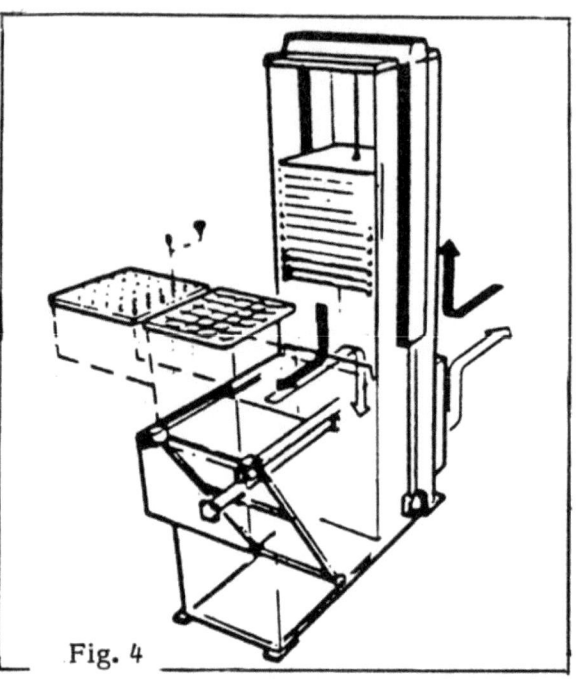

Fig. 4

The ASEA IRBM 1001 Pallet Magazine with a maximum storage of 27 pallets.

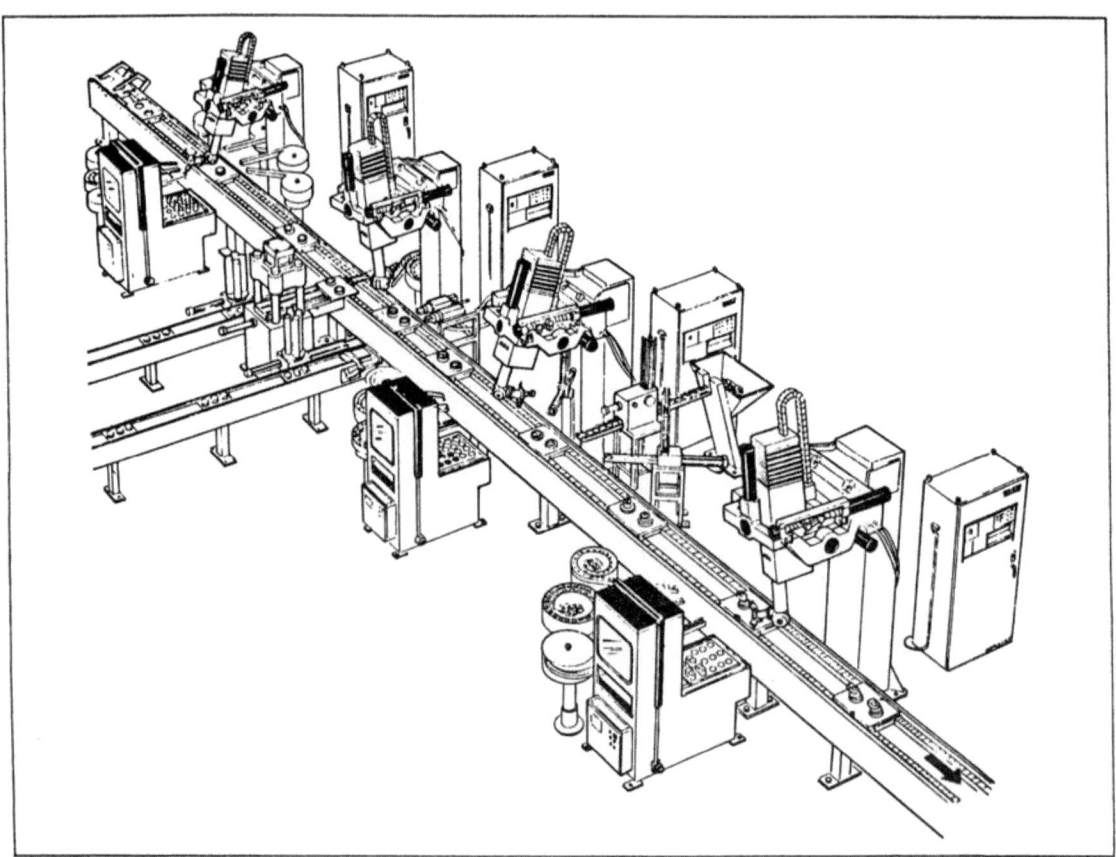

Fig. 5

The flexible assembly line at Schlumberger, consisting of four ASEA IRB 1000 cells in a line.

Evaluating FMS by simulation
O. Arnaud and E. Bloche
Renault Automation, France

INTRODUCTION

In a first part, the authors will point out, using examples, why it is necessary to use simulation for the designing or improving of a flexible manufacturing system.

Then, they will overview several different types of simulators for production unit designing. Through a comparison of these simulators, they will try to show the advantages of specialized simulators for production units.

Finally, the main functions of an FMS specialized simulator will be highlighted.

I. WHY SIMULATE FMS?

A. An example of a flexible production line for the manufacture of mechanical parts.

1. Representation of the production line.

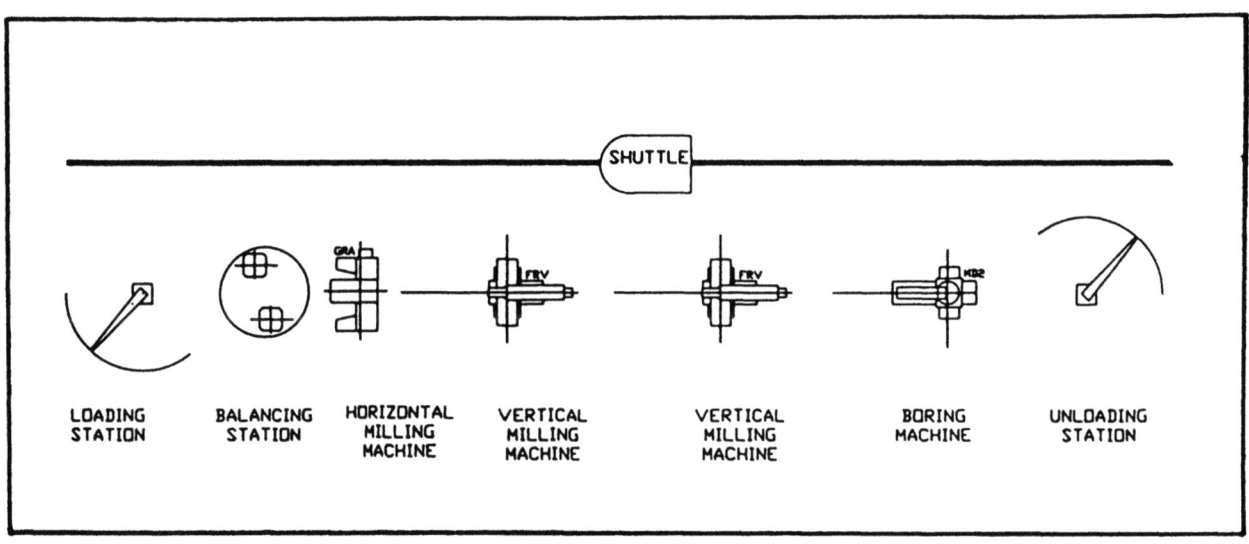

Figure 1. - A flexible line example.

This production line (figure 1.) exists, but is not in such a condition that it can achieve the production programme desired. It is made up of a series of stations, a shuttle to convey the parts from one station to the next and a number of pallets. Four different types of parts pass down the production line as follows:

- Loading
- Balancing
- Horizontal milling
- Milling at one of the vertical milling machines
- Boring
- Unloading, washing

When the operation at one station is completed, the part summons the shuttle, which conveys it to the next station, if this is unoccupied. If it is occupied, the part remains where it is. If the shuttle is summoned at more than one station, priority is given to the part furthest down the production line.

After each production run, an inspector checks the part and initiates finishing.

The desired production capacity is for several dozen parts per day, but in practice, this is not achieved. What methods can be used to understand this phenomenon?

2. Static calculation

An initial rapid calculation of the "station regime" type can be made using:

- The average time taken for the entire operation for each part
- The average time taken for transportation from station to station
- The reliability of the equipment

This calculation gives the average capacity of the production line and the average cycle time. According to the results of this calculation, the production line is capable of achieving the production rate, but it does not take into account:

- Variations in the launching initial of the parts
- Variations in operation times
- The frequency of breakdowns

A second method using statistical calculations can be employed.

3. Statistical calculation

There are several possible methods, such as the CAN-Q algorithm or the MVA (Mean Value Analysis) type algorithm, using the Reiser or Hildebrant heuristics. The CAN-Q method shows us that the production programme could be achieved, but the study is incomplete here as well, as this method uses the following restrictive hypotheses:

- The activity times are distributed according to an exponential distribution;
- Each station has an infinite storage capacity (i.e. there are no blockages).

A simulation of the production line was carried out .

3. Simulation

We have established that the current production line configuration only turns out 50 per cent of the parts envisaged for the production programme.

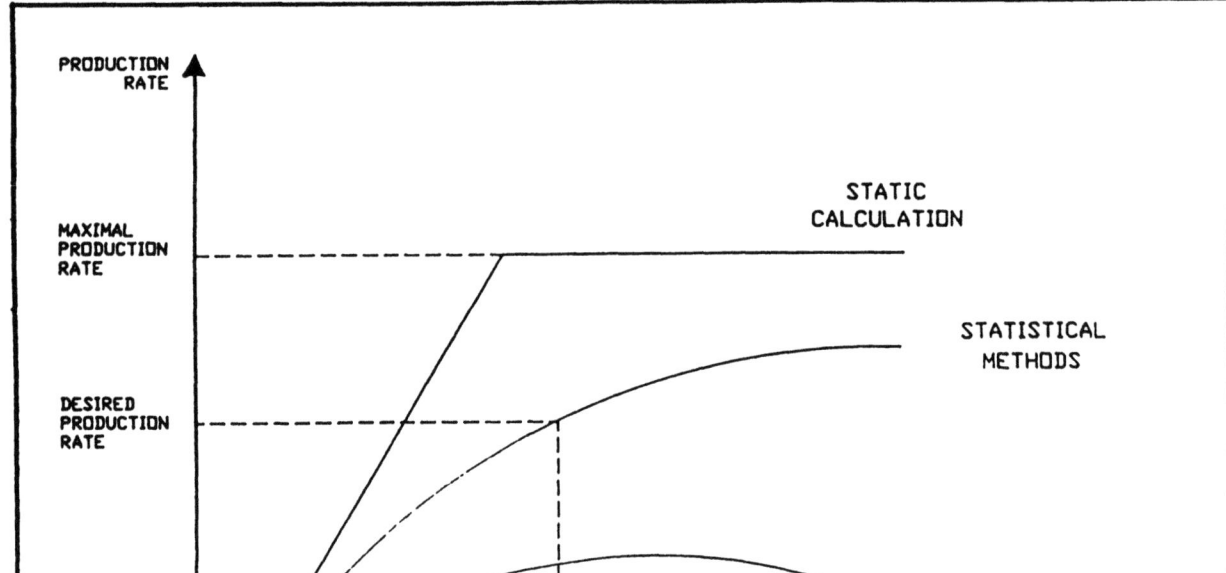

Figure 2. - Comparison of the different methods

This raises certain questions about the insufficient production rate:

- What would be the effect of buffer stocks?
- What effects do breakdowns have?
- What effects does checking and finishing have?
- What effect does the number of shuttles have?
- What effect would the introduction of additional stations have?

By testing several scenarios, it is possible to reproduce the characteristics of the production line.

Firstly, the number of transporters has practically no effect on the flow. Checking and finishing make up 7 per cent of the production. The most important factor is the reduction in the number of breakdowns and production hazards, as, taking the given figures into account, the elimination of all breakdowns would increase production by over 30 per cent. The area of the vertical milling machines was found to be a particular bottleneck.

Finally, if the configuration were changed, a control station and a finishing machine added, an increase of 30 per cent would be attained.

From these conclusions, it is possible to suggest solutions, giving their effects and cost:

- Improvement in the availability of the production line
- Equalisation of time factors
- Better machine linking
- Automation of the workstations
- Introduction of a control station and a finishing machine

From this basis, it is possible to implement the solutions with the most advantageous cost / effect ratio, enabling the production line to sustain the desired flow.

The production line appears on figure 3. with the control station and the finishing machine.

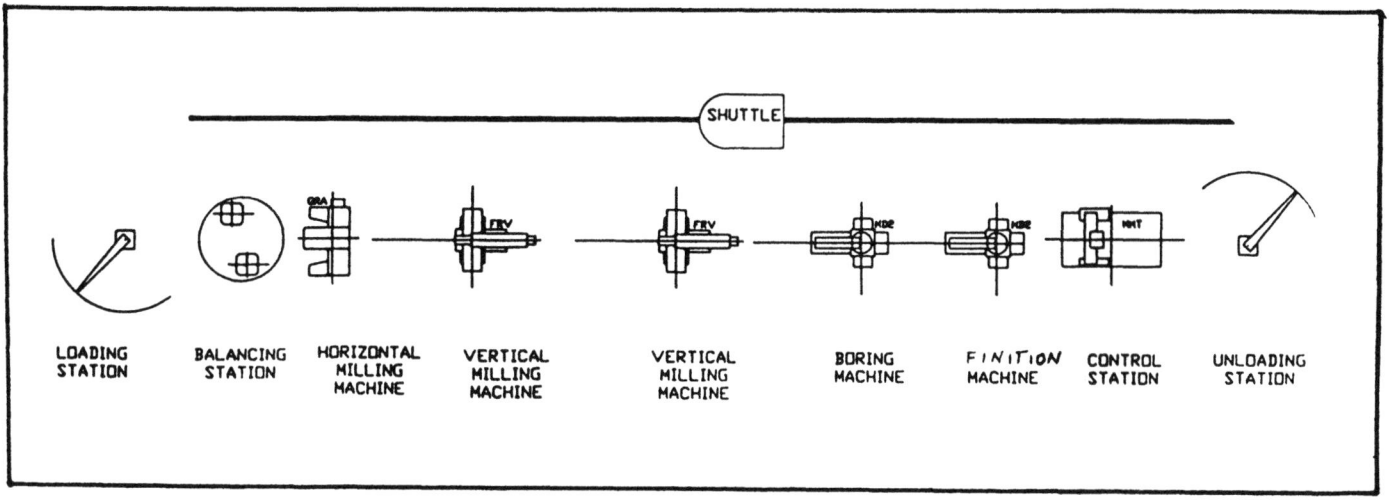

Figure 3. - Flexible line of figure 1. as it appears now

B. An example of a flexible manufacturing system

The workshop represented in figure 4. is made up of:

- 4 machining centers
- 1 palletizing station
- 1 assembly preparation station
- 1 washing machine
- 1 control station
- 1 automatic pallet storage
- 1 automated guided vehicle system

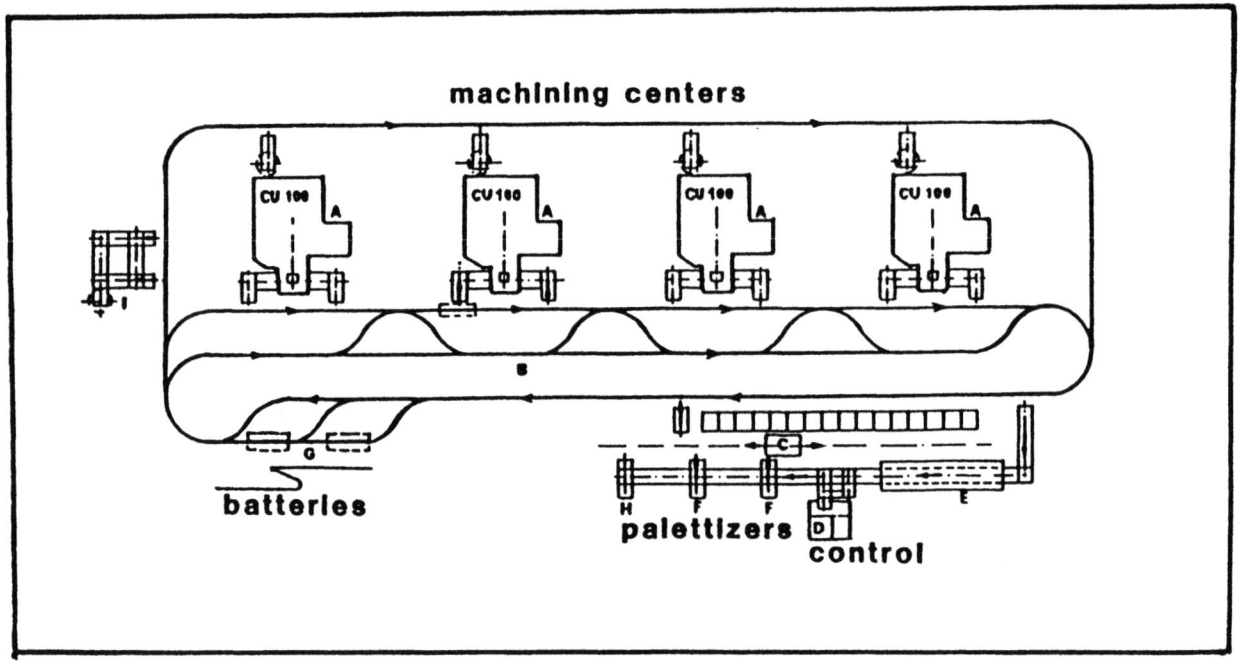

Figure 4. An FMS example

The target is to produce 600 different reference mechanical parts, in small quantities, with machining times varying from 15 minutes to 2 hours. The part goes through a simple process:

- Palletizing
- Machining at one of the machining stations
- Washing
- Dimensional inspection
- Depalletizing

This target should be attained, if a 90 per cent loading rate for the machining stations, excluding breakdowns, is ensured. However, two important constraints have to be taken into account in order to achieve this:

Because of the complexity of the parts to be machined and the differences in their morphology, it was proved to be impossible to find identical clamping tools for the different parts. Therefore, it is necessary to have a particular clamping tool for each part. Considering their high cost, it has been decided that there would be only one clamping tool by type of part.

Each type of part ,requiring specific machining, the operations on the machining station might require the presence of 40 different tools. In fact, several tools could be used for many parts.

In order to ensure a 90 per cent machine usage rate, it was evident that the tool changing time had to be minimized. This was achieved by adding a 40 tool magazine. Thus, the machine has an active tool magazine and a second one prepared by a robot during the manufacturing. It was also necessary to group together the parts requiring the same tools mounted in the machine's magazine.

It is clear that only simulation can solve this problem, answering the following questions:

- How should the parts be regrouped?
- How long in advance should the process be prepared?
- How many shuttles are required in order to ensure handling?
- How does does the workshop behave when a breakdown occurs?

- The simulations undertaken show that the workshop production rate is 93 per cent, thanks to carefull regrouping. They also show that the automated guided vehicle system is the weak point.

- In practice, in order to ensure the flow, it is necessary to employ two autoguided vehicles, but these are only in use 60 per cent of the time (i.e. 1.2 vehicles).

- What happens when a vehicle breaks down?

- A statistical calculation reveals that 20 per cent of the handling capacity is lost and hence 20 per cent of the production. Simulation gives a totally different result - the damage indicated has disasterous results for the process: The production rate of the machinery falls from 93 per cent to 30 per cent and consequently the automatic workshop becomes blocked.

The first example of the flexible production was fairly simple and in this case, only simulation could disperse the bottlenecks. It should be noted that in some cases, the analytical methods are sufficient, however, it is much more accurate to carry out a simulation.

The second example is a flexible cell, where simulation is obviously the only way to ensure its smooth running and thus to make it more economical.

We have seen in two examples how simulation is indespensible in setting out or improving a flexible manufacturing system. We are now going to quickly examine the different types of simulation software available for the design or improvement of production units.

II. SIMULATION LANGUAGES

There exists two types of simulators: general purpose simulators as SLAM and PETRI, and specialized simulators as SAME/AGVS. We will first overview two examples of general purpose simulators.

A. General purpose languages

The most used tools are modelisation and simulation tools for queue systems. These queue systems are represented in either the form of primitives such as "stations","feeders","clients" and "ressources" when using QNAP2 software or in the form of graphical primitives on PETRI networks, or SLAM systems. In this part we will describe the two later systems.

1. PETRI networks

PETRI networks are simple modelisation tools which can easily be applied to simulation. They are based on automation concepts and are similar to GRAFCET.

The advantage of this representation is the small number of symbols (see diagram) ; in fact there are only three symbols:

- Places , corresponding to a notion of state
- Transitions , corresponding to the passage from one state to another
- Arcs , linking places and transitions

Usually, additional properties are associated to the three primitives:

- Places can be temporized (duration of the state).
- Arcs can be valued (number of tokens necessary to validate a transition).
- Transitions can be interpreted. That is that , depending on the systems physical variables, they are given necessary conditions for operation.

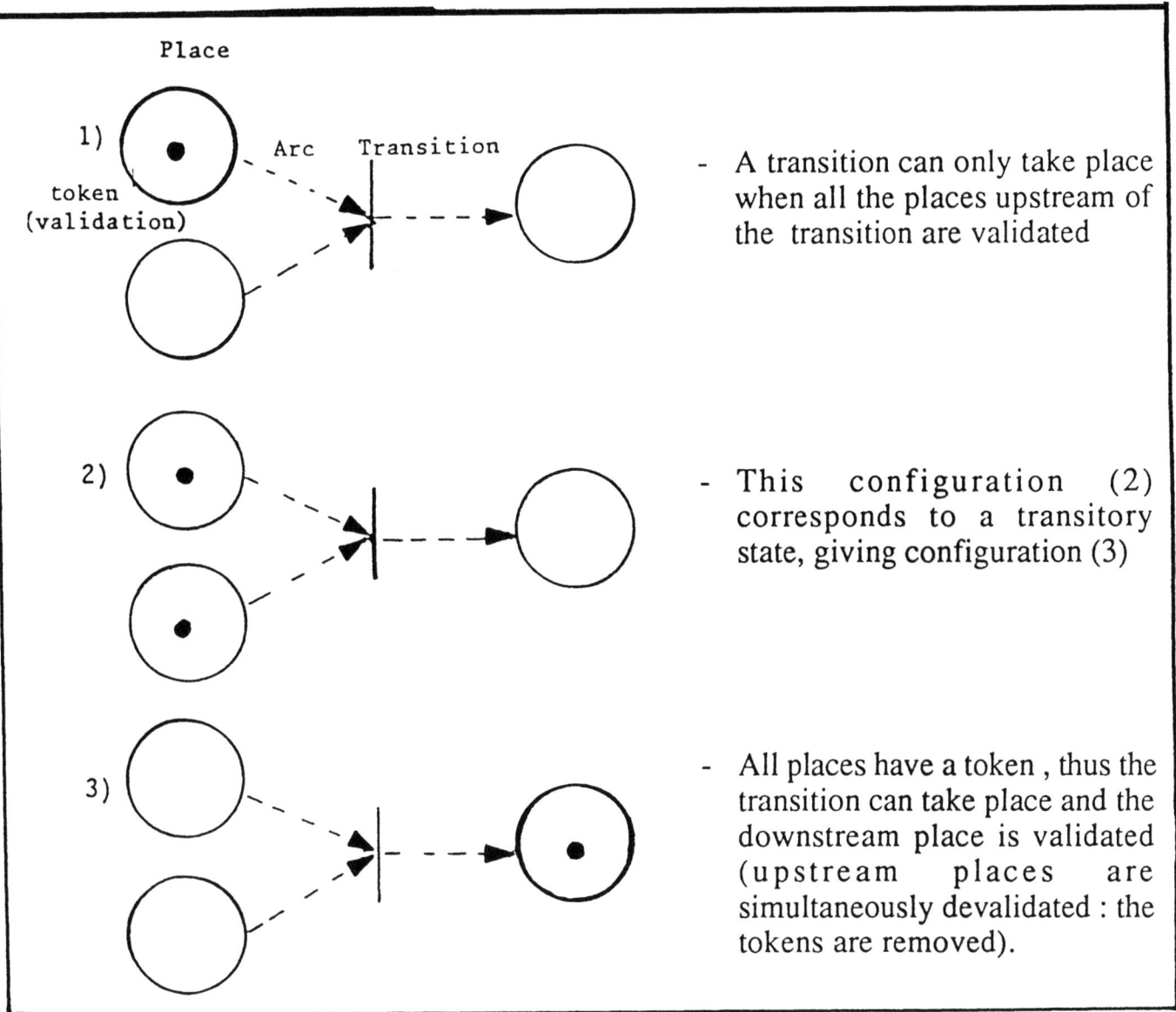

Figure 5. - PETRI representation

The example in figure 6. corresponds to the modelisation of an elevator control system with two positions (upper and lower). This elevator acts as a transfer between two conveyors situated in two different storeys, which, in the example, convey foundry cores (direction is from top to bottom).

The system operates as follows:

- When not in operation the elevator state is the lower position (1)
- When a part arrives on the upper conveyor, the "rising" actuator is activated (2).
- The elevator passes thus in the "rising" state, which lasts t1 (3).
- When the elevator arrives to the upper position, it activates a "stop" activator which stops the engine (4).
- The elevator is now in the upper position. It will remain in this position as long as the "descent" actuator is not activated (for example by the movement of the core).

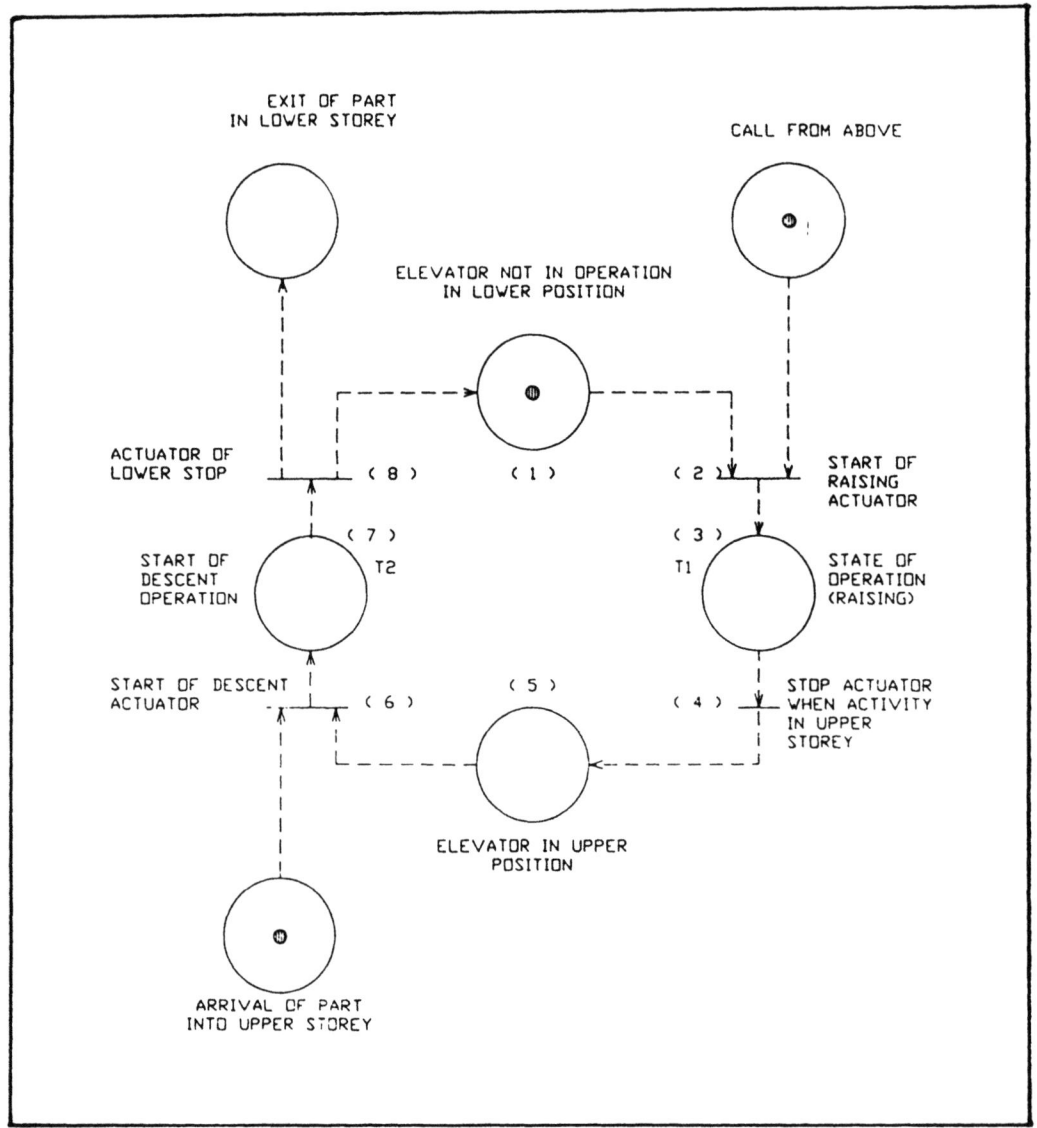

Figure 6. - A PETRI elevator example

The descent operating cycle (5, 6, 7, 8, 1) is the same as the rising cycle.

PETRI networks are particularly well adapted for the analysis and modeling of control systems.

2. SLAM

SLAM, marketed by Pritsker and Associates is a general simulator based on a system of queues and activities. This software uses a high-level graphic language thanks to which the creation of units (parts or information) , the waiting of resources, activities, and the destruction of units can easily be represented.

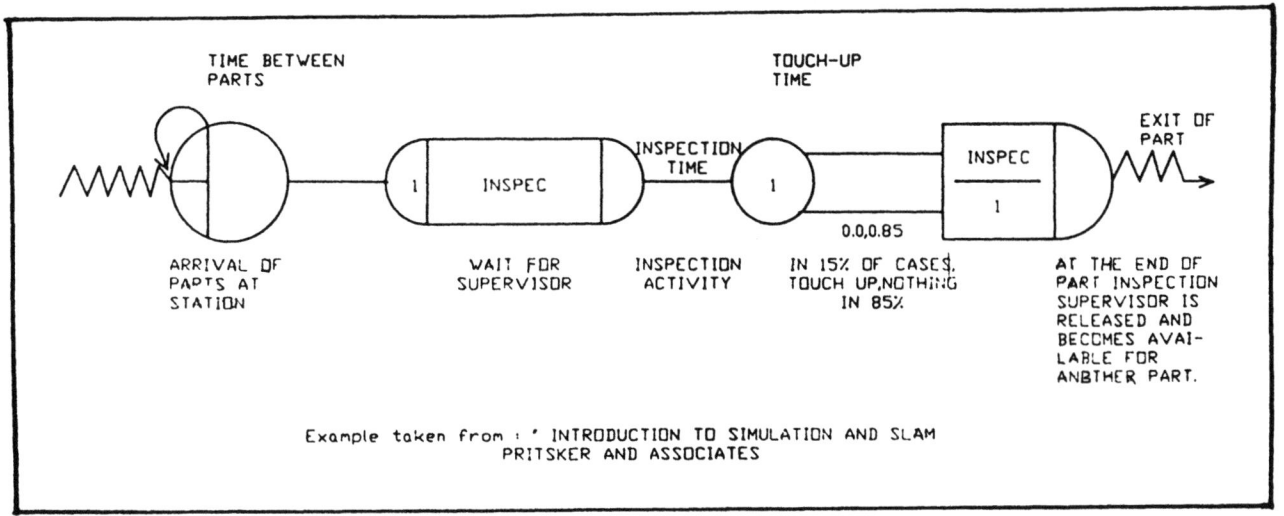

Figure 7. - SLAM example

Figure 7. shows the operation of an inspection station at the exit of a manufacturing workshop :

-   Parts arrive into the workshop at a possibly random rate. They are stored at the foot of the inspection station. There, they wait for the supervisor to check them (they are checked one by one). At the end of inspection 15% of the parts must undergo retouch.

SLAM is an industrial software which is especially well suited for describing physical flows. This explains why general purpose simulators of this type are used to modelise and simulate production units.

B. Specialized simulators

We have seen that general purpose simulators can treat many types of problems but are very abstract. New generations of simulators tend to be more handy and we can find on the market specialized simulators adapted to certain types of production units. We will overview an example of such a simulator for material handling systems developped by Renault Automation.

This simulation package's name is SAME/AGVS which means Simulation Applied to Manufacturing Engineering for Automated Guided Vehicle Systems.

To use SAME/AGVS the conceptor sketchs a network and describes its technical characteristics to the simulator. He can also define the control rules of the AGVS.

Physical description is made using the following components:

- Section
- Diverging switch
- Converging switch
- Crossing
- Source station
- Well station

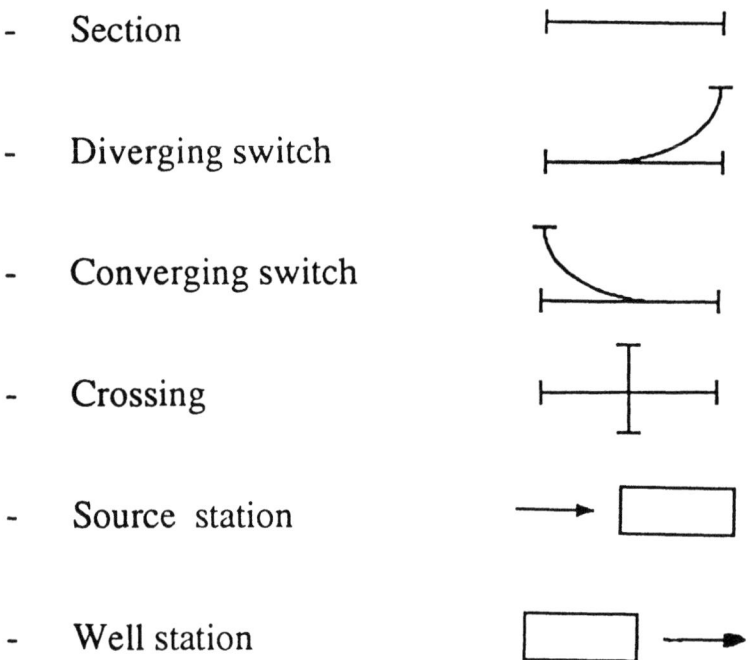

Figure 8. shows the model of an AGVs network using these basic components.

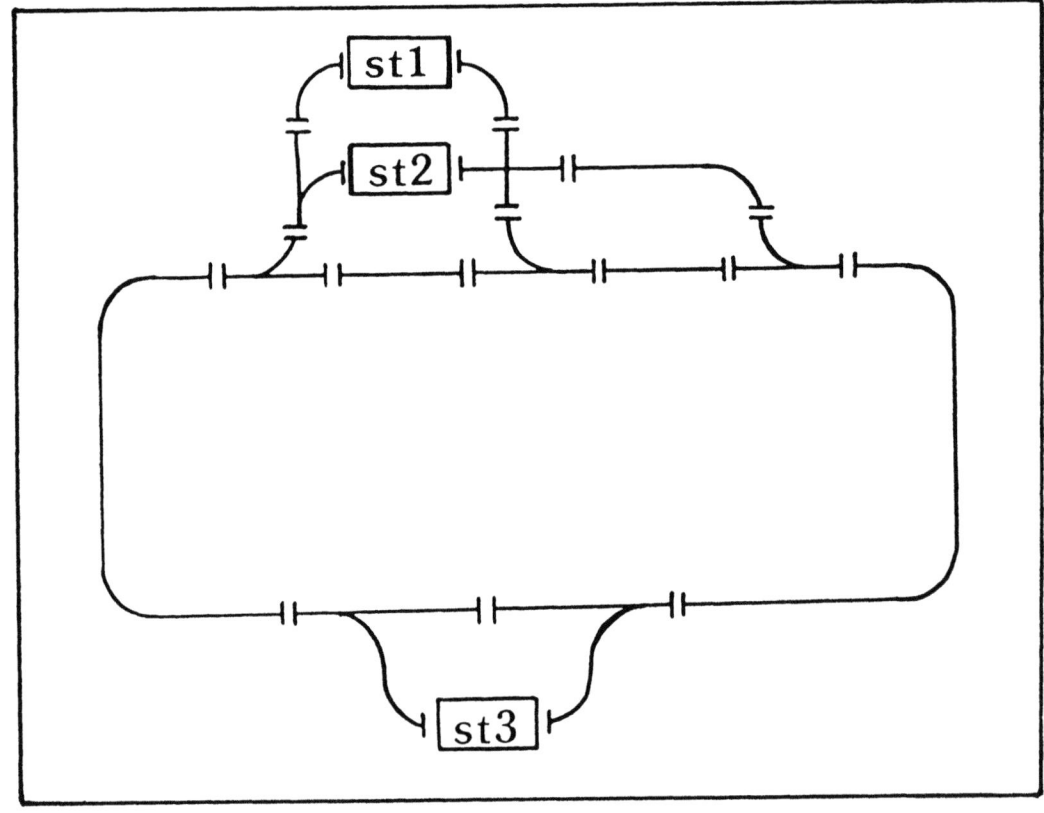

Figure 8. - Model example of AGVS components

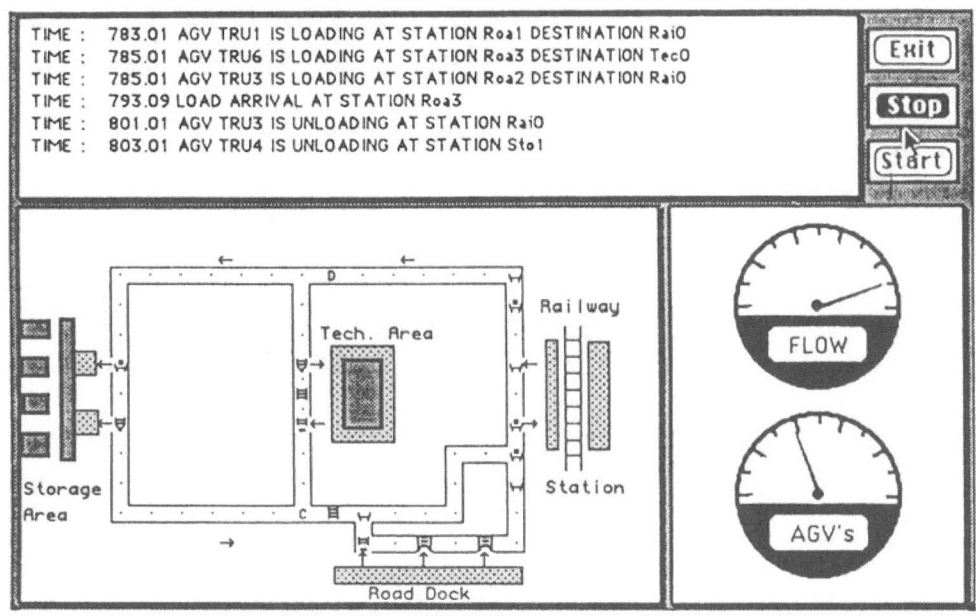

Figure 9. - Graphic display of a simulation run

SAME/AGVS allows the visualization of the behavior of the material handling system (figure 9) in graphic and semi graphic display. SAME/AGVS runs on microcomputers such as IBM PC, Micro Vax II and Apple's Macintosh.

In the next paragraph, the three simulation tools we have just seen will be compared on a simple example.

C. Comparing simulation tools

Figure 10. presents a divergent switch modeled in SLAM; the AGV arrives at the queue H, and awaits the release of the divergent switch. After an unblocking period, it releases the component where it was. At this stage a FORTRAN program decides which exit will be taken in regard to the AGV data : assignement, load,... After a certain period of time (transit time), the returning reply allows the AGV to position itself in the queue I of the left exit (LE), or in J of the right exit (RE). Once this exit is free, it will be possible, after the unblocking period, to release the divergent switch for subsequent traffic.

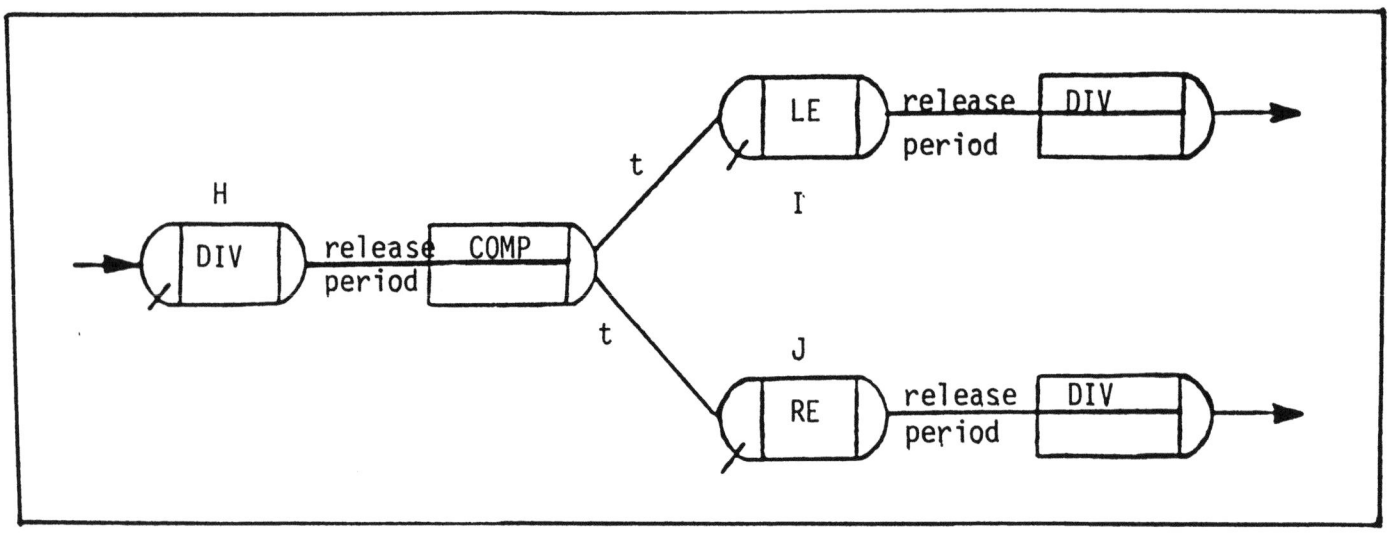

Figure 10. SLAM representation of a divergent switch

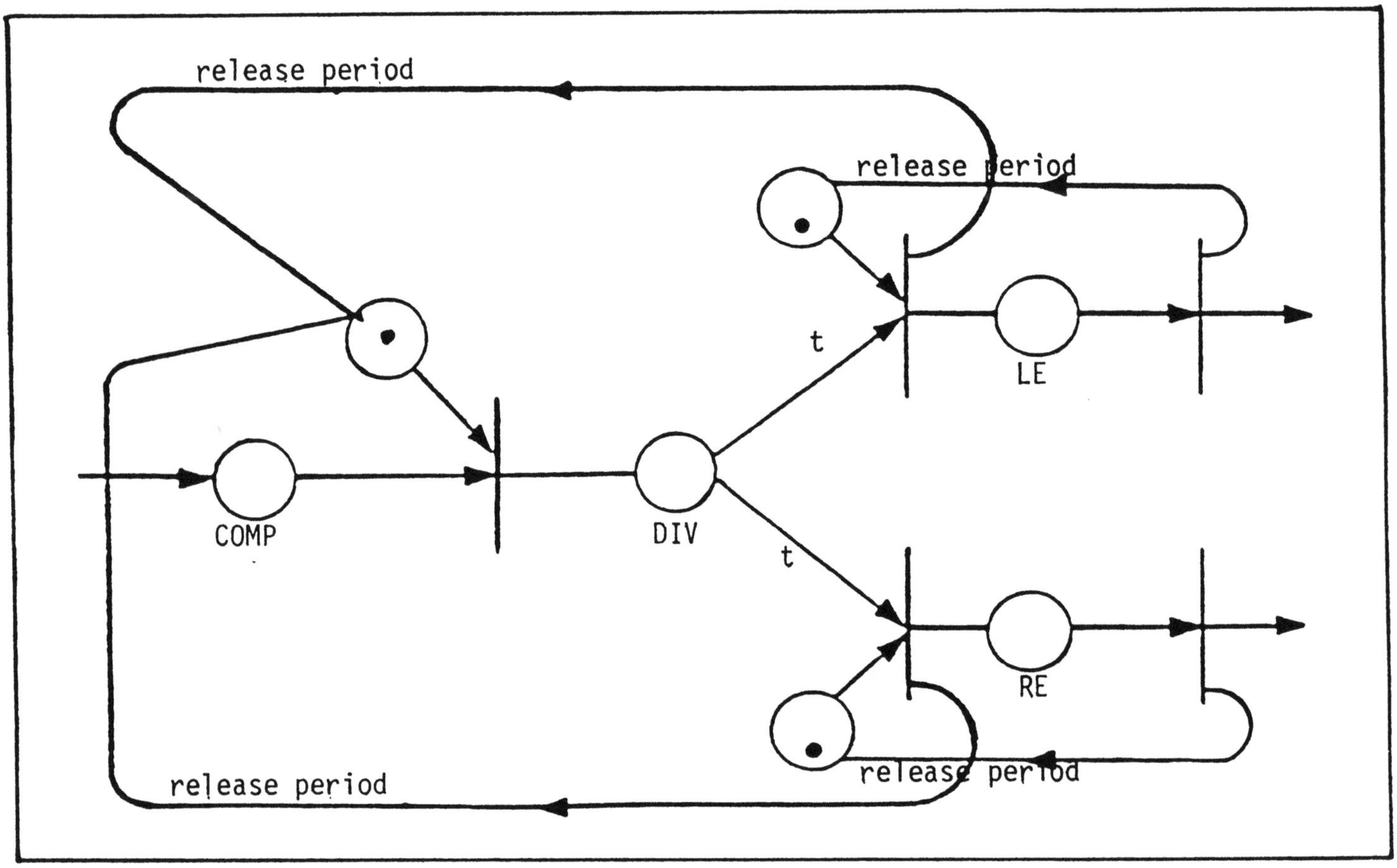

Figure 11. PETRI representation of a divergent switch

The diagram of figure 11. is a model using a PETRI network for the same divergent switch described for SLAM. The AGV will only advance when the tokens mobilized by the passage of the preceding AGV are returned to it; this allowing the PETRI transition.

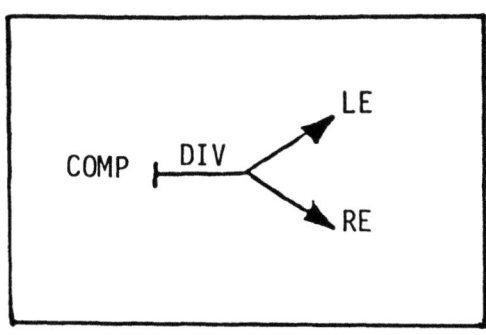

Figure 12. - SAME / AGVS representation of a divergent switch.

The same divergent switch with SAME / AGVS can be seen in figure 12.

Both general and specialized simulation tools are evolving very rapidly. However, as we have seen, some simulators are better suited for representing control operations (PETRI), others for representing flows (SLAM), and yet others for modelising and simulating particular applications (AGV Systems, Flexible Manufacturing Systems, etc...). We will now overview the main functions of FMS.

III. THE FUNCTIONS OF AN FMS ORIENTED SIMULATOR

There are two main functions that define an FMS, production means and piloting rules. We will first see the production means which include all the physical components of an FM workshop.

A. Production means

1. Workstations

There are two general groups of workstations, machining workstations and production workstations. The workstations are logically grouped in types where the same operation can be done by any of the workstations of the catagory.

The workstations of the two groups have common characteristics :

- Their type
- Their utilisation cost
- Their breakdown frequency

However there are differences:

Machining workstations can have at most one input and/or output waiting pad while production workstations can have more than one. On the other hand, production workstations have no tools as machining stations have.

2. Production resources

These workstations need a certain number of production resources. For example a machining workstation can need tools, or parts may need pallets to go through certain workstations. That is why it is necessary to include these ressources in an FMS specialized simulator:

- Workforce (function , number)

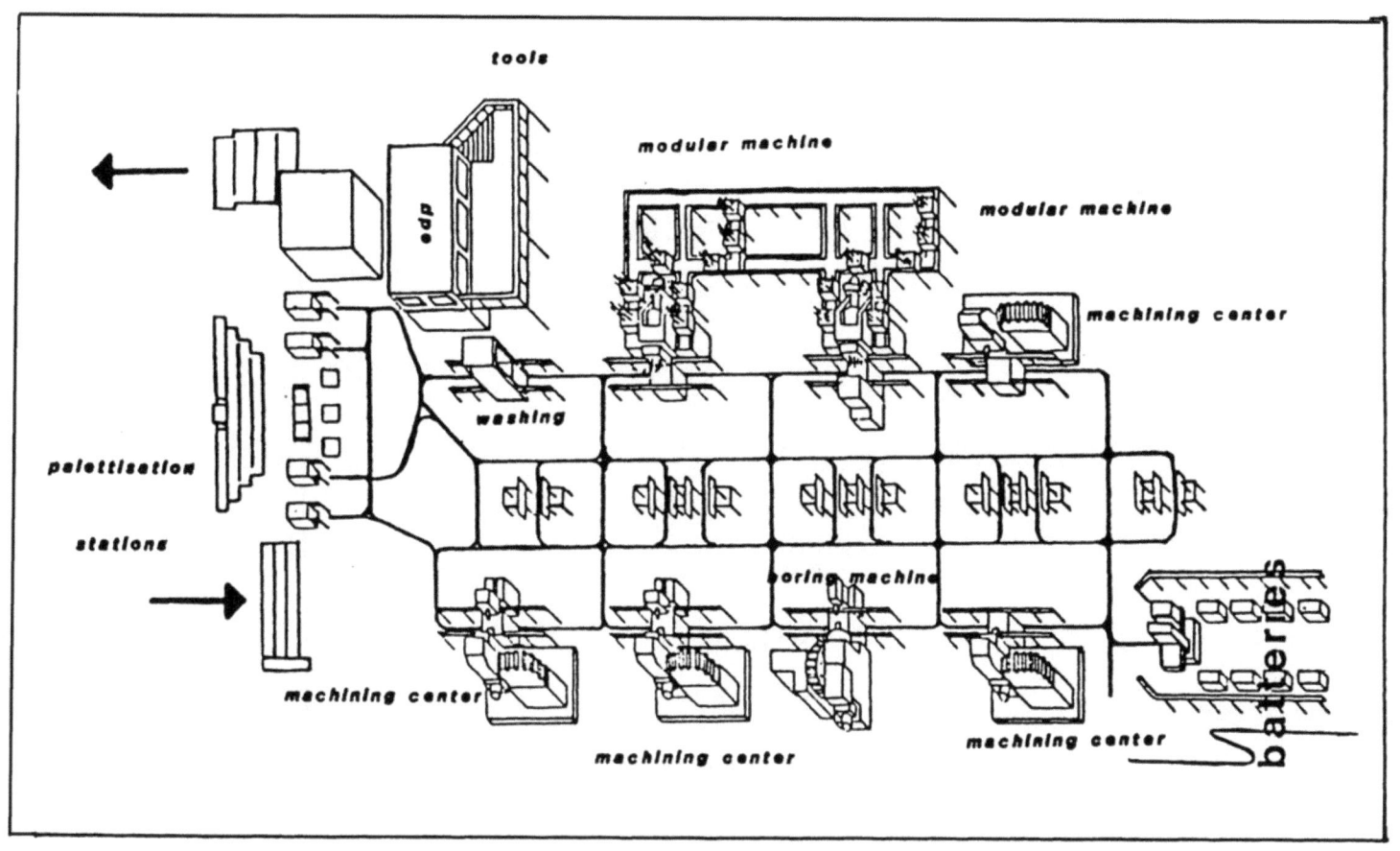

Figure 13. - Layout of Boutheon workshop

- Pallets (type , number)
- Tool batches (type , number)

3. Link with handling and storage systems

The workshop of figure 13. is a Renault FMS that has been settled at Boutheon near Lyon. Every handling in that workshop is made by an AGVS. In order to ensure a good utilization factor of the production unit, it is necessary to know the following characteristics :

- The number of trucks
- Loading and unloading time
- Transport and arrival time
- Breakdown frequency

These parameters are to be taken into account by the FMS simulator. If the conceptor does not know these technical characteristics of the AGVS, it is important to simulate precisely the AGV network. This can be done by using another specialized simulator like SAME/AGVS.

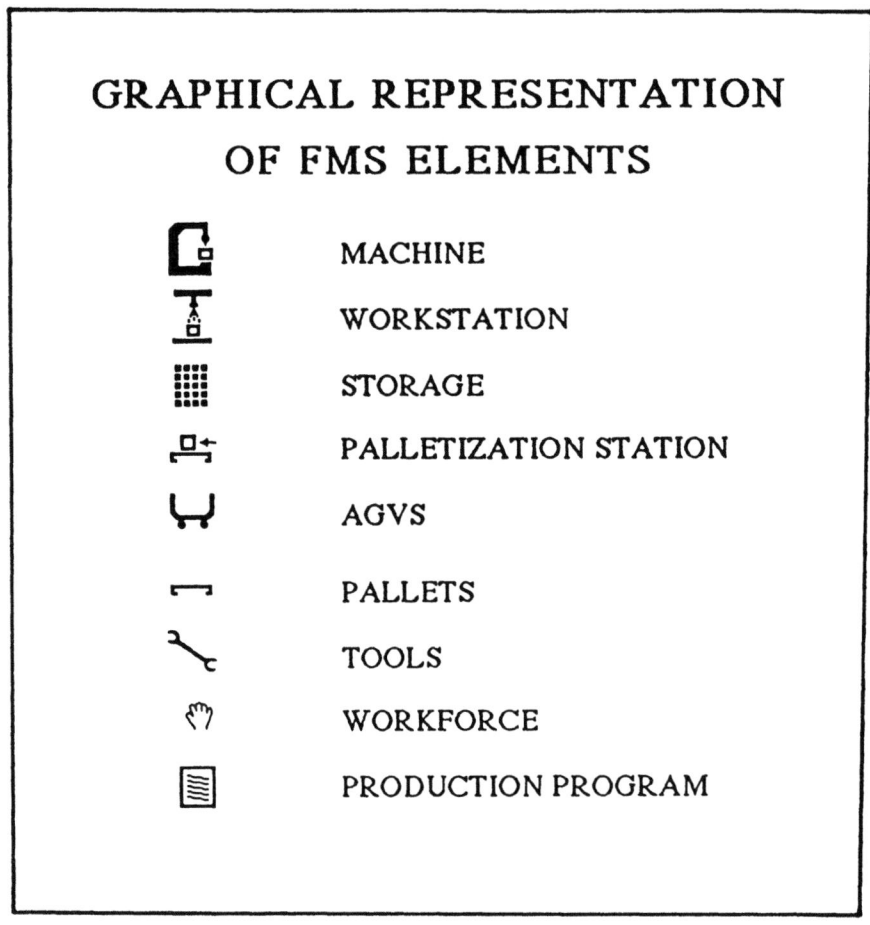

Figure 14. - Main components of FMS.

The problem is the same for storage. The FMS simulator has to take in account storage in a simplified way (the size and the input and output times can be defined). If the conceptor wishes to have a more precise overview of his storage system, he will use yet another simulation package, SAME/ASRS.

4. Operating time sheets

A part's operating time sheet is the logical list of operations that must be done on the part other than handling and storage.

A part can undergo five types of operations :

- Palletization
- Machining
- Production operations (washing, control, etc...)
- Depalletization
- External operations (external to the workshop)

Each operation will be characterized by its type, its duration, the station catagory where it will take place, and eventually the necessary tools, pallets or workforce.

B. Piloting rules

After having described the main production means encountered in FMS workshops and necessary to simulation to point out its behavior, we will now study piloting rules.

1. Scheduling

The goal of a manufacturing workshop is to produce different types of parts. In order to ensure a good utilization rate of the workshop, it is necessary to schedule these parts. The principal scheduling rules are the following:

- At the entrance of the workshop, the parts are either produced in small batches or by unit.

The choice between these two types of rules depends on the tool changing time and the quantity of parts to be produced.

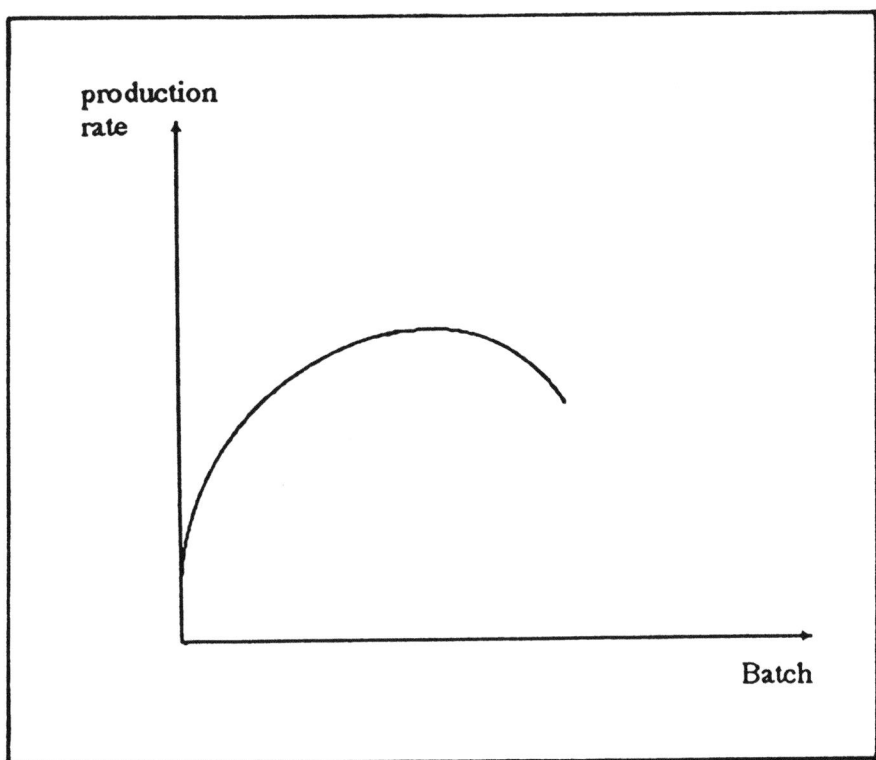

Figure 15. - Batch size influence

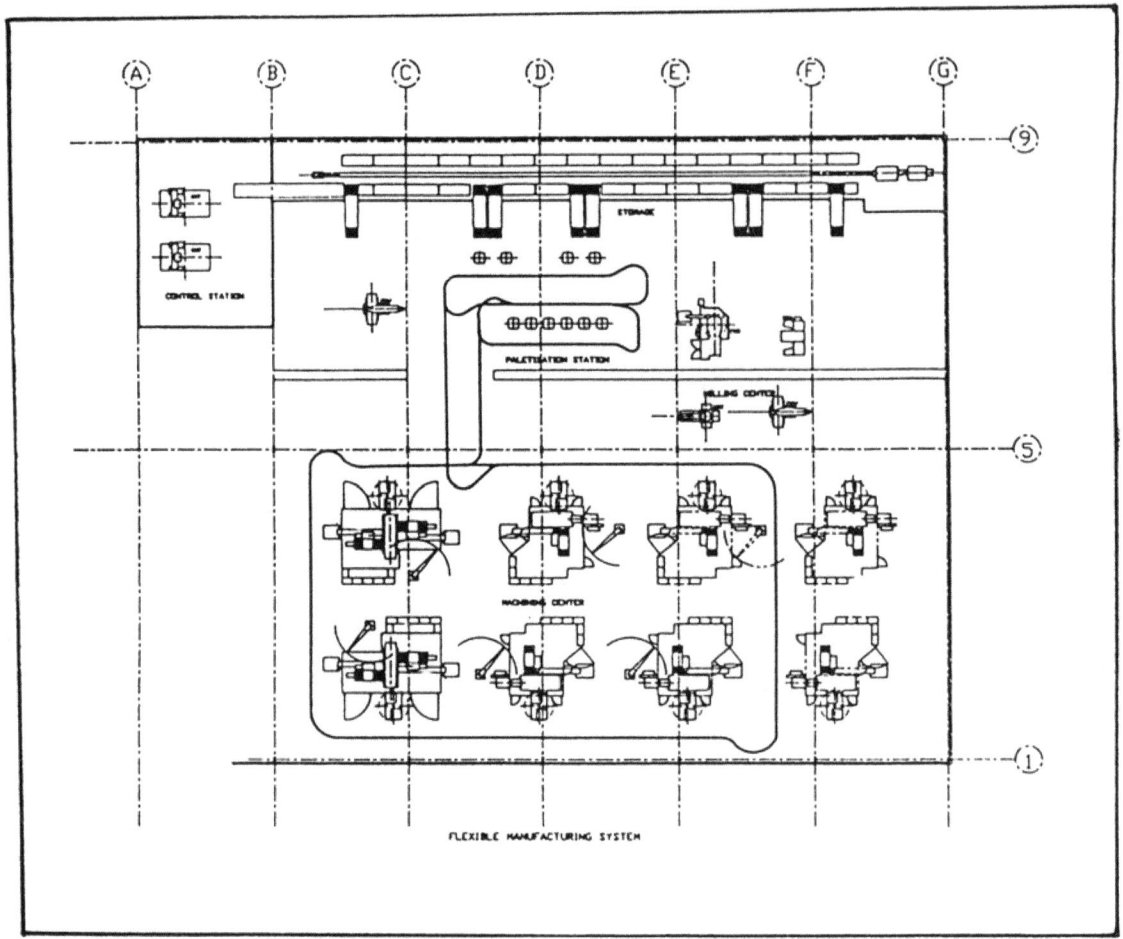

Figure 16. - Layout of EFAB workshop

Within the workshop the choice of the parts to be produced can be determined by a priority affected at the entrance of the part. Generaly these priorities depend on the entrance date, the due date, or simply "First In First Out" or "Last In First Out".

2. Tool changing rules

The layout of figure 16. is a flexible manufacturing system for breeches. The system simultaneously machines two families of highly different parts (16 different parts ranging from 25 to 1085 kg gross weight) performing about 10 operations on each part. This workshop is composed of a group of five CNC machining centers with palletizating and twelve conventional machines and workstations. One of the most important points in the designing of this workshop was the choice of the tool changing rule on machining centers and workstations.

Generaly we can find two main rules for tool managing. (1) The tools are changed at the end of each batch. (2) The tool is changed when worn or not not needed and that its place is necessary for another tool.

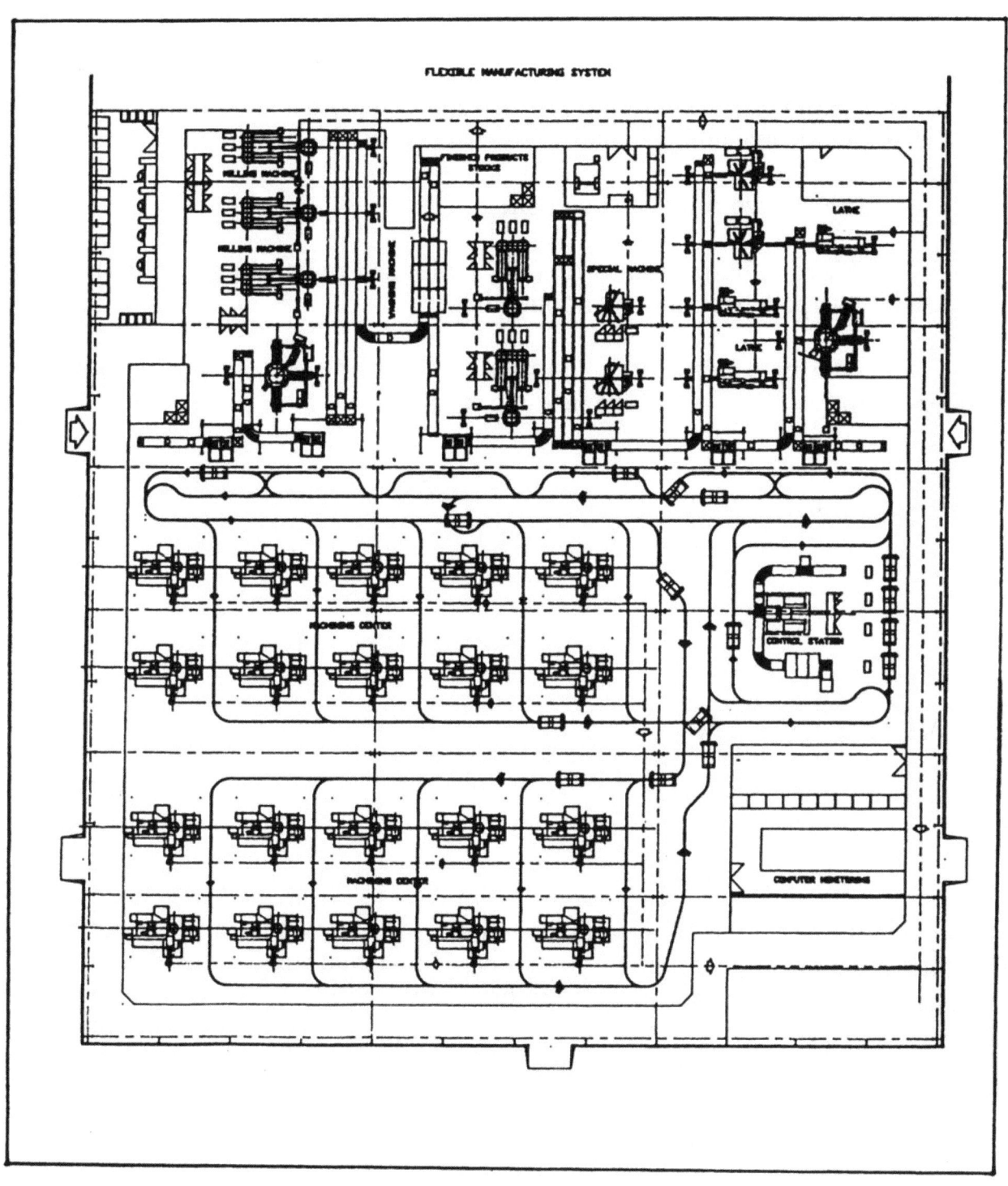

Figure 17. - BALKAN LOVETCH workshop

3. Palletization rules

The workshop of figure 17. manufactures casting for tractors in Bulgaria. The workshop is divided in two areas. The first one is composed of about 20 machining centers. The second area is contains about 10 conventional machines. In order to minimize the working process and to ensure a good utilization factor of the automatic machining center it was necessary to take into consideration the palletization factor.

In general, two rules can be applied:

- Slave palletization. Parts are palletized only when the workstation following the palletization operation is free.

- Master palletization. Parts are palletized when the pallets are free independently of the state of the next workstation.

CONCLUSION

Due to the evolution of personnal computers it is now possible to run high level, low cost simulators on that type of computer. The actual trend is the developpement of specialized simulators to be used without any knowledge of simulation or data processing . In the next few years, we can hope to see these simulators at the workshop level used by foremen and manufacturing technicians.

PCModel  PCModel/GAF

Interactive Graphic Simulation

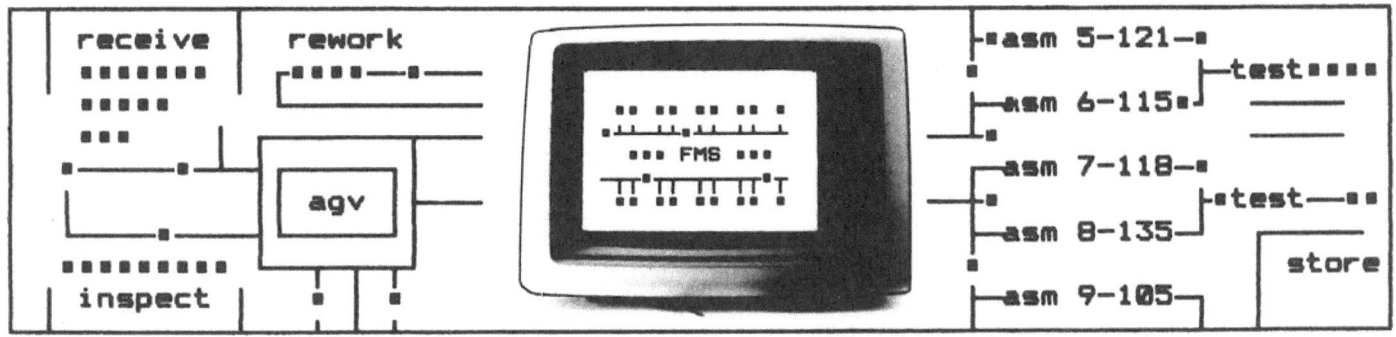

For the IBM PC, XT or AT
(and compatibles)

PCModel — £670, is a highly interactive graphic simulation system for planning and analysing manufacturing and distribution systems. It will satisfy the majority of your simulation needs quickly and cost effectively without the need for extensive programming knowledge. PCModel does not require special hardware or logic and will run with a minimum of 256K in colour or mono. Models, large or small, can be easily communicated to other PC users using the supplied demo disk. PCModel also includes its own training material.

HIGHLIGHTS

- Dynamic character-graphic animation of your process
- Automatic data screens for job performance, utilisation, etc.
- Interface with popular spread sheet programs
- Easy to use Help Menu with quick change to logic/graphics

PCModel/GAF — £6950, is a high resolution simulation system with a drawing facility resembling popular CAD systems. Bit-mapped icon images or backgrounds can be defined in units of measure such as feet, inches, or meters, and moved, rotated etc. during the simulation session. The display can be panned and zoomed to view as much or as little of the model detail as required. Thus you can realistically observe the movement of icons over and under each other. The object path can be defined graphically thereby minimising programming effort. A no-show mode can be selected to bypass animation when only statistical data is required. Models written with the companion system (PCModel) can be run with little modification.

HIGHLIGHTS

- Graphic path definition with rotational and curvilinear motion
- Pan and zoom for positioning all or part of the model onto the screen with varying detail
- No-show mode to bypass animation for maximum model performance
- Custom descriptive help screens and interface with own input/output routines
- Demonstration disks available

Simcon Ltd., (U.K. Sales & Technical Support Centre)
17 Earnley Road, Hayling, Hants PO11 9SU   Tel. (0705) 468908

If you have any concerns about our products,
you can contact us on
ProductSafety@springernature.com

In case Publisher is established outside the EU,
the EU authorized representative is:
**Springer Nature Customer Service Center GmbH
Europaplatz 3, 69115 Heidelberg, Germany**

Printed by Libri Plureos GmbH
in Hamburg, Germany